AF389356

Wireless Power/Data Transfer, Energy Harvesting System Design

Wireless Power/Data Transfer, Energy Harvesting System Design

Editor

Byunghun Lee

MDPI • Basel • Beijing • Wuhan • Barcelona • Belgrade • Manchester • Tokyo • Cluj • Tianjin

Editor
Byunghun Lee
Incheon National University
Korea

Editorial Office
MDPI
St. Alban-Anlage 66
4052 Basel, Switzerland

This is a reprint of articles from the Special Issue published online in the open access journal *Electronics* (ISSN 2079-9292) (available at: https://www.mdpi.com/journal/electronics/special_issues/wpt_eh).

For citation purposes, cite each article independently as indicated on the article page online and as indicated below:

LastName, A.A.; LastName, B.B.; LastName, C.C. Article Title. *Journal Name* **Year**, *Volume Number*, Page Range.

ISBN 978-3-0365-0782-8 (Hbk)
ISBN 978-3-0365-0783-5 (PDF)

Contents

About the Editor

Byunghun Lee (PhD, Assistant Professor)
He received a B.S. degree from Korea University, Seoul, South Korea, and an M.S. degree from the Korea Advanced Institute of Technology (KAIST), Daejeon, South Korea, in 2008 and 2010, respectively. From 2010 to 2011, he worked on wireless power transfer systems for electric vehicles (OLEV) at KAIST as a research engineer. He also received a Ph.D. degree in electrical and computer engineering from Georgia Tech in 2017. Currently, Byunghun Lee is an assistant professor at Incheon National University Department of Electrical Engineering and the director of the INU-Wireless Lab. He served as General Chair for 2020 IEEE PELS Workshop on Emerging Technologies: Wireless Power Transfer (WoW) in Korea, Seoul.

Preface to "Wireless Power/Data Transfer, Energy Harvesting System Design"

In recent decades, wireless power/data and energy harvesting technologies have been developed to provide us with more convenient, comfortable and productive lives than any previous generation, without the burden of physical cables. In the future, wireless power/data and energy harvesting technologies will be completely integrated into our daily lives, supplying power to our personal electronic devices, wearable/implantable electronics, home appliances and electric vehicles. In this Special Issue, we are focusing on emerging technologies in wireless power/data, and energy harvesting applications from a few microwatts to kilowatts with transfer distances from a few millimeters to a few tens of meters. The topics covered in this Special Issue include, but are not limited to, theories and techniques for short- or long-distance wireless/data transfer, RF energy harvesting, various applications of wireless power/data transfer for biomedical/wearable/mobile/IoT/electric vehicle, and system-level implementations. We would appreciate it if researchers would submit related manuscripts of high quality for publication in this Special Issue.

Byunghun Lee
Editor

Article

Inductive Power Transmission for Wearable Textile Heater using Series-None Topology

Hyeokjin Kwon [1], Kang-Ho Lee [2,]* and Byunghun Lee [1,]*

[1] Department of Electrical Engineering, Incheon National University, Incheon 22012, Korea; 2587z@inu.ac.kr
[2] Daegu Research Center for Medical Devices and Rehab Engineering,
Korea Institute of Machinery and Materials, Daejeon 42994, Korea
* Correspondence: kangholee6@kimm.re.kr (K.-H.L.); Byunghun_lee@inu.ac.kr (B.L.);
Tel.: +82-032-835-8766 (B.L.)

Received: 10 February 2020; Accepted: 2 March 2020; Published: 4 March 2020

Abstract: In this paper, an inductive-power-transmission (IPT) system for a wearable textile heater is proposed to comfortably provide heating to a user's body. The conductive thread, which has high electrical resistance, was sewn into a receiver (Rx) coil on clothing to generate high temperature with a low current. The proposed wearable heaters are completely washable thanks to their nonmetallic materials, other than conductive threads in the clothing. We introduced series-none (SN) topology to eliminate a resonant capacitor in the wearable textile heater. A single resonant capacitor in a transmitter (Tx) in SN mode was implemented to resonate both Tx and Rx, resulting in increased power delivered to the load (PDL) while maintaining high-power transfer efficiency (PTE), comparable with conventional series-series (SS) topology. When the supply voltage of the power amplifier was 7 V, while the PTE of the SS and SN modes was 85.2% and 75.8%, respectively, the PDL of the SS and SN modes was 2.74 and 4.6 W, respectively.

Keywords: wearable heater; inductive-power transmission; textile coil

---

1. Introduction

Recently, wearable heaters have attracted a lot of broad attention due to their potential applications in personal heating systems and healthcare management, such as thermotherapy [1–3]. In particular, thermotherapy has been effective in relieving pain and activating muscles or damaged skin, joints, and tissue, resulting in effective physiotherapy [3]. Most research on wearable heaters has typically focused on the development of heating materials for transparent [1], stretchable [1,2], or high-temperature (>100 °C) operations [3]. However, wearable heaters that were studied or commercialized still require electric wires to supply power to the thermal materials, which causes inconvenience and limited action for users. Inductive-power-transmission (IPT) technology could be a potential application in wearable heaters because IPT can eliminate electric wires from thermal materials in clothes.

IPT has been researched in the last few decades because of the convenience of wireless technology and its high applicability with other fields, such as the wireless charging of electric vehicles [4] and implantable biomedical devices [5,6]. In conventional IPT systems, many applications use a capacitive compensation network, such as series-series (SS) [7–9], series-parallel (SP) [8], parallel-series (PS) [10], and parallel-parallel (PP) [11,12], to improve power transfer efficiency (PTE) or power delivered to load (PDL) in given conditions. However, these compensated capacitors in a receiver (Rx) coil can be potentially damaged by repeated washing when it is implemented in a wearable textile heater with IPT technology.

While several IPT topologies were proposed to eliminate the resonant capacitor at the Rx [13] or the Tx [14], they inherently showed less PTE than the maximal achievable PTE of conventional compensation topologies. Series-none (SN) compensation, which removes the resonant capacitor at the

Rx [13], ideally shows approximately the same PTE as that of SS topology, but only when the coupling is strong and quality (Q) factors are large, which is difficult to achieve in a textile coil. Therefore, the proposed IPT with a textile coil focused on the analysis of PTE and PDL losses from the SS topology to find an optimal design point, where PTE loss is not large, but the PDL is increased at the same level as power-amplifier (PA) supply voltage V_{DD}, which implies the alleviation of voltage stress on the PA. Moreover, the SN structure has inherent advantages in terms of wireless efficiency and safety, because the overall system, including the Tx coil, only resonates when the textile Rx coil is properly coupled to the Tx, unlike in a conventional structure.

In the proposed wearable textile heater with an IPT system, a conductive thread (Imbut GmbH 110/f34_PA) with silver-coated polyamide multifilament yarns was utilized, which showed good electrical and thermal conductivity with good washability. We designed a textile coil by sewing the conductive thread onto the clothing. This textile coil simultaneously performs the roles of the Rx coil and thermal material [15]. The proposed wearable heater with an IPT system using a textile coil is shown in Figure 1. The PTE and PDL of the wearable textile heater in the IPT system were analyzed on the basis of circuit theory for conventional SS and proposed SN modes. The system modeling is described in Section 2, and experiment results are shown in Section 3. The conclusion is given in Section 4.

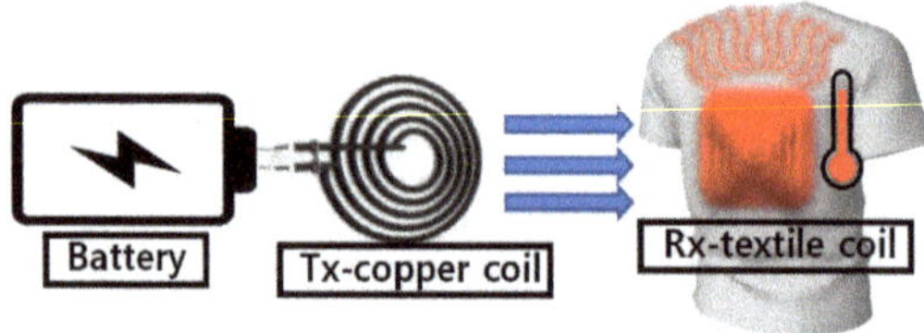

Figure 1. Simplified conceptual representation of proposed wearable heater using inductive power transmission with textile coil.

2. System Modeling

2.1. Circuit Analysis of SS and SN modes

In Figure 2a, the circuit model of SS is presented, which consisted of R_T, C_T, and L_T, where L_T is the inductance of the transmitter (Tx) coil, C_T is a resonant capacitor of Tx, and R_T is the combination of the parasitic resistance of the Tx coil, equivalent series resistance (ESR) of C_T, and the output resistance of PA, represented in V_S. The Rx was composed of R_L, R_R, C_R, and L_R, where L_R is the inductance of the Rx coil, C_R is a resonant capacitor of Rx, R_R is a sum of the parasitic resistance and ESR of C_R, and R_L is load resistance [16].

Since resonant capacitors can compensate for the reactance of inductors from Tx and Rx coils, we could transform the SS circuit to a simple equivalent circuit as shown in Figure 2b. In SS topology, the reflected resistance at Tx is given by

$$R_{reflect,SS} = k^2 Q_T Q_R' R_T, \tag{1}$$

where k is coupling coefficient, and $Q_T = wL_T/R_T$, $Q_R' = wL_R/(R_R + R_L)$, which are the Q factor of Tx and loaded Rx coils, respectively.

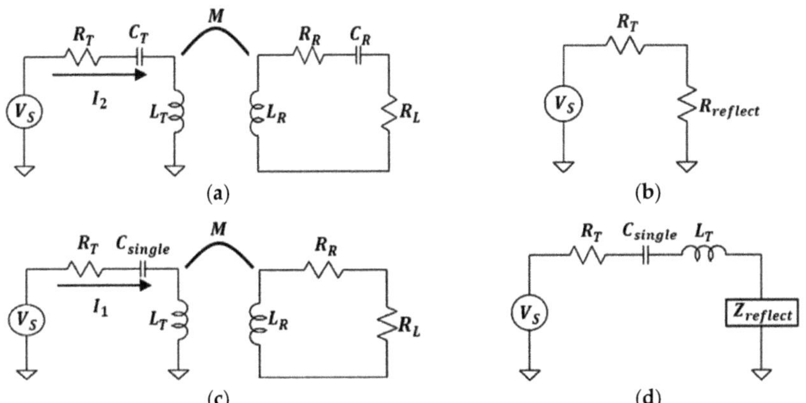

Figure 2. (a) Circuit model of series-series (SS) topology, (b) simplified equivalent SS circuit model, (c) circuit model of proposed series-none (SN) topology, and (d) equivalent SN circuit model using textile coil.

The transmission efficiency of SS topology can be calculated by the ratio of delivered power to the R_T and $R_{reflext,SS}$, resulting in $\eta_{T,SS}$, given by

$$\eta_{T,SS} = \frac{k^2 Q_T Q'_R}{1 + k^2 Q_T Q'_R}. \tag{2}$$

Similarly, Rx efficiency can also be calculated by the ratio of delivered power to the R_R and R_L, resulting in Rx efficiency η_R, given by

$$\eta_R = 1 - \frac{Q'_R}{Q_R}, \tag{3}$$

where $Q_R = \omega L_R / R_R$, which is the Q factor of the Rx coil. Consequently, the total efficiency of the SS mode can be calculated by the multiplication of $\eta_{T,SS}$ and η_R:

$$\eta_{SS} = \frac{k^2 Q_T Q'_R}{1 + k^2 Q_T Q'_R}\left(1 - \frac{Q'_R}{Q_R}\right). \tag{4}$$

In Figure 2a, the source current can be derived as

$$I_{2,pk} = \frac{V_{s,pk}}{R_T + R_{reflect,SS}} = \frac{V_{s,pk}}{R_T\left(1 + k^2 Q_T Q'_R\right)} \tag{5}$$

where $I_{2,pk}$ and $V_{s,pk}$ are the peak current of the SS mode and peak source voltage, respectively. The input power of SS mode is obtained by

$$P_{in,SS} = 0.5 V_{s,pk} I_{2,pk} = \frac{0.5 V_{s,pk}^2}{R_T\left(1 + k^2 Q_T Q'_R\right)}, \tag{6}$$

and the output power of SS mode is defined as the multiplication of $P_{in,SS}$ and η_{SS}:

$$P_{out,SS} = \frac{k^2 Q_T Q'_R}{\left(1 + k^2 Q_T Q'_R\right)^2}\left(1 - \frac{Q'_R}{Q_R}\right)\left(\frac{V_{s,pk}^2}{2 R_T}\right). \tag{7}$$

In SN mode, there is no resonant capacitor in Rx, and the reactance part still remains in the equivalent circuit compared to SS mode, as shown in Figure 2c. In this condition, the reflected impedance can be written as

$$Z_{reflect} = \frac{k^2 Q_T Q_R' R_T}{1 + Q_R'^2} - j\frac{k^2 Q_T Q_R'^2 R_T}{1 + Q_R'^2}. \tag{8}$$

The reactance part of $Z_{reflect}$ can be compensated by a single capacitor at Tx, C_{single}, and it is defined as

$$C_{single} = \frac{1}{w^2 L_T \left(1 - k^2 Q_R'^2 / \left(1 + Q_R'^2\right)\right)}. \tag{9}$$

The Equation (9) implies that the overall inductive link only resonates with C_{single} when the Rx coil is properly coupled with the Tx coil. This is advantageous in terms of wireless power efficiency and safety compared to the conventional structure. Since L_T does not directly resonate with C_{single}, Tx power consumption without the Rx coil is automatically reduced.

Assuming that the reactance part is completely compensated by C_{single} in SN mode, there was only reflected resistance $R_{reflect,SN}$ remaining, as shown in Figure 2d, which can be defined as

$$R_{reflect,SN} = \frac{k^2 Q_T Q_R' R_T}{1 + Q_R'^2}. \tag{10}$$

The Rx efficiency of SN mode was the same as that of SS mode since there was no additional energy consumption in the inductance. Therefore, the remaining procedures to obtain the equations of PTE and PDL are similar to the SS mode:

$$\eta_{T,SN} = \frac{k^2 Q_T Q_R'}{1 + Q_R'^2 + k^2 Q_T Q_R'} \tag{11}$$

$$\eta_{SN} = \frac{k^2 Q_T Q_R'}{1 + Q_R'^2 + k^2 Q_T Q_R'}\left(1 - \frac{Q_R'}{Q_R}\right), \tag{12}$$

$$I_{1,pk} = \frac{V_{s,pk}}{R_T\left(1 + k^2 Q_T Q_R' / \left(1 + Q_R'^2\right)\right)}, \tag{13}$$

$$P_{in,SN} = 0.5 V_{s,pk} I_{1,pk} = \frac{0.5 V_{s,pk}^2}{R_T\left(1 + k^2 Q_T Q_R' / \left(1 + Q_R'^2\right)\right)}, \tag{14}$$

$$P_{out,SN} = \frac{k^2 Q_T Q_R'\left(1 + Q_R'^2\right)}{\left(1 + Q_R'^2 + k^2 Q_T Q_R'\right)^2}\left(1 - \frac{Q_R'}{Q_R}\right)\left(\frac{V_{s,pk}^2}{2 R_T}\right). \tag{15}$$

On the basis of the PTE and PDL from SS and SN modes in Equations (4) and (12), we introduced the ratios of PTE and PDL, n_{PTE} and n_{PDL}, to show the effectiveness of SN mode in a textile coil compared to the conventional SS mode:

$$n_{PTE} = \frac{\eta_{SN}}{\eta_{SS}} = \frac{1 + k^2 Q_T Q_R'}{1 + Q_R'^2 + k^2 Q_T Q_R'}, \tag{16}$$

$$n_{PDL} = \frac{P_{out,SN}}{P_{out,SS}} = \frac{\left(1 + Q_R'^2\right)\left(1 + k^2 Q_T Q_R'\right)^2}{\left(1 + Q_R'^2 + k^2 Q_T Q_R'\right)^2}. \tag{17}$$

When n_{PTE} or n_{PDL} is close to 1, SN mode showed high PTE or PDL, the same as that in conventional SS mode, which means that SN mode could replace SS mode without an additional resonant capacitor in Rx. On the other hand, a small n_{PTE} or n_{PDL} implied that SN mode showed relatively lower achievable PTE or PDL compared to SS mode in the given wireless power conditions.

2.2. PTE and PDL Properties in SS and SN Modes

Figure 3 shows the calculated PTE in SS and SN modes, respectively, as a function of Q_T and Q'_R. Q_T is the effective Q factor of the Tx, including PA output resistance, 0.45 Ω. When the R_L was much larger than R_R, Rx efficiency η_R was close to 1. In the proposed wearable heater using the textile coil, R_R could be considered as zero because the parasitic resistance of the Rx coil is regarded as the load resistance, resulting in the $\eta_R \approx 1$. As shown in Figure 3a, η_{SS} increased when both Q_T and Q'_R increased. However, η_{SN} was maximized at the optimal Q'_R, which could be found from differentiating Equation (12), while higher Q_T was still desirable for the SN mode in Figure 3b. The achievable PTE ratio between SS and SN modes, n_{PTE}, is shown in Figure 3c, and it was used to find the design consideration of Tx and Rx coils in SN mode. When n_{PTE} was close to 1, it was beneficial to use SN mode instead of SS mode in the given inductive link considering the burden of the additional resonant capacitor in the Rx in SS mode.

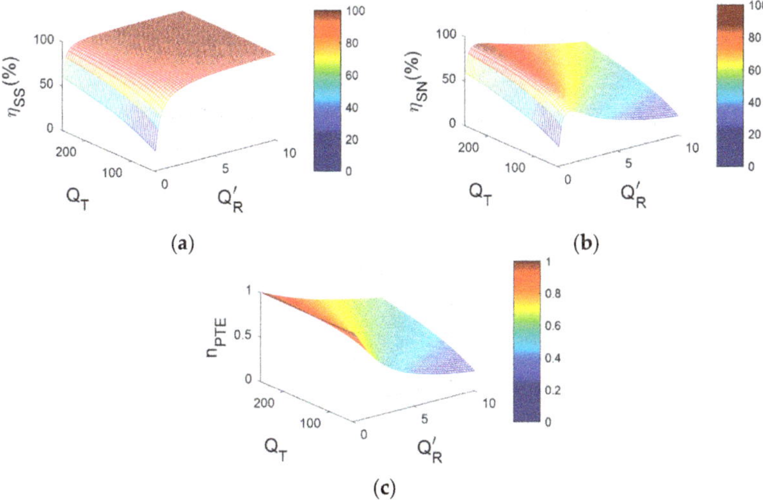

Figure 3. Calculated PTEs of (**a**) SS mode, η_{SS}, (**b**) SN mode, η_{SN}, and (**c**) PTE ratio of SS and SN modes, n_{PTE}, as function of Q_T and Q'_R for $k = 0.24$ and $R_R = 0$ Ω.

By contrast with the PTE properties, the PDL in SS mode was maximized when Q'_R decreased, as shown in Figure 4a. This is because a larger $Z_{reflect}$ is not desirable for a large PDL, while it is beneficial for PTE, based on circuit theory. However, SN mode showed a higher PDL in a wide range of Q'_R compared to SS mode, as shown in Figure 4b, due to $(1 + Q'^2_R)$ terms in Equation (15) compared to Equation (7). Therefore, the PDL ratio between SN and SS modes n_{PDL} could be increased by up to ~60 as Q'_R increased, as shown in Figure 4c. Considering Figures 3c and 4c, we could optimize the textile coil for a wearable heater in SN mode with improved PDL and less PTE loss compared to the conventional SS mode.

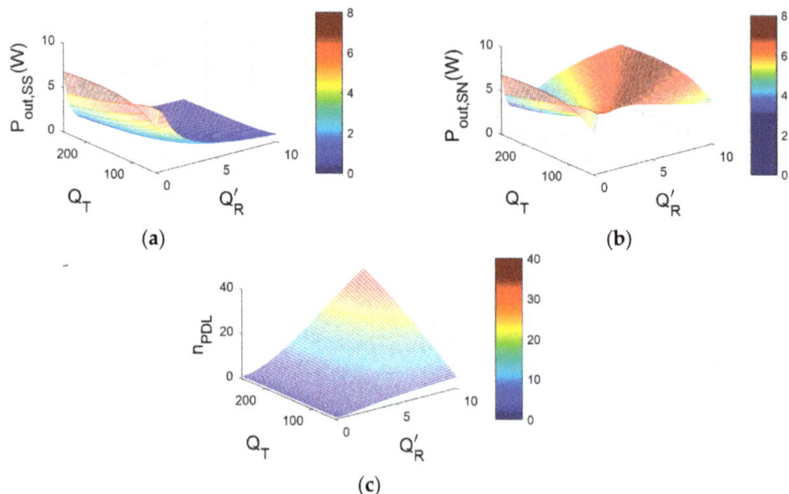

Figure 4. Calculated PDLs of (**a**) SS mode, $P_{out,SS}$, (**b**) SN mode, $P_{out,SN}$ and (**c**) PDL ratio of SS and SN modes, n_{PDL}, as a function of function of Q_T and Q'_R for $k = 0.24$, $R_R = 0\ \Omega$, and $V_{s,pk} = 5$ V.

2.3. Coil Design for Wearable Textile Heater

Considering that the thermal material Rx, attached to the back of a t-shirt, received inductive power from the Tx in a backpack, the maximal outer diameter of the Tx and Rx coils was 16 cm with a gap of 4 cm between coils. In this condition, the coupling coefficient was 0.24, based on the calculation in [16], which was well-matched with the measurement result using the vector network analyzer. On the basis of the derived $nPDL$ and $nPTE$ in Figures 3c and 4c, a higher Q_T is always desirable to achieve better PDL and PTE for fixed PA output resistance. Therefore, the Tx coil was designed to have maximal quality (Q) factor in the experiment conditions considering practical limitations. The Tx coil was designed using 22 American Wire Gauge (AWG) with a 0.3 mm pitch, with a sweep of frequencies and number of turns performed with an electromagnetic (EM) simulator (FEKO, Altair) and measurement, while the lumped-model circuit simulation was performed with LTSpice (Analog Devices). A Tx coil quality factor of 238 without PA output resistance ($Q_T = 153.4$ including PA) was achieved at $N_T = 8$ at 2 MHz carrier frequency, as shown in Figure 5a. From the calculated result from Figures 3c and 4c, $Q'_R = 1$ was selected considering optimal PTE loss (large $nPTE$) and improved PDL (large $nPDL$) in the textile coil. When $Q'_R = 1$, the PTE loss from the conventional SS structure was only 10%, while PDL was improved by 176%, as shown in Figure 5b. Although the figure of merit (FoM) between PDL and PTE could be one of clear standards as described in [17], the design balance between PDL and PTE is still flexible for the design purpose and applications. The Rx coil was designed using conductive thread with a 0.25 mm pitch at 2 MHz carrier frequency. The washable conductive thread (ELITEX®, Art. 110/f34_PA/Ag) had basic 110 dtex/34 filaments yarn counts with a silver coating of 1 μm thickness that showed $70 \pm 20\ \Omega$/m at DC and had a 259 °C melting point based on the datasheet. The number of turns of Rx coil N_R was swept to find the optimal condition of $Q'_R = 1$. As shown in Figure 5c, $N_R = 7$ showed the nearest value to 1. The coil parameters for Tx and Rx are summarized on Table 1. The coil parameters in Table 1 show one of the exemplar designs for the wireless wearable heater for t-shirts. While the outer diameters or thickness of coils could differ by geometrical limitations in practical applications such as gloves, socks, or t-shirts, the optimization procedure is the same as that of the SN topology.

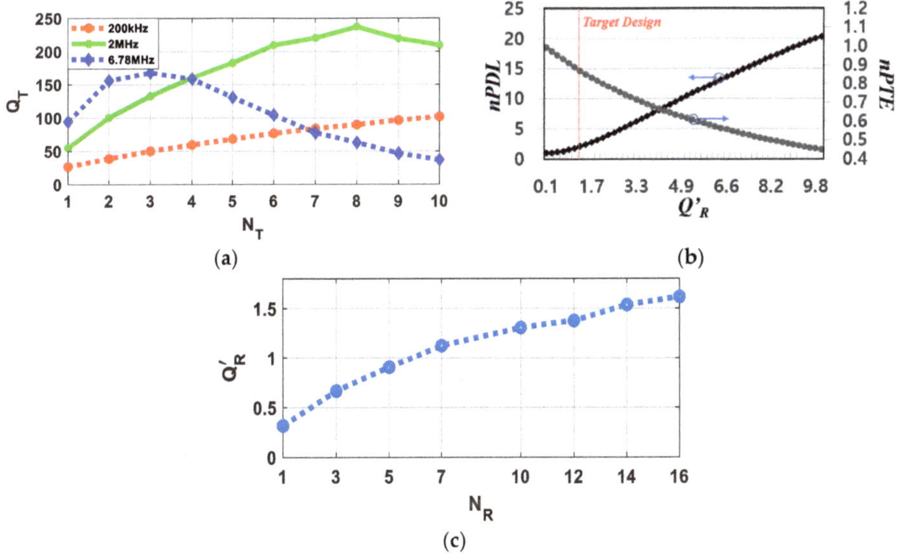

(a)

(b)

(c)

Figure 5. (**a**) Measured Q factor of wire-wound Tx coil for 200 kHz, 2 MHz, and 6.78 MHz carrier frequencies without power-amplifier (PA) output resistance; (**b**) Q'_R optimization for designated Q_T considering n_{PDL} and n_{PTE}; and (**c**) resulting Q factor of conductive-thread textile Rx coil vs. number of turns at 2 MHz carrier frequency.

Table 1. Parameters of transmitter (Tx) and receiver (Rx) coil.

Tx/Rx Coil	Parameters							
	Material	Outer Diameter	Thickness	Pitch (mm)	Turn	Parasitic Res.	Ind. (μH)	Q
Tx coil	Copper	16 cm	22 AWG	0.3	8	0.82 Ω	15.5	238
Rx coil	Coated Textile	16 cm	1 μm [1]	0.25	7	214.3 Ω	18.3	1.1

[1] Thickness of silver layer, 70 ± 20 Ω/m at DC.

3. Experiment Results

Figure 6 shows the experiment setup for the optimized wire-wound Tx coil and conductive-thread Rx coil on the textile with a Class-D power amplifier (PA). The Tx coil was placed 4 cm away from the textile Rx coil, and the PA drove the Tx coil for the inductive heating. The Rx coil received inductive power from the Tx coil, and its internal resistance simultaneously operated as thermal material. The Rx coil made of conductive threads had an electrical resistance of 70 ± 20 Ω/m, allowable current of 0.4 A, and good washability. PA power was controlled by PA supply voltage V_{DD} for the temperature controls in the wearable textile heater. The coupling coefficient of the proposed inductive link was 0.24 at 4 cm distance with $Q_T = 153.4$, including PA output resistance and $Q'_R = 1.1$ as shown in Table 2. All parameters and variables were acquired by the network analyzer (ZND, Rohde and Schwarz) in the experiment.

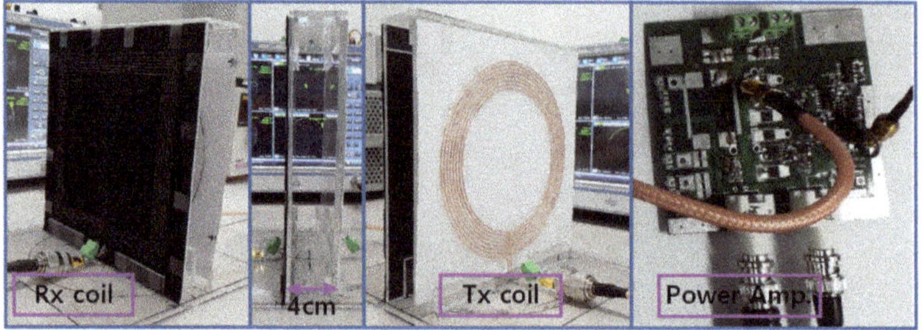

Figure 6. Experiment setup of optimized wire-wound Tx coil and conductive-thread textile Rx coil with power amplifier.

Figure 7 shows the measured temperature of the wearable textile heater depending on the controlled PA power from 3 to 5 W by the adjustment of V_{DD}. Temperature was measured by thermal imaging camera, and was controlled by the input PA power and increased up to 42.5 °C from room temperature when the PA provided the output power of 5 W. For individual PA powers, temperature in the wearable textile heater reached the target temperature within 1 min.

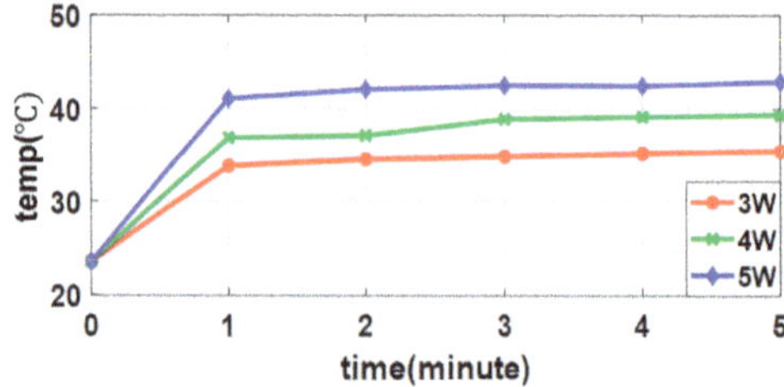

Figure 7. Measured temperature curves of wearable textile heater for different PA power values.

Figure 8 shows the transient waveforms of input voltage and current waveforms in the PA with the temperature of the inductively powered wearable textile heater using the thermal imaging camera. When the textile Rx was properly placed on the Tx coil without any misalignment, the imaginary parts of $Z_{reflect}$ and L_T resonated with C_{single} at 2 MHz carrier frequency from the PA, as shown in Figure 2d. Therefore, the entire inductive link from the Tx to the Rx resonated in-phase between input voltage and current because the real parts of the inductive link remained, and the temperature in the wearable textile heater increased up to 42.5 °C from room temperature, as shown in Figure 8a. The PA did not automatically deliver the inductive power to the wearable textile heater when the Rx coil moved farther from the Tx coil. In this case, the imaginary parts of $Z_{reflect}$ were reduced compared to the nominal condition, and the resonant frequency of the Tx moved farther from the 2 MHz carrier frequency, resulting in an almost 90° phase shift between input voltage and current waveforms, shown in Figure 8b. Therefore, the temperature of the wearable textile heater stayed at room temperature, and the energy was rarely dissipated in this condition.

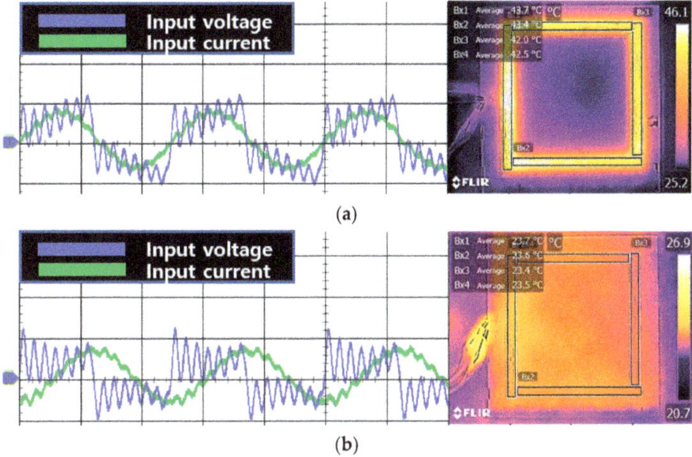

(**a**)

(**b**)

Figure 8. Transient waveforms of input voltage and current with temperature in wearable textile heater for (**a**) aligned and (**b**) misaligned Tx and Rx coils.

PTE and PDL were measured for the proposed system with SN mode, and compared to the conventional SS mode, as summarized on Table 2. The output impedance of the PA, 0.45 Ω, was added in the Q_T. While the measured PTE of the proposed SN mode was reduced by 11%, it showed a 1.6 times higher PDL on average compared to the SS mode without the compensated capacitor in the Rx coil, which showed similar trends in the calculated results. The derived equations proved reliability by showing that the error rate was within 10% between calculated and measured results. The error rates of η_{SS}, η_{SN}, and n_{PDL} were 2.97%, 6.23%, and 10%, respectively. The measured PDLs and the temperatures of the wearable textile heater for SS and SN modes are shown in Figure 9 with respect to controllable PA supply voltage V_{DD}.

Table 2. Calculated and measured results for SS and SN modes in wearable heater.

Fixed Parameters			Topology	
k	Q_T	Q'_R	Series-Series (SS)	Series-None (SN)
0.24	153.37	1.07		
	Calculated PTE		90.7%	81.8%
	Measured PTE		88%	77%
	Calculated n_{PDL}		1.76	
	Measured n_{PDL}		1.6	

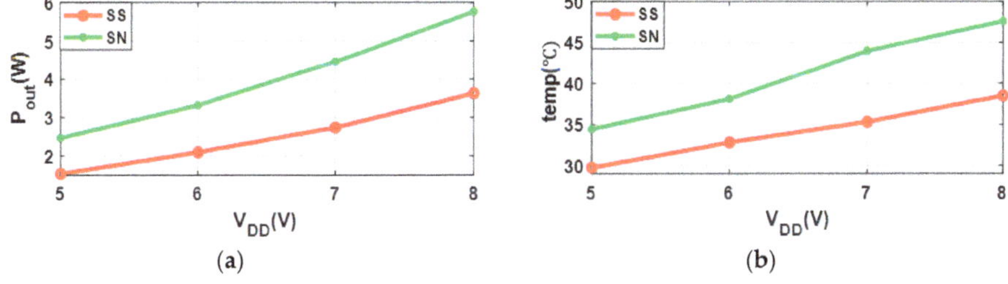

(**a**)   (**b**)

Figure 9. Measured (**a**) power delivered to load (PDL) and (**b**) temperature of wearable textile heater for SS and SN modes with respect to controllable PA supply voltage V_{DD}.

4. Conclusions

We presented the new concept of a wearable heater that combines IPT techniques and a textile coil. The SN mode for the IPT was adapted in the proposed wearable textile heater to eliminate electronic components in the Rx, improving robustness and user convenience. Since the Tx coil only resonates with the textile Rx coil when both coils are properly placed, the system could save energy and ensure safety if the Rx coil is unintentionally misaligned during operation. We optimized the Tx and Rx coils to minimize PTE loss, and maximize the PDL in SN mode through analysis in the IPT system based on SN topology analysis for wireless-wearable-heater applications. The proposed system provides a very robust wearable heater using IPT technology without any electrical components at the Rx, which dramatically increases convenience for the user. Compared to the conventional SS topology in the IPT, the proposed optimized design showed 160% improved PDL, while PDL was only reduced by 11%, which was well-matched with the theoretical results. Furthermore, the proposed design could inherently prevent energy dissipation from the Tx against misalignment between Tx and Rx due to automatic resonant-frequency shifting in the IPT, resulting in improved safety and efficiency with no need for complicated sensing and data-communication circuitries in the Rx. Although the proposed wireless wearable heater showed less sensitivity against temperature variations compared to a wired wearable heater due to the lack of sensing circuitries in the Rx, it can be further improved by a transient observation circuit in the Tx to monitor input voltage and current for estimating the delivered power to the heating material, as shown in Figure 7. To the best of our knowledge, this is the first demonstration of a wireless wearable heater using conductive thread, providing textile-coil optimization for SN topology in an IPT system.

Author Contributions: conceptualization, B.L. and K.-H.L.; methodology, B.L.; validation, H.K. and B.L.; formal analysis, H.K.; investigation, B.L.; data curation, H.K.; writing—original-draft preparation, H.K.; writing—review and editing, B.L.; visualization, H.K.; supervision, project administration, and funding acquisition, B.L. All authors have read and agreed to the published version of the manuscript.

Funding: This work was supported by the Incheon National University (#2018-0165) Research Grant in 2018.

Conflicts of Interest: The authors declare no conflict of interest.

References

1. Hong, S.; Lee, H.; Lee, J.; Kwon, J.; Han, S.; Suh, Y.D.; Cho, H.; Shin, J.; Yeo, J.; Ko, S.H. Highly stretchable and transparent metal nanowire heater for wearable electronics applications. *Adv. Mater.* **2015**, *27*, 4744–4751. [CrossRef] [PubMed]
2. Choi, S.; Park, J.; Hyun, W.; Kim, J.; Kim, J.; Lee, Y.B.; Song, C.; Hwang, H.J.; Kim, J.H.; Hyeon, T.; et al. Stretchable heater using ligand-exchanged silver nanowire nanocomposite for wearable articular thermotherapy. *ACS Nano* **2015**, *9*, 6626–6633. [CrossRef] [PubMed]
3. Zhang, M.; Wang, C.; Liang, X.; Yin, Z.; Xia, K.; Wang, H.; Jian, M.; Zhang, Y. Weft-Knitted Fabric for a Highly Stretchable and Low-Voltage Wearable Heater. *Adv. Electron. Mater.* **2017**, *3*, 1700193. [CrossRef]
4. Liu, C.; Jiang, C.; Qiu, C. Overview of coil designs for wireless charging of electric vehicle. In Proceedings of the 2017 IEEE PELS Workshop on Emerging Technologies: Wireless Power Transfer (WoW), Chongqing, China, 20–22 May 2017; pp. 1–6.
5. Jia, Y.; Mirbozorgi, S.A.; Lee, B.; Khan, W.; Madi, F.; Weber, A.; Li, W.; Ghovanloo, M. A mm-Sized Free-Floating Wirelessly-Powered Implantable Optical Stimulation Device. *IEEE Trans. Biomed. Circuits Syst.* **2019**, *13*, 608–618. [CrossRef]
6. Lee, B.; Ahn, D. Robust Self-Regulated Rectifier for Parallel-Resonant Rx Coil in Multiple-Receiver Wireless Power Transmission System. *IEEE J. Emerg. Sel. Top. Power Electron.* **2019**. [CrossRef]
7. Lee, B.; Yeon, P.; Ghovanloo, M. A multi-cycle Q-modulation for dynamic optimization of inductive links. *IEEE Trans. Ind. Electron.* **2016**, *63*, 5091–5100. [CrossRef] [PubMed]
8. Zhang, Z.; Pang, H.; Georgiadis, A.; Cecati, C. Wireless power transfer—An overview. *IEEE Trans. Ind. Electron.* **2018**, *66*, 1044–1058. [CrossRef]

9.   Sample, A.P.; Meyer, D.T.; Smith, J.R. Analysis, experimental results, and range adaptation of magnetically coupled resonators for wireless power transfer. *IEEE Trans. Ind. Electron.* **2011**, *58*, 544–554. [CrossRef]

10.   Monti, G.; Costanzo, A.; Mastri, F.; Monglardo, M. Optimal design of a wireless power transfer link using parallel and series resonators. *Wirel. Power Transf.* **2016**, *3*, 105–116. [CrossRef]

11.   Hu, A.P.; Hussmann, S. Improved power flow control for contactless moving sensor applications. *IEEE Power Electron. Lett.* **2004**, *2*, 135–138. [CrossRef]

12.   Monti, G.; Mastri, F.; Mongiardo, M.; Corchia, L.; Tarricone, L. Load-Independent Operative Regime for Inductive Resonant WPT Link in Parallel Configuration. *IEEE Trans. Microw. Theory Technol.* **2020**, 1–10. [CrossRef]

13.   Zhang, Y.; Kan, T.; Yan, Z.; Mao, Y.; Wu, Z.; Mi, C.C. Modeling and Analysis of Series-None Compensation for Wireless Power Transfer Systems with a Strong Coupling. *IEEE Trans. Power Electron.* **2018**, *34*, 1209–1215. [CrossRef]

14.   Lee, B.; Kim, H.; Rim, C.T. Resonant power shoes for humanoid robots. In Proceedings of the 2011 IEEE Energy Conversion Congress and Exposition, Phoenix, AZ, USA, 17–22 September 2011; pp. 1791–1794.

15.   Šahta, I.; Baltina, I.; Truskovska, N.; Blums, J.; Deksnis, E. Selection of conductive yarns for knitting an electrical heating element. *High Perform. Optim. Des. Struct. Mater.* **2014**, *137*, 91–102.

16.   Harrison, R.R. Designing efficient inductive power links for implantable devices. In Proceedings of the 2007 IEEE International Symposium on Circuits and Systems, New Orleans, LA, USA, 27–30 May 2007; pp. 2080–2083.

17.   Kiani, M.; Ghovanloo, M. A Figure-of-Merit for Designing High-Performance Inductive Power Transmission Links. *IEEE Trans. Ind. Electron.* **2013**, *60*, 5292–5305. [CrossRef] [PubMed]

Letter

Wireless Power Transfer under Wide Distance Variation Using Dual Impedance Frequency

Woochan Lee and Dukju Ahn *

Department of Electrical Engineering, Incheon National University, Incheon 22012, Korea; wlee@inu.ac.kr
* Correspondence: adjj22@gmail.com; Tel.: +82-32-835-8767

Received: 22 November 2019; Accepted: 5 January 2020; Published: 7 January 2020

Abstract: A dual-impedance operation, where coil impedance is controlled by operating frequency selection, is proposed to maintain optimum reflected impedance across coupling variation. More specifically, this work focuses on how high coupling between coils presents excessively high reflected resistance to transmitter (Tx) inverters, degrading the efficiency and output power of the inverter. To overcome this problem, the proposed system is equipped with dual-impedance coil and selects high- or low-impedance coil based on the ability to operate both at 200 kHz and 6.78 MHz frequencies. The reactive impedances of 6.78 MHz coils are designed to be higher than that of 200 kHz coils. Since the reflected resistance is proportional to the coil impedances and coupling squared, at close distance with high coupling coefficient, 200 kHz coils with low coil impedances are activated to prevent an excessive rise in reflected resistance. On the other hand, at large distance spacing with low coupling coefficient, 6.78 MHz coils with high coil impedances are activated so that sufficient reflected resistance is obtained even under the small coupling. The proposed system's advantages are the high efficiency and the elimination of bulky mechanical relay switches. Measured efficiencies are 88.6–50% across 10 coupling variations.

Keywords: inductive power; dual impedance; dual band; reflected resistance; frequency splitting

---

1. Introduction

One of the major goals in wireless power transfer research today is to accommodate wide distance variations with high efficiency. It has been noted that the distance variation may cause frequency splitting if the transmitter (Tx) and receiver (Rx) coils are brought close together [1–3]. Either delivered voltage or power efficiency experiences a decrease when the system operates at the resonant frequency of Tx coil and Rx coil.

Several techniques have been proposed to solve the frequency splitting problem [4–8]. Works [4,5] propose using special geometry for the resonators. Other solutions include adaptive matching networks which require bulky and mechanically moving relay switches [6–9]. Work [10] proposes the use of a variable inductor at the receiver, this would require an additional buck converter, meaning that the receiver would have two buck converters

For a two-coil system, it has been found that frequency splitting degrades the delivered power whereas coil-to-coil efficiency is not degraded at the center frequency [3]. However, this discussion does not consider Tx inverter efficiency. It is discussed later in this paper that the inverter efficiency can also be degraded at short distance spacing. Therefore, the problems of short distance operation are low output power and the efficiency of Tx inverters.

From the circuit viewpoint, an excessively high reflected resistance is the reason of frequency splitting and consequent degradations. Specifically, the coupling between Tx and Rx coils is abstracted as a reflected resistance, and this hypothetical resistor behaves as the load resistance of a Class-D Tx inverter. Since the efficiency and output power of an inverter depends on its load resistance, the reflected resistance should remain within the appropriate range.

To solve the aforementioned issues, this paper proposes the use of a dual-impedance mode where impedance control is done by operating frequency selection depending on magnetic coupling. One of the advantages is the high efficiency, given that high-impedance coils are designed for long-distance condition while low-impedance coils are optimized for short distance. Moreover, the conventional capacitor-switch matrix used for adaptive impedance matching is not necessary. Therefore, mechanical relay switches, which need complex driving blocks and are susceptible to mechanical damage, are not necessary.

2. Inverter Degradations at Strong Coupling

Figure 1a illustrates the circuit model of a basic wireless power transfer system. From the viewpoint of the transmitter, the coupling with the receiver can be represented by the reflected resistance, R_{refl}, in Figure 1b where:

$$R_{refl} = k^2 \omega L_{TX} \frac{\omega L_{RX}}{R_L + R_{P2}} \tag{1}$$

In other words, the current/voltage characteristic of Figure 1a with receiver and coupling is equivalent to the current/voltage of Figure 1b with the reflected resistance. It is important to recognize that the load resistor of Tx inverter is the reflected resistance.

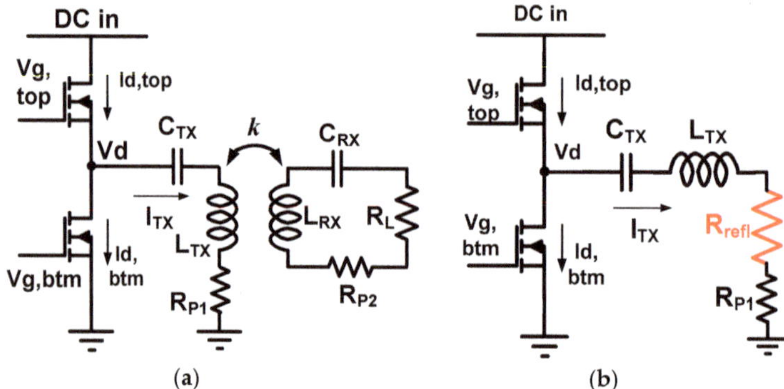

Figure 1. (**a**) Circuit model of a resonant wireless power transfer system. (**b**) From the viewpoint of the transmitter. The coupling with Rx is abstracted as reflected resistance, which serves as the load resistor of a Class-D inverter.

The normal zero-voltage switching (ZVS) waveform of inverter is shown in Figure 2a. After Vg, the upper limit is set to zero and both MOSFETs are off, the coil current (I_{TX}) discharges the parasitic drain capacitors of MOSFETs. This pulls down the drain voltage (Vd) to zero when both MOSFETs are off. After that, the bottom MOSFET is turned on with very low drain voltage (Vd). Therefore, switching loss is small in this case. The MOSFET used in this work is BUK9Y59-60E. Note that although $V_{g,btm}$ goes up at 123 ns, there is ~6 ns turn-on delay. Actual device turn-on point is approximately 128 ns.

However, this zero-voltage switching fails if the reflected resistance (R_{refl}) is too high [11]. High reflected resistance happens when the coupling between Tx and Rx is high at close distance as in Equation (1). This reduces I_{TX} as in Figure 2b. It then takes a longer time for the small I_{TX} to discharge the parasitic drain capacitance of MOSFETs. As a result, the Vd voltage cannot drop to zero before the turning-on of the bottom MOSFET as in Figure 2b, leading to switching loss [11–13].

Another problem of high reflected resistance is the significantly reduced output power of Tx inverters. As the load resistor of a Class-D inverter becomes larger, the output power of the inverter is reduced according to the function V_{DCin}^2 / R_{refl} [13].

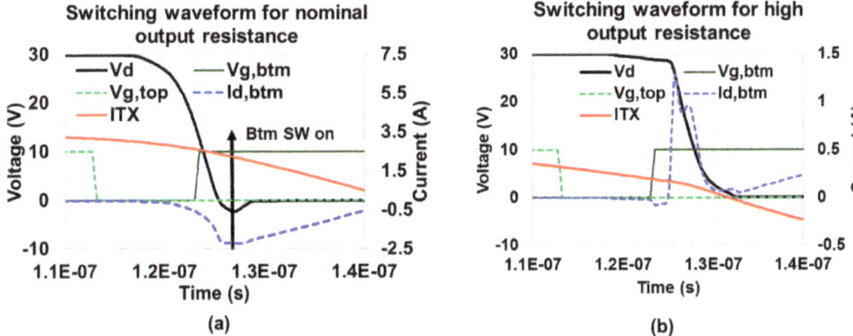

Figure 2. (**a**) Typical normal waveforms when the inverter is in zero-voltage switch mode. (**b**) When Tx and Rx are close together with high coupling. Zero-voltage switching fails due to excessive reflected resistance.

3. Proposed Dual-Impedance-Frequency Operation

The discussion in the previous section indicates that the reflected resistance should be maintained at the appropriate level. Therefore, this paper proposes dual-impedance operation by dual-frequency capability as in Figure 3a.

At a 200 kHz switching frequency, the current cannot flow toward $L_{T(R)XHI}$ because high-Z coils are tuned at ~6.78 MHz and behave as a large capacitive impedance at 200 kHz. Specifically, the impedance looking at a high-Z coil is $Z_{HI} = j\omega L_{HI} - j1/(\omega C_{6.78})$. At 6.78 MHz, the magnitude of ωL_{HI} and $1/(\omega C_{6.78})$ is comparable and thus they cancel each other out. However, for operation with low ω frequency (200 kHz), the ωL_{HI} term disappears whereas the $1/(\omega C_{6.78})$ term becomes enlarged. Hence, the total impedance becomes approximately $Z_{HI} = -j1/(\omega C_{6.78})$, which is capacitive. On the other hand, at a 6.78 MHz switching frequency, the currents cannot flow toward $L_{T(R)XLOW}$ because it behaves as large inductive impedance at 6.78 MHz switching frequency. The impedance looking at a low-Z branch which is tuned for 200 kHz is $Z_{LOW} = j\omega L_{LOW} - j1/(\omega C_{200})$. At 200 kHz, the magnitude of ωL and $1/(\omega C)$ are comparable and cancel each other. However, when ω becomes high (6.78 MHz), the $1/(\omega C_{200})$ term disappears whereas the ωL term becomes enlarged. Hence, the total impedance becomes approximately $Z_{LOW} = j\omega L_{LOW}$, which is inductive.

Therefore, the proposed coil enables dual-band operation without dedicated capacitor-selection switches. Changing the inverter driving frequency is enough for mode selection. The selected frequency bands are compatible with existing standard specification and frequency regulations. 200 kHz belongs to the Qi specification standards [14]. 6.78 MHz is the Industrial, Scientific and Medical (ISM) band which can be freely used [15], an example of which is AirFuel standards [16].

A method of adjusting coil impedance was proposed in [17]. However, this requires the coil to be split into many segments, and each segment should be connected to each relay switch. Another method of impedance reconfiguration is to use multiple capacitors and relay switches [9,18]. The drawback of these schemes is the use of relay switch. The solid-state relay in [17] can withstand only 0.2 W of power. Although mechanical relay switches can withstand higher power, its volume is too large, and its response time is slow. Moreover, a mechanical relay has a short lifetime and generates clicking sounds during operation. MOSFET or IGBT switches that are common in power electronics cannot be easily used to reconfigure resonant networks because the current at the resonant capacitor or coil is bi-directional. The intrinsic antiparallel diode within MOSFET or IGBT allows current flow from source (emitter) to drain (collector) terminal even if gate input is off. Two MOSFETs (IGBTs) should be connected in series with back-to-back configuration to block the bi-directional resonant current [19]. This increases the losses in switch and requires an additional isolated gate driver.

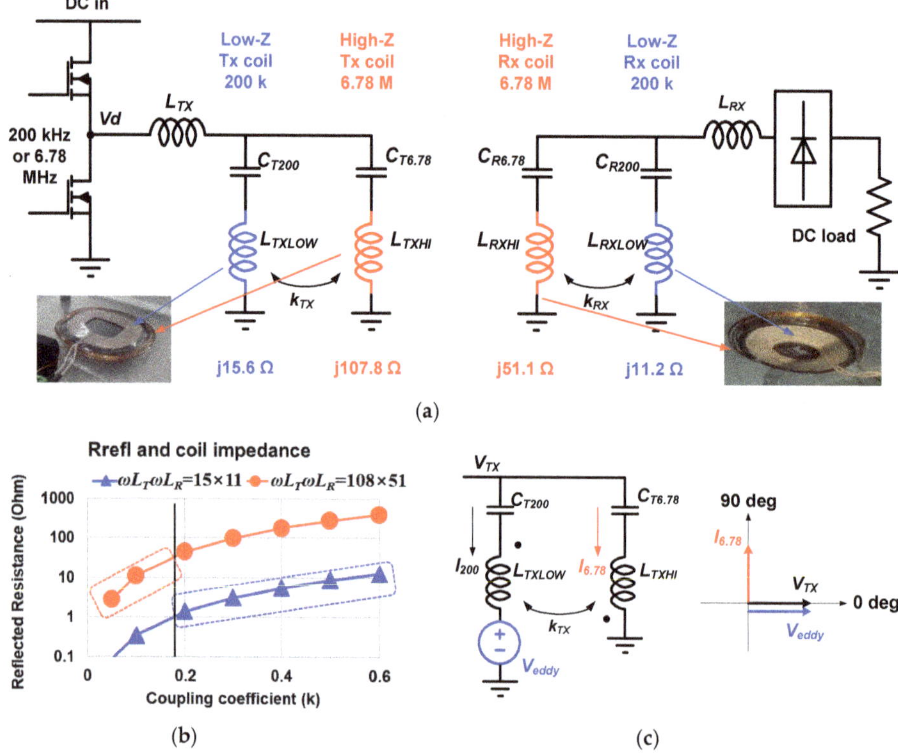

Figure 3. (**a**) Proposed dual-impedance operation by dual-frequency capability. For low-coupling condition, high-impedance coil paths are activated by 6.78 MHz operation. For high-coupling, low-impedance coil paths are activated. This can suppress excessive variations of reflected resistance which would serve as load resistor of Tx inverter, thus stabilizing the inverter efficiency. (**b**) The coil impedances are designed such that the reflected resistance is bounded within 1–30 Ω across the whole coupling coefficient k range. (**c**) Eddy current at $L_{T(R)XLOW}$ during 6.78 MHz mode is prevented by adjusting the magnitude and phase of $I_{6.78}$, so that V_{eddy} and V_{TX} are the same and the voltage across Low-Z path is zero [20].

The proposed method can handle high power without mechanical switches. Although two coils are used, the increment of volume is minimal because one coil is placed within the other coil concentrically.

The coil impedance is designed such that it is high at 6.78 MHz mode and low at 200 kHz. Hence, the coil impedance is adjusted by simply selecting the operating frequency. When the coupling k becomes low, the reflected resistance drops as in Equation (1). However, Equation (1) implies that the reflected resistance can be high if $\omega L_{TXHI} \omega L_{RXHI}$ is high even with the small coupling k. Therefore, the high-Z coils are activated by operating at 6.78 MHz. When the coupling k becomes too high, the reflected resistance becomes excessively high, resulting in the degradation of inverter efficiency as in Section 2. To prevent excessive rise of reflected resistance, small $\omega L_{TXLOW} \omega L_{RXLOW}$ is activated by 200 kHz switching frequency mode.

Referring to Figure 3b, the target range of reflected resistance, R_{refl}, is set to 1–30 Ω because, within this range, the inverter efficiency remains at least 84% or higher. This range of R_{refl} should be maintained across the whole operating distance of 0.5–4.4 cm, within which the coupling coefficient k varies from 0.057 to 0.6. For the blue triangular trace (low-Z coil) of Figure 3b, its reflected resistance drops below 1 Ω when the coupling k is lower than 0.17. At this low-coupling region, the red circular trace (high-Z) can lift the reflected resistance to 30 Ω. To this end, the high-Z coil is designed such that the multiplication of TX impedance and RX impedance of high-Z is 30 times higher than the multiplication of low-Z TX and RX impedances. The TX sector and the RX sector can have different impedance values—this does not affect the proposed design method. The RX is usually a movable device, and hence it is designed with smaller geometry and resultant lower impedance.

One practical issue is that, because $L_{T(R)XHI}$ and $L_{T(R)XLOW}$ are placed concentrically each other, high coupling between them exists, k_{TX} and k_{RX}. This coupling may cause an eddy current at $L_{T(R)XLOW}$ during 6.78 MHz switching frequency mode. To avoid this eddy current, $C_{T(R)6.78}$ is designed to be smaller than a value which would completely cancel the $L_{T(R)XHI}$ impedance [20]. Referring to Figure 3c, phasor diagram, the $I_{6.78}$ phase is faster than V_{TX} by 90 degree if $C_{T6.78}$ is chosen to be smaller than a value that resonates L_{TXHI}. From the dot convention and mutual inductance equation, the V_{eddy} is given as $V_{eddy} = -j\omega k_{TX} \sqrt{L_{TXHI}L_{TXLOW}}$, which becomes in-phase with V_{TX}. Therefore, the voltage difference across L_{TXLOW}-C_{T200} is zero, which prevents eddy current. This allows closer placement of low-frequency coil and high-frequency coil each other than the coils of [21].

4. Measurement Result

Figure 4 shows measurement setup. The L_{TXHI} and L_{RXHI} are 2.53 µH and 1.2 µH and the L_{TXLOW} and L_{RXLOW} are 12.4 µH and 8.92 µH, respectively. The C_{T200} and $C_{T6.78}$ are 55 nF and 130 pF, respectively. The C_{R200} and $C_{R6.78}$ are 64.1 nF and 252 pF, respectively.

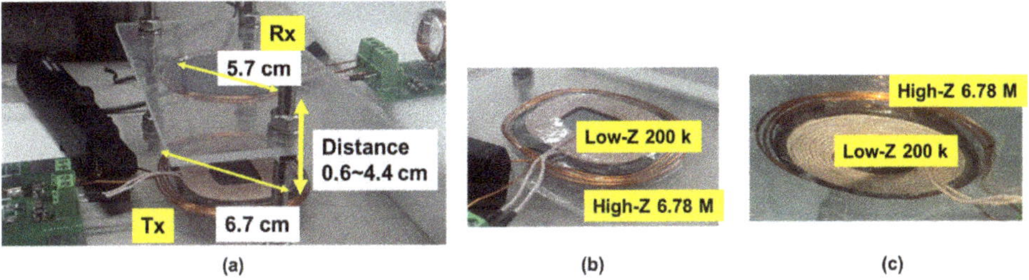

Figure 4. (**a**) Measurement setup. (**b**) Tx coils. Low-Z coils are within the high-Z coil. High-Z coil is wound around low-Z coil (**c**) Rx coils have similar configurations.

The distance between Tx and Rx varies from 0.6 cm to 4.4 cm. The corresponding coupling variation is from 0.6 to 0.057. The receiver output is 12 V–2 A, 24 W.

Figure 5 demonstrates the problems of excessively high reflected impedance. Figure 5a is the measured voltage and current waveform when the distance is only 1.1 cm and the reflected resistance is too high. The I_{TX} is suppressed due to high impedance, thereby causing longer delay for V_D to reach zero. The bottom MOSFET is turned on before V_D drops to zero, causing severe switching loss. On the other hand, the I_{TX} at 2.9 cm distance of Figure 5b is moderate and V_D can drop to zero before the bottom MOSFET is turned on. Zero-voltage switching is achieved in this case.

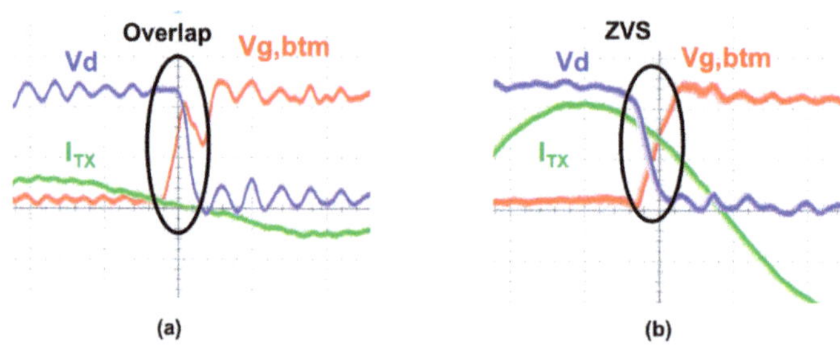

Figure 5. (**a**) Measured 6.78 MHz inverter waveforms at 1.1 cm distance (excessive reflected resistance). Gate is turned on before Vd drops to zero, causing switching losses. (**b**) 2.9 cm separation (moderate reflected resistance). Zero-voltage switching (ZVS) is achieved.

Figure 6a shows the measured efficiency and received voltage for the standalone single impedance system. The standalone single impedance system refers to the conventional single-frequency single-coil system. It is apparent that a single configuration cannot accommodate full distance range. Specifically, low-Z mode is better for 2 cm and below while high-Z performs better above 2 cm. The high-Z mode cannot output the required 12 V because of the excessive reflected resistance seen by the inverter.

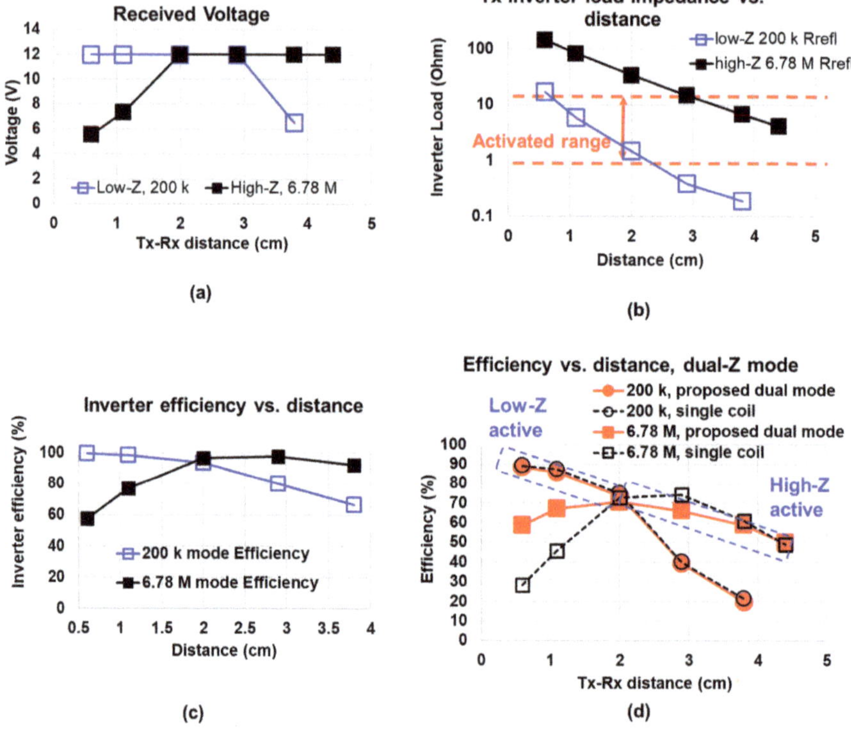

Figure 6. (**a**) Received voltage for conventional standalone single coil. A single configuration cannot accommodate the full distance range. (**b**) Measured reflected resistance for standalone low-Z and high-Z coils. (**c**) Simulated inverter efficiency. (**d**) Measured efficiency. The degradation from conventional single coil to proposed dual mode is minimal, which proves that Figure 3 successfully combines different standalone impedances into one dual-impedance system.

Figure 6b presents measured reflected resistance (i.e., inverter load) for standalone 200 kHz low-Z and 6.78 MHz high-Z configuations. At 1.1 cm and below, the inverter load resistor is too high. By introducing low-Z mode, the inverter load can be managed within 30 Ω. Figure 6c shows inverter efficiencies. At 3–4 cm 200 k mode, the inverter load (reflected resistance) is too small, causing efficiency to drop. At 0.6–1.1cm 6.78 MHz mode, inverter efficiency is degraded due to excessively high reflected resistance and ZVS failure. This is supported by Figure 5a,b.

In Figure 6d, it is proved that the two different standalone impedances are successfully merged into one dual-impedance system using Figure 3, because the efficiency degradation from standalone circuit to dual-impedance system is minimal. In a real usage scenario, the low-Z is activated simply by driving the Tx inverter to 200 kHz when reflected impedance becomes high, while the high-Z is activated by 6.78 MHz driving the inverter when reflected resistance becomes low.

One possible shortcoming of the proposed method is the parasitic resistance from L_{TX} and L_{RX} inductors. However, other matching techniques such as mechanical relays also cause parasitic resistance. Moreover, in Figure 6, at the longest distance (4.5 cm), the efficiency with the proposed method is the same as the efficiency of a single-mode coil. That means, the efficiency did not degrade due to the dual-mode capability.

5. Conclusions

A frequency-controlled dual-impedance operation which prevents excessive variation of inverter load is proposed. The low-impedance coils or the high-impedance coils are selected depending on coupling condition simply by frequency selection. This contrasts with the conventional impedance matching which relies on mechanical relay switch arrays to select capacitor bank or coil snippet.

The increment of coil volume is kept minimized by packing one coil within the other coil. The proposed method obviates the necessity of any impedance tuning switches—such as mechanical relays and back-to-back series semiconductor switches—which cause losses and high volumes. In summary, a compact, high efficiency, and mechanical-part-free method of impedance reconfiguration in response to coupling variation is realized.

Increasing the number of coils and frequency bands is not feasible at this moment because of the eddy current. For two coils in this paper, it is easy to prevent the eddy current following the scheme of Figure 3c. However, it is not directly applicable to more than 3 coils. Triple-band systems with eddy current blocking capability will be a topic for further research.

Author Contributions: Conceptualization, W.L.; Methodology, D.A.; Validation, D.A. and W.L.; Formal Analysis, W.L.; Investigation, W.L.; Resources, D.A.; Data Curation, D.A.; Writing—Original Draft Preparation, W.L.; Writing—Review & Editing, D.A.; Visualization, D.A.; Supervision, D.A.; Project Administration, D.A.; Funding Acquisition, D.A. All authors have read and agreed to the published version of the manuscript.

Funding: This work is supported by Incheon National University Grant (#2018-0482).

Conflicts of Interest: The authors declare no conflict of interest.

References

1. Liao, Z.-J.; Sun, Y.; Ye, Z.-H.; Tang, C.-S.; Wang, P.-Y. Resonant analysis of magnetic coupling wireless power transfer systems. *IEEE Trans. Power Electron.* **2019**, *33*, 5513–5523. [CrossRef]
2. Aditya, K.; Williamson, S. Design guidelines to avoid bifurcation in a series–series compensated inductive power transfer system. *IEEE Trans. Ind. Electron.* **2019**, *66*, 3973–3982. [CrossRef]
3. Ahn, D.; Hong, S. Effect of coupling between multiple transmitters or multiple receivers on wireless power transfer. *IEEE Trans. Ind. Electron.* **2013**, *60*, 2602–2613. [CrossRef]
4. Lee, W.-S.; Son, W.-I.; Oh, K.-S.; Yu, J.-W. Contactless energy transfer systems using antiparallel resonant loops. *IEEE Trans. Ind. Electron.* **2013**, *60*, 350–359. [CrossRef]
5. Lyu, Y.-L.; Meng, F.-Y.; Yang, G.-H.; Che, B.-J.; Wu, Q.; Sun, L.; Erni, D.; Li, J. A method of using nonidentical resonant coils for frequency splitting elimination in wireless power transfer. *IEEE Trans. Power Electron.* **2015**, *30*, 6097–6107. [CrossRef]

6. Lee, J.; Lim, Y.-S.; Yang, W.-J.; Lim, S.-O. Wireless power transfer system adaptive to change in coil separation. *IEEE Trans. Antennas Propagat* **2014**, *62*, 889–897. [CrossRef]

7. Kim, J.; Jeong, J. Range-adaptive wireless power transfer using multiloop and tunable matching techniques. *IEEE Trans. Ind. Electron.* **2015**, *62*, 6233–6241. [CrossRef]

8. Dang, Z.; Qahouq, J. Extended-Range Two-Coil Adaptively Reconfigurable Wireless Power Transfer System. In Proceedings of the IEEE Applied Power Electronics Conference and Exposition, Charlotte, NC, USA, 15–19 March 2015; pp. 1630–1636.

9. Jeong, S.; Lin, T.-H.; Tentzeris, M. A real-time range-adaptive impedance matching utilizing a machine learning strategy based on neural networks for wireless power transfer systems. *IEEE Trans. Microw. Theory Tech.* **2019**, *67*, 5340–5347. [CrossRef]

10. Mai, R.; Yue, P.; Liu, Y.; Zhang, Y.; He, Z. A dynamic tuning method utilizing inductor paralleled with load for inductive power transfer. *IEEE Trans. Power Electron.* **2018**, *33*, 10924–10934. [CrossRef]

11. Czarkowski, D.; Kazimierczuk, M. ZVS class D series resonant inverter—discrete-time state-space simulation and experimental results. *IEEE Trans. Circuits Syst. I* **1998**, *45*, 1141–1147. [CrossRef]

12. Erickson, R.W. Resonant Conversion. In *Fundamentals of Power Electronics*; Kluwer Academic Publishers: New York, NY, USA, 2001; pp. 723–726.

13. Grebennikov, A.; Sokal, N.O. *Switchmode RF Power Amplifiers*; Elsevier: Amsterdam, The Netherlands, 2007; pp. 78–79.

14. Wireless Power Consortium. *The Qi Wireless Power Transfer System Power Class 0 Specification*; Part 4: Reference designs, ver. 1.2.2, April. 2016; Wireless Power Consortium: Piscataway, NJ, USA, 2016.

15. International Telecommunication Union (ITU). *Wireless Power Transmission Using Technologies other than Radio Frequency Beam*; Report ITU-R SM.2303-1, 2015; International Telecommunication Union (ITU): Geneva, Switzerland, 2015.

16. AirFuel Alliance. Available online: https://airfuel.org/ (accessed on 19 December 2019).

17. Mercier, P.; Chandrakasan, A. Rapid wireless capacitor charging using a multi-tapped inductively-coupled secondary coil. *IEEE Trans. Circuit Syst. I* **2013**, *60*, 2263–2272. [CrossRef]

18. Li, Y.; Hu, J.; Chen, F.; Liu, S.; Yan, Z.; He, Z. A new-variable-coil-structure-based IPT system with load-independent constant output current or voltage for charging electric bicycles. *IEEE Trans. Power Electron.* **2018**, *33*, 8226–8230. [CrossRef]

19. Erickson, R.W. Switch Realization. In *Fundamentals of Power Electronics*; Kluwer Academic Publishers: New York, NY, USA, 2001; p. 72.

20. Ahn, D.; Kim, S.; Kim, S.-W.; Moon, J.; Cho, I.-K. Wireless power transmitter and receiver supporting 200-kHz and 6.78-MHz dual-band operation without magnetic field canceling. *IEEE Trans. Power Electron.* **2017**, *32*, 7068–7082. [CrossRef]

21. Riehl, P.; Satyamoorthy, A.; Akram, H.; Yen, Y.-C.; Yang, J.-C.; Juan, B.; Lee, C.-M.; Lin, F.-C.; Muratov, V.; Plumb, W.; et al. Wireless power systems for mobile devices supporting inductive and resonant operating modes. *IEEE Trans. Microw. Theory Tech.* **2015**, *63*, 780–790. [CrossRef]

 *electronics*

Article

Duty-Cycled Wireless Power Transmission for Millimeter-Sized Biomedical Implants

Muhammad Abrar Akram [1], **Kai-Wen Yang** [1] **and Sohmyung Ha** [1,2,*]

[1] Division of Engineering, New York University Abu Dhabi, Abu Dhabi 129188, UAE; ma5844@nyu.edu (M.A.A.); kwy225@nyu.edu (K.-W.Y.)
[2] Tandon School of Engineering, New York University, New York, NY 11201, USA
*   Correspondence: sohmyung@nyu.edu

Received: 14 October 2020; Accepted: 20 November 2020; Published: 12 December 2020

Abstract: Wireless power transmission (WPT) using an inductively coupled link is one of the most popular approaches to deliver power wirelessly to biomedical implants. As the electromagnetic wave travels through the tissue, it is attenuated and absorbed by the tissue, resulting in much weaker electromagnetic coupling than in the air. As a result, the received input power on the implant is very weak, and so is the input voltage at the rectifier, which is the first block that receives the power on the implant. With such a small voltage amplitude, the rectifier inevitably has a very poor power conversion efficiency (PCE), leading to a poor power transfer efficiency (PTE) of the overall WPT system. To address this challenge, we propose a new system-level WPT method based on duty cycling of the power transmission for millimeter-scale implants. In the proposed method, the power transmitter (TX) transmits the wave with a duty cycle. It transmits only during a short period of time and pauses for a while instead of transmitting the wave continuously. In doing so, the TX power during the active period can be increased while preserving the average TX power and the specific absorption rate (SAR). Then, the incoming voltage becomes significantly larger at the rectifier, so the rectifier can rectify the input with a higher PCE, leading to improved PTE. To investigate the design challenges and applicability of the proposed duty-cycled WPT method, a case for powering a 1×1-mm^2-sized neural implant through the skull is constructed. The implant, a TX, and the associated environment are modeled in High-Frequency Structure Simulator (HFSS), and the circuit simulations are conducted in Cadence with circuit components in a 180-nm CMOS process. At a load resistor of 100 kΩ, an output capacitor of 4 nF, and a carrier frequency of 144 MHz, the rectifier's DC output voltage and PCE are increased by 300% (from 1.5 V to 6 V) and by 50% (from 14% to 64%), respectively, when the duty cycle ratio of the proposed duty-cycled power transmission is varied from 100% to 5%.

Keywords: wireless power transmission (WPT); power conversion efficiency (PCE); mm-sized implant; duty cycle; pulsed power transmission; power transfer efficiency (PTE); rectifier

---

1. Introduction

Wireless power transfer (WPT) has become one of the most promising power transmission schemes for various biomedical implants as it obviates the need for bulky batteries and, consequently, invasive battery replacement surgeries [1–4]. Among multiple WPT methods for biomedical implants such as inductive powering [5], capacitive powering [6], optical powering [7], and ultrasonic energy transferring [8], inductive powering is one of the most developed and popular approaches to deliver power wirelessly over a-few-cm distances [1,4,9,10]. However, this technique suffers from low power transfer efficiency

Electronics **2020**, *9*, 2130; doi:10.3390/electronics9122130   www.mdpi.com/journal/electronics

(PTE) when used for millimeter-scale implants because of the poor electromagnetic coupling between the external power transmitter and the implant's coil antenna [11]. The PTE largely depends on the coupling coefficient k and the quality factor Q of the coils. Both of them are drastically degraded as the coil antenna size is scaled down to millimeter scales [11,12]. For millimeter-scale free-floating implants, an achievable Q factor on the implant side is typically around or below 10. Thus, an achievable k is around or less than 0.005, and the resulting PTE is only 2% or less [13,14]. Hence, low achievable PTEs remain one of the most limiting bottlenecks for further miniaturization of implants.

In addition to such poor electromagnetic coupling, the absorption of electromagnetic waves by biological tissues confines the maximum transmittable power, thus further limiting the power delivered to the load (PDL) [15]. The international safety standards on specific absorption rate (SAR) limit the transmitted power of TX to 1.6 W/kg according to the IEEE guidelines [16]. SAR is the parameter commonly used to quantify absorptive and exposure levels of biological tissue to electromagnetic waves. Due to these reasons, therefore, the input power at the implant is typically very weak.

1.1. PTE-Improvement Techniques for Inductively Coupled WPT Systems

To improve the amount of power delivered to the biomedical implants and the efficiency, various techniques have been proposed and demonstrated. As one of the most basic techniques, impedance matching should be properly achieved for high PTEs. To obtain a proper impedance matching, an inductance tapping circuit and a voltage doubler circuit were proposed [17]. For the robust operation of the inductive link, various tuning techniques were also demonstrated [17–19]. As it is difficult to achieve a high Q for the coil on the miniature biomedical implant, some techniques to increase the inductance on the implant side were proposed [20–23]. Besides, techniques to optimize the design parameters of the planar printed spiral coils were demonstrated, achieving a PTE of 41.2% with a link operating frequency of 1 MHz [22]. In [20], a micro-coil array was implemented for a multi-channel wireless implantable system to achieve a high PTE. Similarly, a three-coil inductive link was proposed to achieve both high PTE and PDL [23]. However, an extra coil generally requires a larger space on the implant side and more complex designs, posing several design constraints on the inductor geometries [24].

1.2. Techniques to Improve the Power Conversion Efficiency of Rectifiers

To improve the overall PTE, many recent studies have investigated the rectifier circuits. Because the rectifier is the first power-receiver block on the implant side, the rectifier should extract as much power as possible from the RF input. Rectifiers convert the incoming RF input to a DC voltage ($V_{DC,OUT}$) at the output. The power conversion efficiency (PCE) of a rectifier is the ratio of the DC output power to the input power of the rectifier [25]. The PCE largely depends on its circuit topology, diode-device parameters, RF input power, and output loading conditions. Comparators can be used to turn on and off the MOSFETs to improve the PCE [26]. However, they typically work at relatively low frequencies as 13.56 MHz, and they are in-efficient at high frequencies. For millimeter-sized implants, typical carrier frequencies are over 100 MHz and even above 1 GHz [12,27,28]. The rectifiers for such applications are typically made up with passive components. For most passive rectifier topologies, the threshold voltage (V_{LTH}) of the diode devices is the major limiting factor of the PCE [29], especially when the input signal's amplitude is small. Diode-based rectifiers typically suffer from a huge voltage drop across the diodes, leading to a poor PCE and wide dead zone at low input levels. As compared to typical pn-junction diodes and MOSFET-based diodes, using Schottky diodes in a rectifier can enhance its PCE, especially at low input power, because of their very small threshold voltage and fast switching speeds [30–32]. The fully cross-coupled rectifier [25] can alleviate the dead-zone problem (the input sensitivity problem) to some extent and operate at a relatively low input power by applying the input signal deferentially across the four rectifying transistors. However, it suffers from a poor PCE due to the presence of reverse leakage currents.

Even with recent improvements in rectifier circuits [29,33–35], the improvement in the PCE and eventually in the PTE is still not enough. The input sensitivity issue caused by the dead zone is not resolved. To address the aforementioned challenges of low PCE and PTE at weak input power and the dead-zone issue, we present a new system-level WPT method utilizing a duty-cycled WPT technique that significantly improves the PCE and wireless power delivery to the miniature biomedical implants at low input power levels. Similar preliminary works were presented previously [36–38]. Compared to these, our paper includes more systematic analyses on the method from the rectifier to the whole system, mathematical derivations on the optimal load condition, and practical considerations in various aspects. Details of the proposed method are presented in the next section. The validation setup, including the design specifications of the HFSS circuit models, is described in Section 3. Section 4 shows the results of simulations using a 1×1-mm^2 implant model and discusses main practical constraints of the proposed schemes. Section 5 concludes the paper.

2. The Proposed Scheme

Figure 1 shows an overall circuit diagram of a general WPT system, which includes a transmitter (TX), receiver (Rx), and biological tissue in between. The two coils, L_1 and L_2, are inductively coupled with a coupling coefficient k. The LC tanks on both sides are tuned to resonate at the same resonant frequency with C_1 and C_2, respectively. The LC tank on the secondary side is connected to a rectifier, which generates a DC voltage $V_{DC,OUT}$ to the load R_L. The figure also illustrates voltage waveforms of major nodes (VL_1, V_{LL} and $V_{DC,OUT}$) for the conventional and the proposed power transmission methods. As compared to the conventional WPT method where the power is continuously transmitted, in the proposed duty-cycled power transmission, the RF power signal is transmitted during a designated portion of time only and remains dormant for the rest of the duration. The active and dormant periods are iterated with a periodic interval of T_I, as shown in Figure 2. Figure 2 shows the relationship between T_I and the active duty period (T_D) of the duty-cycled waveform. The ratio of these two is the duty cycle ratio D, which is given as follows:

$$D = \frac{T_D}{T_I}. \tag{1}$$

As illustrated in Figure 1, the instantaneous TX power for the proposed method can be larger than that of the conventional continuous power transmission at the same RF input. The TX power for the proposed method is increased up to the point where the average TX power is the same as that of the continuous powering case. Then, the average SAR levels of both cases are at the same level. As a result, the rectifier's voltage input is also increased, so a greater proportion of the input can more easily overcome the threshold voltage of the rectifier. Hence, more RF power is converted into the output DC power with a higher PCE of the rectifier, effectively increasing the overall PCE of the rectifier circuit and the output DC voltage $V_{DC,OUT}$ level and the PTE of the WPT system.

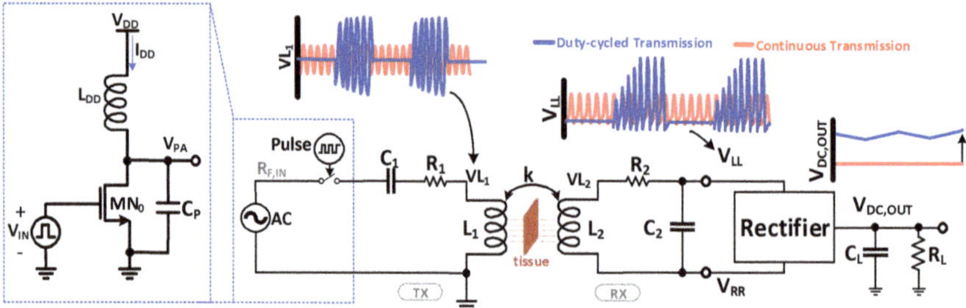

Figure 1. Circuit diagram of the general wireless power transmission (WPT) system with waveforms of major voltage nodes for the conventional continuous power transmission (in red) and the proposed duty-cycled power transmission (in blue). Inset: Implementation scenario with a class-E amplifier for the input voltage control.

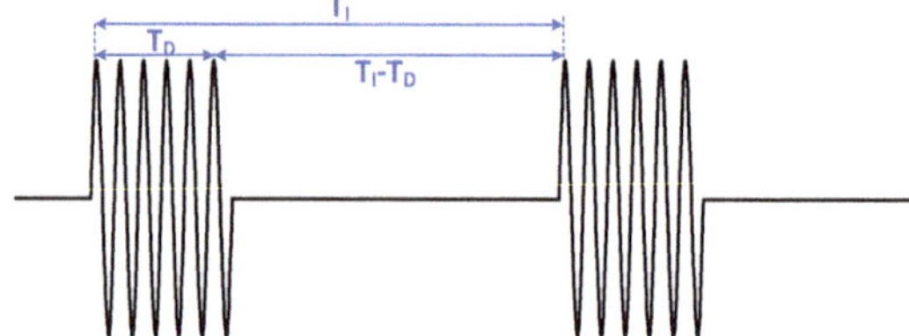

Figure 2. Duty-cycled waveform with indications of the duty cycle interval T_I and the active duty period T_D.

3. Validation Setup

3.1. HFSS Simulation Model

To validate the proposed scheme, we constructed a three-dimensional finite element method (FEM) model in Ansys HFSS, as shown in Figure 3. The model models multi-layered brain tissues, including the skull, based on frequency-dependent dielectric properties [12]. An 1×1-mm²-sized implant, which is assumed to be a 200-μm-thick semiconductor chip fabricated in a CMOS Silicon-on-Insulator (SOI) process (XFAB XT018), is covered with Parylene C and placed on the grey matter under the skull at a depth of 9.7 mm. The top view of the RX coil is shown in Figure 3. The 7-turn RX coil occupies the peripheral space of 1×1 mm² of the implant. The widths of the metal coil and gap are kept at 1:1 while the center part (the area not occupied by the coil) is 0.5×0.5 mm². A single-turn power TX coil is positioned concentrically with the implant coil above the skin at an air-gap of 1 mm. The TX coil has a radius of 12 mm. The detailed geometry of the TX coil is also shown in Figure 3. Using this HFSS model, simulations were performed at the carrier frequency 144 MHz. With the two-port vector network analyzer in HFSS, Z parameters were obtained from the simulations. R_i, L_i, C_i, and k values were calculated with the following equations:

$$R_i = Re(Z_{i,i}) \tag{2}$$

$$L_i = \frac{Im(Z_{i,i})}{2\pi f} \tag{3}$$

$$C_i = \frac{1}{L_i \times (2\pi f)^2} \tag{4}$$

$$k = \sqrt{\frac{Im(Z_{1,2}) \times Im(Z_{2,1})}{Im(Z_{1,1}) \times Im(Z_{2,2})}}. \tag{5}$$

The following circuit parameters for both the TX and RX coils were acquired and used for the circuit simulations in Cadence: L_1 = 18.3 nH, R_1 = 0.285 Ω, L_2 = 57.6 nH, R_2 = 6.78 Ω, C_1 = 38.3 pF, C_2 = 12.2 pF, k = 0.00136. The resonance frequency is set at 144 MHz.

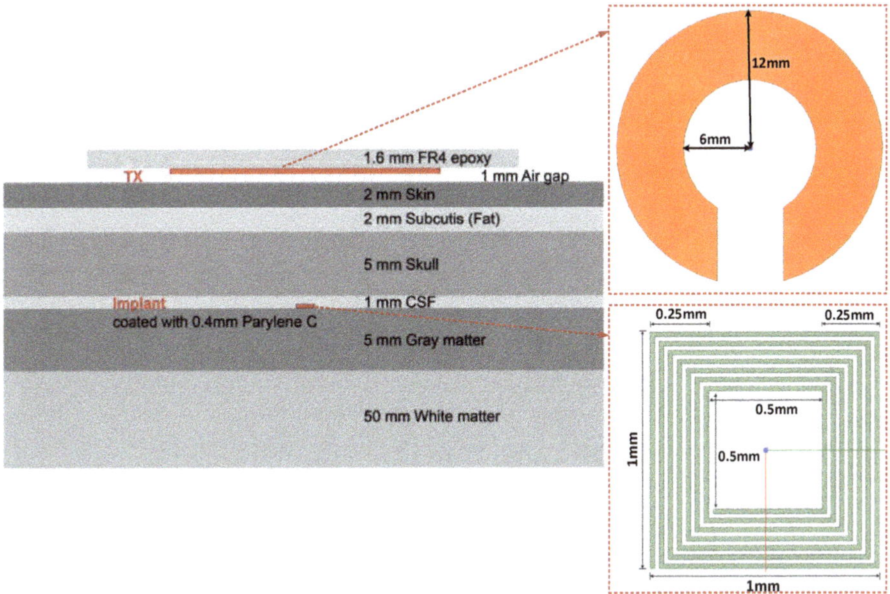

Figure 3. Cross-sectional configuration of the implemented finite element method (FEM) model in High-Frequency Structure Simulator (HFSS). The 1 $\times$ 1 mm^2-sized seven-turn-coil implant is placed at 9.7-mm depth of the brain. The one-tern circular transmitter (TX) is placed above the skin.

3.2. Rectifier Circuit and Characterization Results

On the RX side, the LC tank is connected to a rectifier. The rectifier in Figure 4 is chosen for validation of the proposed scheme, while the proposed method can be applied to most types of rectifiers in general. The rectifier is composed of a pair of Schottky diodes (D_1 and D_2) and a pair of cross-coupled NMOS transistors (MN_1 and MN_2). All of these circuit components are in a 180-nm CMOS SOI process. In the rectifier design, we selected Schottky diodes because of their low threshold voltage of 150–350 mV, much lower than typical pn junction diodes. Sizes of the NMOS transistors and Schottky diodes were optimized to acquire the maximum possible PCE at the given conditions.

Operational waveforms of the rectifier are shown in Figure 5. The input power to the rectifier is received through V_{LL} and V_{RR} during the active period of the duty-cycled powering. V_{LL} and V_{RR} are rectified during the active period, resulting in an increase of $V_{DC,OUT}$. During the dormant period of duty-cycled powering, $V_{DC,OUT}$ droops slightly. To observe the rectifier operation more in detail during the positive and the negative cycles of V_{LL} and V_{RR}, these waveforms are zoomed in and shown in the inset of Figure 5. When V_{RR} is high and V_{LL} is low, it corresponds to the forward-biased condition for MN_1

and D_2. Because V_{RR} the gate voltage of MN_1 is positive and high, MN_1 has a small turn-on resistance. At the same time, V_{RR} overcomes the threshold voltage of D_2, allowing the forward current I_{D2} to $V_{DC,OUT}$. The rectifier works for the other polarity vice versa.

Before investigating the impacts of the duty-cycled power transmission on the PCE, PTE, and $V_{DC,OUT}$, the rectifier is characterized first. The amplitude of the input voltage V_{IN} on the primary side is varied from 0.1 V to 1.2 V at a fixed load resistor R_L of 100$k\Omega$ and output capacitor C_L of $4nF$. The RF signal is transferred wirelessly to the secondary side and arrives at the input of the rectifier. The resulting $V_{DC,OUT}$, PCE, and PTE over $V_{LL,ptp}$ the peak-to-peak voltage at the input of the rectifier are plotted in Figure 6. Here the PCE is the ratio of DC output power to the input power of the rectifier (PCE = $P_{DC,OUT}/P_{IN}$ of the rectifier), and the PTE of a WPT system is the ratio of the DC output power to the input power of TX coil (PTE = $P_{DC,OUT}/P_{IN}$ on the TX side). As shown in Figure 6a, $V_{DC,OUT}$ linearly increases after a dead-zone of about 300 mV as $V_{LL,ptp}$ increases. As shown in Figure 6b,c, the PCE and PTE also have a similar dead zone and increases monotonically afterwards as $V_{LL,ptp}$ increases. When $V_{LL,ptp}$ is 2 V, the resulting $V_{DC,OUT}$, PCE and PTE are 1.48 V, 14% and 0.0011%, respectively. As clearly shown in the results, this continuous power transmission suffers not only from its small outputs ($V_{DC,OUT}$, PCE, and PTE) but also from the dead-zone, which is $\approx$300 mV. When the RF input is within the dead zone, all $V_{DC,OUT}$, PTE, and PCE are stuck at zero.

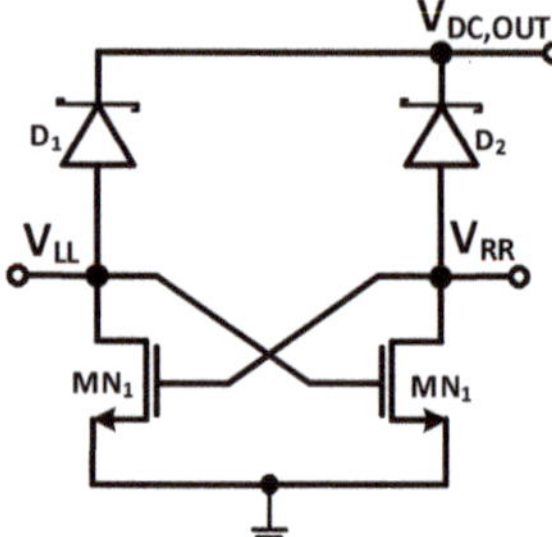

Figure 4. Circuit diagram of the rectifier used in the validation.

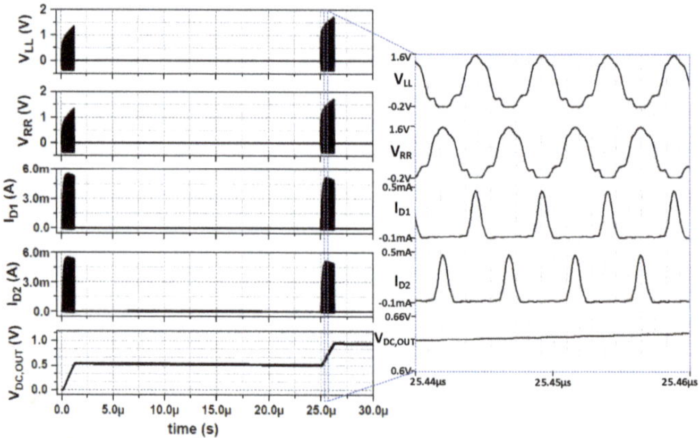

Figure 5. Operational waveforms of the rectifier.

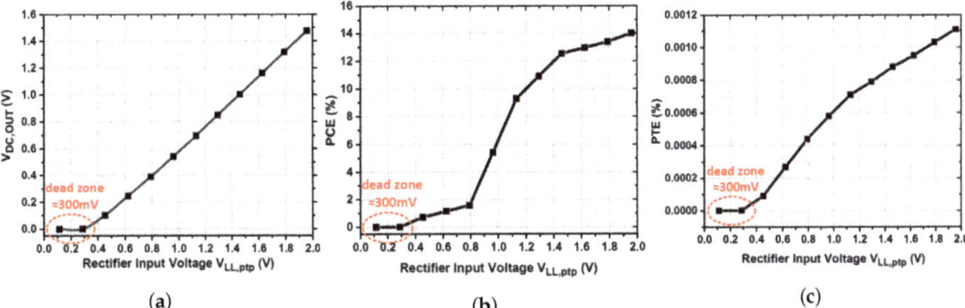

Figure 6. Characterization results of the rectifier shown in Figure 4. (**a**) $V_{DC,OUT}$, (**b**) power conversion efficiency (PCE), and (**c**) power transfer efficiency (PTE) of the rectifier with respect to $V_{LL,ptp}$ the peak-to-peak amplitude of the input voltage of the rectifier.

4. Results and Discussion

4.1. Circuit Simulation Setup

The circuit in Figure 1 was constructed in Cadence using the 180-nm CMOS SOI process. For each run of transient simulations, the circuit was simulated for a long enough duration to settle down at a steady state. For each simulation, $V_{DC,OUT}$, PCE of the rectifier, and PTE of the whole WPT system are observed. We simulated with the target circuit over V_{IN} from 0.1 V to 1.2 V for various duty cycle ratios, i.e., 100%, 50%, 10%, and 5%. For each duty cycle, the TX power should be maintained at the same level, so is the specific absorption rate (SAR). Since k is very small as 0.00136, the transmitted power is directly proportional to the power from the input source on the primary side. Thus, the input voltage V_{IN} required for each case of different duty cycle ratio is determined as follows:

$$V_{IN} = \frac{V_{IN,D=100\%}}{\sqrt{D}}, \tag{6}$$

where $V_{IN,D=100\%}$ is V_{IN} for the case with 100% of duty cycle, and D is the duty cycle ratio. For instance, assume that the RF power is transmitted at 0.5 V of V_{IN} for a 100% duty cycle ratio. For $D = 50\%$, the RF signal is transmitted only during half of the time, so the TX power can be doubled while maintaining the same SAR. Therefore, the required V_{IN}, in this case, is 0.707 V.

4.2. Simulation Results

The simulation results at $R_L = 100$ kΩ and $C_L = 4$ nF for $D = 100\%$, 50%, 10%, and 5% are shown in Figure 7. The results for 100% of the duty is the same as the conventional continuous power transmission. For all duty-cycle cases including the continuous transmission case, the average TX power is maintained at 25.5 dBm. Figure 7a clearly shows that $V_{DC,OUT}$ significantly increases as D reduces for the all range of $V_{IN,D=100\%}$. $V_{DC,OUT}$ increases by more than four times as D decreases from 100% to 5% for $V_{IN,D=100\%}$ of 0.3 to 1.2 V. The increase in $V_{DC,OUT}$ is attributed to the effective improvement in the rectifier's PCE shown in Figure 7b. The maximum PCE is drastically increased from 14% to 64% at $V_{IN,D=100\%}$ of 1.2 V. This inevitably leads to improvement in the overall PTE of the WPT system, as shown in Figure 7c. Moreover, as compared to the continuous power transmission ($D = 100\%$), the proposed duty-cycled power transmission technique eliminates or significantly reduces the dead-zone for low-RF-input cases, as shown in the figures. In Figure 7d, the PCE of the rectifier is plotted with respect to the duty cycle ratio for various $V_{IN,D=100\%}$. As shown, significant improvements in the PCE are made by reducing the duty cycle ratio for every $V_{IN,D=100\%}$ case. In addition, the proposed duty-cycled WPT system achieves 33% and 39% better PCE performances as compared to the previous PWM-based WPT techniques [37,38], respectively, at $R_L = 100$ kΩ.

To evaluate the effect of the duty cycle interval T_I on the performance of the proposed method, simulations were done by sweeping T_I from 1 µs to 100 µs for D of 100%, 50%, 10%, and 5%, at $V_{IN,D=100\%}$ = 0.5 V, $R_L = 100$ kΩ, and $C_L = 4$ nF. Figure 8 shows the resulting $V_{DC,OUT}$ and PCE. $V_{DC,OUT}$ is maintained for $D = 50\%$, but it decreases as T_I decreases for shorter duty cycle ratios. The decreasing slope is steeper for the case with $D = 5\%$. It is because it requires some recovery time to reach the maximum amplitude due to the resonance whenever the active transmission (T_D) starts. Figure 9 shows simulated V_{L1} and V_{LL} at at $T_I = 1$ µs and $D = 5\%$ ($T_D = 0.05$ µs). The active duty period of the waveforms is zoomed in and shown on the right side of Figure 9. As shown, only several cycles of the RF carrier signal are sent for each duty interval (T_D). Since T_D is too short, the amplitude of the signals cannot reach the maximum level within T_D. Due to this effect, $V_{DC,OUT}$ decreases as T_I decreases for small D (Figure 8a). On the contrary, the PCE is less affected by T_I, as shown in Figure 8b.

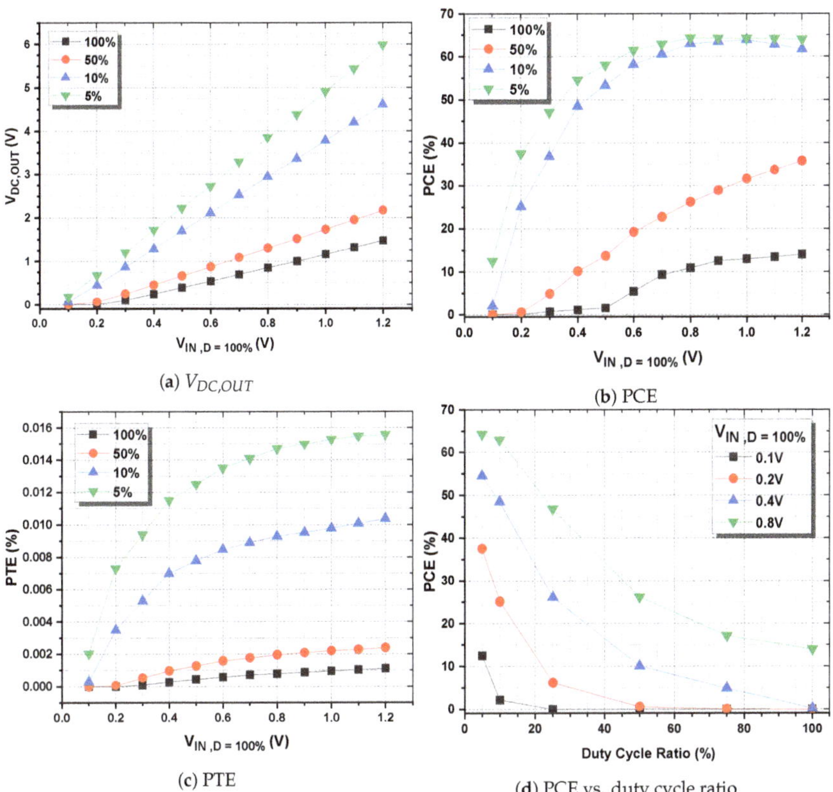

(**a**) $V_{DC,OUT}$

(**b**) PCE

(**c**) PTE

(**d**) PCE vs. duty cycle ratio

Figure 7. Simulation results of the rectifier and the WPT system for various duty cycle ratios for the same amount of RF input power. (**a**) $V_{DC,OUT}$, (**b**) PCE, and (**c**) PTE over $V_{IN,D=100\%}$. (**d**) PCE vs. duty cycle ratio for various $V_{IN,D=100\%}$.

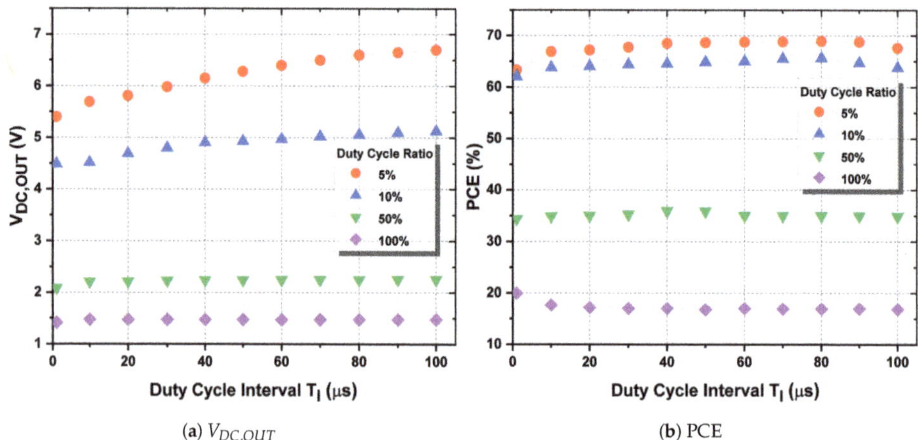

(a) $V_{DC,OUT}$

(b) PCE

Figure 8. (a) $V_{DC,OUT}$ and (b) PCE of the rectifier with respect to T_I the duty cycle interval.

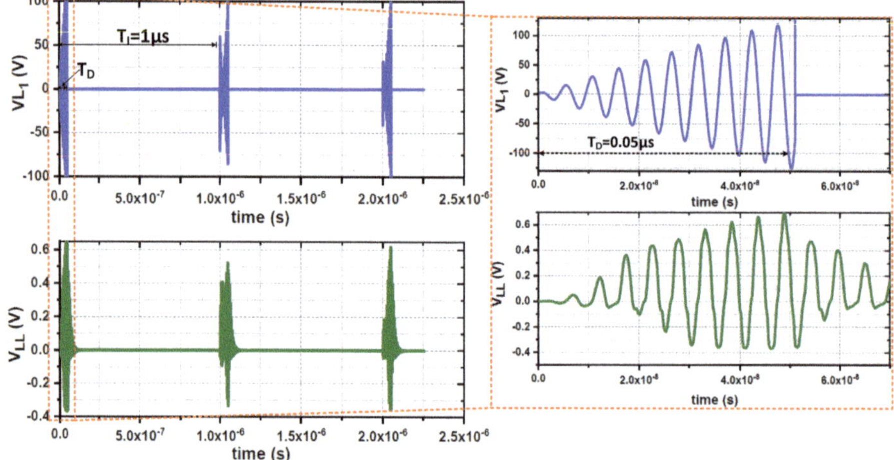

Figure 9. Simulated V_{L1} and V_{LL} waveforms at T_I = 1 µs and D = 5%.

The performance gain by the duty-cycled power transmission depends on the load condition. The overall PTE of a WPT system strongly depends on the load, and the maximum power condition is differed by the load. Because the duty cycling effectively changes the load in view of the secondary LC tank and the overall WPT system, the duty cycle ratio and the load are interrelated. Hence, depending on the load condition, the PTE may not increase as the duty cycle ratio decreases. Therefore, the load conditions should also be considered to achieve the peak PTE.

The PTE of the WPT system is evaluated under various load conditions by sweeping the duty cycle ratios from 1% to 100%, as shown in Figure 10a. While the PTE is at its peak when D = 1% for R_L of 100 kΩ and 1 MΩ, the PTE reaches to the maximum at D = 3% for R_L = 40 kΩ, at D = 4% for R_L = 20 kΩ, and at D = 8% for R_L = 10 kΩ. The maximum point becomes D of 1% for R_L of 1 kΩ, again. To further evaluate the optimum condition for the maximum PTE, the PTE is observed by sweeping R_L from 1 kΩ to 10 MΩ

for fixed duty cycle ratios of 1%, 5%, 10%, 25%, 50%, 75%, and 100%. The resulting PTEs are shown in Figure 10. As shown, the load condition should be considered together to find the optimal duty cycle ratio.

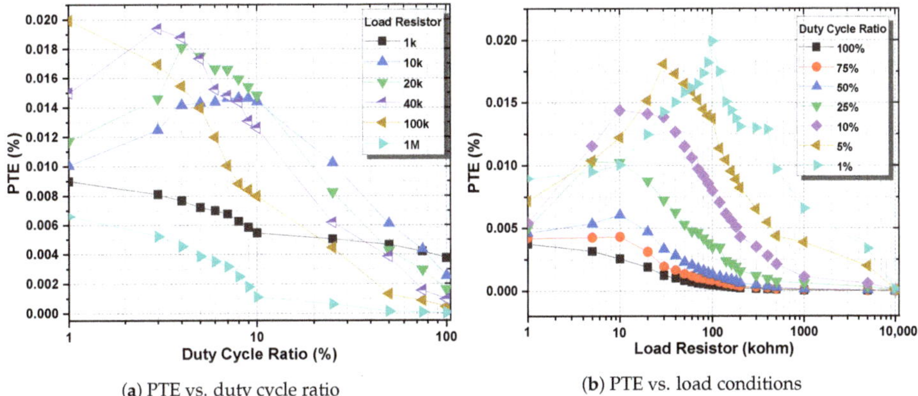

(a) PTE vs. duty cycle ratio   (b) PTE vs. load conditions

Figure 10. (**a**) PTE over the duty cycle ratio for various load resistances. (**b**) PTE over the load resistance for various duty cycle ratios.

4.3. Optimal Load Conditions

As presented in the previous section, the PTE depends on the load and the duty cycle ratio. The PTE of an inductive link shown in Figure 1 is given by:

$$\text{PTE} = \frac{k^2 Q_1 Q_{2,loaded}}{1 + k^2 Q_1 Q_{2,loaded}} \cdot \frac{R_{P,2}}{R_{L,ac} + R_{P,2}}, \tag{7}$$

where k is the coupling coefficient, Q_1 and Q_2 are the quality factors of the TX and RX coils, and R_1 and R_2 are the equivalent series resistances, which model the losses of the TX and RX coils. $R_{L,ac}$, which is connected in parallel to the secondary LC tank, is the AC load resistance that models the rectifier and the load resistor R_L as shown in Figure 11. $Q_{2,loaded}$ and $R_{P,2}$ are given as follows:

$$Q_{2,loaded} = \eta_2 Q_2 \tag{8}$$

$$R_{P,2} = R_2(Q_2^2 + 1), \tag{9}$$

where $\eta_2 = R_{L,ac}/(R_{L,ac} + R_{P,2})$ [39]. In our work, k is very small, so $1 + k^2 Q_1 Q_{2,loaded} \cong 1$. Equation (7) can be simplified as follows:

$$\text{PTE} = k^2 Q_1 Q_{2,loaded} \cdot \frac{R_{P,2}}{R_{L,ac} + R_{P,2}}. \tag{10}$$

By putting Equation (8) into Equation (10), the following equation is obtained:

$$\text{PTE} = k^2 Q_1 Q_2 R_{P,2} \cdot \frac{R_{L,ac}}{(R_{L,ac} + R_{P,2})^2}. \tag{11}$$

To find the maximum condition of the PTE with respect to $R_{L,ac}$, we need to take its partial derivative as follows:

$$\frac{\partial PTE}{\partial R_{L,ac}} = k^2 Q_1 Q_2 R_{P,2} \cdot \frac{R_{P,2} - R_{L,ac}}{(R_{L,ac} + R_{P,2})^3} = 0.$$ (12)

Hence, the optimal $R_{L,ac}$ can be found as follows:

$$R_{L,ac,opt} = R_{P,2} = R_2((Q_2)^2 + 1).$$ (13)

Since $Q_2^2 \gg 1$ typically, Equation (13) can be more simplified to:

$$R_{L,ac,opt} \cong Q_2^2 R_2 = \frac{L_2}{C_2 R_2}.$$ (14)

Therefore, $R_{L,ac,opt}$ is the output impedance of the parallel-tuned LC tank on the secondary side (Figure 1).

$R_{L,ac}$ can be related to the output load (R_L) of the rectifier by using the functions of PCE and voltage conversion efficiency (VCE). The PCE is the relationship between the output DC power of the rectifier (P_L) and the input power ($P_{in,rec}$) of the rectifier. The VCE is the relationship between the output DC voltage of the rectifier ($V_{DC,OUT}$ and the input voltage amplitude of the rectifier. The PCE and VCE are given as follows:

$$PCE = P_L / P_{in,rec}$$ (15)

$$VCE = V_{DC,OUT} / V_{L2},$$ (16)

where $P_L = (V_{DC,OUT})^2 / R_L$, and $P_{in,rec} = D \cdot (V_{L2})^2 / (2R_{L,ac})$. From Equations (14)–(16), the overall optimal load resistance ($R_{L,opt}$) can be given as follows:

$$R_{L,ac} = D \cdot \frac{PCE}{2VCE^2} \cdot R_L,$$ (17)

Thus, the optimal load resistance $R_{L,opt}$ is:

$$R_{L,opt} = \frac{1}{D} \cdot \frac{2VCE^2}{PCE} \cdot R_{L,ac,opt} = \frac{1}{D} \cdot \frac{2VCE^2}{PCE} \cdot \frac{L_2}{C_2 R_2}$$ (18)

Here, note that PCE and VCE are functions of the rectifier input voltage, which is a function of D.

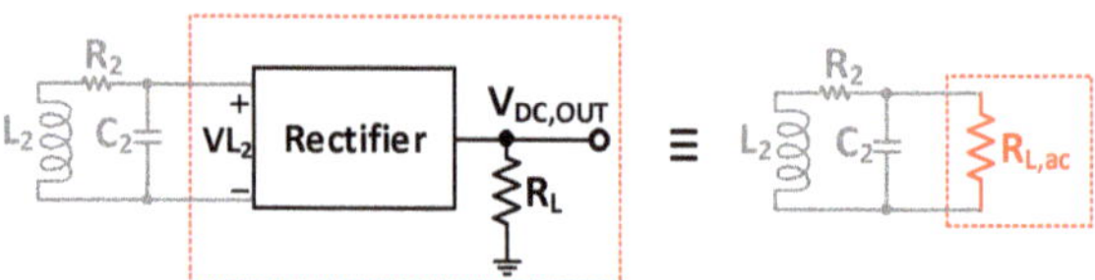

Figure 11. Illustration for $R_{L,ac}$, which is the equivalent parallel resistor for the rectifier and R_L.

Using the derived Equation (18), we calculated the optimal load where the proposed WPT system achieves maximum PTE over the duty cycle. Furthermore, we compared the calculated optimal loads with the simulated ones from the results shown in Figure 10b. As shown in Figure 12, the simulated and calculated results are well-matched over the duty cycle.

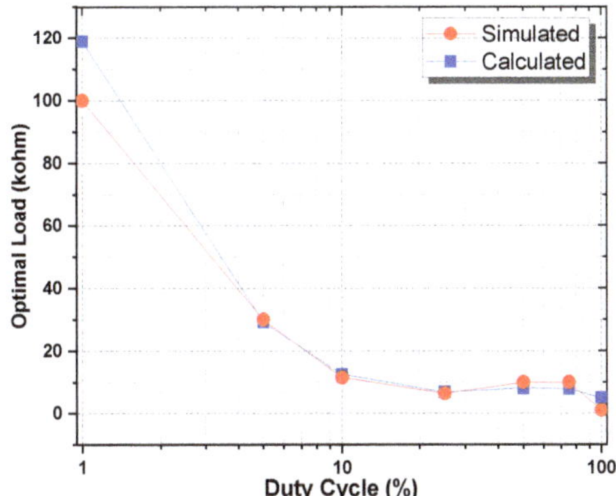

Figure 12. Simulated and calculated optimized load conditions for the maximum PTE over the duty cycle.

4.4. Practical Constraints

Although the PCE and output DC voltage $V_{DC,OUT}$ can be significantly enhanced with the proposed duty-cycled power transmission, there are two practical constraints of the proposed technique: (1) the voltage ripple at $V_{DC,OUT}$ and (2) the maximum instantaneous TX power.

Firstly, as the power is transmitted during the active phase only (T_D in Figure 2), $V_{DC,OUT}$ droops during the dormant phase (T_I–T_D in Figure 2), inducing voltage ripples on $V_{DC,OUT}$. The ripple magnitude increases as the duty cycle phase gets shorter for given input power, load condition, and output capacitor.

Figure 13a shows the ripple ratio (RR) in respect to the duty cycle ratio at T_I = 1 µs, R_L = 100 kΩ, and C_L = 4 nF. RR is defined as the ratio between $V_{R,pp}$ the peak-to-peak amplitude of the ripple voltage and $\overline{V_{DC,OUT}}$ the average dc output voltage:

$$\text{RR} = \frac{V_{R,pp}}{\overline{V_{DC,OUT}}}. \tag{19}$$

As shown in Figure 13a, RR is at the minimum when the power is continuously transmitted (D = 100%). RR increases as D decreases, and it reaches approximately 7% and 15% at D = 5% and 1%, respectively. To further investigate RR, T_I the duty cycle interval is varied from 1 µs to 100µ for five fixed duty cycle ratios. As shown in Figure 13b, RR can be decreased by decreasing T_I, while conserving the same duty cycle ratio. It is because the dormant time decreases as T_I decreases even with the same duty cycle ratio.

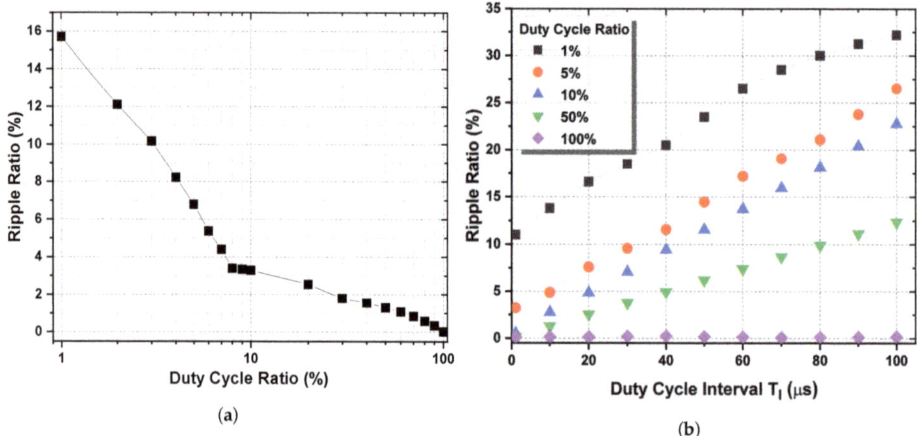

Figure 13. (a) Ripple ratios (RR) vs. the duty cycle ratio (D) at T_I the duty cycle interval of 1µs. (b) RR vs. T_I at 100%, 50%, 10%, and 1% of the duty cycle ratios

To reduce the undesirable ripples further, the size of the load capacitor (C_L) can be increased. Nevertheless, there is a limit on the C_L size that can be integrated on the semiconductor chip or in the miniaturized millimeter-scale implant. Due to this limit, 4 nF of C_L is chosen as the maximum achievable capacitance and is used in the simulations in this paper. Although RR is high for shorter duty cycle ratios, it can be subdued by high power-supply rejection (PSR) of the low-dropout regulator (LDO) [40,41], which is typically implemented after the rectifier circuit in WPT systems. Moreover, the frequency components of the ripples are typically out of the frequency range of the biological signals of interest. For instance, assume an implant to measure action potentials from neurons. Frequency ranges of the signal and the recording channel are typically from 100 Hz to 10 kHz, so ripples around 1 kHz may affect the signal reading adversely. Such ripples can be avoided easily by choosing T_I outside of the range between 0.1 ms and 10 ms.

In addition to the ripple effect, there are also some instrumental limits in the duty-cycled power transmission. In practice, the instantaneous TX power cannot be increased indefinitely to implement the case for short duty cycle ratios. This is mainly because of limited power supply levels on the TX side. Within a limit of the maximum TX power, the duty cycle ratio can be shortened to increase the effective PDL and PTE.

5. Conclusions

This paper proposed a new system-level duty-cycled WPT technique, which can be applied to miniature biomedical implants. The proposed duty-cycled WPT technique enhances the overall power-transfer efficiency (PTE) to the implants by effectively increasing the power conversion efficiency (PCE) of the rectifier at weak and attenuated input power. Based on an FEM model of a WPT system in HFSS for a 1×1-mm^2 implant on the brain surface, circuit simulations of the proposed method were performed in a 180-nm CMOS SOI process. The proposed technique increases the output voltage and PCE of the rectifier by $\times 4$ and $\times 4.5$, respectively, and hence the overall PTE is improved by $\times 15$, as compared to typical continuous WPT techniques. The proposed technique is more effective for the cases with a low k and a low input voltage for the rectifier. Thus, its potential applications include millimeter-sized implants for brain signal recording and stimulation and other miniature sensors inside the body. This technique can also be applied to wireless power transmission for other environmental sensors that are placed at a low k. In addition, this technique is not limited to the inductively coupled systems only, but it can also be applied to ultrasonic powering. In the future, a fully adaptive input-power-aware duty-cycled WPT system is likely to be proposed. It is expected to be able to decide its duty cycle automatically according to the received power at the implant. Such an adaptive scheme can further optimize the PTE and power delivery under various conditions.

Author Contributions: Conceptualization, K.-W.Y. and S.H.; HFSS modeling, K.-W.Y. and S.H.; Cadence circuit modeling, M.A.A. and K.-W.Y.; simulations, M.A.A. and K.-W.Y.; analysis, M.A.A., K.-W.Y., and S.H.; resources, S.H.; writing—original draft preparation, M.A.A.; writing—review and supervision, S.H. All authors have read and agreed to the published version of the manuscript.

Funding: This research received no external funding.

Conflicts of Interest: The authors declare no conflict of interest.

References

1. Khan, S.R.; Pavuluri, S.K.; Cummins, G.; Desmulliez, M.P.Y. Wireless Power Transfer Techniques for Implantable Medical Devices: A Review. *Sensors* **2020**, *20*, 3487.
2. Hui, S.Y.R.; Zhong, W.; Lee, C.K. A Critical Review of Recent Progress in Mid-Range Wireless Power Transfer. *IEEE Trans. Power Electron.* **2014**, *29*, 4500–4511.
3. Wang, W.X.; Dai, H.A. A Critical Review of Wireless Power Transfer via Strongly Coupled Magnetic Resonances. *Energies* **2014**, *7*, 4316–4341.
4. Kim, H.; Hirayama, H.; Kim, S.; Han, K.J.; Zhang, R.; Choi, J. Review of Near-Field Wireless Power and Communication for Biomedical Applications. *IEEE Access* **2017**, *5*, 21264–21285.
5. Basar, M.R.; Ahmad, M.Y.; Ibrahim, F.; Cho, J. Resonant inductive power transfer system for freely moving capsule endoscope with highly uniform magnetic field. In Proceedings of the 2016 IEEE Industrial Electronics and Applications Conference (IEACon), Kota Kinabalu, Malaysia, 20–22 November 2016; pp. 393–397.
6. Kline, M.; Izyumin, I.; Boser, B.; Sanders, S. Capacitive power transfer for contactless charging. In Proceedings of the 2011 Twenty-Sixth Annual IEEE Applied Power Electronics Conference and Exposition (APEC), Fort Worth, TX, USA, 6–11 March 2011; pp. 1398–1404.
7. Rahman, M.M.; Adalian, D.; Chang, C.; Scherer, A. Optical power transfer and communication methods for wireless implantable sensing platforms. *J. Biomed. Opt.* **2015**, *20*. [CrossRef]
8. Ozeri, S.; Shmilovitz, D. Ultrasonic transcutaneous energy transfer for powering implanted devices. *Ultrasonics* **2010**, *50*, 556–566.
9. Lu, Y.; Ki, W. *CMOS Integrated Circuit Design for Wireless Power Transfer*; Springer: New York, NY, USA, 2018; pp. 13–32.
10. Lee, B.; Jia, Y.; Mirbozorgi, S.A.; Connolly, M.; Tong, X.; Zeng, Z.; Mahmoudi, B.; Ghovanloo, M. An Inductively-Powered Wireless Neural Recording and Stimulation System for Freely-Behaving Animals. *IEEE Trans. Biomed. Circuits Syst.* **2019**, *13*, 413–424.
11. Ahn, D.; Hong, S. Effect of Coupling Between Multiple Transmitters or Multiple Receivers on Wireless Power Transfer. *IEEE Trans. Ind. Electron.* **2013**, *60*, 2602–2613.
12. Yang, K.; Oh, K.; Ha, S. Challenges in Scaling Down of Free-Floating Implantable Neural Interfaces to Millimeter Scale. *IEEE Access* **2020**, *8*, 133295–133320.
13. Feng, P.; Yeon, P.; Cheng, Y.; Ghovanloo, M.; Constandinou, T.G. Chip-Scale Coils for Millimeter-Sized Bio-Implants. *IEEE Trans. Biomed. Circuits Syst.* **2018**, *12*, 1088–1099.
14. Kim, C.; Park, J.; Ha, S.; Akinin, A.; Kubendran, R.; Mercier, P.P.; Cauwenberghs, G. A 3 mm × 3 mm Fully Integrated Wireless Power Receiver and Neural Interface System-on-Chip. *IEEE Trans. Biomed. Circuits Syst.* **2019**, *13*, 1736–1746. [PubMed]

15. Ahn, D.; Ghovanloo, M. Optimal Design of Wireless Power Transmission Links for Millimeter-Sized Biomedical Implants. *IEEE Trans. Biomed. Circuits Syst.* **2016**, *10*, 125–137. [PubMed]

16. Lin, J.C. IEEE Standard for Safety Levels with Respect to Human Exposure to Radio Frequency Electromagnetic Fields, 3 kHz to 300 GHz. *IEEE Antennas Propag. Mag.* **2006**, *48*, 157–159.

17. Jouran, M.A.; Yang, X.; Chen, W. Magnetically coupled resonance WPT: Review of compensation topologies, resonator structures with misalignment, and EMI diagnostics. *Electronics* **2018**, *7*, 296.

18. Hochmair, E.S. System Optimization for Improved Accuracy in Transcutaneous Signal and Power Transmission. *IEEE Trans. Biomed. Eng.* **1984**, 177–186. [CrossRef]

19. Zierhofer, C.M.; Hochmair, E.S. High-efficiency coupling-insensitive transcutaneous power and data transmission via an inductive link. *IEEE Trans. Biomed. Eng.* **1990**, *37*, 716–722.

20. Mou, Z.; Zheng, X.; Hou, W.; Wang, X.; Wu, X.; Jiang, Y. An Analytical Model for Inductively Coupled Multichannel Implantable System With Micro-Coil Array. *IEEE Trans. Magn.* **2012**, *48*, 2421–2429.

21. Gong, C.; Liu, D.; Miao, Z.; Li, M. A Magnetic-Balanced Inductive Link for the Simultaneous Uplink Data and Power Telemetry. *Sensors* **2017**, *17*, 1768.

22. Jow, U.; Ghovanloo, M. Design and Optimization of Printed Spiral Coils for Efficient Transcutaneous Inductive Power Transmission. *IEEE Trans. Biomed. Circuits Syst.* **2007**, *1*, 193–202.

23. Kiani, M.; Jow, U.; Ghovanloo, M. Design and Optimization of a 3-Coil Inductive Link for Efficient Wireless Power Transmission. *IEEE Trans. Biomed. Circuits Syst.* **2011**, *5*, 579–591.

24. Zargham, M.; Gulak, P.G. Maximum Achievable Efficiency in Near-Field Coupled Power-Transfer Systems. *IEEE Trans. Biomed. Circuits Syst.* **2012**, *6*, 228–245. [PubMed]

25. Kotani, K.; Sasaki, A.; Ito, T. High-Efficiency Differential-Drive CMOS Rectifier for UHF RFIDs. *IEEE J. Solid-State Circuits* **2009**, *44*, 3011–3018.

26. Lee, H.-M.; Ghovanloo, M. An Integrated Power-Efficient Active Rectifier with Offset-Controlled High Speed Comparators for Inductively Powered Applications. *IEEE Trans. Circuits Syst. I Regul. Pap.* **2011**, *58*, 1749–1760. [PubMed]

27. Karimi, Y.; Khalifa, A.; Montlouis, W.; Stanaćević, M.; Etienne-Cummings, R. Coil array design for maximizing wireless power transfer to sub-mm sized implantable devices. In Proceedings of the 2017 IEEE Biomedical Circuits and Systems Conference (BioCAS), Torino, Italy, 19–21 October 2017; pp. 1–4.

28. Khalifa, A.; Karimi, Y.; Huang, Y.; Stanaćevic, M.; Etienne-Cummings, R. The Challenges of Designing an Inductively Coupled Power Link for µm-sized On-Chip Coils. In Proceedings of the 2018 IEEE Biomedical Circuits and Systems Conference (BioCAS), Cleveland, OH, USA, 17–19 October 2018; pp. 1–4.

29. Almansouri, A.S.; Ouda, M.H.; Salama, K.N. A CMOS RF-to-DC Power Converter With 86% Efficiency and −19.2-dBm Sensitivity. *IEEE Trans. Microw. Theory Tech.* **2018**, *66*, 2409–2415.

30. Karthaus, U.; Fischer, M. Fully integrated passive UHF RFID transponder IC with 16.7-µW minimum RF input power. *IEEE J. Solid-State Circuits* **2003**, *38*, 1602–1608.

31. Barnett, R.; Balachandran, G.; Lazar, S.; Kramer, B.; Konnail, G.; Rajasekhar, S.; Drobny, V. A Passive UHF RFID Transponder for EPC Gen 2 with −14 dBm Sensitivity in 0.13µm CMOS. In Proceedings of the 2007 IEEE International Solid-State Circuits Conference. Digest of Technical Papers, San Francisco, CA, USA, 11–15 February 2007; pp. 582–623.

32. Wu, K. *Introduction To SCHOTTKY Rectifier and Application Guidelines*; Taiwan Semiconductor: Hsinchu, Taiwan, 2003.

33. Kim, C.; Ha, S.; Park, J.; Akinin, A.; Mercier, P.P.; Cauwenberghs, G. A 144-MHz Fully Integrated Resonant Regulating Rectifier with Hybrid Pulse Modulation for mm-Sized Implants. *IEEE J. Solid-State Circuits* **2017**, *52*, 3043–3055.

34. Xue, Z.; Fan, S.; Li, D.; Zhang, L.; Gou, W.; Geng, L. A 13.56 MHz, 94.1% Peak Efficiency CMOS Active Rectifier with Adaptive Delay Time Control for Wireless Power Transmission Systems. *IEEE J. Solid-State Circuits* **2019**, *54*, 1744–1754.

35. Lee, B.; Ahn, D. Robust Self-Regulated Rectifier for Parallel-Resonant Rx Coil in Multiple-Receiver Wireless Power Transmission System. *IEEE J. Emerg. Sel. Top. Power Electron.* **2019**. [CrossRef]

36. Lo, C.; Yang, Y.; Tsai, C.; Lee, C.; Yang, C. Novel wireless impulsive power transmission to enhance the conversion efficiency for low input power. In Proceedings of the 2011 IEEE MTT-S International Microwave Workshop Series on Innovative Wireless Power Transmission: Technologies, Systems, and Applications, Kyoto, Japan, 12–13 May 2011; pp. 55–58.
37. Yang, Y.; Yang, C.; Tsai, C.; Lee, C. Efficiency improvement of the impulsive wireless power transmission through biomedical tissues by varying the duty cycle. In Proceedings of the 2011 IEEE MTT-S International Microwave Workshop Series on Innovative Wireless Power Transmission: Technologies, Systems, and Applications, Kyoto, Japan, 12–13 May 2011; pp. 175–178.
38. Yang, Y.; Tsai, C.; Yang, C.; Yang, C. Using pulse width and waveform modulation to enhance power conversion efficiency under constraint of low input power. In Proceedings of the 2012 Asia Pacific Microwave Conference Proceedings, Kaohsiung, Taiwan, 4–7 December 2012; pp. 400–402.
39. Park, J.; Kim, C.; Akinin, A.; Ha, S.; Cauwenberghs, G.; Mercier, P.P. Wireless powering of mm-scale fully-on-chip neural interfaces. In Proceedings of the 2017 IEEE Biomedical Circuits and Systems Conference (BioCAS), Torino, Italy, 19–21 October 2017; pp. 1–4.
40. Yun, S.J.; Yun, J.S.; Kim, Y.S. Capless LDO Regulator Achieving -76 dB PSR and 96.3 fs FOM. *IEEE Trans. Circuits Syst. II Express Briefs* **2017**, *64*, 1147–1151.
41. Lu, Y.; Ki, W.; Yue, C.P. 17.11 A 0.65 ns-response-time 3.01 ps FOM fully-integrated low-dropout regulator with full-spectrum power-supply-rejection for wideband communication systems. In Proceedings of the 2014 IEEE International Solid-State Circuits Conference Digest of Technical Papers (ISSCC), San Francisco, CA, USA, 9–13 February 2014; pp. 306–307.

 *electronics*

Article

A Novel Metal Foreign Object Detection for Wireless High-Power Transfer Using a Two-Layer Balanced Coil Array with a Serial-Resonance Maxwell Bridge

Sunhee Kim [1], Haeyong Jung [2], Youngjun Ju [3] and Yongseok Lim [2,*]

[1] Department of System Semiconductor Engineering, Sangmyung University, Cheonan-si 31066, Korea; happyshkim@smu.ac.kr

[2] Korea Electronics Technology Institute, Seoul 03924, Korea; sunwater4@keti.re.kr

[3] EMF Safety Inc., Yongin-si, Gyeonggi-do 16890, Korea; youngjunju@emfsafety.co.kr

* Correspondence: busytom@keti.re.kr; Tel.: +82-2-6388-6669

Received: 31 October 2020; Accepted: 1 December 2020; Published: 4 December 2020

Abstract: In a wireless high-power transfer system with a distance of several tens of centimeters apart between the transmitter and receiver coils, one of the most challenging issues is to detect metallic foreign objects between the transmitter and receiver coils. The metallic foreign objects must be detected and removed since these reduce the transmission efficiency and cause heat generation of the transmitter and receiver. This paper presents two-layer symmetric balanced coil array so that if there are metallic foreign objects, it can be detected through the change of the inductance of the balanced coils. Since the balanced coil is composed of coils that are in a symmetrical relationship in position, there is no need for a reference coil, and interference between coils is reduced by dividing the coil into two layers. In addition, a novel serial-resonance Maxwell bridge circuit to improve the inductance change detection performance is presented in this paper. The proposed metallic foreign object detection system is implemented using two-layer balanced coil array with a serial-resonance Maxwell bridge and the experimental results show that voltage changes of hundreds of mV to several V occur when a metallic foreign object is inserted, so that even small metals such as clips can be detected.

Keywords: balanced coil; foreign object detection; Maxwell bridge; metal object detection; wireless power transfer

---

1. Introduction

Recently, research on wireless power transfer (WPT) has been focused on wireless charging of large mobile devices such as electric vehicles, automatic guided vehicles, and robots beyond wireless small devices such as mobile phones, smart watches and wearable devices [1–3]. Since it requires thousands of times more power than conventional small devices, high-power WPT has additional requirements for safety during charging as well as the requirements for charging efficiency [4,5]. Therefore, technologies such as unwanted emission electromagnetic field reduction, heat control, foreign object detection (FOD), position detection (PD), and human safety have attracted more attention [6,7]. In particular, technology that detects metallic foreign objects (MO) such as coins, keys, and foil not only reduces transmission efficiency, but can also cause circuit damage or fire, and even life-threatening if creatures such as cats and dogs get in between. Therefore, technologies such as FOD including metal object detection (MOD) and live object detection (LOD) are defined as very important technologies in the application of the product and are also defined as essential elements that must be observed in industry standardization organizations such as the Wireless Power Consortium (WPC) and Society of Automotive Engineers (SAE) International.

The Qi Wireless Power Transfer System Power Class 0 Specification version 1.2.3 proposes two FOD methods [8]. The first method monitors the temperature change at the interface surface of the transmitter. Another method monitors power loss between the transmitting interface and the receiving interface. However, in the case of high-power WPT, it is not appropriate to apply the methods proposed by Qi WPT. This is because the power loss caused by a metallic foreign object (MO) is very small compared to the output power in high-power WPT systems, making it difficult to detect. In addition, if there is a separation distance between the transmitter coil and receiver coil unlike the low-power WPC system, the distance between the coils is not constant and the power transfer efficiency varies greatly by misalignment [9,10]. Although SAE International's J2954 wireless power transfer for light-duty plug-in/electrical vehicles and alignment methodology does not propose specific FOD methods, it is regulated to conduct MOD test including coins, paper clips, steel sheet, beverage cans, etc. [6,11].

For FOD methods in high-power WPT, wave-based detection methods and field-based detection methods are considered [6]. Wave-based detection methods require additional sensing devices such as imaging cameras, thermal cameras, or radar sensors to detect the presence of foreign objects. Thus the method has a high cost and is difficult to integrate with the power transfer system. The field-based detection method is a method of detecting a change in a magnetic field generated by the metal objects. To this end, a detection coil other than the transmitter coil and the receiver coil is added, and placed between the transmitter coil and the receiver coil to detect change in the characteristics of the detection coil. However, since the characteristics of the detection coil are also affected by the location of the receiver coil and the surrounding conditions such as temperature, techniques using a balanced coil have recently been introduced.

The balanced coil is also basically located between the transmitter coil and the receiver coil. In addition, a reference coil is added to detect the changes in the characteristics of the balanced coil when metal foreign objects are inserted. The balanced coil and the reference coil are arranged symmetrically with respect to the transmitter coil, so that the magnitude of the magnetic flux induced in the two coils is the same. When a metal object is inserted, the balance of magnetic flux induced in the two coils is broken. Then, the presence of a metal object is detected by detecting that the amount of current flowing through the two coils is different. When there is no metal foreign object, the current does not flow in the coil [12].

The size of the MO that can be detected depends on the size of the balanced coil, so the balanced coil must be small enough to detect general metal objects. In addition, depending on the shape or arrangement of balanced coils, there may be areas where it is difficult to detect MOs. In other words, when a relatively small MO is inserted or a MO is inserted in an area that is difficult to detect, the change in inductance of the balanced coil is very small and it is difficult to detect it. Therefore, studies have been proposed to amplify a voltage or current generated by the change of inductance with an amplifier, but they have a problem of amplifying noise generated by power transfer signals. Therefore, in this paper, we propose a balanced coil structure with a serial-resonance Maxwell bridge circuit that can detect small MOs in wireless power transfer systems with a certain distance between the transmitter coil and the receiver coil.

2. Materials and Methods

2.1. Effect of Metallic Foreign Object on Wireless Charging

Figure 1 shows the basic principle of transferring electric energy over the air in wireless power transfer using magnetic field coupling. In general, the transmitter system generates AC power through an inverter-type power conversion circuit and transmits power by forming a magnetic field over the air through the power transfer coil. The receiver system converts the power received from the power reception coil to DC through reception circuits including a rectifier and transmitted to the load. As the distance between the coils increases, the induced magnetic flux drastically decreases, and the power transfer efficiency decreases rapidly. Therefore, when there is a separation distance of 10 cm or more, the magnetic resonance method is generally used among the magnetic field coupling methods in order to overcome the problem of reducing efficiency according to the distance.

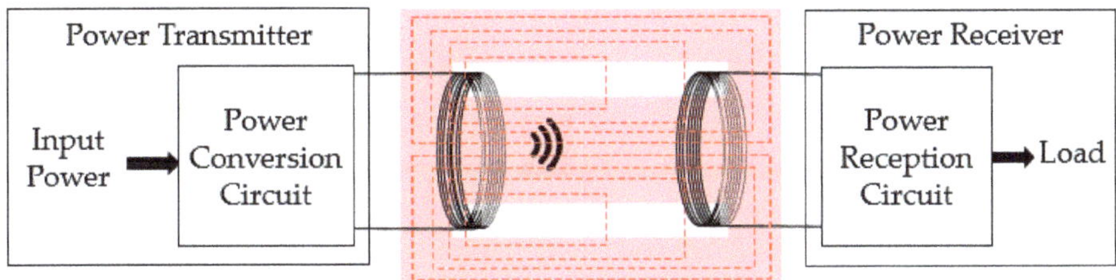

Figure 1. The basic principle of transferring electric energy over the air.

Figure 2 shows the current and magnetic field generation, and the equivalent circuit when a MO is placed on the transmitter coil. When the current I_{tx} flows through the transmitter coil, it generates the magnetic field H_{tx}. When the magnetic field H_{tx} passes through an area with the MO, an electromotive force is generated inside the metal object, thereby generating a current I_{mo}. This current I_{mo} is generated in the form of a loop on a plane perpendicular to the magnetic field H_{tx} and is called an eddy current. Then, it generates a new magnetic field H_{mo}. The magnetic field H_{mo} is generated in the opposite direction to the change of the original magnetic field H_{tx}, thereby interfering with power transfer at the transmitter. In addition, heat is generated in the metal object by the eddy current flowing through the resistance component of the metal object.

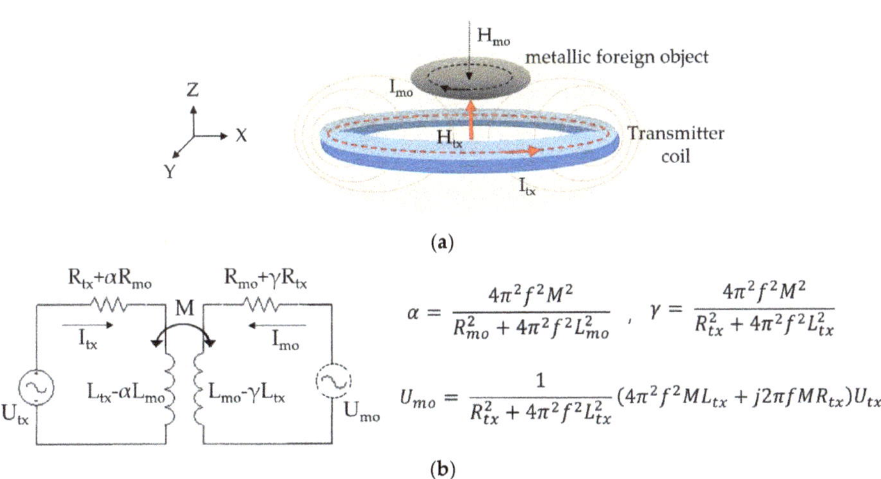

$$\alpha = \frac{4\pi^2 f^2 M^2}{R_{mo}^2 + 4\pi^2 f^2 L_{mo}^2} \quad , \quad \gamma = \frac{4\pi^2 f^2 M^2}{R_{tx}^2 + 4\pi^2 f^2 L_{tx}^2}$$

$$U_{mo} = \frac{1}{R_{tx}^2 + 4\pi^2 f^2 L_{tx}^2}(4\pi^2 f^2 M L_{tx} + j2\pi f M R_{tx})U_{tx}$$

Figure 2. (**a**) The current and magnetic field generation, and (**b**) the equivalent circuit when a metallic foreign object (MO) is placed on the top of the coil.

In the equivalent circuit as shown in Figure 2b, L_{tx} and R_{tx} are the inductance and resistance of the transmitter coil respectively, L_{mo} and R_{mo} are the inductance and resistance of the MO, respectively. M is the mutual inductance between the transmitter coil and the metal object and U_{tx} is the AC power source at the transmitter system. When there is no MO, the impedance Z_{tx} of the transmitter circuit can be expressed as follows.

$$Z_{tx} = R_{tx} + j2\pi f L_{tx} \tag{1}$$

where f is the frequency of the AC power and I_{tx} is the current flowing through the transmitter coil. When a MO is inserted, Kirchhoff's voltage law theory is applied to the transmitter and the MO as shown in Equations (2) and (3), respectively.

$$U_{tx} = (R_{tx} + j2\pi f L_{tx})I'_{tx} - j2\pi f M I_{mo} \tag{2}$$

$$R_{mo}I_{mo} + j2\pi f L_{mo}I_{mo} - j2\pi f M I'_{tx} = 0 \tag{3}$$

where I_{tx}' is the current of the transmitter coil changed by the MO. If these two equations are combined as in Equation (4), the impedance of the transmitter circuit is changed by mutual inductance as Equation (5). In addition, the changed impedance of the MO and induced voltage are as Equations (6) and (7), respectively.

$$U_{tx} = \left(R_{tx} + \frac{4\pi^2 f^2 M^2}{R_{mo}^2 + 4\pi^2 f^2 L_{mo}^2} R_{mo}\right)I'_{tx} + j2\pi f\left(L_{tx} - \frac{4\pi^2 f^2 M^2}{R_{mo}^2 + 4\pi^2 f^2 L_{mo}^2} L_{mo}\right)I'_{tx} \tag{4}$$

$$Z'_{tx} = (R_{tx} + \alpha R_{mo}) + j2\pi f(L_{tx} - \alpha L_{mo}), \quad \alpha = \frac{4\pi^2 f^2 M^2}{R_{mo}^2 + 4\pi^2 f^2 L_{mo}^2} \tag{5}$$

$$Z'_{mo} = (R_{mo} + \gamma R_{tx}) + j2\pi f(L_{mo} - \gamma L_{tx}), \quad \gamma = \frac{4\pi^2 f^2 M^2}{R_{tx}^2 + 4\pi^2 f^2 L_{tx}^2} \tag{6}$$

$$U_{mo} = \frac{1}{R_{tx}^2 + 4\pi^2 f^2 L_{tx}^2}\left(4\pi^2 f^2 M L_{tx} + j2\pi f M R_{tx}\right)U_{tx} \tag{7}$$

where Z_{tx}' are the impedance of the transmitter coil changed by the MO. As you can see from the Equation (5), the real part of the transmitter coil impedance is increasing, while the reactive part is decreasing. As a result, when a MO is inserted between the transmitter coil and the receiver coil, the inductance of the transmitter coil decreases and the resonant frequency of the transmitter changes, resulting in a sharp drop in power transfer efficiency. In addition to the heat hazard caused by eddy current in MOs, heat is generated by the increase of the conductance of the transmitter coil, causing serious damage such as failure of the transmitter circuit or a fire. For this reason, MOD technology is very important in high-power wireless power transfer.

2.2. Two-Layer Balanced Coil Array Architecture

As shown in Figure 3, a balanced coil array is placed between the transmitter and receiver coils. That is, it is placed on the transmitter coil and designed to cover the entire cross section of the transmitter coil. As shown in Figures 1 and 2, since the transmitter coil is a form of symmetry, the generated magnetic field is symmetric not only with respect to the x-y plane of the transmitter coil, but also with respect to the z-axis. In other words, although the intensity of the field is not uniform on the x-y plane, it is formed symmetrically with respect to the center point. Therefore, the balanced coil is constructed with two coils in a symmetrical relationship in position. Therefore, since the two coils act as reference coils with each other, separate reference coils are not added.

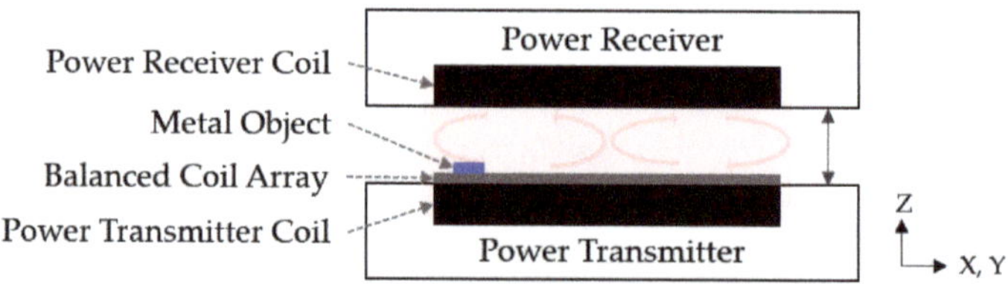

Figure 3. A cross-section of the system.

Figure 4 shows a proposed two-layer balanced coil array architecture. Each balanced coil array consists of $(N \times M)/2$ rectangular loop coils. In the first layer, $N/2$ coils arranged along the Y-axis are connected to form a total of M coil sets. Then, $M/2$ balanced coil arrays B$\{X_m{:}X_{M-m+1}\}$ are formed between coil sets that are symmetrical to the Y-axis. In the second array, $M/2$ coils arranged along the X-axis are connected to form a total of N coil sets. Likewise, $N/2$ balanced coil arrays B$\{Y_n{:}Y_{N-nm+1}\}$ are formed between coil sets that are symmetrical to the X-axis. However, as shown in Figure 4, in this paper, the coils are not arranged completely symmetrically. This is because if a balanced coil set is constructed with coils in exactly symmetrical positions, when a MO spans $X_{M/2}$ and $X_{M/2+1}$ (or $Y_{N/2}$ and $Y_{N/2+1}$), the degree of unbalance between the two coils is insignificant.

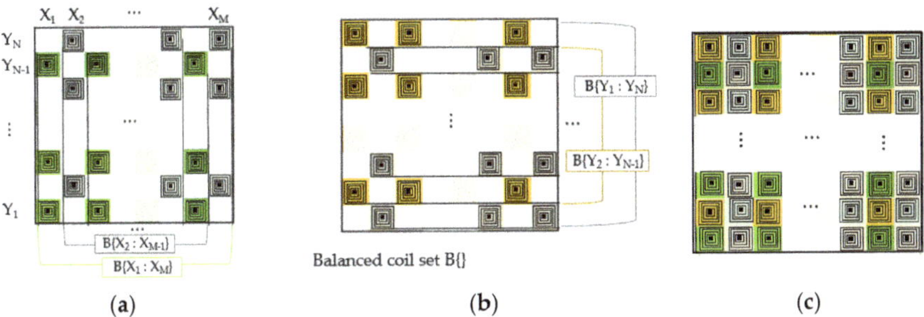

Figure 4. A two-layer balanced coil array architecture. (**a**) Balanced coil array in the first; (**b**) balanced coil array in the second layer; and (**c**) overlapped layers to cover the entire plane.

One layer does not cover the entire plane, and the coils are arranged across one by one in the $(N \times M)$ grid structure. These two layers are overlapped to cover the entire plane. If $N \times M$ coils are placed in one plane to cover the entire plane, the LC resonance circuit is formed by the parasitic parallel capacitance produced between adjacent coils. Thus, when a MO is inserted, the inductance of the coil at the position where the MO is placed changes, and a series of current change occurs in neighboring coils, leading to the change of current flowing through the all coils. This phenomenon attenuates the effect of comparing the signal difference using the balanced coil. Thus, like the proposed structure, the coils are arranged one by one to minimize the effect of the surrounding coils

If a MO is inserted on top of X_2, the inductance of X_2 is reduced by the MO as shown in Equation (7). Thus, the balance of X_2 and X_{M-1} is broken. When you detect which coil pair is out of balance, you know the presence of a metal foreign object.

In a balanced coil array, if the area of the rectangular loop coil is narrowed and the number of coils arranged is increased, the size of the detectable foreign object decreases and the place resolution increases. However, the magnetic flux passing through the cross-section of the coil also decreases, which reduces the amount of induced current. In addition, since systems such as wireless charging for vehicles use high-power AC signals, noise generated in the system is large, making it more difficult to detect the induced current change. Therefore, it is common to add an amplifier in order to improve the foreign object detection sensitivity [12]. However, there is a problem that even noise is amplified in this case. Therefore, we propose a balanced coil array with a serial-resonance Maxwell bridge circuit that can detect a small change of inductance.

2.3. Serial-Resonance Maxwell Bridge Circuit

The Maxwell bridge as shown in Figure 5 is generally used to measure an unknown inductance using a calibrated resistance, a known resistance and a known inductance [13]. The value of the unknown inductance is detected by measuring the voltage between two nodes A and B at the center of the bridge circuit.

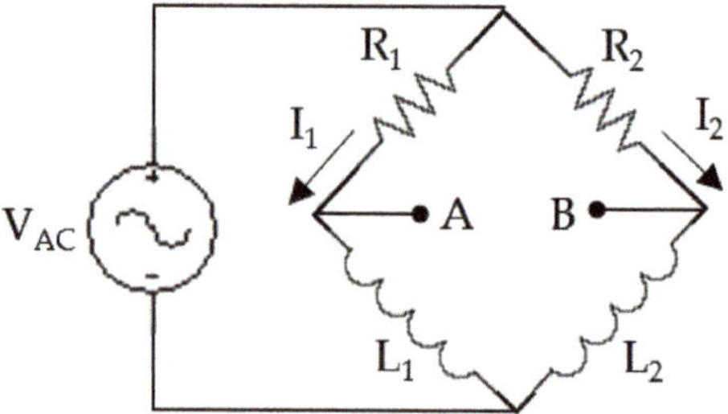

Figure 5. Maxwell bridge circuits.

When AC voltage V_{AC} with constant frequency f is supplied, the voltage between two nodes A and B, V_{AB}, is as

$$V_{AB} = (V_{AC} - R_1 I_1) - (V_{AC} - R_2 I_2) = R_2 I_2 - R_1 I_1 \tag{8}$$

where, if the following equations are satisfied, the voltage between the two nodes becomes zero.

$$L_1 I_1 = L_2 I_2 \text{ and } R_1 I_1 = R_2 I_2 \text{ , or } \frac{R_1}{L_1} = \frac{R_2}{L_2} \tag{9}$$

In this paper, as shown in Figure 6, each coil of the balanced coil is used as a known inductor and an unknown inductor, respectively, of Maxwell bridge. Therefore, the calibrated resistance R_2 is initially changed to set the voltage between the two nodes to be zero. When a MO is inserted, the inductance of L_1 or L_2 is reduced and the voltage difference occurs between the two nodes.

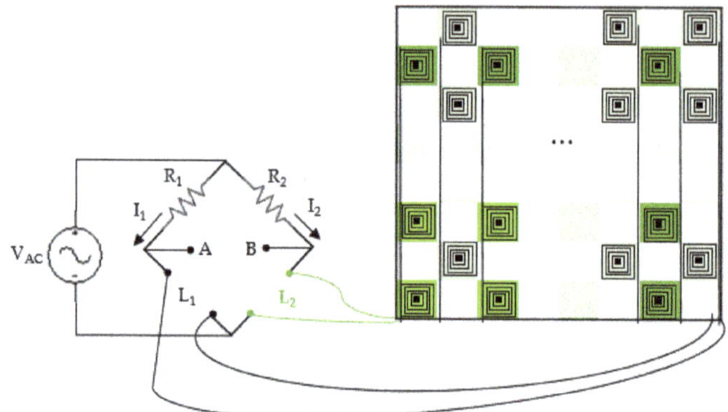

Figure 6. Structure that combines a balanced coil array and Maxwell bridge circuits.

When the voltage difference according to the change of inductance is measured using a Maxwell bridge, it is larger than when the Maxwell bridge is not used. However, the amount of change of inductance due to a small metal foreign object is absolutely small. Therefore, in order to make the voltage or current change larger, we propose an LC resonance circuit by connecting capacitors to the balanced coil in series as shown in Figure 7.

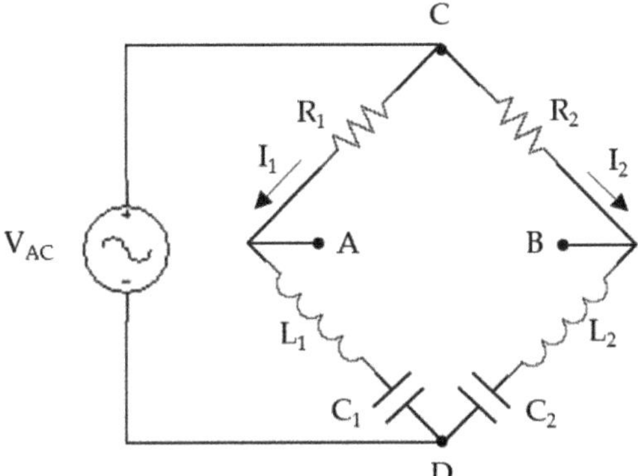

Figure 7. Proposed Maxwell bridge circuit with serial-resonance capacitors.

Similar to the Equation (9) above, the voltage difference between both nodes of the serial-resonance Maxwell bridge circuit is zero when the following conditions are satisfied.

$$R_1 I_1 = R_2 I_2 \text{ and } L_1 I_1 = L_2 I_2 \text{ and } \frac{C_1}{I_1} = \frac{C_2}{I_2}, \text{ or } L_1 C_1 = L_2 C_2 \tag{10}$$

In this paper, since balanced coil pair and resonant capacitors are used, theoretically, L_1 and L_2, C_1 and C_2 are identical respectively. Therefore, I_1 and I_2 become the same in the absence of a MO, and the voltage V_{AB} becomes zero. When a MO is inserted, the inductance of the coil at that location decreases and the resistance increases. In addition, the capacitance is changed by the capacitance of the MO itself and the parasitic capacitance generated in parallel. Therefore, the voltage V_{AB} of the serial-resonance Maxwell bridge circuit has a non-zero value as shown in the following Equation (11).

$$V_{AB} = (R_2 + \alpha R_{mo})I'_2 - R_1 I_1 \tag{11}$$

where a metal object is on L_2 coil and I'_2 is the current flowing through L_2 which means it is no longer the same as I_1. Since the potential drops across the arm CA and AD are equal the potential drops across the arm CB and BD, the following Equations (12)–(14) are established.

$$\left(R_1 + j2\pi f L_1 - j\frac{1}{2\pi f C_1}\right)I_1 = \left(R_2 + \alpha R_{mo} + j2\pi f(L_2 - \alpha L_{mo}) - j\frac{1}{2\pi f C'_2}\right)I'_2 \tag{12}$$

$$I'_2 = \frac{R_1 + j2\pi f L_1 - j\frac{1}{2\pi f C_1}}{R_2 + \alpha R_{mo} + j2\pi f(L_2 - \alpha L_{mo}) - j\frac{1}{2\pi f C'_2}} I_1 = \beta I_1 \tag{13}$$

$$V_{AB} = \{\beta(R_2 + \alpha R_{mo}) - R_1\} I_1 \tag{14}$$

where β is the ratio of the current flowing through L_1 and the current flowing through L_2. Both resistance and inductance are changed by the MO, so the denominator of β and the numerator of β have different phases. That is, the phases of I_1 and I_2 are different. Therefore, V_{AB} has a non-zero value due to a voltage phase difference according to the degree of change of resistance, inductance and capacitance caused by a MO.

3. Results

A wireless power transfer system including balanced coil arrays with serial-resonance Maxwell bridges was constructed and tested as shown in Figure 8. The AC generator supplies 800 kHz AC to the Maxwell Bridge circuit for MOD. The transmitter generates and passes 70 kHz AC signals up to 2 kW to the power transfer coil that is placed under the balanced coil array.

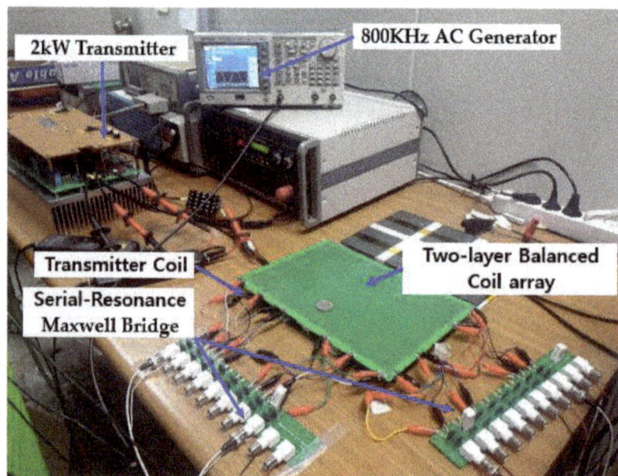

Figure 8. Proposed system and test environment.

For the MOD experiment, as shown in Table 1 and Figure 9, a 500 Korean Won(KRW) coin, a 10 KRW coin, a clip, and a screw were used.

Table 1. Comparison of metal foreign objects used in the experiment.

	500 KRW Coin	10 KRW Coin	Clip	Screw
size	diameter 2.65 cm	diameter 1.8 cm	3.5 cm × 0.8 cm	0.5 cm × 0.5 cm × 1.2 cm
weight (component)	7.9 g (copper 75%, nickel 25%)	1.2 g (copper 48%, aluminum 52%)	-	-

Figure 9. Metal foreign objects used in the experiment.

The transfer coil was configured in a 30 cm × 20 cm rectangular shape and the loop coil of the balanced coil was made of 5 cm × 5 cm. Therefore, as shown in Figure 10, a two-layer balanced coil array consists of a total of 4 × 6 rectangular loop coils. Here, X_n means a set of loop coils connected along the Y axis in the first layer. Y_m refers to the set of loop coils connected along the X axis in the second layer. And (x_n, y_m) means the location where X_n and Y_m intersect. Since the loop coil is arranged one space apart, the loop coil at the intersection is connected to only one of the X_n and Y_m. The following Equations (15) and (16) summarize the loop coils connected to X_n and Y_m for each layer.

$$X_n \ni \begin{cases} (x_n, y_{2k-1}), \ where \ n = 1, \ 3, \ 5 \ and \ k = 1, \ 2 \\ (x_n, \ y_{2k}), \ where \ n = 2, \ 4, \ 6 \ and \ k = 1, \ 2 \end{cases} \tag{15}$$

$$Y_m \ni \begin{cases} (x_{2k}, y_m), \ where \ m = 1, \ 3 \ and \ k = 1, \ 2, \ 3 \\ (x_{2k-1}, \ y_m), \ where \ m = 2, \ 4 \ and \ k = 1, \ 2, \ 3 \end{cases} \tag{16}$$

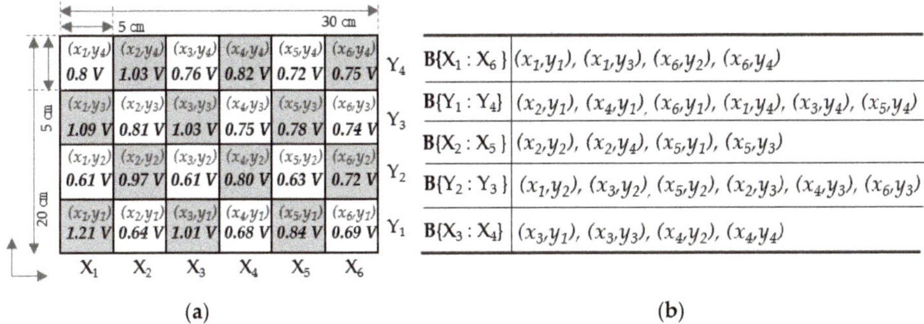

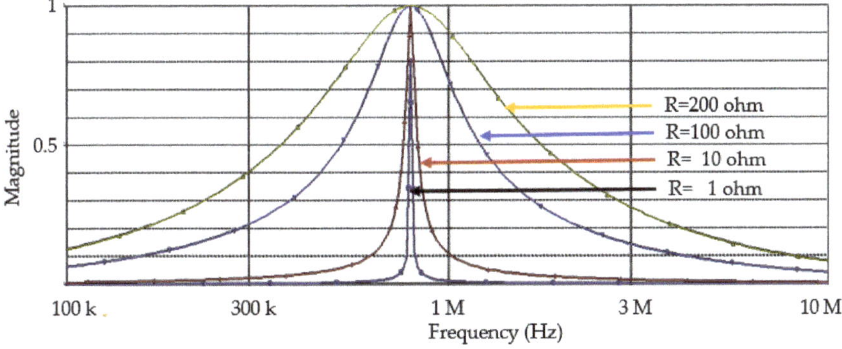

(a) (b)

Figure 10. (**a**) Coordinates to indicate the position of the coin and the zero-to-peak value of V_{AB} measured at each position, and (**b**) the balanced coil set and the position of the loop coil belonging to the set.

As shown in Equations (15) and (16), X_n consists of two loop coils and Y_m consists of three loop coils. Therefore, the coil inductance of the balanced coil array of the first and second layers is different. The inductance of X and Y is 40 µH and 60 µH, respectively. The 1000 pF and 680 pF capacitors were used, respectively, to achieve resonance at 800 kHz with the balanced coil.

The quality factor (Q) of the serial resonant circuit was simulated as shown in Figure 11 to determine the resistance R_1 and R_2 of the proposed Maxwell bridge. The smaller the resistors, the larger the Q. However, if the resistor is too small, the current flows excessively and it takes a long time for the circuit to stabilize, so it is not suitable for a circuit that measures impedance. If the resistor is too large, the current flows less, so it makes difficult to detect the impedance change by a foreign metal object. Therefore, the following simulations and experiments were repeated to determine the magnitude of R_1 and R_2.

Figure 11. Quality factor of the serial resonant circuit according to resistances.

The serial-resonance Maxwell bridge circuit has a V_{AB} of zero when the Equation (10) is satisfied. However, there are errors in the circuit components. Figure 12a shows the V_{AB} when the values of R_1 and R_2 do not match. When the resistors, R_1 and R_2, are selected as 100 Ω and the resistors have a tolerance of 5%, V_{AB} is about 120 mV at the maximum. Similarly, when the capacitors, C_1 and C_2, have a tolerance of 5%, V_{AB} is about 800 mV at the maximum as shown in Figure 12b.

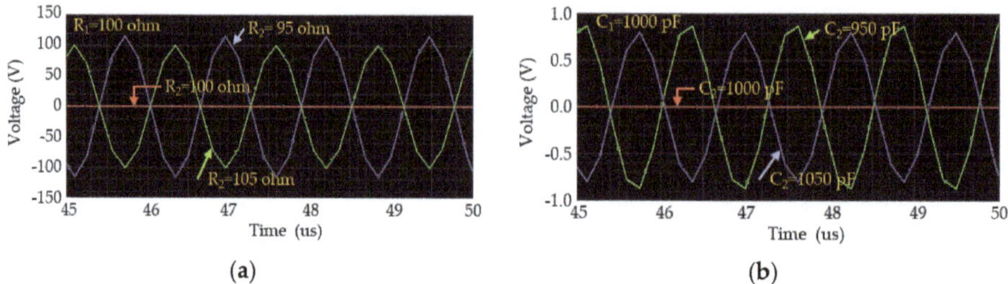

Figure 12. V_{AB} of the Maxwell bridge (**a**) when R_1 and R_2 do not match and (**b**) when C_1 and C_2 do not match.

When the capacitor changes, the resonance frequency of the circuit changes, and when the resistor changes, the bandwidth of Q changes. If the resistors are 10 Ω, the bandwidth of Q becomes narrower than when the resistors are 100 Ω and V_{AB} is about 2 V for a 5% tolerance change of the capacitor. Conversely, if the resistors are 200 Ω, the bandwidth of Q becomes wider, and V_{AB} is about 450 mV. That is, when two capacitors are mismatched, if the resistors are small, V_{AB} increases, and if the resistors are large, V_{AB} decreases. Therefore, it is preferable that the resistors are appropriately large. However, when the equivalent capacitor slightly changes due to a small MO, a change of V_{AB} must occurs, so it is preferable that the resistors are appropriately small. Therefore, in this paper, R_1 and R_2 are determined as 100 Ω.

In the absence of MO, V_{AB} can be made to zero by adjusting the resistors. In this paper, the resistors were not fine-tuned to remove the offset. Instead, the offset does not affect the MOD by adjusting the threshold to determine MO.

Figure 10b shows the balanced coil set whose balance is broken according to the position of the MO. When a MO is at position (x_1, y_1), the MO is above coil set X_1, not Y_1. Only the inductance of X_1 is changed by MO. Since X_1 constitutes a balanced coil with X_6, the balance between X_1 and X_6 is broken, and V_{AB} for the balanced coil B{X_1:X_6}, B(X_1,X_6), changes. Conversely, if B(X_3, X_4) changes, according to Equation (15), it can be seen that MO is placed on one of the loop coils of (x_3, y_1), (x_3, y_3), (x_4, y_2) and (x_4, y_4), which belong to X_3 or X_4. If a MO is at (x_1, y_2), the balance of B{Y_2:Y_3} is broken, so B(Y_2, Y_3) changes. Similarly, if B(Y_1, Y_4) changes, according to Equation (16), it can be seen that MO is placed on one of the loop coils of (x_2, y_1), (x_4, y_1), (x_6, y_1) and (x_1, y_4), which belong to Y_1 or Y_4. The voltage for each position in Figure 10a is a value measured when 500 KWN coin is placed in the center of each position and is about 700–1200 mV$_{\text{zero-to-peak}}$.

Figure 13 shows the voltage difference between when Maxwell bridge is used and when not used for MOD. Figure 13a shows the case where a 500 KWN coin is at (x_1, y_1). The actual voltage difference between the coil X_1 and X_6 is about 0.5 V. However, as in this paper, if the voltage difference is measured using the Maxwell bridge, the phase difference is also reflected, so it is measured as about 2.2 V peak-to-peak. In other words, the voltage difference occurs about four times more than when the Maxwell bridge circuit is not used.

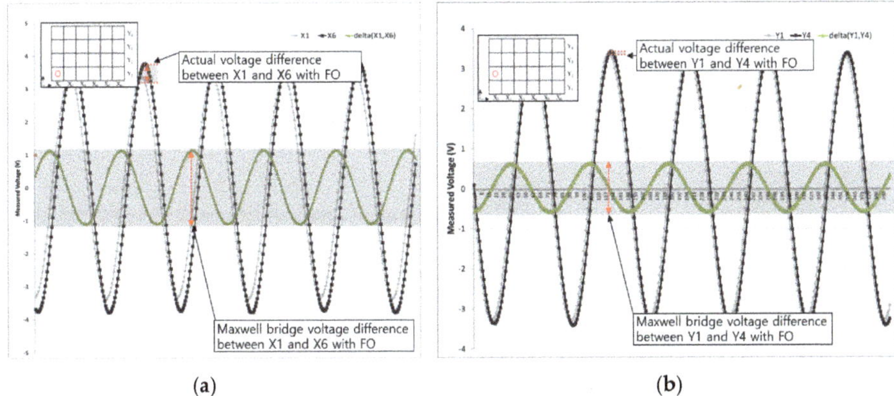

(**a**)   (**b**)

Figure 13. Voltage difference between when Maxwell bridge is used and when not used. (**a**) When a 500 KWN coin is at (x_1, y_1), and (**b**) at (x_1, y_2).

As shown in Figure 13b, when a 500 KWN coin is at (x_1, y_2), the actual voltage difference is about 0.05 V and the voltage difference using the Maxwell bridge is about 1.2 Vpeak-to-peak. The reason why $B(X_1, X_6)$ and $B(Y_2, Y_3)$ are different for the same MO is that the inductance and capacitance of the balanced coil set $B\{X_1:X_6\}$ and $B\{Y_2:Y_3\}$ are different. As shown in Equations (14) and (15), the voltage difference V_{AB} occurs according to the phase difference between the currents flowing through the two nodes, and the current is determined by resistance R, inductance L and capacitance C in each path. Even with the same MO, the degree of phase change is different because the R, L, and C values of $B\{X_1:X_6\}$ and $B\{Y_2:Y_3\}$ are different. Therefore, the larger the phase change by MO, the larger the value of V_{AB}.

When each MO was placed in the center of (x_1, y_1), the zero-to-peak V_{AB} in the serial-resonance Maxwell bridge was measured and summarized in Table 2. When there is no MO, V_{AB} was 0.197 V. Although it was not possible to compare the components of the MOs, the voltage difference tended to decrease as the cross-sectional area was small. In this experiment, we confirmed that the clips were sufficiently detectable by increasing the voltage difference by approximately 1.5 times compared to when there were not MOs.

Table 2. Measured V_{AB} for various MOs.

	no MO	500 KRW Coin	10 KRW Coin	Clip	Screw
V_{AB} (zero-to-peak)	0.197	1.1	0.675	0.3	0.249

To find the hidden area, the position of the 500 KRW coin was moved as shown in Figure 14 and the voltage difference $B(X_1, X_6)$ and $B(Y_1, Y_4)$ were measured. The measured results are summarized in Figure 14b and Table 3. When the coin is in the 'a' position, i.e., the coin is at the center of (x_1, y_1), the impedance of X_1 changes. Therefore, as a result of measurement, $B(X_1, X_6)$ is about 1.1 V. As the coin is away from the center of the X_1 coil, the influence of the coin on the loop coil at the (x_1, y_1) position decreases, so the difference between the impedance of X_1 and the impedance of X_6 gradually decreases. That is, $B(X_1, X_6)$ gradually decreases. When there is no overlap with the X_1 coil, i.e., when the MO is in the 'd' and 'e' positions, the coin hardly affects the impedance of X_1. So, $B(X_1, X_6)$ is almost the same level as when there is no coin. Conversely, when the coin is on the loop coil of (x_1, y_1), it hardly affects the impedance of Y_m. As the coin is placed on the loop coil of (x_2, y_1), the impedance of Y_1 starts to change. Therefore, when the coin is placed in the 'e' position, which is the center of (x_2, y_1), the impedance difference between Y_1 and Y_4 becomes the greatest. That is, $B(Y_1, Y_4)$ was greatest when the coin was in the 'e' position, and it decreased toward the 'a' position.

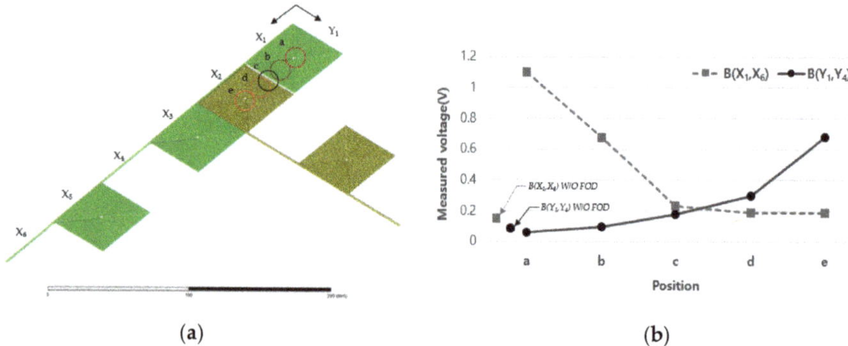

(a) (b)

Figure 14. V_{AB} according to MO's position (**a**) MO position in a balanced coil, (**b**) measured V_{AB}.

Table 3. V_{AB} according to MO's position.

	Without MO	a	b	c	d	e
$B(X_1, X_6)$ (V)	0.197	1.1	0.675	0.301	0.19	0.187
$B(Y_1, Y_4)$ (V)	0.095	0.06	0.096	0.18	0.3	0.68

When there was a MOD on the coil of layer 1, $B(X_n, X_{N-n+1})$ was increased several hundred mV or more. The changed value that is about 3 to 5 times larger than the value measured when there is no MO. When it was on the coil of layer 2, $B(Y_m, Y_{M-m+1})$ was increased relatively small. However, the changed value that is about 3 to 7 times larger than the value measured when there is no MO. Since the loop coils are arranged regularly, the same phenomenon occurs at other locations. Therefore, it is possible to detect MO over the entire area.

The process and results of MOD using the proposed Maxwell bridge with an independent power of 800 kHz are described. However, since this circuit is included in the wireless power transfer system, the following experiment was conducted to find out the effect between the wireless power transfer operation and the MOD operation. As shown in the Figure 15, there is a balanced coil array between the transmitter and receiver coils. While transmitting the power of 70 kHz, the V_{AB} of the Maxwell bridge was measured while changing the distance between the transmitter and receiver coils to 3 cm, 5 cm, and 7 cm.

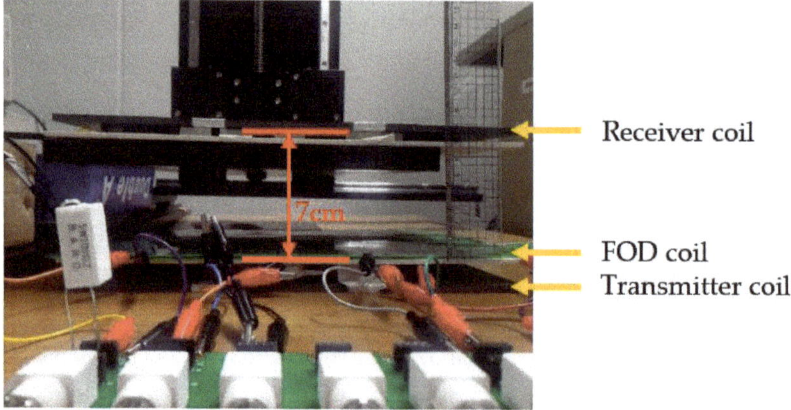

Figure 15. Experimental environment to test the effect between the wireless power transfer operation and the metal object detection (MOD) operation.

Figure 16 shows the results of measuring V_A, V_B, and V_{AB} of the proposed Maxwell Bridge circuit at 7 cm distance. When 70 kHz high power is radiated from the transmitter coil to transmit power in the middle of using detecting MOs at 800 kHz, it is also induced in the balanced coil. As a result, it can be seen that the 800 kHz signal has an envelope of 70 kHz. Table 4 summarizes the measured values of $V_{AB\_peak\_to\_peak}$ and V_{ENV_70k} at each distance during power transferring. Since the resonance between the transmitter and receiver coils is matched when it is 5 cm or more, the maximum power is transmitted when the distance is 7 cm, and the V_{ENV_70k} is also maximum. However, $V_{AB\_peak\_to\_peak}$ is almost constant regardless of power transmission as well as distance.

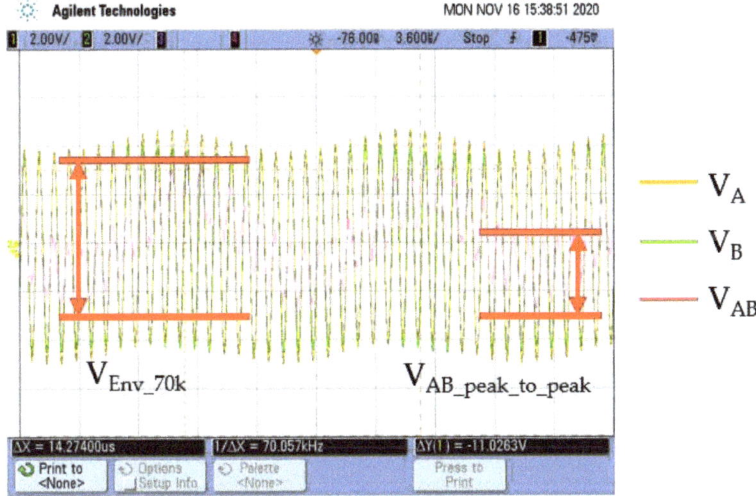

Figure 16. Measured voltages, V_A, V_B, and V_{AB} at 7 cm apart in powering.

Table 4. Measured $V_{AB\_peak\_to\_peak}$ and V_{ENV_70k} according to distance between the transmitter and receiver coils.

	No Powering	3 cm Apart in Powering	5 cm Apart in Powering	7 cm Apart in Powering
$V_{AB\_peak\_to\_peak}$	1.4 V	1.5 V	1.5 V	1.6 V
V_{ENV_70k}	1.4 V	2.8 V	3.3 V	4.9 V

In order to increase the accuracy of V_{AB} signal, a filter to remove signals in the 70 kHz band can be used. Figure 17a,b show magnitude in time domain and in frequency domain of V_{AB} during power transmitting, respectively. Figure 17c,d show magnitude in time domain and in frequency domain of V_{AB} after removing 70 kHz band signals, respectively. As shown in Figure 16, there are 800 kHz band signals and 70 kHz band signals during power transmitting, but the 70 kHz band signals are removed after using the filter. In this paper, however, the slotted Q factor method during power transmitting is used. A slot to detect MOs in the middle of the in-power is added to make it as less sensitive to the power of the transmitter coil as possible.

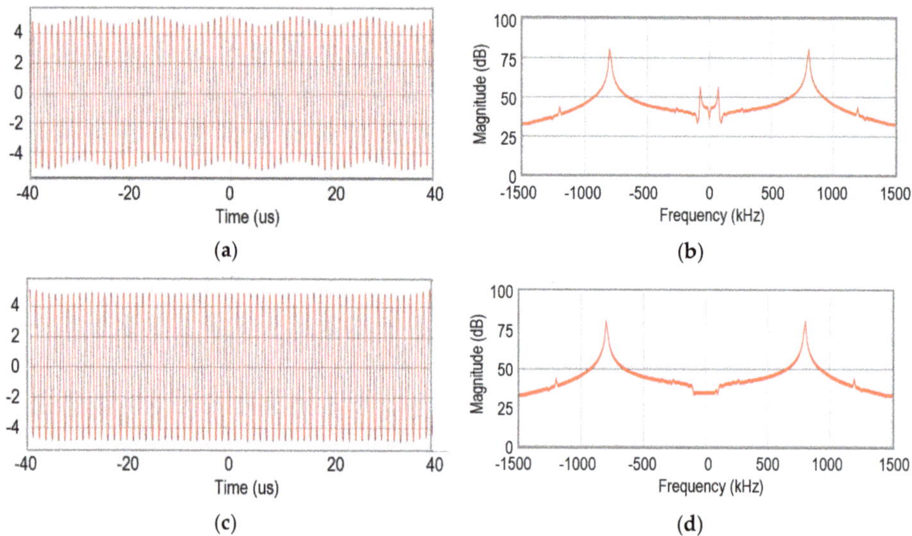

Figure 17. Magnitude of V_{AB} (**a**) in time domain and (**b**) in frequency domain during power transmitting (70 kHz + 800 kHz), and (**c**) in time domain and (**d**) in frequency domain after removing 70 kHz band signals.

4. Conclusions

In this paper, we have presented a novel MOD method based on two-layer balanced coil array with serial-resonant Maxwell bridge circuits, which can detect even small MOs through high voltage change sensing. The balanced coil arrays are divided into two layers and placed one by one at each layer to reduce interference by the surrounding coils other than the MO to be detected. When a MOs are inserted, it is detected that the balance of the current induced in the two coils is broken. The smaller the size of the MO, the smaller the induced current difference. To solve this problem, we also proposed a Maxwell bridge circuit and serial-resonant capacitors, which result in a dramatically improved detection sensitivity. In the proposed system, the balanced coil arrays of the first and second layers have inductance of about 60 μH and 40 μH respectively, and serial-resonance Maxwell bridges have about approximately 100 Ω resistors and 1000 pF and 680 pF capacitors, respectively, for each layer. As a result of testing for 800 kHz AC generator, the proposed system can detect even coins and clips with a voltage difference of several hundred mV. It can be seen that the detection performance is improved compared to conventional circuits detecting MO with a difference of several μV~mV. In addition, the magnitude of the voltage is different depending on the location where the MO was placed, but it shows about 3 to 5 times larger than that without the MO. Therefore, the proposed method in this paper is expected to be useful as a MOD technology for high-power wireless charging with a distance of several tens of centimeters apart between the transmitter and receiver coils such as a robot or an electric vehicle.

5. Patents

Yongseok Lim, Haeyong Jung, and YongJu Park. Metal foreign object detection method in wireless power transmission. KR 10-2020-0111944.

Author Contributions: Conceptualization, S.K. and Y.L.; methodology, S.K., H.J. and Y.L.; software, S.K.; validation, H.J.; formal analysis, Y.J.; investigation, S.K. and Y.L.; resources, H.J.; writing—original draft preparation, S.K.; writing—review and editing, S.K and Y.L.; supervision, Y.L.; project administration, Y.L. All authors have read and agreed to the published version of the manuscript.

Funding: This work was supported by Institute for Information and Communications Technology Promotion (IITP) grant funded by the Korean Government (MSIP) (No. 2020-0-00148, development of ultra-compact/high efficiency wireless charging technology for 1 kW class robot).

Conflicts of Interest: The authors declare no conflict of interest.

References

1. Yang, Y.; Cui, J.; Cui, X. Design and Analysis of Magnetic Coils for Optimizing the Coupling Coefficient in an Electric Vehicle Wireless Power Transfer System. *Energies* **2020**, *13*, 4143. [CrossRef]
2. Seong, J.Y.; Lee, S.-S. A Study on Precise Positioning for an Electric Vehicle Wireless Power Transfer System Using a Ferrite Antenna. *Electronics* **2020**, *9*, 1289. [CrossRef]
3. Wen, F.; Chu, X.; Li, Q.; Gu, W. Compensation Parameters Optimization of Wireless Power Transfer for Electric Vehicles. *Electronics* **2020**, *9*, 789. [CrossRef]
4. TDK Developing Technologies, Achieves Wireless Power Transfer to Mobile Objects with High Efficiency Can Also Be Used for Rotating Parts Such as a Robot Arm. Available online: https://product.tdk.com/info/en/techlibrary/developing/wireless/index.html (accessed on 26 October 2020).
5. Lim, Y.; Tang, H.; Lim, S.; Park, J. An Adaptive Impedance-Matching Network Based on a Novel Capacitor Matrix for Wireless Power Transfer. *IEEE Trans. Power Electron.* **2014**, *29*, 4403–4413. [CrossRef]
6. Zhang, Y.; Yan, Z.; Zhu, J.; Li, S.; Mi, C. A review of foreign object detection (FOD) for inductive power transfer systems. *eTransportation* **2019**, *1*, 100002. [CrossRef]
7. Gyun, W.D. Optimal Design and Control Strategy of Inductive Power Transfer Charging System for Electric Vehicles. Ph.D. Thesis, Sungkyunkwan University, Seoul, Korea, December 2015.
8. Wireless Power Consortium. Foreign Object Detection. In *Qi Wireless Power Transfer System Power Class 0 Specification*; Parts 1 and 2: Interface Definitions, Version 1.2.3; Wireless Power Consortium: Piscataway, NJ, USA, 2017; pp. 127–139.
9. Jeong, S.Y.; Kwak, H.G.; Jang, G.C.; Choi, S.Y.; Rim, C.T. Dual-Purpose Nonoverlapping Coil Sets as Metal Object and Vehicle Position Detections for Wireless Stationary EV Chargers. *IEEE Trans. Power Electron.* **2017**, *33*, 7387–7397. [CrossRef]
10. Jeong, S.Y.; Thai, V.X.; Park, J.H.; Rim, C.T. Self-Inductance-Based Metal Object Detection With Mistuned Resonant Circuits and Nullifying Induced Voltage for Wireless EV Chargers. *IEEE Trans. Power Electron.* **2018**, *34*, 748–758. [CrossRef]
11. Xiang, L.; Zhu, Z.; Tian, J.; Tian, Y. Foreign Object Detection in a Wireless Power Transfer System Using Symmetrical Coil Sets. *IEEE Access* **2019**, *7*, 44622–44631. [CrossRef]
12. Zhou, B.; Liu, Z.Z.; Chen, H.X.; Zeng, H.; Hei, T. A New Metal Detection Method Based on Balanced Coil for Mobile Phone Wireless Charging System. In Proceedings of the International Conference on New Energy and Future Energy System, Beijing, China, 19–22 August 2016. [CrossRef]
13. Maxwell's Bridge. Available online: https://circuitglobe.com/maxwells-bridge.html (accessed on 30 October 2020).

Article

Optimum Receiver-Side Tuning Capacitance for Capacitive Wireless Power Transfer

Sungryul Huh [1] and Dukju Ahn [2,*]

[1] The Cho Chun Shik Graduate School for Green Transportation, Korea Advanced Institute of Science and Technology, Daejeon 34141, Korea; tjdfuf2397@kaist.ac.kr
[2] Dept. of Electrical Engineering, Incheon National University, Incheon 22012, Korea
*   Correspondence: adjj22@gmail.com; Tel.: +82-32-835-8767

Received: 12 November 2019; Accepted: 9 December 2019; Published: 13 December 2019

Abstract: This paper reveals the optimum capacitance value of a receiver-side inductor-capacitor (LC) network to achieve the highest efficiency in a capacitive power-transfer system. These findings break the usual convention of a capacitance value having to be chosen such that complete LC resonance happens at the operating frequency. Rather, our findings in this paper indicate that the capacitance value should be smaller than the value that forms the exact LC resonance. These analytical derivations showed that as the ratio of inductor impedance divided by plate impedance increased, the optimum Rx capacitance decreased. This optimum capacitance maximized the *TX-to-RX* transfer efficiency of a given set of system conditions, such as matching inductors and coupling plates.

Keywords: wireless power transfer; capacitive power transfer; parallel-plate contactless power

---

1. Introduction

Capacitive wireless power-transfer systems wirelessly transmit electrical energy without the use of actual wire coils. Instead, thin metallic plates form a capacitor through which current can flow. Such a system has previously been investigated for biomedical applications [1], electric vehicles [2–5], mobile devices [6], and constant-current applications [7]. Although a variety of circuit topologies are available for capacitive power system [1–8], an inductor-capacitor (LC) section in a receiver (RX) is the simplest topology for systems with small coupling capacitances [2,3]. Additional matching inductors, such as those in [5], require a large inductance value (~240 µH), which is too bulky. Parasitic capacitances due to nearby metals can also be merged with a parallel tuning capacitor [2] to form another type of LC section.

Although much work has been done on the LC matching network design, only a few works have focused on efficiency maximization, which is an important key requirement in an effective wireless power-transfer system. Reference [3] proposes operating near the resonance frequency of an inductor and capacitor, either for constant current or constant voltage operation. Their operating frequency would slightly deviate from the self-resonant frequency of an LC matching network for constant voltage or current operation, where the amount of deviation is determined by the strength of the capacitive coupling. At weak coupling, the operating frequency would approach the LC resonant frequency.

While the LC matching design of [3] successfully achieves either constant voltage or current operation, this design does not focus on efficiency maximization. Reference [9] analyzes the effect of matching detuning and proposes a design method to operate over a wide frequency bandwidth. Although this design successfully operated over this wider frequency range via inverter soft-switching, the optimum matching capacitance for maximum efficiency has not yet been discussed.

Reference [8] proposed a resonance-matching network to improve the power factor. This is equivalent to enhancing the real part of Z_{RX}, as seen in Figure 1. The large resistive impedance of the receiver increased the power factor here because the CP impedances were highly imaginary. A matching network design [10] also aimed toward power-factor maximization. Unfortunately, as will be discussed later in this paper, power-factor maximization does not necessarily maximize efficiency. Hence, any design method that focused on achieving maximum efficiency would be different from the methods in [8,10].

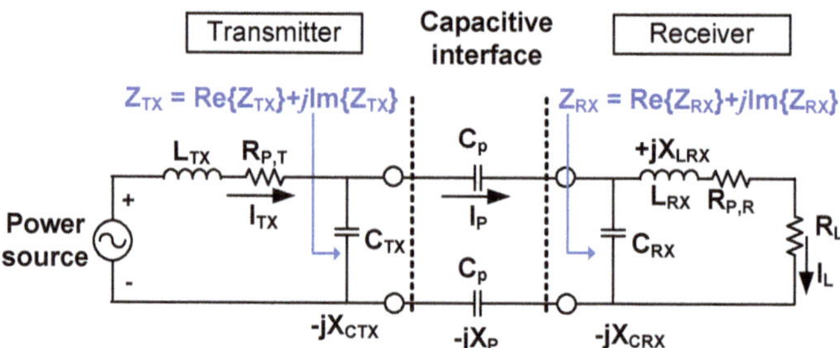

Figure 1. Equivalent circuit of a capacitive wireless-power transfer system. The Re{Z_{TX}} resistance at the transmitter should be high compared to $R_{P,T}$ in order to achieve high Tx-to-Rx efficiency.

Reference [11] proposed that the matching capacitor should be small in order to reduce sensitivity to parameter variations and voltage stress. Through this method, the drift of system performance against load or component variation would be minimized.

While the various design methods mentioned above aimed to achieve different goals, such as constant output, wide bandwidth, high power factor, or reduced sensitivity, none of them explicitly defined the optimum RX capacitance for maximum efficiency. This paper investigates exactly what optimum capacitance value could maximize the transfer efficiency for a given set of system parameters, such as coupling plates, load, and matching inductors. The results showed that the optimum RX capacitor value should be smaller than that of a value that achieves complete LC resonance. A quantitative closed-form equation predicted the optimum capacitor value as a function of coupling plates, load, and inductors.

2. Optimum Resonant Matching Capacitor

2.1. Circuit Topology of Capacitive Power System

Figure 1 shows the equivalent circuit of capacitive power transfer. This work used a differential Class-E amplifier that produced a sinusoid output voltage and current. Its schematic and measured waveform is presented in Figure 8. However, this mathematical derivation was also applicable to a square-wave voltage source as well (e.g., voltage-mode Class-D inverter) because the first-harmonic approximation was valid due to the high-Q of the matching network. In other words, due to the high selectivity of L_{TX}—C_{TX} resonance, higher frequency components of I_{TX} were suppressed and I_{TX} became sinusoidal. Since the impedance of capacitive interface, $X_P = (\omega C_P/2)^{-1}$, is extremely high in noncontact applications, it is common to boost the receiver load R_L using an L_{RX}–C_{RX} network. The parallel resonance of L_{RX}–C_{RX}–R_L increases the real part of Z_{RX} while minimizing the reactance of Z_{RX}.

For the given L_{RX} inductance, the C_{RX} capacitance would normally be chosen such that L_{RX}–C_{RX} was resonant at the operating frequency. This was to maximize the real part of Z_{RX} (i.e., Re{Z_{RX}}) while minimizing the reactive part of Z_{RX} (Im{Z_{RX}}). In this paper, however, we revealed that an exact LC resonance was not the optimum design for efficiency maximization. Rather, the capacitance should be slightly smaller, such that there exists significant inductive impedance in Z_{RX}. This Im{Z_{RX}} partially cancels out the larger capacitive X_P. Although the Re{Z_{RX}} obtained via the proposed shifted resonance was lower than that obtained with exact LC resonance, it is also analyzed in this paper that a higher Re{Z_{RX}} is not always beneficial: there was an optimum Re{Z_{RX}}.

2.2. Analytical Derivation for Optimum C_{RX}

The Tx-to-Rx efficiency of Figure 1 was defined as

$$\eta_{TX\text{-}to\text{-}RX} = \frac{\text{Load power}}{\text{Supplied power by source}}$$
$$= \frac{I_L^2 R_L}{I_{TX}^2 \left(\text{Re}\{Z_{TX}\} + R_{P,T}\right)} \tag{1}$$

The Z_{TX} is the equivalent impedance with regard to the C_{TX} capacitance, and $R_{P,T}$ and $R_{P,R}$ are the parasitic resistance of the L_{TX} and L_{RX} inductors, respectively. Equation (1) can be separated into a two-stage equation. The first stage, which was the transmitter efficiency, consisted of the power entering the capacitive interface $I_{TX}^2\text{Re}\{Z_{TX}\}$, divided by the total power supplied from our power source $I_{TX}^2(\text{Re}\{Z_{TX}\} + R_{P,T})$. The second stage, which was the receiver efficiency, consisted of the power dissipation at final load divided by the power dissipation across the whole receiver. Hence, Equation (1) could be written as

$$\eta_{TX\text{-}to\text{-}RX} = \frac{I_{TX}^2\text{Re}\{Z_{TX}\}}{I_{TX}^2\left(\text{Re}\{Z_{TX}\} + R_{P,T}\right)} \times \frac{I_L^2 R_L}{I_L^2(R_L + R_{P,R})} \tag{2}$$

thereby arriving at the impedance ratio equation of

$$\eta_{TX\text{-}to\text{-}RX} = \frac{\text{Re}\{Z_{TX}\}}{\text{Re}\{Z_{TX}\} + R_{P,T}} \times \frac{R_L}{R_L + R_{P,R}} \tag{3}$$

Since the power delivered to the receiver was equal to the power dissipated at the $\text{Re}\{Z_{TX}\}$, it was important that we obtained a large value of $\text{Re}\{Z_{TX}\}$ to maximize transmission efficiency. In other words, the power delivered to RX is $P = |I_{TX}|^2\text{Re}\{Z_{TX}\}$, whereas the power dissipated at TX parasitic is $P = |I_{TX}|^2 R_{P,T}$. Hence, the $\text{Re}\{Z_{TX}\}$ should have been higher than $R_{P,T}$. The $\text{Re}\{Z_{TX}\}$ is defined found as follows:

$$\text{Re}\{Z_{TX}\} = \frac{X_{CTX}^2\text{Re}\{Z_{RX}\}}{\text{Re}\{Z_{RX}\}^2 + \left(X_{CTX} + X_P - \text{Im}\{Z_{RX}\}\right)^2} \tag{4}$$

where $\text{Re}\{Z_{RX}\}$ and $\text{Im}\{Z_{RX}\}$ is

$$\text{Re}\{Z_{RX}\} = \frac{X_{CRX}^2 R_L}{R_L^2 + \left(X_{LRX} - X_{CRX}\right)^2} \tag{5}$$

$$\text{Im}\{Z_{RX}\} = -\frac{X_{CRX}\left\{X_{LRX}(X_{LRX} - X_{CRX}) + R_L^2\right\}}{R_L^2 + \left(X_{LRX} - X_{CRX}\right)^2} \tag{6}$$

and $X_{CTX} = (\omega C_{TX})^{-1}$, $X_{CRX} = (\omega C_{RX})^{-1}$, $X_P = (\omega C_P/2)^{-1}$, and $X_{LRX} = \omega L_{RX}$.

After substituting Equations (5) and (6) into Equation (4), Equation (4) became a function of the receiver parameters, such as R_L, X_{LRX}, and X_{CRX}. The typical complete resonance, $\omega = 1/\sqrt{L_{RX}C_{RX}}$, almost cancelled out the $\text{Im}\{Z_{RX}\}$, whereas the opposite was true for the $\text{Re}\{Z_{RX}\}$, which was maximized.

In this paper, we tested the theory that there may be an optimum $C_{RX,opt}$ to maximize the efficiency of Equation (3) for any given set of system parameters. Our efficiency maximization was realized by a maximum $\text{Re}\{Z_{TX}\}$ resistance and a corresponding minimum I_{TX} current, thereby suppressing power losses at the inverter and TX passive components thanks to a minimum of I_{TX} current.

Differentiating Equation (4) with respect to X_{CRX} and setting this differentiation to zero, i.e., $\partial\text{Re}\{Z_{TX}\}/\partial X_{CRT} = 0$, the optimum X_{CRX} was derived:

$$X_{CRT,opt} = \frac{((R_L + R_{P,R})^2 + X_{LRX}^2)(X_P + X_{CTX})}{X_{LRX}(X_P + X_{CTX}) - (R_L + R_{P,R})^2 - X_{LRX}^2} \qquad (7)$$

Note that the derivation of Equation (7) does not involve any approximations and therefore was generally applicable for any given set of system parameters e.g., load, L_{RX}, C_P, C_{TX} etc. Equation (7) can be simplified because the coupling plate impedance, X_P, is usually a much higher value than the X_{CTX}. [2,3]. Moreover, the Rx inductor reactance, X_{LRX}, is also usually designed as a much higher value than R_L in order to boost a small R_L into a large Re{Z_{RX}}, generally because Re{Z_{RX}} $\approx X_{LRX}^2/R_L$. Under these conditions, Equation (7) was simplified as follows:

$$X_{CRX,opt} \cong X_{LRX}\left(1 - \frac{X_{LRX}}{X_P}\right)^{-1} \qquad (8)$$

Equation (8) indicated that the optimum X_{CRX} impedance, which maximized the Re{Z_{TX}} and our efficiency, should be higher than the inductor impedance X_{LRX}. The ratio between inductor impedance and coupling plate impedance, i.e. X_{LRX}/X_P, determined the level of deviation from the complete LC canceling condition of $X_{CRX} = X_{LRX}$. Equation (8) indicated that a higher ratio of X_{LRX}/X_P required a larger deviation of X_{CRX} from the X_{LRX}.

2.3. Discussion

Figure 2b is the Z_{RX} representation of Figure 2a at a conventional resonance. Conventional RX cancelled the Im{Z_{RX}} while maximizing the R_L into a high Re{Z_{RX}} so that the power factor of Z_{CAP} was maximized. However, higher Re{Z_{RX}} was not always beneficial for TX-to-RX efficiency. As seen in Figure 1, the I_{TX} supplied from the inverter was directed toward two separate paths: one was through C_{TX} (which did not contribute to power delivery), and the other was through I_P flowing into the receiver. If Re{Z_{RX}} was too high, then most of the I_{TX} was circulated to C_{TX} and only limited current could flow through I_P, which resulted in a reduced power efficiency. The bottom graph of Figure 2d shows that at conventional resonance the I_{TX} required to deliver a specified I_L should have been increased.

However, the proposed C_{RX} detuning in Figure 2c did not maximize the Re{Z_{RX}} and, at the same time, intentionally generated +Im{Z_{RX}}. This partially cancelled X_P by detuning L_{RX}–C_{RX}. Its impedance, as seen in Figure 2d, was a frequency of 7.1 MHz. The overall impedance |Z_{CAP}| = Re{Z_{RX}} + j(Im{Z_{RX}} − X_P) was significantly reduced compared to conventional L_{RX}–C_{RX}. As a result, the bottom graph of Figure 2d shows that the I_{TX} current required to deliver a given load current I_L could be minimized, which in turn could reduce the losses in the transmitter.

The exact amount of detuning of L_{RX}–C_{RX} was quantitatively obtained from Equations (7) and (8). Figure 3 illustrates the design trade-off. In Figure 3a, while Re{Z_{RX}} should have been high to maximize the load power per unit I_P of current, the Re{Z_{RX}} should not have been so excessively high that the I_P current per unit I_{TX} could not be maintained. At the same time, in Figure 3b, jIm{Z_{RX}} − jX_P was minimized by maximizing the +jIm{Z_{RX}} so that the I_P was increased per given I_{TX}. However, as seen in Figure 3c, excessively high Im{Z_{RX}} may have compromised the achievable Re{Z_{RX}}. The proposed Equations (7) and (8) optimized the trade-offs of Figure 3 and produced an optimum Re{Z_{RX}} and Im{Z_{RX}} that maximized power efficiency.

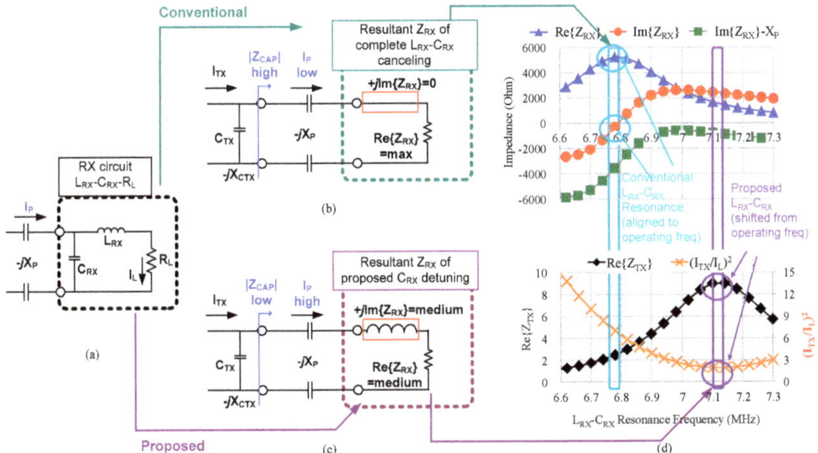

Figure 2. Comparison between conventional resonance and proposed C_{RX}. (**a**) *RX* circuit consisting of L_{RX}–C_{RX} and load R_L. (**b**) Typical complete LC canceling causes high Z_{CAP} and low I_P. (**c**) Proposed C_{RX} condition yields a high $+j\mathrm{Im}\{Z_{RX}\}$ that partially cancels the high $-jX_P$ of the coupling plates. Moreover, $\mathrm{Re}\{Z_{RX}\}$ was moderate. The two improvements of Z_{RX} allowed a higher I_P current toward RX. (**d**) The *x*-axis was L_{RX}–C_{RX} resonance frequency. Operating frequency was fixed at 6.78 MHz. The proposed C_{RX} of Equation (7) partly cancels the $-jX_P$ impedance and yields an appropriate value of $\mathrm{Re}\{Z_{RX}\}$, both of which increased the current I_P and maximized the load power. This maximized $\mathrm{Re}\{Z_{TX}\}$ and minimized the I_{TX} required to deliver a given load current I_L. C_{TX} = 168.5 pF, C_P = 14.5 pF.

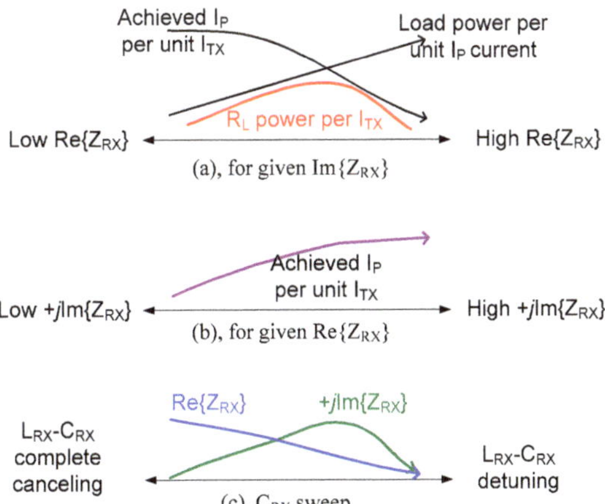

Figure 3. Design considerations. Equation (7) optimized C_{RX} tuning, when considering all the trade-offs. (**a**) $\mathrm{Re}\{Z_{RX}\}$ optimization to maximize R_L power per given I_{TX}. (**b**) High $+j\mathrm{Im}\{Z_{RX}\}$ improved the I_P per given I_{TX}. However, (**c**) excessively high $\mathrm{Im}\{Z_{RX}\}$ compromised the achievable $\mathrm{Re}\{Z_{RX}\}$.

Figure 4 compares the conventional and the proposed methods. The proposed method surpassed the upper limit imposed by conventional *RX* tuning.

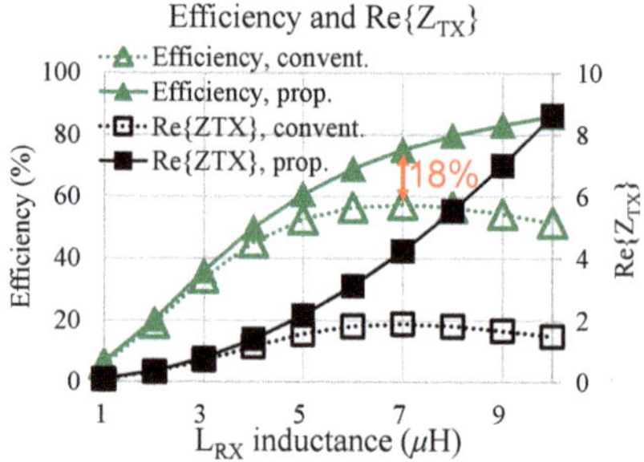

Figure 4. Calculated efficiency and Re{Z_{TX}}. The proposed method always achieves higher efficiency than conventional tuning method for every value of L_{RX}.

The high Re{Z_{TX}} might also be obtained by using a small C_{TX}, as in Equation (4). However, a small C_{TX} demands a large L_{TX}, which increases inductor volume and parasitic $R_{P,T}$. As an example, in Figure 2d the bottom graph typical resonance still gives the same Re{Z_{TX}} = 9 Ω if the C_{TX} was reduced from 168.5 to 87.5 pF. However, then the required L_{TX} should have increased from 3.8–6.8 μH. Due to the increased parasitic $R_{P,T}$, the spice-simulated efficiency degraded from 77.4% to 69.9%. Hence, an optimum $C_{RX,opt}$ becomes important in order to produce the highest Re{Z_{TX}} under the constraint of L_{TX} volume and parasitic resistance.

3. Results

Figure 5 shows the measurement setup using wireless charging of an unmanned aerial vehicle (Drone) prototype can be seen in Figure 5. The load condition was 36 V–1.8 A and resulted in a value of 64.8 W. A differential Class-E inverter and full-bridge rectifier were used. Efficiency in this paper was defined as from DC source to DC load.

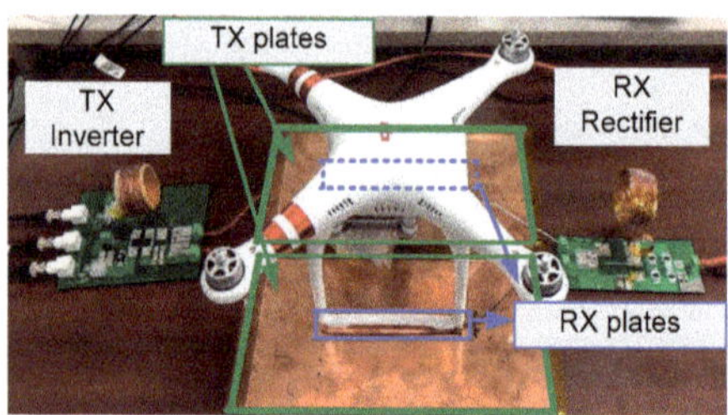

Figure 5. Measurement validation using the unmanned aerial vehicle prototype. The *TX* plates were protected by a 2–4 mm thick acrylic sheet to prevent hazardous electrical shorts caused by collision with unexpected foreign objects.

A 0.2 mm thick copper plate was used for each plate. A transmit plate of 30 × 30 cm^2 was placed underneath the landing pad and a receiver plate of 13 × 1.5 cm^2 was attached under the landing foot of the UAV. The C_P was 23 and 14 pF for a 2 and 4 mm distance, respectively. These distances were due to electrical isolation by way of an acrylic sheet to prevent electrical shorts and mechanical damage of the *TX* plates that may have resulted from a collision with foreign objects. L_{TX}, L_{RX}, and C_{TX} were 3.8 μH, 7.13 μH, and 165 pF, respectively. A GS66508T FET and PMEG6045 diode were used as our inverter and rectifier, respectively. Please note that Equations (7) and (8) are generally applicable to different systems with different component parameters. Table 1 provides circuit parameters.

Table 1. Circuit parameters.

Parameter	Value	Parameter	Value
C_p	14~23 pF	Load	36 V, 64.8 W
L_{TX}	3.8 μH	TX plate	30×30 cm^2
L_{RX}	7.13 μH	RX pad	13×1.5 cm^2
C_{TX}	165 pF	Switching freq.	6.78 MHz

Figure 6a presents the DC-to-DC efficiency for each *RX* capacitor value. A L_{RX} value of 7.13 μH was chosen because, as can be seen from Figure 4, efficiency could be maximized near ~7 μH at a Cp of 10 pF (worst coupling) using typical LC resonance. The C_{RX} of 77.3 pF corresponded to typical LC resonance, whose resonance frequency coincided with an operating frequency of 6.78 MHz. The optimum $C_{RX,opt}$ was predicted by Equation (8) for different C_P coupling plates. As expected, our proposed $C_{RX,opt}$ values achieved the highest efficiency for the given set of system constraints. Figure 6b presents the I_{TX} current required to deliver the given load power, which was minimized at the proposed $C_{RX,opt}$ capacitor tuning. This result was expected because Equations (7) and (8) maximized the Re$\{Z_{TX}\}$, and therefore the power delivered to the receiver, which was $P = |I_{TX}|^2 \text{Re}\{Z_{TX}\}$.

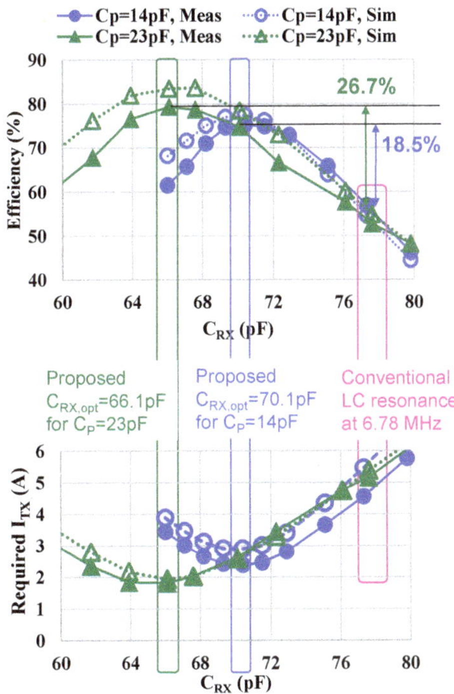

Figure 6. Proposed tuning that achieved higher efficiency. (**a**) Efficiency vs. C_{RX} for $L_{RX} = 7.13$ μH. The proposed C_{RX} values achieved higher efficiency than the conventional exact LC resonance. (**b**) I_{TX} current required to deliver a given load power. The proposed C_{RX} condition lowered the required I_{TX} current, thereby reducing the power loss of the transmitter.

Figure 7 presents the loss analysis for the same load power. The proposed $C_{RX,opt}$ greatly reduced the power loss of the transmitter. This was because the C_{RX} affected the Re$\{Z_{TX}\}$, which in turn determined the magnitude of current through the transmitter. The waveform in Figure 8 shows that the inverter achieved zero-voltage switching.

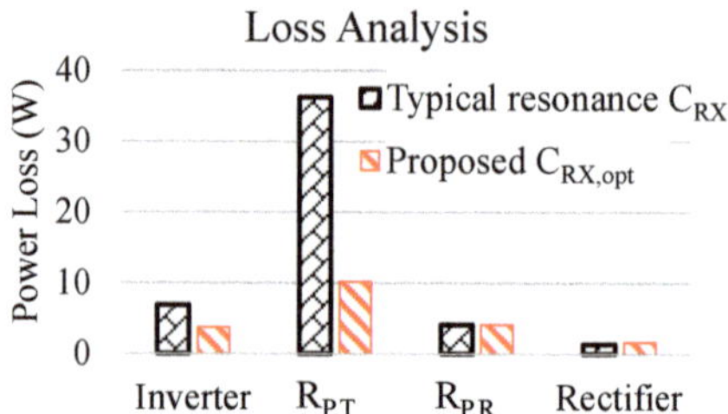

Figure 7. Loss breakdown analysis for the same load power. The proposed method improved the losses in transmission while not affecting the receiver-loss characteristics.

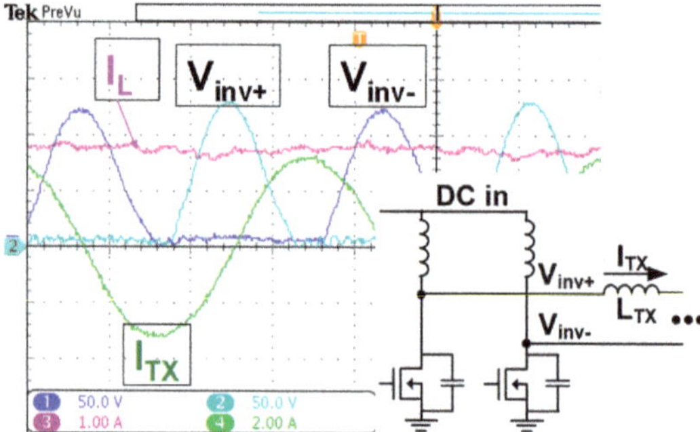

Figure 8. Measured waveforms. The inverter was operated at 6.78 MHz using zero-voltage switching.

4. Conclusions

This paper thoroughly reveals the optimum parallel capacitance value of a receiver for a given set of system parameters. Our finding showed that a complete LC resonance at operating frequency did not result in the highest efficiency. Rather, the *RX* capacitor should have been be of a smaller capacitance value than the nominal resonance-tuning value. The optimal deviation from nominal resonance should have been proportional to the ratio between the *RX* inductor impedance and the coupling plate impedance, as formulated in Equation (8). This minimized our I_{TX} value and its associated losses in the transmitter, thereby increasing overall efficiency while not affecting receiver loss characteristics.

Author Contributions: Conceptualization, S.H. and D.A.; Formal analysis, S.H. and D.A.; Funding acquisition, D.A.; Methodology, S.H. and D.A.; Project administration, D.A.; Supervision, D.A.; Validation, S.H.

Funding: This work is supported by Incheon National University Grant (#2018-0156).

Conflicts of Interest: The authors declare no conflict of interest.

References

1. Jegadeesan, R.; Agarwal, K.; Guo, Y.-X.; Yen, S.-C.; Thakor, N.V. Wireless power delivery to flexible subcutaneous implants using capacitive coupling. *IEEE Trans. Microwave Theory Tech.* **2017**, *65*, 280–292. [CrossRef]

2. Regensburger, B.; Kumar, A.; Sinha, S.; Doubleday, K.; Pervaiz, S.; Popovic, Z.; Afridi, K. High-Performance Large Air-Gap Capacitive Wireless Power Transfer System for Electric Vehicle Charging. In Proceedings of the 2017 IEEE Transportation Electrification Conference and Expo (ITEC), Chicago, IL, USA, 26–28 June 2017; pp. 638–643.

3.   Lu, F.; Zhang, H.; Hofmann, H.; Mi, C. A double-sided LC compensation circuit for loosely-coupled capacitive power transfer. *IEEE Trans. Power Electron.* **2018**, *33*, 1633–1643. [CrossRef]

4.   Zhang, H.; Lu, F.; Hofmann, H.; Liu, W.; Mi, C. A four-plate compact capacitive coupler design and LCL-compensated topology for capacitive power transfer in electric vehicle charging application. *IEEE Trans. Power Electron.* **2016**, *31*, 8541–8551.

5.   Zhang, H.; Lu, F.; Hofmann, H.; Liu, W.; Mi, C. Six-plate capacitive coupler to reduce electric field emission in large air gap capacitive power transfer. *IEEE Trans. Power Electron.* **2017**, *33*, 665–675. [CrossRef]

6.   Dai, J.; Ludois, D. Biologically Inspired Coupling Pixilation for Position Independence in Capacitive Power Transfer Surfaces. In Proceedings of the 2015 IEEE Applied Power Electronics Conference and Exposition (APEC), Charlotte, NC, USA, 15–19 March 2015.

7.   Su, Y.-G.; Xie, S.-Y.; Hu, A.; Tang, C.-S.; Zhou, W.; Huang, L. A capacitive power transfer system with a mixed-resonant topology for constant-current multiple-pickup applications. *IEEE Trans. Power Electron.* **2017**, *32*, 8778–8786. [CrossRef]

8.   Theodoridis, M. Effective capacitive power transfer. *IEEE Trans. Power Electron.* **2012**, *27*, 4906–4913. [CrossRef]

9.   Zhang, H.; Lu, F. An improved design methodology of the double-sided LC-compensated CPT system considering the inductance detuning. *IEEE Trans. Power Electron.* **2019**, *34*, 11396–11406. [CrossRef]

10.   Li, S.; Liu, Z.; Zhao, H.; Zhu, L.; Shuai, C.; Chen, Z. Wireless power transfer by electric field resonance and its application in dynamic charging. *IEEE Trans. Ind. Electron.* **2016**, *63*, 6602–6612. [CrossRef]

11.   Mostafa, T.; Bui, D.; Muharam, A.; Hattori, R.; Hu, A. A Capacitive Power Transfer System with a CL Network for Improved System Performance. In Proceedings of the 2018 IEEE Wireless Power Transfer Conference (WPTC), Montreal, QC, Canada, 3–7 June 2018.

 *electronics*

Article

High-PSRR Wide-Range Supply-Independent CMOS Voltage Reference for Retinal Prosthetic Systems

Ruhaifi Bin Abdullah Zawawi [1], Hojong Choi [2],* and Jungsuk Kim [3],*

[1] Department of Health Science and Technology, Gachon Advanced Institute for Health Sciences & Technology, Incheon 21999, Korea; ruhaifi@bme.gachon.ac.kr
[2] Department of Medical IT Convergence Engineering, Kumoh National Institute of Technology, 350-27, Gum-daero, Gumi 39253, Korea
[3] Department of Biomedical Engineering, Gachon University, 191, Hambakmoe-ro, Incheon 21936, Korea
*   Correspondence: hojongch@kumoh.ac.kr (H.C.); jungsuk@bme.gachon.ac.kr (J.K.)

Received: 25 October 2020; Accepted: 27 November 2020; Published: 30 November 2020

Abstract: This paper presents a fully integrated voltage-reference circuit for implantable devices such as retinal implants. The recently developed retinal prostheses require a stable supply voltage to drive a high-density stimulator array. Accordingly, a voltage-reference circuit plays a critical role in generating a constant reference voltage, which is provided to a low-voltage-drop regulator (LDO), and filtering out the AC ripples in a power-supply rail after rectification. For this purpose, we use a beta-multiplier voltage-reference architecture to which a nonlinear current sink circuit is added, to improve the supply-independent performance drastically. The proposed reference circuit is fabricated using the standard 0.35 μm technology, along with an LDO that adopts an output ringing compensation circuit. The novel reference circuit generates a reference voltage of 1.37 V with a line regulation of 3.45 mV/V and maximum power-supply rejection ratio (PSRR) of −93 dB.

Keywords: reference circuit; inductive link; implantable device; line regulation; wireless power telemetry; supply independence

---

1. Introduction

Short-distance wireless communication for retinal prosthetic systems plays a critical role in delivering a radio-frequency (RF) power carrier and data from the external world to an implant inside the eyeball. Communication techniques based on inductively coupled coils have been widely used, owing to their high power-transfer efficiency and hardware simplicity [1–4]. The recent advances in submicron complementary metal oxide semiconductor (CMOS) technologies have also facilitated the reduction of the size of the implanted hardware such as an inductive coil receiver, digital controller, or high-density stimulator array. As a result, the device can be implanted above the ganglion cells in the case of epiretinal prostheses [5,6] or below the bipolar cells in subretinal prostheses [7–9].

Figure 1 shows a retinal prosthetic system architecture wherein a dual half-wave rectifier is used to recover dual-rail DC power from a received RF power carrier, through inductively coupled coils. The external device is composed of a commercial class-E power amplifier, an amplitude-shift-keying (ASK) modulator circuit, and a current-sense circuit for back-telemetry data recovery. The implanted device consists of a rectifier, regulator, over-voltage–protection circuit, demodulator, reverse telemetry controller, global digital controller, and stimulator array. The class-E amplifier is driven by an RF carrier signal of 13.56 MHz, which is allowed for industry, science, and medical purposes, containing command data of 1.356 Mbps, which is transmitted to the implanted device through inductively coupled coils. The RF signals received in the implanted device are rectified and fed to the regulator to generate dual-rail DC supply voltages, $+V_{1,2}$ and $-V_{1,2}$. The connection of reference circuits (+REF, −REF) is shown in Figure 1, where +REF and −REF, respectively generate positive and negative reference voltages for internal LDO. LDO1 and LDO2 supply constant voltages to analog and digital circuits in the implanted devices respectively. The command data modulated on the RF carrier are recovered using the demodulator and sent to the global digital controller that decodes the demodulated data in order to activate the stimulator array. Here, a single-pixel stimulator, which is composed of a photosensor, current amplifier, and pulse shaper [10], generates a biphasic current pulse, which is delivered to the bipolar cells through a microelectrode. Back telemetry is utilized to observe the operating status of the implanted device. A load-shift keying (LSK) technique is used for back telemetry, which is fully controlled by the reverse telemetry controller in the retinal prosthesis. The over-voltage-protection circuit in the implanted device is activated when the rectified signal exceeds the allowable voltage limit. This circuit allows to shift the reactance of the implanted secondary coil and capacitor away from its resonance, and as a result, the received signal is attenuated to a safe voltage level.

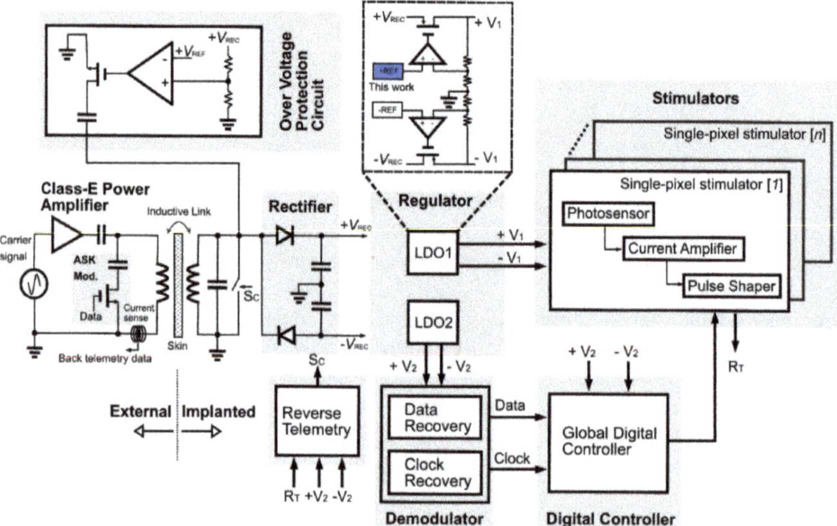

Figure 1. Retinal prosthetic system architecture.

In this retinal prosthetic implant, it is important to provide stable DC voltages to each functional block such as the demodulator, digital controller, and stimulator array. If the DC voltage level varies during stimulation, it will affect the biphasic pulse amplitude, resulting in more or less charge injection to the bipolar cells. However, the amplitude of the rectified DC voltage that determines the regulated DC supply voltage level often fluctuates for the following two reasons. First, natural eye rolling may cause axial and/or lateral misalignment of the implanted coil. This angular misalignment decreases the coupling coefficient, leading to low power-transfer efficiency and DC supply voltage drop [11]. Second, a high-modulation-index waveform of the received RF carrier, based on the ASK scheme, causes a deep ripple in the rectified voltage. This is because the on-chip small capacitor for the rectifier is not sufficient to eliminate the rapid change in charging and discharging. Therefore, it is indispensable for the reference circuit to provide a stable DC voltage to the regulator, regardless of the deep ripple variation after rectification.

For a retinal implant, the reference circuit must meet three design requirements. First, the reference circuit requires a wide-range supply-independent reference voltage. As mentioned above, eye rolling affects the rectified voltage level shift, and can vary the reference voltage output. This also causes a regulator output biasing drop. Second, a high-power supply rejection ratio (PSRR) is necessary for the reference circuit to lessen the deleterious effect of the deep AC ripple. An abrupt change in the ripple after rectification produces a high-frequency noise that can affect the output voltage. Finally, the reference circuit for the implantable device should be fully integrated on a single chip to reduce the active area. Because the eyeball is not large enough to accept external components, all components should be miniaturized on a single chip. Motivated by this, in this paper, we propose an integrated reference circuit that has a nonlinear current sink circuit (NSC) to improve the supply dependency and ripple rejection performances. This novel circuit is designed, fabricated using an SK Hynix 0.35-μm CMOS technology, and demonstrated on a benchtop.

The current paper is an extension of our previous work in [12]. Our last work utilizes the temperature compensation technique, which had been proven to reduce the variation of the reference voltage at higher temperatures. However, the proposed temperature circuit limits the line regulation performance as the biasing voltage of the compensation circuit varies proportionally to the supply changes. As a result, some current from the output path sinks out through the compensation circuit, which eventually deteriorates the output reference voltage. Besides that, the performance of the operational amplifier drops as the power supply increases because the input voltage of the amplifier exceeding the allowable input common-mode range, set for the current source of the amplifier to be kept in the saturation region. Consequently, the amplifier becomes unstable, which leads to the poor performances of line regulation and PSRR. The new architecture of the reference circuit resolves this issue to fulfil all requirements highlighted in this paper. With the proposed circuit, the improvements have been made in terms of its supply rejection and wide-range supply independent performance. The proposed circuit was also evaluated in the full-path retinal prosthetic system shown in Section 3. This system includes on-chip digital controller and stimulator, which is not implemented in [12].

2. Methods

2.1. Static Analysis

The newly proposed circuit and its small-signal equivalent circuit are illustrated in Figure 2a,b respectively. The proposed circuit is based on a V_{GS} reference supply independent current-reference circuit [13]. The loop around M_{N1} and M_{N2} has negative feedback; consequently, the output voltage at node V_{REF} is unchanged even when the voltage across the resistor R_1 varies. However, the nonlinear current I_1 in the current mirror, produced by the square-low behavior of the transistor and channel length modulation that results in a nonzero slope for I_D/V_{DS}, deteriorates the DC level of the reference voltage. The second-order factor in the reference voltage due to the nonlinear current can be eliminated if a linear current flows into M_{P3}. Therefore, the NSC formed by transistors M_{P4} and M_{N3} is proposed in Figure 2a, whereby a linear current can be produced in the output path. The slope of the output voltage with respect to V_{DD} is also controlled by the NSC. As a result, improvements have been made in terms of its supply dependency and ripple rejection performance.

If I_4 and V_{REF} are constant, the gate–source voltages of M_{N1} and M_{N2} will also be constant. V_{REF} can be expressed as follows:

$$V_{REF} = V_{GS,N1} + V_{GS,N2}. \tag{1}$$

A supply-independent reference voltage can be achieved if:

$$\frac{\partial V_{REF}}{\partial V_{DD}} = 0 \tag{2}$$

or

$$\frac{\partial V_{GS,N1}}{\partial V_{DD}} + \frac{\partial V_{GS,N2}}{\partial V_{DD}} = 0 \tag{3}$$

The gate–source voltage of M_{N1} can be obtained by first allowing the current in M_{N1} to saturate; hence,

$$I_4 = \frac{1}{2}k_{n,N1}(V_{GS,N1} - V_{TH,N1})^2(1 + \lambda_{N1}V_{DS,N1}). \tag{4}$$

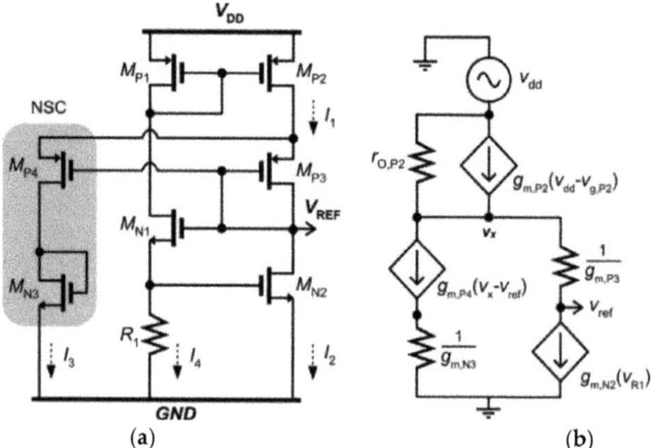

Figure 2. (a) Proposed reference circuit and (b) its small-signal equivalent circuit.

Rearranging Equation (4) gives:

$$V_{GS,N1} = \sqrt{\frac{2I_4}{K_{n,N1}}} \cdot \frac{1}{(1 + \lambda_{N1} V_{DS,N1})^{\frac{1}{2}}} + V_{TH,N1} \tag{5}$$

and taking the derivative of $V_{GS,N1}$ with respect to $V_{DS,N1}$ results in:

$$\frac{\partial V_{GS,N1}}{\partial V_{DS,N1}} = -\sqrt{\frac{2I_4}{K_{n,N1}}} \cdot \frac{\lambda_{N1}}{2(1 + \lambda_{N1} V_{DS,N1})}. \tag{6}$$

By approximating $\frac{\partial V_{DS,N1}}{\partial V_{DD}} = 1$, we obtain, from Equation (6),

$$\frac{\partial V_{GS,N1}}{\partial V_{DS,N1}} = \frac{\partial V_{GS,N1}}{\partial V_{DD}} \tag{7}$$

From Figure 2a, we know that:

$$I_2 = I_1 - I_3. \tag{8}$$

The NSC senses the voltage variation at the source terminals of M_{P3} and M_{P4} by assuming that V_{REF} is constant and that the nonlinear current in I_1 sinks into M_{P4} leaving a linear current in M_{P3}. By assuming that I_1 and I_3 grow linearly with respect to V_{DD}, we can rewrite I_2 in Equation (8) as follows:

$$\begin{aligned} I_2 &= m_1 V_{DD} + C_1 - (m_3 V_{DD} + C_3) \\ &= V_{DD}(m_1 - m_3) + (C_1 - C_3), \end{aligned} \tag{9}$$

where m_1 and m_3 are the slopes of I_1 and I_3, respectively, and C_1 and C_3 are constant values. I_2 can also be written as:

$$I_2 = \frac{1}{2} k_{n,N2} (V_{GS,N2} - V_{TH,N2})^2. \tag{10}$$

By substituting Equation (9) into (10), Equation (10) becomes:

$$V_{DD}(m_1 - m_3) + (C_1 - C_3) = \frac{1}{2} k_{n,N2} (V_{GS,N2} - V_{TH,N2})^2. \tag{11}$$

Rearranging Equation (11) for $V_{GS,N2}$ gives:

$$V_{GS,N2} = \sqrt{\frac{2}{K_{n,N2}}} \cdot \sqrt{(m_1 - m_3)V_{DD} + (C_1 - C_3)} + V_{TH,N2} \tag{12}$$

and taking the derivative of $V_{GS,N2}$ with respect to V_{DD} results in:

$$\frac{\partial V_{GS,N2}}{\partial V_{DD}} = \sqrt{\frac{2}{K_{n,N2}}} \cdot \frac{(m_1 - m_3)}{2\sqrt{(m_1 - m_3)V_{DD} + (C_1 - C_3)}} \cdot \tag{13}$$

Substituting Equations (6) and (13) into (3) produces:

$$-\sqrt{\frac{2I_4}{K_{n,N1}}} \cdot \frac{\lambda_{N1}}{2(1 + \lambda_{N1}V_{DS,N1})} + \sqrt{\frac{2}{K_{n,N2}}} \cdot \frac{(m_1 - m_3)}{2\sqrt{(m_1 - m_3)V_{DD} + (C_1 - C_3)}} = 0. \tag{14}$$

A key parameter of interest in Equation (14) is m_3. To satisfy the condition in Equation (2), m_3 in Equation (14) can be controlled by adjusting the size of transistor M_{P4}. The slope of m_3 should be less than m_1, so that the first term of Equation (14) can be subtracted to zero. Trimming the transistor's width and length can be performed by the Cadence Spectre simulation to optimize the output voltage variation.

2.2. Dynamic Analysis

The dynamic behavior of the proposed circuit can be determined by analyzing the equivalent circuit shown in Figure 2b. M_{N1} and R_1 form a source follower. For simplifying the design equations, we presume that $g_{m,N1}R_1 \gg 1$; as a result, v_{R1} is close to v_{ref} and $v_{g,p2}$ can be approximately equal to v_{dd}. From Figure 2b, by applying Kirchhoff's current law at node v_x, we can derive the following equation:

$$\frac{v_{dd} - v_x}{r_{o,P2}} = g_{m,P4}v_x + v_{ref}(g_{m,N2} - g_{m,P4}). \tag{15}$$

v_x can be derived from the circuit given as:

$$v_x = v_{ref}\left(\frac{g_{m,N2} + g_{m,P3}}{g_{m,P3}}\right) \tag{16}$$

By substituting Equation (16) into (15), the ratio of v_{ref} to v_{dd} can be obtained as follows:

$$\frac{v_{ref}}{v_{dd}} \approx \frac{g_{m,P3}}{r_{o,P2}(g_{m,N2} \times g_{m,P4} + g_{m,N2} \times g_{m,P3})} \tag{17}$$

Accordingly, Equation (17) shows that v_{ref} becomes decoupled from small variations in v_{dd} when the output resistance of $r_{o,P2}$ is sufficiently high. The NSC produces the $g_{m,P4}$ in the denominator of Equation (17), which can help improve the PSRR performance.

Table 1 shows all the parameters used in the proposed reference circuit in Figure 2b.

Table 1. Component parameters.

Component	Parameter	Component	Parameter
M_{P1}, M_{P2}	$W = 4\ \mu m, L = 1\ \mu m, m = 50$	M_{N2}	$W = 25\ \mu m, L = 3\ \mu m, m = 4$
M_{P3}	$W = 1.15\ \mu m, L = 1\ \mu m, m = 1$	M_{N3}	$W = 4\ \mu m, L = 1\ \mu m, m = 20$
M_{P4}	$W = 4\ \mu m, L = 1\ \mu m, m = 9$	R_1	$70\ k\Omega$
M_{N1}	$W = 4\ \mu m, L = 1\ \mu m, m = 10$		

3. Simulation and Measurement Results

The proposed supply independent voltage reference circuit was fabricated using SK Hynix 0.35 μm CMOS technology. The micrograph of the proposed reference circuit is displayed in Figure 3a, and it occupies an active area of 0.0131 mm^2. Figure 3b illustrates the transient response of the output voltage V_{REF} when V_{DD} is ramped from 0 to 5 V. When the current starts flowing in the circuit, the self-biasing reference circuit drives itself towards the desired stable state, according to the measurement, the output voltage starts settling at 1.37 V when V_{DD} reaches over 2 V.

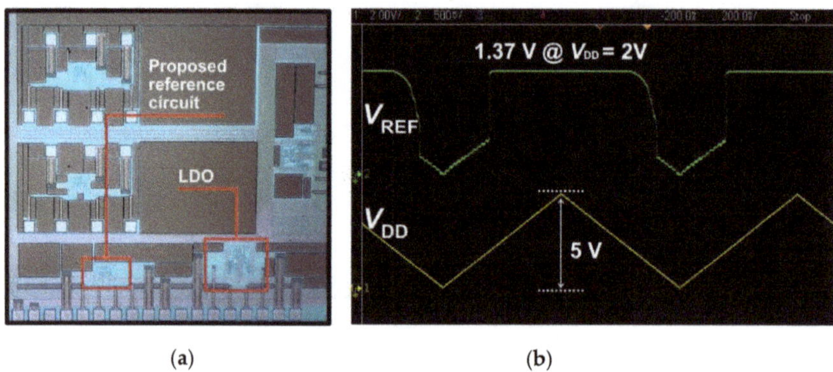

(a)   (b)

Figure 3. (a) Micrograph of the proposed reference circuit, (b) measured transient response.

The measured output voltage variation with respect to V_{DD} is plotted in Figure 4a. The inset graphs show the measured and simulated results when V_{DD} varies from 2 V to 5 V and 0 to 30 V, respectively. The observed output voltage increases linearly by 10 mV for V_{DD} variation from 2.1 V to 5 V. This results in a line regulation of 3.45 mV/V. This dependency can be further optimized by adjusting M_{P4}, as indicated in Equation (14). The simulation shows a stable reference voltage even when V_{DD} reaches up to 30 V. However, the measurement was stopped by the breakdown voltage of ~6 V for transistors. The variation of V_{REF} between 2.4 V to 30 V for V_{DD} is only 11 mV, which results in a line regulation of 0.4 mV/V. The measured and simulated PSRR of the proposed reference circuit is displayed in Figure 4b, where we obtain the simulated PSRR of −67 dB without the NSC block and −112 dB with one, respectively. This shows that the nonlinear current sink circuit we proposed in this work works as expected. In reality, however, a maximum PSRR of −93 dB is observed for frequencies lower than 1 kHz. The difference of −19 dB between the simulation and measurement is probably due to the parasitic capacitance that arises from the metal lines and pads in the fabricated chip, where our proposed reference circuit shares the power supply rails of other test blocks.

Figure 5a illustrates 500 Monte Carlo simulation results of the reference voltage. The average reference voltage, μ, and the standard deviation, σ, are 1.3955 V and 66.15 m, respectively. The line regulation results in different corners are shown in Figure 5b. The results indicate that the effectiveness of NSC produces stable reference voltage in all corners. The worst case occurs in SS condition in which the line regulation of 34.4 mV/V is obtained for V_{DD} variation from 2.1 V to 5 V.

The overall electrical performance of the CMOS reference circuit designed in this work is summarized in Table 2, where it is also compared with the performances of the prior designs presented in [14–17].

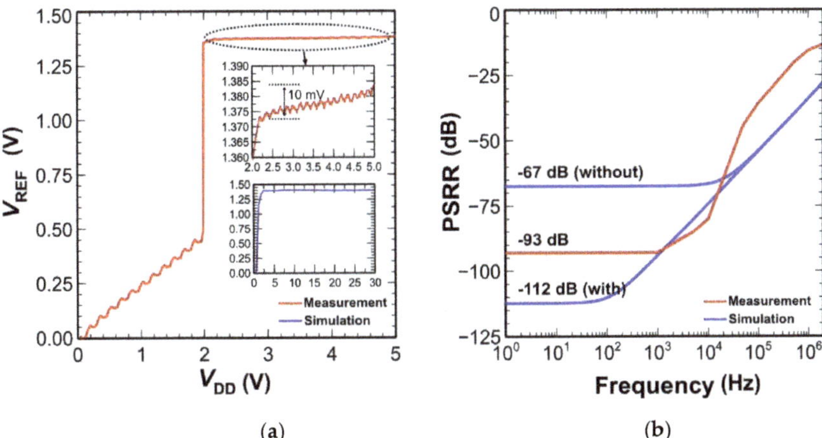

Figure 4. Measured results of (**a**) line regulation and (**b**) power supply rejection ratio (PSRR).

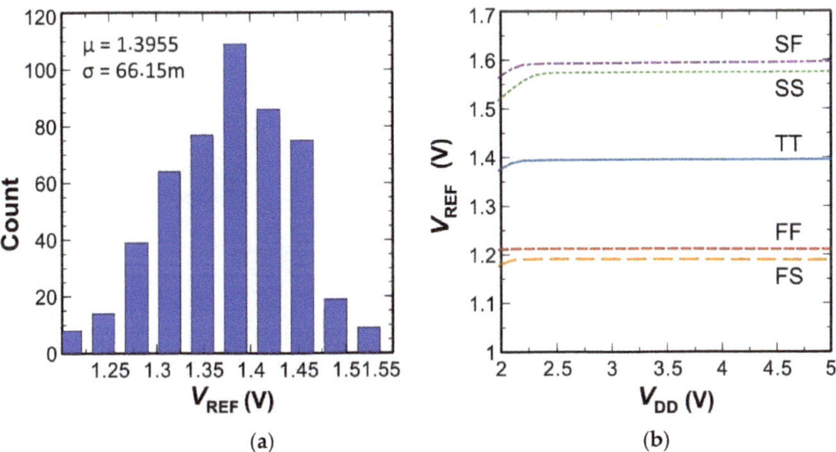

Figure 5. (**a**) 500 Monte Carlo simulation results at 3.3 V supply, 27 °C; (**b**) line regulation results in different corners.

Table 2. Electrical performance summary and comparison of reference circuit with previous designs.

Parameter	[14]	[15]	[16]	[17]	This Work
Supply voltage range (V)	2 to 5	0.5 to 1	1.3 to 1.8	2.6 to 12 (2.4 to 20) *	2.1 to 5 (2.4 to 30) *
Reference output voltage (V)	1.14055	0.495	1.17	1.6	1.37
Line regulation (mV/V)	2	3.2	0.35	0.957 (0.511) *	3.45 (0.39) *
PSRR (dB)	−61	−50	−52	−60 (−59.2) *	−93 (−112) *
Chip Area (mm²)	0.0396	0.0522	0.082	-	0.0131
Technology (µm)	0.35	0.065	0.18	1.6	0.35

* Simulation result.

To verify the real performance of the proposed reference circuit, we applied it to the low-voltage-drop regulator (LDO) circuit plotted in Figure 6a. This LDO circuit that utilizes an output ringing compensation circuit was fabricated in the same chip displayed in Figure 3a. The output of the LDO provides a positive supply voltage to 64 pixels stimulator circuit which requires maximum current of 10 mA. This requirement is based on our previous work which can be found in [10]. In the experimental setup to evaluate the line-transient responses of LDO, a 100 Ω resistor was chosen as a load for the LDO output terminal so that the load current is larger than 10 mA. When we applied a ramp voltage varying from 0 to 5 V, the LDO still output a constant voltage of 1.62 V, as shown in Figure 6b, while the reference circuit output 1.37 V. While maintaining a load current of 16.2 mA, the LDO output variation was observed as 17.6 mV in the supply range from 2.5 V to 5 V. The overshoot and undershoot voltages were 80 mV and 100 mV, respectively, as illustrated in Figure 6c. This LDO also had a fast recovery time of 240 ns, corresponding to the rapid change of V_{DD} from 5 V to 5.2 V in 20 ns. Presently, we evaluated the full-path system using separate chips for LDO, digital controller and stimulator shown in Figure 7.

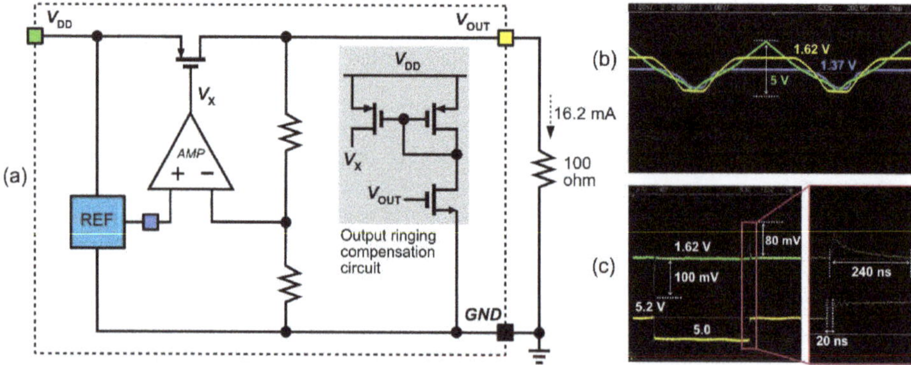

Figure 6. (**a**) Low-voltage-drop regulator (LDO) circuit incorporating the proposed reference circuit; (**b**) measured line-transient responses; and (**c**) measured overshoot and undershoot voltages.

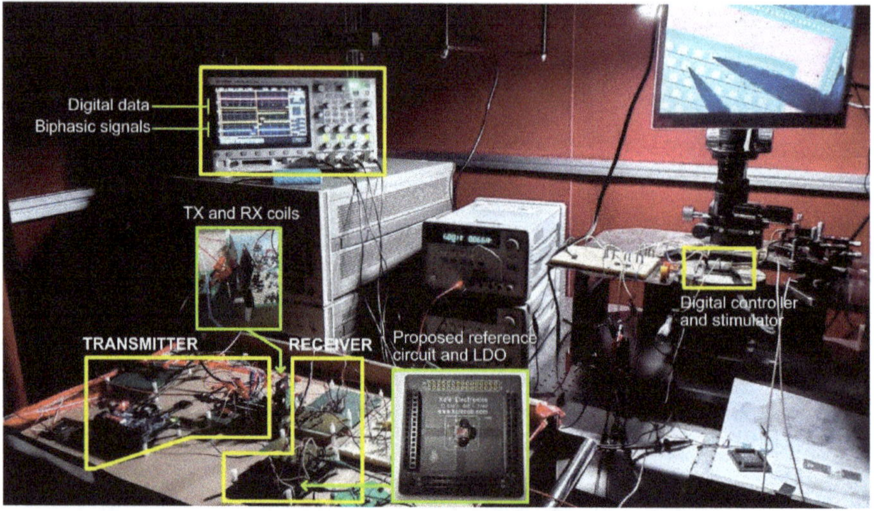

Figure 7. Full-path prosthetic system experimental setup.

In the near future, the proposed voltage-reference circuit working with the LDO can be applied to fully integrated retinal prosthetic systems. Our proved concept prototype system could be further miniaturized and potentially transformed into stretchable devices which is beyond the flatland constrain of the traditional wafer-based system. Currently, we have developed a stretchable receiving coil and are progressively working on flexible microelectrode arrays.

4. Conclusions

In this paper, a novel CMOS voltage-reference circuit with high PSRR and wide-range supply independence was proposed and designed for a subretinal prosthetic system. The proposed NSC reduced the supply ripples significantly and sank out undesired signals in the reference voltage. Experimental results exhibited good agreement with the proposed concept and demonstrated better performance in supply independence and PSRR, when compared to the previous works. The proposed circuit provided a constant output voltage of 1.37 V and exhibited 10 mV variations over the supply range from 2.1 V to 5 V, resulting in a line regulation of 3.45 mV/V. The maximum PSRR was observed to be −93 dB for frequencies below 1 kHz. The current proof-of-concept prototype was implemented on a single chip with the LDO circuit, using a standard 0.35 µm CMOS process. The proposed reference design occupied an active area of 0.0131 mm^2. Considering the high precision and small size design, this novel reference circuit can be used for implantable devices such as retinal prostheses and cochlear implants. As a future work, we will integrate the proposed reference circuit and LDO onto a single chip along with other functional blocks such as a demodulator, digital control, and high-density stimulator array.

Author Contributions: Conceptualization, R.B.A.Z., and J.K.; methodology, R.B.A.Z. and J.K.; formal analysis, R.B.A.Z., H.C., and J.K.; writing—original draft preparation, R.B.A.Z., H.C., and J.K.; writing—review and editing, R.B.A.Z., H.C., and J.K.; supervision, H.C., and J.K.; project administration, J.K.; funding acquisition, J.K. All authors have read and agreed to the published version of the manuscript.

Funding: This research was partially supported by the National Research Foundation of Korea (Grant No. NRF-2017M3A9E2056461) and the Gachon University Research Fund (2018-0324). This work was supported by the National Research Foundation of Korea grant funded by the government (MSIT) (No. 2020R1A2C4001606).

Acknowledgments: The authors would like to express their sincerest appreciation to the IC Design Education Center for chip fabrication.

Conflicts of Interest: The authors declare no conflict of interest.

References

1.   Lee, B.; Kiani, M.; Ghovanloo, M. A Triple-Loop Inductive Power Transmission System for Biomedical Applications. *IEEE Trans. Biomed. Circuits Syst.* **2016**, *10*, 138–148. [CrossRef] [PubMed]

2.   Jegadeesan, R.; Nag, S.; Agarwal, K.; Member, S. Enabling Wireless Powering and Telemetry for Peripheral Nerve Implants. *IEEE J. Biomed. Health Inform.* **2015**, *19*, 958–970. [CrossRef] [PubMed]

3.   Lo, Y.-K.; Chen, K.; Gad, P.; Liu, W. An On-Chip Multi-Voltage Power Converter with Leakage Current Prevention Using 0.18 um High-Voltage CMOS Process. *IEEE Trans. Biomed. Circuits Syst.* **2016**, *10*, 163–174. [CrossRef] [PubMed]

4.   Ha, S.; Khraiche, M.L.; Akinin, A.; Jing, Y.; Damle, S.; Kuang, Y.; Bauchner, S.; Lo, Y.-H.; Freeman, W.R.; Silva, G.A.; et al. Towards high-resolution retinal prostheses with direct optical addressing and inductive telemetry. *J. Neural Eng.* **2016**, *13*, 056008. [CrossRef] [PubMed]

5.   Goetz, G.A.; Palanker, D.V. Electronic approaches to restoration of sight. *Rep. Prog. Phys.* **2016**, *79*, 096701. [CrossRef] [PubMed]

6.   Lin, T.-C.; Chang, H.-M.; Hsu, C.-C.; Hung, K.-H.; Chen, Y.-T.; Chen, S.-Y.; Chen, S.-J. Retinal prostheses in degenerative retinal diseases. *J. Chin. Med. Assoc.* **2015**, *78*, 501–505. [CrossRef] [PubMed]

7.   Wu, C.-Y.; Tseng, C.-K.; Liao, J.-H.; Chiao, C.-C.; Chu, F.-L.; Tsai, Y.-C.; Ohta, J.; Noda, T. CMOS 256-Pixel/480-Pixel Photovoltaic-Powered Subretinal Prosthetic Chips with Wide Image Dynamic Range and Bi/Four-Directional Sharing Electrodes and Their Ex Vivo Experimental Validations With Mice. *IEEE Trans. Circuits Syst. I Regul. Pap.* **2020**, *67*, 3273–3283. [CrossRef]

8. Kuo, P.H.; Wong, O.-Y.; Tzeng, C.-K.; Wu, P.-W.; Chiao, C.C.; Chen, P.-H.; Tsai, Y.-C.; Chu, F.-L.; Ohta, J.; Tokuda, T.; et al. Improved Charge Pump Design and Ex Vivo Experimental Validation of CMOS 256-Pixel Photovoltaic-Powered Subretinal Prosthetic Chip. *IEEE Trans. Biomed. Eng.* **2020**, *67*, 1490–1504. [CrossRef] [PubMed]

9. Özmert, E.; Arslan, U. Retinal Prostheses and Artificial Vision. *Turk. J. Ophthalmol.* **2019**, *49*, 213–219. [CrossRef] [PubMed]

10. Kang, H.; Abbasi, W.H.; Kim, S.-W.; Kim, J. Fully Integrated Light-Sensing Stimulator Design for Subretinal Implants. *Sensors* **2019**, *19*, 536. [CrossRef] [PubMed]

11. Kim, J.; Basham, E.; Pedrotti, K.D. Geometry-based optimization of radio-frequency coils for powering neuroprosthetic implants. *Med. Biol. Eng. Comput.* **2013**, 123–134. [CrossRef] [PubMed]

12. Zawawi, R.B.A.; Abbasi, W.H.; Kim, S.-H.; Choi, H.; Kim, J. Wide-Supply-Voltage-Range CMOS Bandgap Reference for In Vivo Wireless Power Telemetry. *Energies* **2020**, *13*, 2986. [CrossRef]

13. Gray, P.; Meyer, R. *Analysis and Design of Analog Integrated Circuits*, 5th ed.; John Wiley and Sons: Hoboken, NJ, USA, 2010.

14. Zhou, Z.-K.; Shi, Y.; Wang, Y.; Li, N.; Xiao, Z.; Wang, Y.; Liu, X.; Wang, Z.; Zhang, B. A Resistorless High-Precision Compensated CMOS Bandgap Voltage Reference. *IEEE Trans. Circuits Syst. I Regul. Pap.* **2019**, *66*, 428–437. [CrossRef]

15. Chi-Wa, U.; Zeng, W.-L.; Law, M.-K.; Lam, C.-S.; Martins, R.P. A 0.5-V Supply, 36 nW Bandgap Reference With 42 ppm/°C Average Temperature Coefficient Within −40 °C to 120 °C. *IEEE Trans. Circuits Syst. I Regul. Pap.* **2020**, *67*, 3656–3669.

16. Kim, M.; Cho, S.H. A 0.0082-mm², 192-nW Single BJT Branch Bandgap Reference in 0.18-μm CMOS. *IEEE Solid State Circuits Lett.* **2020**, *3*, 426–429.

17. Sodagar, A.; Najafi, K. A wide-range supply-independent CMOS voltage reference for telemetry-powering applications. In Proceedings of the 9th International Conference on Electronics, Circuits and Systems, Dubrovnik, Croatia, 15–18 September 2002; Institute of Electrical and Electronics Engineers (IEEE): Piscataway, NJ, USA, 2003; Volume 1, pp. 401–404.

Publisher's Note: MDPI stays neutral with regard to jurisdictional claims in published maps and institutional affiliations.

 *electronics*

Article

Analysis of Fundamental Differences between Capacitive and Inductive Impedance Matching for Inductive Wireless Power Transfer

Yelzhas Zhaksylyk *, Einar Halvorsen, Ulrik Hanke and Mehdi Azadmehr

Department of Microsystems, University of South-Eastern Norway, Campus Vestfold, NO-3184 Horten, Norway; Einar.Halvorsen@usn.no (E.H.); Ulrik.Hanke@usn.no (U.H.); Mehdi.Azadmehr@usn.no (M.A.)
* Correspondence: Yelzhas.Zhaksylyk@usn.no

Received: 21 February 2020; Accepted: 6 March 2020; Published: 13 March 2020

Abstract: Inductive and capacitive impedance matching are two different techniques optimizing power transfer in magnetic resonance inductive wireless power transfer. Under ideal conditions, i.e., unrestricted parameter ranges and no loss, both approaches can provide the perfect match. Comparing these two techniques under non-ideal conditions, to explore fundamental differences in their performance, is a challenging task as the two techniques are fundamentally different in operation. In this paper, we accomplish such a comparison by determining matchable impedances achievable by these networks and visualizing them as regions of a Smith chart. The analysis is performed over realistic constraints on parameters of three different application cases both with and without loss accounted for. While the analysis confirms that it is possible to achieve unit power transfer efficiency with both approaches in the lossless case, we find that the impedance regions where this is possible, as visualized in the Smith chart, differ between the two approaches and between the applications. Furthermore, an analysis of the lossy case shows that the degradation of the power transfer efficiencies upon introduction of parasitic losses is similar for the two methods.

Keywords: impedance matching network; parasitic resistance; power loss; reflection coefficient; Smith chart; wireless power transfer

---

1. Introduction

Recent demand on mobility and accessibility of devices is pushing the development of wireless technology to new levels. There are good solutions for data transfer such as WiFi and Bluetooth, whereas, power is still delivered by either batteries or cable, the main bottleneck in the strive for cutting all the wires and limiting the mobility of devices.

For daily-life applications, inductive wireless power transmission has drawn increasing attention from researchers as it offers the highest power transfer efficiency (PTE) among other alternatives such as capacitive, microwave, laser, and acoustic [1,2]. Various products such as electric toothbrushes and mobile chargers using this technique are already commercially available. This technique provides two advantages compared to others: transfer of high power and low-frequency operation, making it less hazardous to the human body [3]. The main issue with this type of inductive wireless power transfer (WPT) is the mobility as the sender and receiver need to be close to each other, less than a few centimetres. Magnetic resonant (MR) WPT, an inductive technique based on highly coupled high-Q resonators, addresses this issue and offers a reasonable distance of power transfer (up to 2 m) [4]. However, a considerable challenge of MR-WPT is to maintain high power transfer throughout a range of distances between resonators and for variations in load value, as these will cause a mismatch between the source and input impedances [5].

In order to solve this challenge, different types of impedance matching techniques have been developed in the last decades. The simplest and most popular ones use capacitive or inductive impedance matching networks (IMNs). The capacitive method uses variable capacitors to tune the transmitter to the resonant frequency or a predefined capacitor sequence for different distances [6,7]. There is a variety of adaptive frequency tuning systems where L, T and Π-type impedance matching networks contain capacitors [7,8]. Matching can potentially also provide power to multiple device WPT by using only a single transmitter [9].

The four-coil MR-WPT system presented by the MIT group in 2007 [4] has become a recognized solution [10–14] for highly resonant WPT systems for medium distances. The system consists of two or more high-Q resonating coils which are driven by a low-Q coil connected to the power source. The load is also connected to a low-Q coil. The coupling between resonator and the driver coils (or the load coil) can be considered as parts of a matching network, where tuning of the impedance can be achieved by changing the coupling between them. In our study, this method is referred to as inductive matching. In a previous work, we showed that these two matching techniques, i.e., the capacitive and inductive matching could potentially achieve a similar level of matching in certain cases [15]. Among the many different capacitive compensation circuitries, we chose the parallel-series compensation according to reference [8] for comparison to the inductive method. This network offers sufficient degrees of freedom to match perfectly if there is no loss and no restriction on parameter ranges, hence it is sufficient to give an insight into the effect of these limitations. The presented comparison method can also be used to identify matchable regions of other compensation structures.

The aforementioned matching techniques, i.e., capacitive and inductive can be applied to any mismatches in a WPT system to improve the power transfer efficiency [16]. However, there is a challenge in the direct comparison of their matching performances because they have different circuit topologies. The paper describes a method that makes a systematic comparison of their performance possible. The proposed method is based on comparison of the conjugate impedance of the matchable load, displayed in the Smith chart. The conjugate matching method has been analysed in [17], where all concepts of conservation and amplification of power by two port network was defined. This method is used by [18,19] to describe efficiency of the WPT system with inductive IMN. Our work presents a comparison of the matchable loads offered by the inductive and capacitive matching networks over a full range of realistic parameter ranges for three different applications distinguished by their operating frequency and power level [20–22]. The operating frequencies are within the allowed Industrial, Scientific and Medical (ISM) bands [9], which also limits the frequency range in the analysis. For the ease of simulation and calculations we choose to keep the distance constant and match the different load values. We assume the coil sizes are such that the systems operate in near field and the inductances of sender and receiver coils are equal, as in the [18–22].

This paper is organized as follows. In Section 2, impedances are analysed by derivation of reflection coefficients. We intentionally exclude the parasitic components in the system in order to have a clear comparison between these matching techniques in the ideal case. Subsequently, Section 3 visualizes the matchable reflection coefficients in the Smith chart, which graphically illustrates all of the possible complex impedances that are obtained by sweeping matching parameters. Therefore, it demonstrates which method offers the wider area of impedances that can be matched. Furthermore, the impact of parasitic loss to the matchable region is analyzed and optimized power simulation is given in the Section 4 and Section 5 discusses the outcome of the comparison.

2. Reflection Coefficients

In order to map and compare the tunable impedances of WPT systems, suitable circuit models and corresponding impedance expressions should be established. A generic WPT system with impedance matching networks can be represented as shown in Figure 1. The driving source consists of an ideal voltage source (V_S) and a series resistance R_S. The two-port network consists of resonators, and here we will consider capacitive and inductive impedance matching circuits. The network is terminated by load impedance Z_L at the output. Here, i_S and i_L are currents through R_S and Z_L, respectively.

Matching networks are necessary to obtain a match between input impedance Z_{in} and source impedance R_S. They affect the Power Transfer Efficiency (PTE), defined as ratio between power delivered to the load and input power of the two-port network [18]. This research focuses on the comparison of the matchable loads offered by the capacitive and inductive IMNs. Therefore, the two-port network is redrawn as in Figure 2 to get value of an effective impedance Z_{out} at the output, which is a complex conjugate Z_L^* form of load impedance that can be perfectly matched.

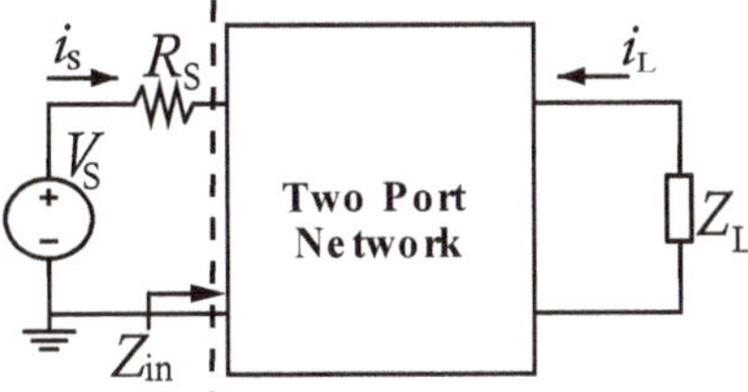

Figure 1. Two-port network representation of a highly resonant WPT system.

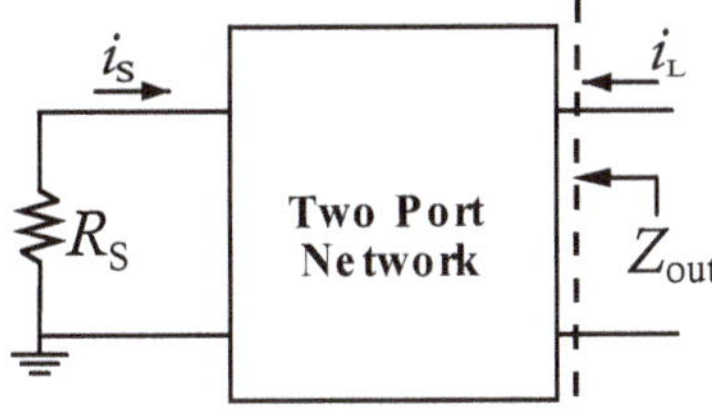

Figure 2. Two-port network representation when source is terminated.

2.1. Capacitive Matching Network

A lossless model of the inductive resonant WPT system with capacitive impedance matching is shown in Figure 3. The circuit elements are ideal, i.e., inductors and capacitors do not have parasitics. Matching networks consist of series-parallel connection of capacitors C_{ts}, C_{tp} and C_{rp}, C_{rs} at the transmitter (Tx) and receiver (Rx) sides. The source impedance is considered as resistance R_S.

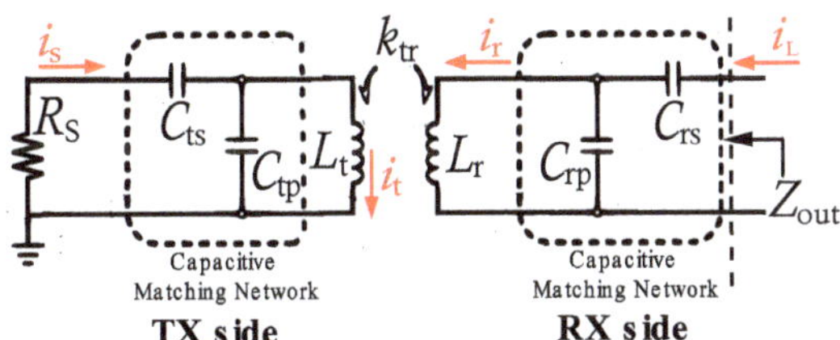

Figure 3. Equivalent circuit of a WPT system with capacitive impedance matching network.

Applying Kirchhoff's voltage law (KVL) to the circuit in Figure 3, the voltage-current relations can be written in an impedance matrix form

$$\begin{bmatrix} V_{out} \\ 0 \\ 0 \\ 0 \end{bmatrix} = \begin{bmatrix} Z_{11} & Z_{12} & 0 & 0 \\ Z_{21} & Z_{22} & Z_{23} & 0 \\ 0 & Z_{32} & Z_{33} & Z_{34} \\ 0 & 0 & Z_{43} & Z_{44} \end{bmatrix} \begin{bmatrix} i_L \\ i_r \\ i_t \\ i_s \end{bmatrix} \tag{1}$$

where

$$Z_{11} = \frac{1}{j\omega C_{rs}} + \frac{1}{j\omega C_{rp}}, \ Z_{12} = -\frac{1}{j\omega C_{rp}}, \tag{2}$$

$$Z_{21} = -\frac{1}{j\omega C_{rp}}, \ Z_{22} = j\omega L_r + \frac{1}{j\omega C_{rp}}, \ Z_{23} = j\omega M_{12}, \tag{3}$$

$$Z_{32} = j\omega M_{tr}, Z_{33} = j\omega L_t + \frac{1}{j\omega C_{tp}}, Z_{34} = -\frac{1}{j\omega C_{tp}}, \tag{4}$$

$$Z_{43} = -\frac{1}{j\omega C_{tp}}, \ Z_{44} = R_S + \frac{1}{j\omega C_{ts}} + \frac{1}{j\omega C_{tp}}. \tag{5}$$

Here,

$$M_{tr} = k_{tr}\sqrt{L_t L_r}, \ 0 \le k_{tr} \le 1 \tag{6}$$

is the mutual inductance between inductors L_t and L_r. The coefficient k_{tr} represents the coupling between them and its value is inversely proportional to the cube of their distance [15]. The distance change and variation of load impedance can be controlled by adjusting the capacitances C_{ts}, C_{tp} and C_{rs}, C_{rp} in the matching networks.

The effective impedance Z_{out} at the output of two-port network is

$$Z_{\text{out}} = jX_r + \frac{Z_{12}^2(Z_{34}^2 - Z_{33}Z_{44})}{(Z_{22}Z_{33} - Z_{32}^2)Z_{44} - Z_{34}^2 Z_{22}}, \tag{7}$$

where

$$X_r = -\frac{1}{\omega C_{rs}} - \frac{1}{\omega C_{rp}}. \tag{8}$$

The real and imaginary parts of the impedance are

$$Re\{Z_{\text{out}}\} = \Delta R_S, \ Im\{Z_{\text{out}}\} = \Delta A - \frac{1}{\omega C_{rs}} - \frac{1}{\omega C_{rp}}, \tag{9}$$

where

$$\Delta = \frac{\omega^2 M_{tr}^2 Z_{12}^2 Z_{34}^2}{B^2 R_S^2 + (X_t B - Z_{34}^2 Z_{22})^2}, \tag{10}$$

$$B = Z_{22}Z_{33} - Z_{32}^2, \ X_t = -\frac{1}{\omega C_{ts}} - \frac{1}{\omega C_{tp}}, \tag{11}$$

$$A = \frac{(Z_{22}Z_{34}^2 - X_t\omega^2 M_{tr}^2)Z_{34}^2 - Z_{33}B(X_t^2 + R_S^2)}{\omega^2 M_{tr}^2 Z_{34}^2}. \tag{12}$$

The impedance Z_{out} can be seen at the output of the two-port network, which is complex conjugate form of load impedance Z_L. This impedance is used to derive the reflection coefficient (Γ), which can be seen from the load side

$$\Gamma = (Z_{\text{out}} - Z_0)/(Z_{\text{out}} + Z_0) \tag{13}$$

where Z_0 is reference impedance equal to R_S.

Equation (13) is used to draw Γ in the Smith chart to visualize graphically and estimate the values of load impedance that can be perfectly matched. Furthermore, a derivation of the reflection coefficient expression for the inductive matching is discussed in next section, and numerical results are given in Section 3.

2.2. Inductive Matching Network

Inductive coupling is another method widely exploited to match the input impedance for different distances between resonators or load variation. This system uses additional magnetically coupled coils at the transmitter or receiver, or both, to enhance the PTE. These coils do not need as high Q as the resonator coils. The most popular one is a four-coils system with source coil L_S, high-Q transmitter coil L_t, high-Q receiver coil L_r, and load coil L_L [4,15]. The matching can be controlled by varying couplings between the source/load coils and high-Q coils—k_S, k_L. The equivalent lossless model of such a system is shown in Figure 4. The high-Q coils are connected to series external capacitors to form resonators.

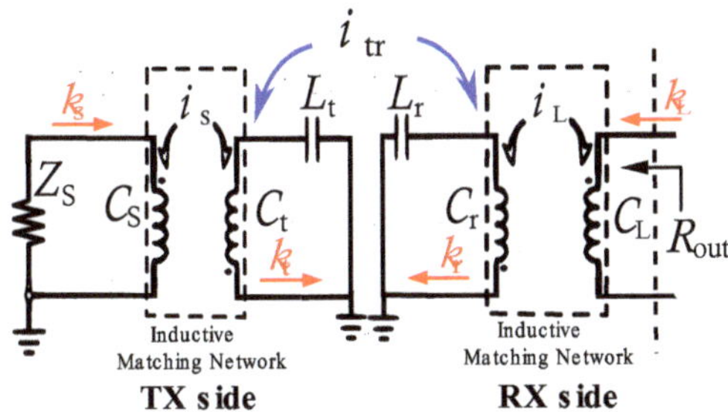

Figure 4. Equivalent circuit of WPT system with inductance matching network.

For such a circuit, the same voltage-current relations can be used as in Equation (1), where impedances are

$$Z_{11} = j\omega L_L, \ Z_{12} = -j\omega M_L, \tag{14}$$

$$Z_{21} = -j\omega M_L, Z_{22} = j\omega L_t + \frac{1}{j\omega C_t}, Z_{23} = j\omega M_{tr}, \tag{15}$$

$$Z_{32} = j\omega M_{tr}, Z_{33} = j\omega L_r + \frac{1}{j\omega C_r}, Z_{34} = -j\omega M_S, \tag{16}$$

$$Z_{43} = -j\omega M_S, \ Z_{44} = R_S + j\omega L_S, \tag{17}$$

and by neglecting cross-coupling:

$$Z_{13} = Z_{14} = Z_{24} = Z_{31} = Z_{41} = Z_{42} = 0. \tag{18}$$

Here, $M_{S/L}$ is a mutual inductance between source/load coil ($L_{S/L}$) and high-Q coil ($L_{t/r}$)

$$M_{S/L} = k_{S/L}\sqrt{L_{S/L}L_{t/r}}, \ 0 \le k_{S/L} \le 1. \tag{19}$$

and M_{tr} is the mutual inductance between resonator coils

$$M_{tr} = k_{tr}\sqrt{L_tL_r}, \ 0 \le k_{tr} \le 1. \tag{20}$$

If we assume that the two resonators have same resonance frequency $\omega = 1/\sqrt{L_{t/r}C_{t/r}}$, then

$$Z_{22} = Z_{33} = 0. \tag{21}$$

From impedance Equations (14)–(18) the real and imaginary parts of output impedance Z_{out} become

$$Re\{Z_{out}\} = \Delta R_S, \ Im\{Z_{out}\} = \omega(L_L - \Delta L_S), \tag{22}$$

where

$$\Delta = \frac{\omega^2 M_S^2 M_L^2}{M_{tr}^2 (R_S^2 + \omega^2 L_S^2)}. \tag{23}$$

Furthermore, Equations (13), (22) and (23) are used to calculate the reflection coefficient in the following section. These results conclude the theoretical analysis that is required to compare the two techniques.

3. Matchable Regions of Lossless Model

This section presents graphs of reflection coefficients in the Smith Chart, based on the equations derived in the previous section for both capacitive and inductive matching. These graphs help us to estimate the matchable loads and to select the proper matching network at the transmitter and receiver side. In the capacitive method the input impedance is controlled via capacitances C_{ts}, C_{tp}, C_{rs}, C_{rp}, whereas in the inductive matching it is controlled by coupling coefficients k_S, k_L between the source/load coils and high-Q coils.

In this section, the parasitic components of the system are intentionally excluded to have a clear and ideal case comparison between these two matching techniques. Reflection coefficients were examined for three specific applications, and parameter values used for the cases given in Table 1. The presented cases have been chosen so that they cover a wide range of WPT applications with different specifications for power level and operating frequency [20–22].

Table 1. Applications and parameters.

Application	Operation Frequency	$L_t = L_r$	$C_t = C_r$	k_{tr}	References
Case A—Car charging	85 kHz	60 μH	58.4 nF	0.01	[20]
Case B—Tablet charging	6.78 MHz	6 μH	91.8 pF	0.01	[21]
Case C—High Frequency	100 MHz	2.5 μH	1.01 pF	0.01	[22]

Figures 5–10 show the realizable reflection coefficient values in the Smith Chart. The matchable regions are indicated by bold black borders. Each figure consists of three impedance regions, where each region corresponds to different resulting impedances in the circuit: Z_{tx}—impedance at the transmitter, Z_{tr}—impedance after transmission, Z_{out}—impedance at the output. They are obtained by sweeping the impedance matching network parameters over the realistic range of values, which are given in Table 2. Chosen constraint for the inductive IMN is based on an assumption that the driving and load loops have an inductance equal or smaller than the resonator inductances [4]. The bottom limit for the capacitance variance in the capacitive IMN is the lowest value of capacitance in the market, which is approximately 500 fF (ignoring the possibilities of series connection), whereas the upper limit was chosen sufficient for the application choice. As we can see further from the results the upper limit in the inductive method and lower limit in the capacitive method decides the final shape in the Smith chart. The circuits are equivalent models of an inductive WPT, which consists of source resistance R_S, matching networks (capacitive or inductive), and lossless coils for transmission and reception.

Case A application in Table 1 is an Electric Vehicle (EV) charging station for transmission of high power. It is designed for low-frequency operation, in our case at 85 kHz frequency, and designed for coils around $L_t = L_r = 60$ μH. Figure 5a shows the impedance at the transmitter. This impedance region agrees well with known results for L-type networks in [23]. It is controlled by varying the capacitances C_{ts} and C_{tp} within the range given in Table 2. Since the L-type capacitive matching network at the transmitter (Tx) cannot match all impedances, a matching network is required for the receiver part as well. Second stage, in Figure 5b, illustrates reflection graph after the resonator coils L_t and L_r, which are coupled at $k_{tr} = 0.01$. The inductances change the region into a circle—smaller than Smith chart, which means there is still a limitation in the matchable area. Finally, in Figure 5c, third stage gives impedances that can be matched by the complete network consisting of matching networks at the Tx (C_{ts} and C_{tp}) and Rx (C_{rs} and C_{rp}) sides. All the capacitances in this example are varied from 0.1 pF to 200 pF (Table 2). The impedance matching network at the Rx side greatly improves the matchable area, which now practically fills the Smith chart. It means that for case A without losses any load is matchable by the capacitive matching method at $k_{tr} = 0.01$, but it does not mean that this still holds for other coupling coefficient values.

The result of following a similar procedure for case A with the inductive impedance matching network is shown in Figure 6. In this case, matching works by adjusting source/load inductances L_S/L_L and coupling coefficients between these and resonator coils (k_S, k_L). Inductance and coupling constant ranges are given in Table 2. In Figure 6a, we can notice that the inductive method gives an extremely limited region of impedances that can be matched, hence, matching at the receiver becomes crucial. One thing that can be noticed from Figure 6b impedance range of Z_{tr} after the coupling ($k_{tr} = 0.01$) between resonators L_t, L_r is even smaller, and it shows that three-coil system is not suitable for applications where load values are diverse. However, the resulting matchable impedance of the complete system with four coils, where impedance can be matched at both sides (Tx and Rx), fills around 80% of the Smith chart and is shown in Figure 6c. According to the figure, for case A without losses there is roughly 20% of the impedances that cannot be perfectly matched by the inductive approach.

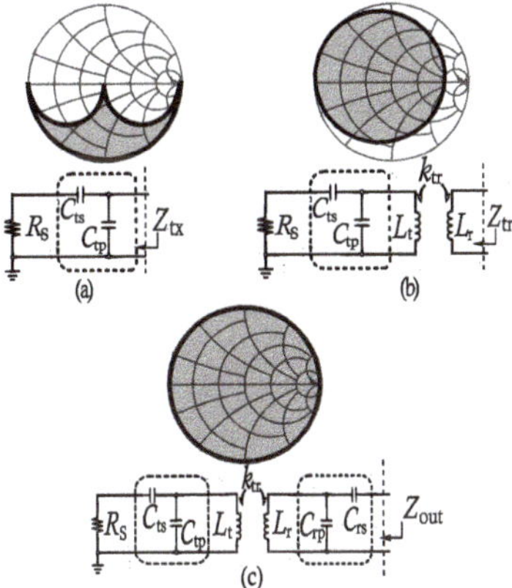

Figure 5. Reflection coefficient graphs show impedances that can be obtained by capacitive matching networks placed into WPT system in case A: (**a**) Z_{tx}—impedance at the transmitter, (**b**) Z_{tr}—impedance after transmission, (**c**) Z_{out}—impedance at the output.

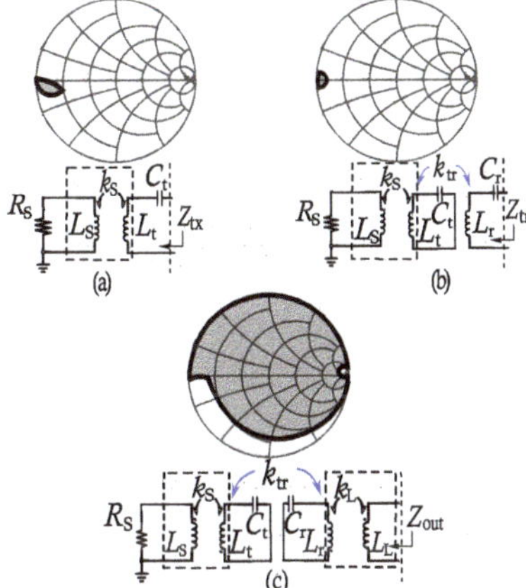

Figure 6. Reflection coefficient graphs show impedances that can be obtained by inductive coupling in four coiled WPT system in case A: (**a**) Z_{tx}—impedance at the transmitter, (**b**) Z_{tr}—impedance after transmission, (**c**) Z_{out}—impedance at the output.

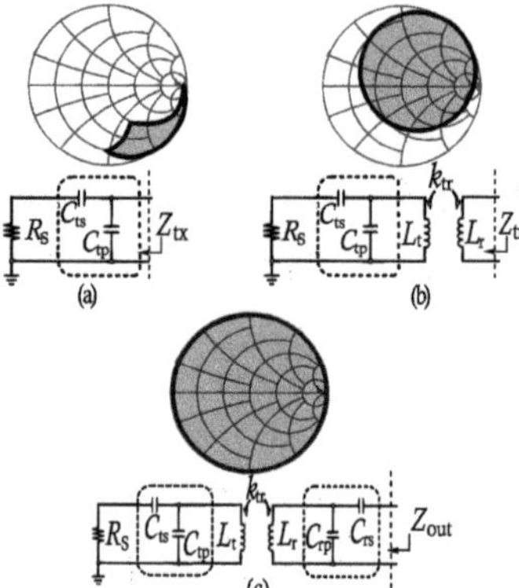

Figure 7. Reflection coefficient graphs show impedances that can be obtained by capacitive matching networks placed into WPT system in case B: (**a**) Z_{tx}—impedance at the transmitter, (**b**) Z_{tr}—impedance after transmission, (**c**) Z_{out}—impedance at the output.

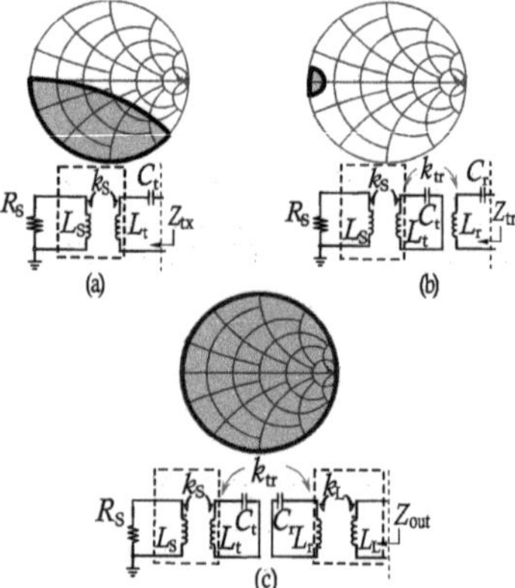

Figure 8. Reflection coefficient graphs show impedances that can be obtained by inductive coupling in four coiled WPT system in case B: (**a**) Z_{tx}—impedance at the transmitter, (**b**) Z_{tr}—impedance after transmission, (**c**) Z_{out}—impedance at the output.

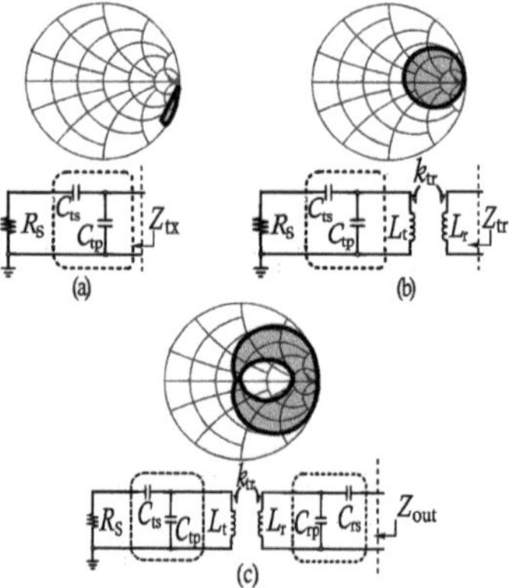

Figure 9. Reflection coefficient graphs show impedances that can be obtained by capacitive matching networks placed into WPT system in case C: (**a**) Z_{tx}—impedance at the transmitter, (**b**) Z_{tr}—impedance after transmission, (**c**) Z_{out}—impedance at the output.

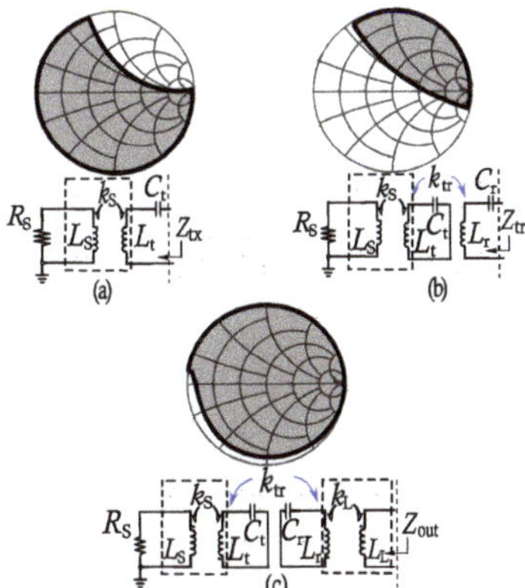

Figure 10. Reflection coefficient graphs show impedances that can be obtained by inductive coupling in four coiled WPT system in case C: (**a**) Z_{tx}—impedance at the transmitter, (**b**) Z_{tr}—impedance after transmission, (**c**) Z_{out}—impedance at the output.

Table 2. Matching network parameters.

Application	L_S, L_L	X_S, X_L	k_S, k_L	$C_{ts/rs}$, $C_{tp/rp}$	$X_{ts/rs}$, $X_{ts/rs}$
Case A	0.1–60 µH	8.5 mΩ–32.0 Ω	0.001–1	0.1–200 nF	187 kΩ–9 Ω
Case B	0.1–6 µH	0.7 Ω–40.7 Ω	0.001–1	0.5–200 pF	47 kΩ–117 Ω
Case C	0.1–2.5 µH	10 Ω–1.6 kΩ	0.001–1	0.5–2 pF	3.2 kΩ–796 Ω

Case B is a low power and high-frequency system, a mobile phone charging device from Airfuel Alliance, which has standard parameters as 6.78 MHz operational frequency and around $L_t = L_r = 6$ µH coils for Tx and Rx [21]. Results are given in Figures 7 and 8. Here, the inductive method again shows low performance for a three-coil system since the matchable area fills only about 10% of the Smith chart. However, both methods are able to match all loads with the complete network.

Finally, we consider application case C, which is a high-frequency device with $L_t = L_r = 2.5$ µH for Rx and Tx coils. The inductive matching network has larger matchable region than the capacitive method, see Figures 9 and 10. In a complete network, around 90% of the load impedances are matchable by the inductive method, whereas the capacitive approach can match only around 30% of the loads. One thing to note is that the matchable region for the capacitive network has a hole, which appears because of the minimum constraint 0.5 pF in the parameter range. Consequently, the capacitive matching network with these constraints is less versatile in this type of application.

4. Performance of Lossy System

In this section, system performance is examined in the presence of loss. Therefore, the lossless inductors L_t and L_r in the previous circuits is replaced by a model of a non-ideal inductor shown in Figure 11. The model consists of an ideal inductor with a series parasitic resistance and a parallel capacitance. We only consider the effect of the parasitic resistance since the parasitic capacitance can be taken care of by compensation circuits. The parasitic resistance is a combination of ohmic and radiative losses of the coil. Other circuit parameters are kept the same as in the previous section for calculation of reflection coefficients. The realizable reflection coefficients over the chosen range of parameters are shown in Figures 12–14.

The inductances of the coils in case A are larger than for the other two cases, so the parasitic resistance $R = 1$ Ω of each coil is higher than for case B and comparable to case C where the skin effect matters. R is estimated assuming a copper coil made from a 30-m long wire of diameter 0.8 mm [20]. The result for case A is shown in Figure 12, where Figure 12a presents realizable reflection coefficients for capacitive IMN of Figures 3 and 12b presents the results for inductive IMN of Figure 4. Solid lines show borders of matchable regions for lossles networks determined in the previous section, whereas dashed lines bound matchable regions for lossy networks.

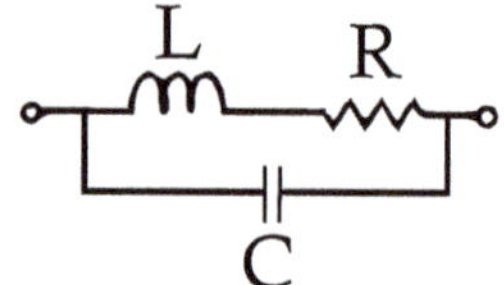

Figure 11. Model of non-ideal inductor.

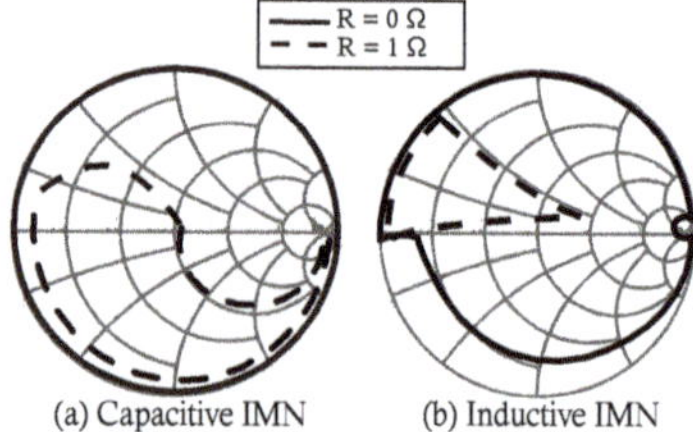

(a) Capacitive IMN   (b) Inductive IMN

Figure 12. Matchable regions in lossless and lossy model for case A.

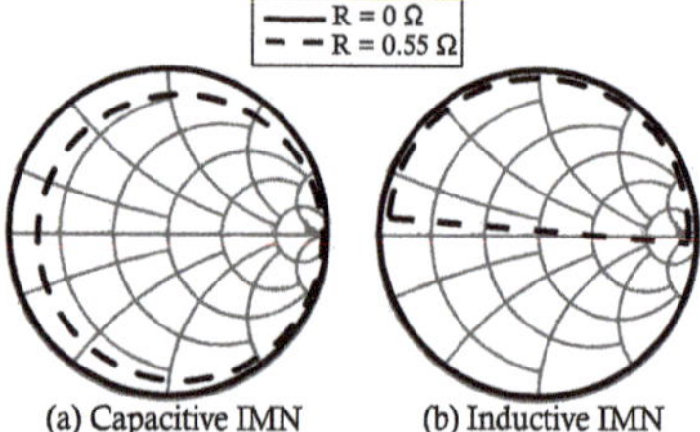

(a) Capacitive IMN   (b) Inductive IMN

Figure 13. Matchable regions in lossless and lossy model for case B.

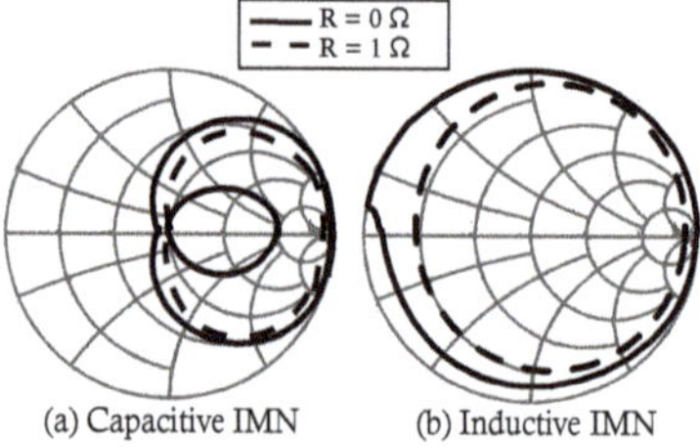

(a) Capacitive IMN   (b) Inductive IMN

Figure 14. Matchable regions in lossless and lossy model for case C.

The matchable area has been dramatically reduced from 100% to approximately 40% of the Smith chart for capacitive and from 80% to 20% for the inductive technique. It shows that range of matchable impedances is sensitive to loss for both methods and that the inductive method is slightly sensitive to the loss than the capacitive method.

The comparison of matchable regions, after the introduction of parasitic resistance $R = 0.55\ \Omega$ ([21]) in the coils for case B, is shown in Figure 13. The reflection coefficient graphs of capacitive and inductive methods are shown in Figure 13a,b, respectively. The change in the matchable region is again more dramatic for the inductive method since it has about 55% reduction from the ideal case compared to the capacitive method's 20%.

Matchable regions for case C are given in Figure 14. Since the resonator coils operate at the highest frequency in the comparison, we considered skin effect as well. Therefore, ohmic loss at the coils is chosen as a combination of parasitic resistance and skin effect loss, which is $R = 1\ \Omega$. R is estimated assuming a copper coil made from a 75-cm long wire of diameter 0.8 mm [22]. As a peculiarity in Figure 14a, we note that the matchable region for the lossless capacitive network has a hole in it and that this vanishes when losses are introduced. The overall numbers are 2% reduction in area for the capacitive and 15% for the inductive approach. While the inductive method in this case still has a matchable region more sensitive to loss than capacitive IMN, the inductive method covers a larger area of matchable impedances both with and without loss.

As mentioned before in Section 2, we considered regions of matchable impedances. It can also be interesting to see how well the network performs for any given load impedance. Here we consider this question by calculating the delivered power to load impedances sampled form the entire Smith chart. It can be obtained in several ways, but we used SPICE AC analysis here. We focused on cases A and B because their matchable regions are most affected by the introduction of ohmic loss. The results are shown in Figures 15 and 16 and demonstrate that capacitive and inductive IMNs can be comparable in matching various load impedances, when the networks are optimized for each particular impedance. Device in case A has more power loss than in case B. Overall, power level is distributed between -4.2 dBm and -4.8 dBm for case A, and -3.5 dBm and -4 dBm for case B.

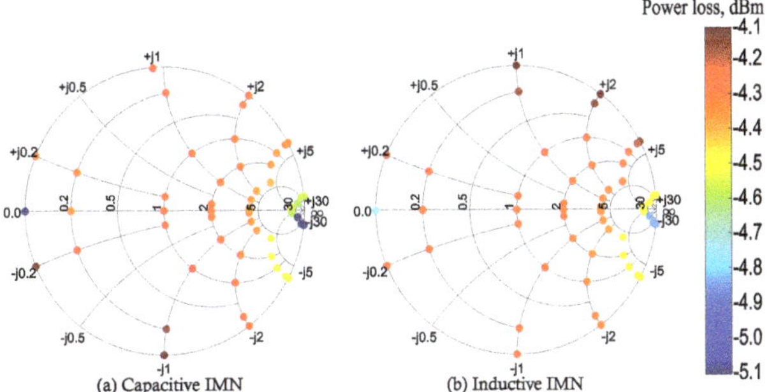

Figure 15. Comparison of optimal power delivered to the various loads for case A.

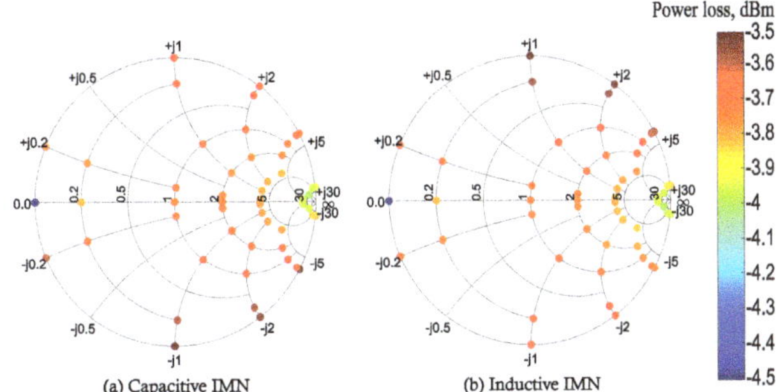

Figure 16. Comparison of optimal power delivered to the various loads for case B.

5. Discussion

Section 3 presented three impedance areas obtained for different stages of ideal circuits (Figures 5–10), which is summarized in Figure 17. In this figure, each trace represents specific application (case A, B, C) and matching network. The impedances at the different stage of circuits, in Figures 5–10, are given in x-axis, whereas y-axis shows a share of Smith chart in percentage. All the areas have been estimated from the graphs. According to the figure, Z_{out} fills a larger area than Z_{tr} for all cases, which shows that the impedance matching network at the Rx side is crucial in obtaining large tunable area of load impedance. There is a degradation from 100 to 30% in the area of output impedance Z_{out} by the capacitive method from case A to C. An improvement from 10 to 75% in the area of impedance at the transmitter side Z_{tx} can be seen for the inductive technique from case A to C. This is due to difference in the operation frequency of the specific applications since for capacitive IMN $X_C \propto 1/f$, while for inductive IMN $X_L \propto f$. Therefore, the region of inductive method shrinks in case C (85 kHz—low frequency) and widens in case A (100 MHz—high frequency), and it is vice versa for capacitive matching. At low operation frequencies, the available range of the capacitance values is enough to match most of the load impedances, whereas at high frequency, the matching is limited by the smallest capacitance value in the parameter range, which is 0.5 pF. In the inductive approach, the ranges of load and source inductance values are the same as for the resonator coils. For case C, covering most of the Smith chart can be achieved with a reasonable inductance range, whereas for case A, the inductance range is not large enough to match all the load impedances.

The limitation discussed above also gives a better understanding of the parasitic effects in the lossy WPT systems. For capacitive method, it is crucial to avoid parasitic capacitances of resonator coils at high frequencies since they are comparable with 0.5 pF, whereas for low frequency applications they can be ignored.

The matchable areas of the lossless model and the circuit with parasitics are compared in Figure 18. Here, blue/yellow colors correspond to matching network type and solid/hatched patterns of charts represent ideal and lossy scenarios of examination. The figure shows that the circuit with inductive IMN is more sensitive to the loss than capacitive IMN. In case A, the reductions in matchable areas of inductive and capacitive IMNs are comparable. In case B and case C, the reduction is higher for inductive IMN than for capacitive IMN. Case C is least affected by the parasitics due to high operation frequency. It should be noted that a larger parasitic resistance shrinks the matchable region, whereas a smaller parasitic resistance increases it closer to the ideal case.

It is clear that matching networks provide quite different areas of perfectly matchable impedances. However, the comparison in Figures 15 and 16 showed that the methods have comparable power transfer when the networks are optimized for each load.

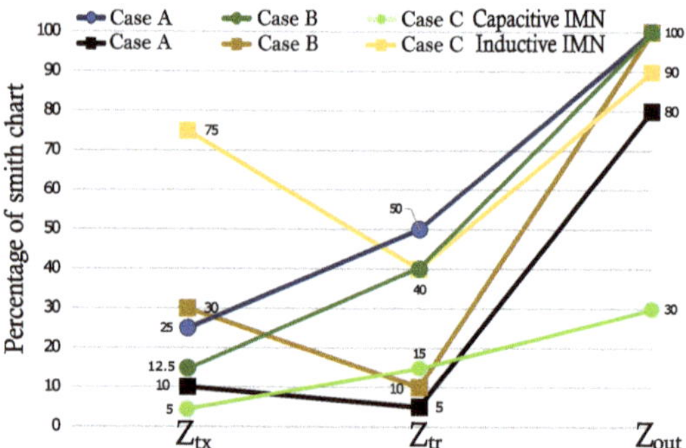

Figure 17. Matchable area comparison for different stages of the circuit in the ideal case.

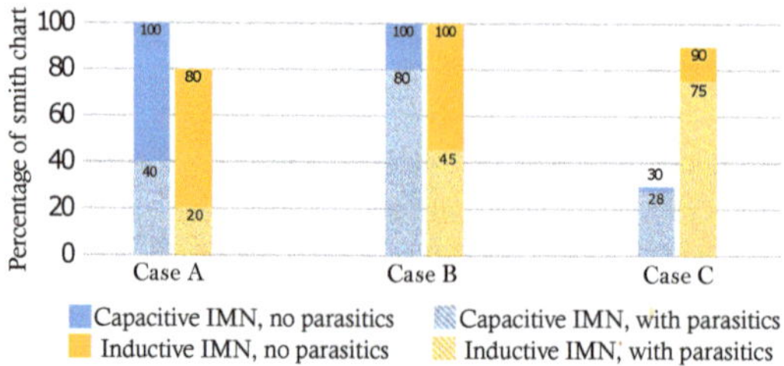

Figure 18. Area of matchable loads before and after introducing the parasitics.

6. Conclusions

In this paper, we compare areas of matchable loads for capacitive and inductive impedance matching networks (IMNs), which are the common matching techniques in the magnetic resonant wireless power transfer system. Graphical visualization of the impedances by Smith Chart is used for effortless comparison of the IMNs performances. An analytic expression for effective output impedance is derived and used to display the conjugate-image impedance of the load. Without any limitations in the parameter values and frequency range, it is always possible to match any load, i.e., any point in the Smith chart. Therefore, three different applications were considered with constraints: case A—car charging operating at 85 kHz, case B—mobile phone charging at 6.78 MHz, and case C— a high-frequency charging device at 100 MHz.

For the lossless system, the capacitive circuit's matchable area fills the Smith chart for case A, whereas around 20% of the chart that cannot be matched by the inductive IMN. For case B both methods could match any load impedance. On the other hand, the inductive IMN has shown about 60% larger area than the capacitive IMN for case C. The matching network at the receiver improves the area in all cases.

Finally, the impact of parasitic resistance of the resonator coils to the matchable area has been examined. In case A, the reduction of matchable area is 60 percentage points for both methods. The matchable area by the inductive IMN is more sensitive to the parasitic resistance than capacitive IMN for cases B and C. However, simulation of power transfer has shown that both matching networks can be equally effective in matching different load values.

Author Contributions: All authors contributed to problem formulation, conceptualization and choice of methods. Y.Z. performed all calculations, simulations and original draft preparation. All authors contributed to interpretation of results, and to editing into the final manuscript. All authors have read and agreed to the published version of the manuscript.

Funding: Y.Z. benefits from a PhD scholarship from University of South-Eastern Norway funded by the Norwegian Ministry of Education and Research. Article processing charge (APC) was funded by the University of South-Eastern Norway.

Acknowledgments: The Research Council of Norway is acknowledged for the support to the Norwegian Ph.D. Network on Nanotechnology for Microsystems, Nano-Network (221860/F40).

Conflicts of Interest: The authors declare no conflict of interest.

References

1. Aqeel, J.M.; Nordin, R.; Gharghan, S.K.; Jawad, H.M.; Ismail, M. Opportunities and Challenges for Near-Field Wireless Power Transfer: A Review. *Energies* **2017**, *10*, 1022. [CrossRef]
2. Akhtar, F.; Rehmani, M.H. Energy replenishment using renewable and traditional energy resources for sustainable wireless sensor networks: A review. *Renew. Sustain. Energy Rev.* **2015**, *45*, 769–784. [CrossRef]
3. Vázquez-Leal, H.; Gallardo, A.; González-Martínez, F.J.; Castañeda-Sheissa, R. *The Phenomenon of Wireless Energy Transfer: Experiments and Philosophy*; INTECH Open Access Publisher: London, UK, 2012.
4. Kurs, A.; Karalis, A.; Moffatt, R.; Joannopoulos, J.D.; Fisher, P.; Soljacic, M. Wireless power transfer via strongly coupled magnetic resonances. *Science* **2007**, *317*, 83–86. [CrossRef] [PubMed]
5. Sample, A.P.; Meyer, D.A.; Smith, J.R. Analysis, experimental results, and range adaptation of magnetically coupled resonators for wireless power transfer. *IEEE Trans. Ind. Electron.* **2011**, *58*, 544–554. [CrossRef]
6. Eplett, B.K. On-Chip Impedance Matching Using a Variable Capacitor. U.S. Patent 2008/0211598 A1, 4 September 2008.
7. Lee, W.; Lee, H.; Oh, K.; Yu, J. Switchable distance-based impedance matching networks for a tunable HF system. *Progr. Electromagn. Res.* **2012**, *128*, 19–34. [CrossRef]
8. Lim, Y.; Tang, H.; Lim, S.; Park, J. An Adaptive Impedance-Matching Network Based on a Novel Capacitor Matrix for Wireless Power Transfer. *IEEE Trans. Power Electron.* **2014**, *29*, 4403–4413. [CrossRef]
9. Kim, J.; Kim D.H.; Park, Y.J. Analysis of Capacitive Impedance Matching Networks for Simultaneous Wireless Power Transfer to Multiple Devices. *IEEE Trans. Ind. Electron.* **2015**, *62*, 2807–2813. [CrossRef]

10. Chen, J. A Study of Loosely Coupled Coils for Wireless Power Transfer. *IEEE Trans. Circuits Syst. II Express Briefs* **2010**, *57*, 536–540. [CrossRef]
11. Kim, J. Coil Design and Shielding Methods for a Magnetic Resonant Wireless Power Transfer System. *Proc. IEEE* **2013**, *101*, 1332–1342. [CrossRef]
12. Kim, J.; Jeong, J. Range-Adaptive Wireless Power Transfer Using Multi-loop and Tunable Matching Techniques. *IEEE Trans. Ind. Electron.* **2015**, *62*, 6233–6241. [CrossRef]
13. Kesler, M. *Highly Resonant Wireless Power Transfer: Safe, Efficient, and Over Distance*; WiTricity Corporation: Watertown, MA, USA, 2013.
14. Hui, S.Y.R.; Zhong, W.; Lee, C.K. A Critical Review of Recent Progress in Mid-Range Wireless Power Transfer. *IEEE Trans. Power Electron.* **2014**, *29*, 4500–4511. [CrossRef]
15. Zhaksylyk, Y.; Azadmehr, M. Comparative Analysis of Inductive and Capacitive feeding of Magnetic Resonance Wireless Power Transfer. In Proceedings of the 2018 IEEE PELS Workshop on Emerging Technologies: Wireless Power Transfer (Wow), Montréal, QC, Canada, 3–7 June 2018.
16. Li, Y.; Dong, W.; Yang, Q.; Zhao, J.; Liu, L.; Feng, S. An Automatic Impedance Matching Method Based on the Feedforward-Backpropagation Neural Network for a WPT System. *IEEE Trans. Ind. Electron.* **2019**, *66*, 3963–3972. [CrossRef]
17. Roberts, S. Conjugate-Image Impedances. *Proc. IRE* **1946**, *34*, 198–204. [CrossRef]
18. Wang, Q.; Che, W.; Monti, G.; Mongiardo, M.; Dionigi M.; Mastri, F. Conjugate image impedance matching for maximizing the gains of a WPT link. In Proceedings of the 2018 IEEE MTT-S International Wireless Symposium (IWS), Chengdu, China, 6–10 May 2018.
19. Dionigi, M.; Mongiardo, M.; Perfetti, R. Rigorous Network and Full-Wave Electromagnetic Modeling of Wireless Power Transfer Links. *IEEE Trans. Microw. Theory Tech.* **2015**, *63*, 65–75. [CrossRef]
20. Zhou, S.; Chris Mi, C. Multi-Paralleled LCC Reactive Power Compensation Networks and Their Tuning Method for Electric Vehicle Dynamic Wireless Charging. *IEEE Trans. Ind. Electron.* **2016**, *63*, 6546–6556. [CrossRef]
21. Uchida, A.; Shimokawa, S.; Oshima, H. Effect of load dependence of efficiency in a multi-receiver WPT system. In Proceedings of the 2017 IEEE Wireless Power Transfer Conference (WPTC), Taipei, Taiwan, 10–12 May 2017.
22. Gernsback, H. *Lighting lamp by S-W- Radio*; Short Wave & Television: New York, NY, USA, 1937, pp. 166–191.
23. Rhea, R. The Yin-Yang of Matching: Part 1—Basic Matching Concepts. *High Freq. Electron.* **2006**, *5*, 16–25.

Article

Study on Battery Charging Converter for MPPT Control of Laser Wireless Power Transmission System

Seongjun Lee [1], Namgyu Lim [2], Wonseon Choi [3], Yongtak Lee [3], Jongbok Baek [4] and Jungsoo Park [1,*]

[1] Department of Mechanical Engineering, Chosun University, Gwangju 61452, Korea; lsj@chosun.ac.kr
[2] Department of Mechanical System & Automotive Engineering, Chosun University, Gwangju 61452, Korea; lnk9100@chosun.kr
[3] Gwangju Institute of Science and Technology, Gwangju 61452, Korea; bourne.ws.choi@gmail.com (W.C.); ytlee@gist.ac.kr (Y.L.)
[4] Energy ICT Convergence Research Department, Korea Institute of Energy Research, Daejeon 34129, Korea; jongbok.baek@kier.re.kr
* Correspondence: j.park@chosun.ac.kr; Tel.: +82-62-230-7057

Received: 24 September 2020; Accepted: 19 October 2020; Published: 21 October 2020

Abstract: Herein, the voltage and current output characteristics of a laser photovoltaic (PV) module applied to a wireless power transmission system using a laser beam are analyzed. First, an experiment is conducted to obtain the characteristic data of the voltage and current based on the laser output power of the laser PV module, which generates the maximum power from the laser beam at a wavelength of 1080 nm; subsequently, the small-signal voltage and current characteristics of the laser PV module are analyzed. From the analysis results, it is confirmed that the laser PV module has a characteristic in which the maximum power generation point varies according to the power level of the laser beam. In addition, similar to the solar cell module, it is confirmed that the laser PV module has a current source and a voltage source region, and it shows a small signal resistance characteristic having a negative value as the operating point goes to the current source region. In addition, in this paper, by reflecting these electrical characteristics, a method for designing the controller of a power converter capable of charging a battery while generating maximum power from a PV module is proposed. Since the laser PV module corresponds to the input source of the boost converter used as the power conversion unit, the small-signal transfer function of the boost converter, including the PV module, is derived for the controller design. Therefore, by designing a controller that can stably control the voltage of the PV module in the current source, the maximum power point, and voltage source regions defined according to the output characteristics of the laser PV module, the maximum power is generated from the PV module. Herein, a systematic controller design method for a boost converter for laser wireless power transmission is presented, and the proposed method is validated based on the simulation and experimental results of a 25-W-class boost converter based on a microcontroller unit control.

Keywords: laser wireless power transmission; PV module; maximum power point; battery charging

---

1. Introduction

Recently, studies regarding wireless charging technology for supplying electric power to electric vehicles, various IoT (Internet of Things) devices, and unmanned moving objects have been actively conducted. Hitherto, wireless charging technology has been mainly studied based on magnetic induction, magnetic resonance, and electromagnetic wave methods. The magnetic induction method is mainly used when the distance between the transmitting and receiving coils is short (1–2 cm for electronic products or 0.15 m for electric vehicles), and the magnetic resonance method is applied when induction, magnetic resonance, and electromagnetic wave methods. The magnetic induction method is mainly used when the distance between the transmitting and receiving coils is short (1–2 cm for electronic products or 0.15 m for electric vehicles), and the magnetic resonance method is applied when the transmission distance is longer than the magnetic induction method. In this case, the main target is a distance of less than 10 m. The electromagnetic wave method enables power to be transmitted up to a transmission distance of several kilometers compared with the other two methods; however, its transmission efficiency is low, and it is harmful to the human body [1–4].

On the other hand, the wireless charging method using the laser is also being studied in military and space applications [2,5]. Figure 1 shows a schematic diagram of an unmanned aerial vehicle (UAV) wireless power transmission system using a laser beam proposed in [5]. The light energy of the laser beam irradiated from the ground is transferred to the laser photovoltaic (PV) module installed inside the UAV and converted into electric energy. The converted electrical energy is used to charge the battery through a power converter or as energy for components mounted on the UAV. Research on the laser wireless charging system to date is as follows.

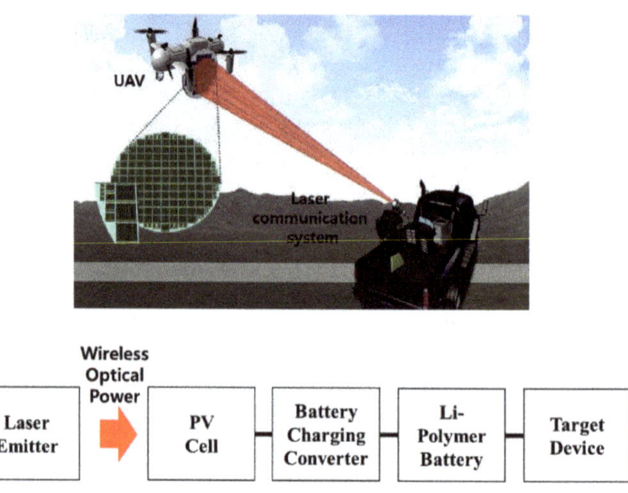

Figure 1. Laser wireless charging system configuration of unmanned aerial vehicle (UAV) [5].

Results pertaining to the voltage, current, and efficiency based on the material type used and the laser wavelength in a laser PV cell are presented in [5,6]. However, in addition to these results, a method to control a converter that can charge a battery when a laser PV module is applied to a UAV is necessitated. Results pertaining to the wavelength and temperature output characteristics of PV panels applied to charging systems using laser beams as well as the mathematical modeling of PV panels are presented in [7]. In this study, the modeling of the laser PV module proposed in [7] was used to design a PV module applied to a simulation model; however, the output characteristics were analyzed through testing a prototype laser PV module. In addition, a control design method that can generate the maximum power from a laser PV module using a battery charging converter was investigated. In [8], the concept of the maximum power point tracking (MPPT) algorithm applied with a neural network (NN) algorithm considering the characteristics of a PV module, and a laser beam was presented. However, in that paper, the improvement compared with perturb-and-observe and incremental conductance methods, as well as the specific design method and experimental results of NN, were not presented.

Similar to the previous studies mentioned above, studies regarding laser wireless charging systems mainly focus on improving the efficiency based on the material of the laser PV module and the introduction of research areas where the laser charging system is applicable. However, for the practical use of the laser charging system, studies regarding the control of the battery charging system using laser PV modules is required. There has been a lot of research on battery charging systems using solar cell modules, but there are no research articles on the power converter design and control of using solar cell modules, but there are no research articles on the power converter design and control of laser wireless charging systems using laser beams. Therefore, this paper deals with a controller design method based on the experimental results of laser PV output characteristics for controlling the power converter of a laser wireless charging system.

To extract the maximum power from a laser PV module, the operating point of the PV module must be controlled. To design a controller for controlling the operating point of the input source, a small-signal model of the PV module is required in the current source, voltage source, and maximum power point regions based on the laser output. First, we conducted an experiment to investigate voltage and current characteristics based on the laser output power of a laser PV module to generate the maximum power from a laser of a specific wavelength; subsequently, we derived the resistance values for small-signal fluctuations of voltage and current. After deriving the transfer function of the boost converter, including the small-signal resistance of the laser PV source, we designed a controller that can satisfy the dynamic point in the entire operational area of the laser PV. As described above, we present herein a design method for small-signal transfer functions that can reflect the small-signal characteristics of a laser PV and a laser PV controller design method that can generate the maximum PV power based on the MPPT algorithm. The proposed scheme was validated based on the simulation and experimental results of a PV-class laser of rate 25-W voltage boost converter prototype.

2. Modeling and Controller Design of Laser Wireless Power Transmission System

2.1. Laser PV Module Characteristics

The PV module, which is the input source of the laser wireless charging system, has a configuration in which 16 PV cells are connected in series, as shown in Figure 2; it was manufactured by the Korea Advanced Nano Fabrication Center [5]. Using a high-power continuous fiber laser MFSC-200, laser beams with a wavelength of 1080 nm were irradiated onto the laser PV module at 2.478 and 2.874 W/cm² [9]. Figure 3 shows the voltage–current and voltage–power characteristics based on the laser beam power used in the experiment.

Figure 2. Laser photovoltaic (PV) module with 16 cells connected in series.

As shown in Figure 3, the PV module presents nonlinear output characteristics, in which the short-circuit current, open-circuit voltage, and location of the maximum power generation point vary according to the laser power. Since the maximum power generated from the PV module is the point at which the slope of the voltage–power curve, confirmed that it reaches the maximum power generation point the supplied in the experiment is approximately 7 W. To charge quickly the battery, the laser PV battery from the battery charging the battery charging controller must use an MPPT algorithm to identify the maximum power point of the laser PV module as well as stably control the identify the maximum power point of the laser PV module as well as stably control the operating point of the input source determined by the algorithm. Therefore, the small-signal characteristics of the laser PV module must be analyzed to control the operating point of the input source; the small-signal resistance (denoted as r_s) for the increment of the measured voltage and current of the laser PV module is shown in Figure 4. As shown, the laser PV module exhibits negative small-signal resistance characteristics, and the small-signal resistance value increases as it approaches the current source region. In the next section, a method for deriving the small-signal transfer function of a boost converter that reflects the small-signal resistance characteristics of a laser PV module and a controller design method are presented:

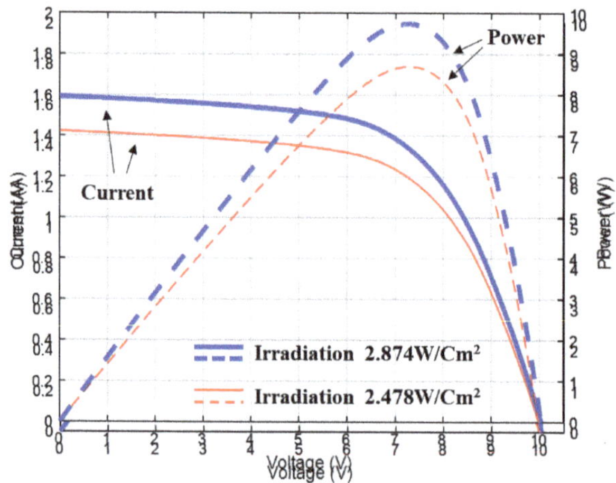

Figure 3. Voltage–current and voltage–power characteristics based on laser power of PV module.

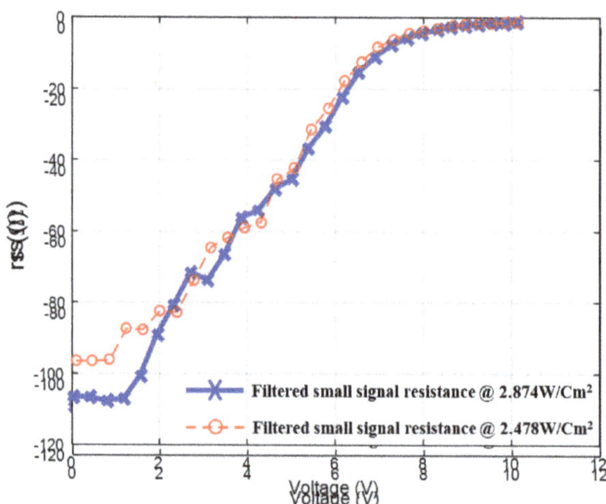

Figure 4. Small-signal resistance characteristics of laser PV module.

2.2. Modeling and Controller Design of Boost Converter

Figure 5 shows the boost converter topology of the battery charging system to which the laser PV module was applied. In Figure 5, L is the inductor of the boost converter, C is the input capacitor, r_c is the capacitor's equivalent series resistance (ESR), and V_b is the voltage source representing the battery connected to the output. The boost converter must have an operating point control function to generate the maximum power from the PV module before the battery reaches its full voltage [10–12]. In this study, the boost converter was set to control the voltage of the PV module such that the generate the maximum power from the PV module before the battery reaches its full voltage [10–12]. laser PV module can form a stable operating point in a large signal. For the controller design and stability analysis of the boost converter, a small-signal modeling technique using the state-space averaging method was applied, which is a method of obtaining the transfer function from the input to the output of the system by linearizing the system based on the operating point in the steady state [13,14].

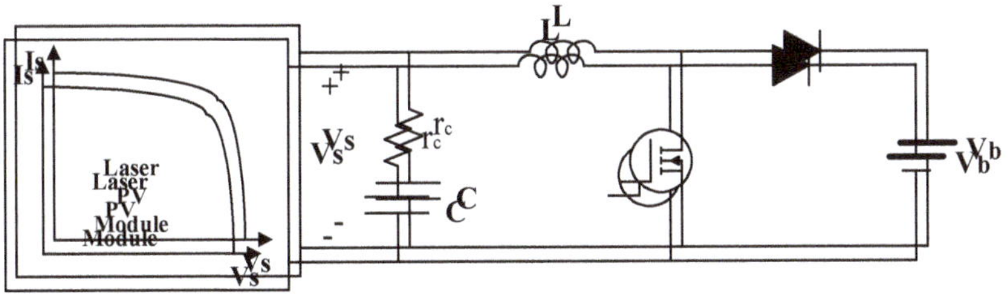

Figure 5. Boost converter for battery charging with laser PV module.

The small-signal transfer function required for designing the voltage controller of the laser PV module of the boost converter was derived in the following two steps. First, as shown in Figure 6, a small-signal model of the unterminated model for the boost converter modeling of the PV module as a current source was developed. Subsequently, the small-signal transfer function of the boost converter to which the laser PV module was applied was derived by including the relationship between the small-signal voltage and current of the laser PV module mentioned in Section 2.1.

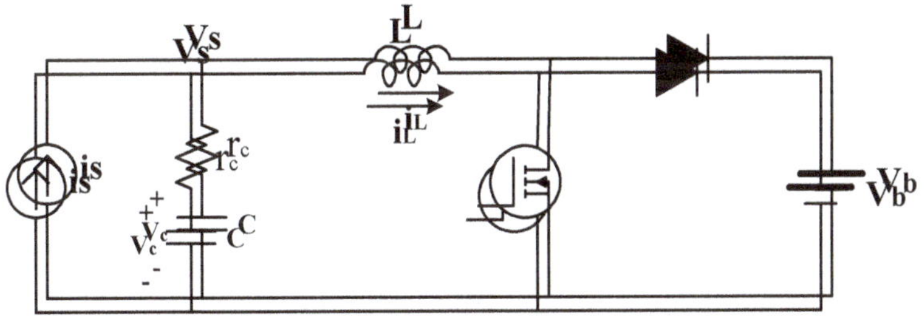

Figure 6. Unterminated model of boost converter with input as the current source.

The state equation obtained by averaging the differential equations of the voltage and current based on the switching states of the power switch of the boost converter in Figure 6 is shown in Equation (1), where d denotes the ratio of turning ON during one period of the switching frequency, and d' is defined as $1 - d$.

$$\begin{bmatrix} \dot{i}_L \\ \dot{v}_C \end{bmatrix} = \begin{bmatrix} -\frac{r_c}{L} & \frac{1}{L} \\ -\frac{1}{C} & 0 \end{bmatrix} \begin{bmatrix} i_L \\ v_c \end{bmatrix} + \begin{bmatrix} \frac{r_c}{L} & -\frac{1}{L}d' \\ \frac{1}{C} & 0 \end{bmatrix} \begin{bmatrix} i_s \\ v_b \end{bmatrix} \quad (1)$$

To develop the small-signal model, the state and control variables of the boost converter were defined as having small perturbations at the steady-state operating point, which is indicated by capital letters as follows: $v_c = V_c + \hat{v}_c$, $i_L = I_L + \hat{i}_L$, $d = D + \hat{d}$, $v_b = V_b + \hat{v}_b$ and $i_s = I_s + \hat{i}_s$. Since the small perturbation excluding the steady-state operating point yields a small-signal model, the small-signal transfer function of the boost converter is as shown in Equation (2). In addition, the small-signal block diagram of the output to the input can be represented as shown in Figure 7.

$$
\left\{
\begin{aligned}
G_4 &= \frac{\hat{v}_s}{\hat{v}_b} = D'\frac{(1+sr_cC)}{\Delta} \\[4pt]
G_5 &= \frac{\hat{i}_L}{\hat{d}} = \frac{sCV_b}{\Delta} \\[4pt]
G_6 &= \frac{\hat{v}_s}{\hat{d}} = -V_b\frac{(1+sr_cC)}{\Delta}
\end{aligned}
\right.
\tag{2}
$$

transfer function of the boost converter is as shown in Equation (2). In addition, the small-signal block diagram of the output to the input can be represented as shown in Figure 7.

The characteristic equation Δ, Q-factor, and resonance frequency of the transfer function are expressed as shown in Equation (3).

$$
\left\{
\begin{aligned}
G_1 &= \frac{\hat{i}_L}{\hat{i}_s} = \frac{(1+sr_cC)}{\Delta} \\[4pt]
G_2 &= \frac{\hat{i}_L}{\hat{v}_b} = \frac{sCD'}{\Delta} \\[4pt]
G_3 &= \frac{\hat{v}_s}{\hat{i}_s} = \frac{1+sr_cC}{\Delta} \\[4pt]
G_4 &= \frac{\hat{v}_s}{\hat{v}_b} = D'\frac{(1+sr_cC)}{\Delta} \\[4pt]
G_5 &= \frac{\hat{i}_L}{\hat{d}} = \frac{sCV_b}{\Delta} \\[4pt]
G_6 &= \frac{\hat{v}_s}{\hat{d}} = -V_b\frac{(1+sr_cC)}{\Delta}
\end{aligned}
\right.
\tag{3}
$$

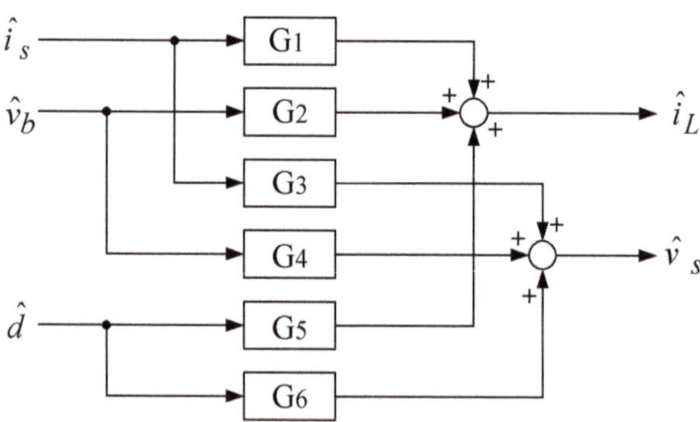

Figure 7. Small-signal block diagram of unterminated model.

The characteristic equation Δ, the Q-factor, and the resonance frequency of the transfer function are expressed as shown in Equation (3).

$$
\left\{
\begin{aligned}
\Delta &= 1 + \frac{s}{Q\omega_0} + \frac{s^2}{\omega_0^2} \\[4pt]
Q &= \frac{1}{r_c}\sqrt{\frac{L}{C}} \\[4pt]
\omega_0 &= \frac{1}{\sqrt{LC}} \quad (r_c < 0)
\end{aligned}
\right.
\tag{4}
$$

As mentioned in Section 2, the laser PV module represents nonlinear voltage and current characteristics; therefore, a small-signal model can be obtained through the Taylor series for the operating point, as shown in Equation (4).

$$
\left\{
\begin{aligned}
i_s &= f(v_s) \\[4pt]
I_s + \hat{i}_s &= f(V_s + \hat{v}_s) = f(V_s) + \dot{f}(V_s)\cdot\hat{v}_s \\[4pt]
\hat{i}_s &= r_s\hat{v}_s \quad (r_s < 0)
\end{aligned}
\right.
\tag{4}
$$

In the unterminated model, the relationship between $\hat{v}_s$ and $\hat{i}_s$ corresponds to the small-signal resistance r_s of the laser PV module; therefore, the small-signal block diagram in Figure 7 can be reorganized as shown in Figure 8. As depicted in Figure 8, when the laser PV module is connected to the input of the boost converter, a loop gain T is formed where $\hat{v}_s$ is fed back to $\hat{i}_s$; therefore, the final

final small-signal transfer function of the boost converter considering the laser PV module can be derived from Equation (5).

Electronics **2020**, *9*, 1745

small-signal transfer function of the boost converter considering the laser PV module can be derived from Equation (5).

$$\begin{cases} \dfrac{\hat{v}_s}{\hat{v}_b} = \dfrac{G_4}{1+T} \\[2mm] \dfrac{\hat{v}_s}{\hat{d}} = \dfrac{G_6}{1+T} \\[2mm] \dfrac{\hat{i}_L}{\hat{v}_b} = \dfrac{G_6}{1+T} + \dfrac{G_1}{r_s}\dfrac{G_4}{1+T} \\[2mm] \dfrac{\hat{i}_L}{\hat{d}} = G_2 + \dfrac{G_1}{r_s}\dfrac{G_4}{1+T} \\[2mm] \dfrac{\hat{i}_L}{\hat{d}} = G_5 + \dfrac{G_1}{r_s}\dfrac{G_6}{1+T} \end{cases} \tag{5}$$

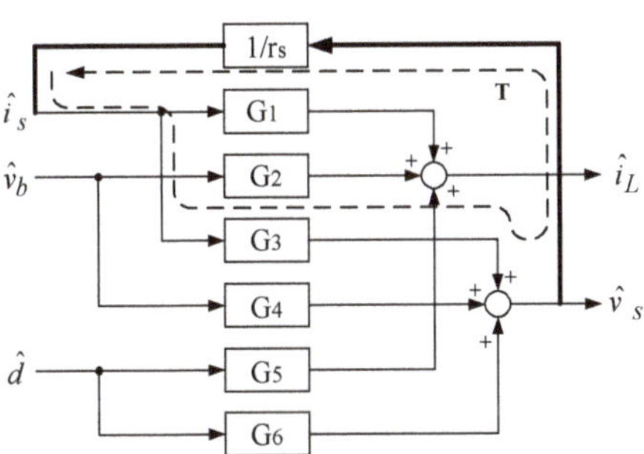

Figure 8. Small-signal block diagram with added PV module.

Therefore, the final transfer function of the boost converter considering the laser PV module is expressed as in Equation (5), and the small-signal block diagram including the inductor current of the boost converter and the input voltage of the laser PV module is shown in Figure 8. The input voltage was the voltage of the PV module, and the MPPT algorithm tracks the maximum point of the laser PV module. A two-loop control was adopted, in which a voltage control was applied to the outer loop while a current loop was applied to the inner loop. Since the converter operated through the digital controller (MCU), the transfer function (MCU), block transfer function expressed as a discrete-time transfer function.

The open-loop current gain (Ti) for controlling the inductor current of the boost converter is shown in Figure 9. At this time, for digital control, the continuous-time small-signal transfer function obtained previously was converted into a discrete-time transfer function using the zero-order hold (ZOH) method, and the conversion equation is as shown in Equation (10). Here, $Z\{\cdot\}$ represents the z-transform, $(1-e^{-sT_s})/s$ represents the transfer function of the ZOH, and $G_p(s)$ represents the transfer function of Equations (6)–(9).

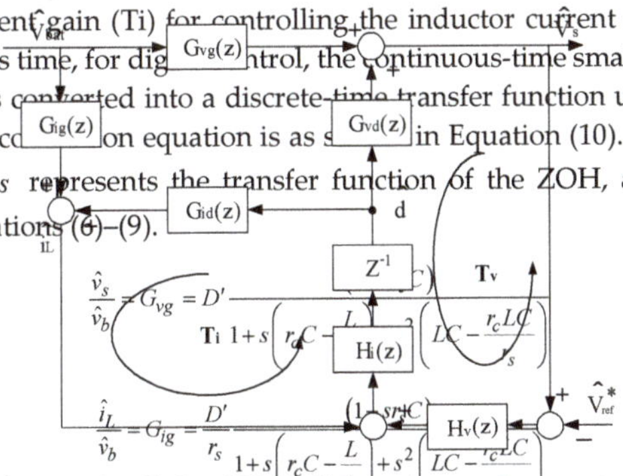

$$G_p(z) = Z\left\{\frac{1-e^{-sT_s}}{s}G_p(s)\right\} \tag{10}$$

$$\frac{\hat{v}_s}{\hat{v}_b} = G_{vg} = D' \frac{(1+sr_cC)}{1+s\left(r_cC-\dfrac{L}{r_s}\right)+s^2\left(LC-\dfrac{r_cLC}{r_s}\right)} \tag{6}$$

$$\frac{\hat{i}_L}{\hat{v}_b} = G_{ig} = \frac{D'}{r_s} \frac{}{1+s\left(r_cC-\dfrac{L}{r_s}\right)+s^2\left(LC-\dfrac{r_cLC}{r_s}\right)} \tag{7}$$

Figure 9. Small-signal block diagram for two-loop control.

$$\frac{\hat{v}_s}{\hat{d}} = G_{vd} \frac{(1+sr_cC)}{1+s\left(r_cC-\dfrac{L}{r_s}\right)+s^2\left(LC-\dfrac{r_cLC}{r_s}\right)} \tag{8}$$

The open-loop proportional gain for controlling that was used for the boost converter is shown in Figure 9. At this time, for digital control, the boost converter's small-signal transfer function obtained previously was considered a time delay until the calculated data is reflected, and it was considered as a (ZOH) sampling delay. In this study, Figure 10 shows the Bode diagram of the designed current controller. CSR denotes the current-source region of the laser PV module, MPP denotes the region near the maximum power point, and VSR denotes the voltage-source region. Since the laser PV module has nonlinear characteristics, it must be designed to ensure stability in the CSR, VSR, and MPP of the laser PV module, as shown in the Bode diagram.

$$H_i(z) = 0.075\frac{(z-0.93)}{(z-1)} \tag{11}$$

$$T_i = G_{id}(z) \cdot H_i(z) \cdot z^{-1} \tag{12}$$

$$\frac{\hat{i}_L}{\hat{d}} = G_{id} = -\frac{V_b}{r_s}\frac{(1-sr_sC)}{1+s\left(r_cC-\frac{L}{r_s}\right)+s^2\left(LC-\frac{r_cLC}{r_s}\right)} \tag{9}$$

Electronics 2020, 9, 1745

z-transform, $(1-e^{-sT_s})/s$ represents the transfer function of the ZOH, and $Gp(s)$ represents the transfer function of Equations (6)–(9).

$$\frac{\hat{v}_s}{\hat{v}_b} = G_{vg} = D'\frac{(1+sr_cC)}{1+s\left(r_cC+\frac{L}{r_s}\right)+s^2\left(LC-\frac{r_cLC}{r_s}\right)} \tag{6}$$

$$\frac{\hat{v}_s}{\hat{v}_b} = G_{ig} = \frac{D'}{r_s}\frac{sr_sC}{1+s\left(r_cC-\frac{L}{r_s}\right)+s^2\left(LC-\frac{r_cLC}{r_s}\right)} \tag{7}$$

$$\frac{\hat{v}_s}{\hat{d}} = G_{vd} = -V_b\frac{(1+sr_cC)}{1+s\left(r_c\right)+s^2\left(LC-\frac{r_cLC}{r_s}\right)} \tag{8}$$

$$\frac{\hat{i}_L}{\hat{d}} = G_{id} = -\frac{V_b}{r_s}\frac{sr_sC}{1+s\left(r_cC-\frac{L}{r_s}\right)+s^2\left(LC-\frac{r_cLC}{r_s}\right)} \tag{9}$$

$$G_p(z) = Z\left\{\frac{(1-e^{-sT_s})}{s}G_p(s)\right\} \tag{10}$$

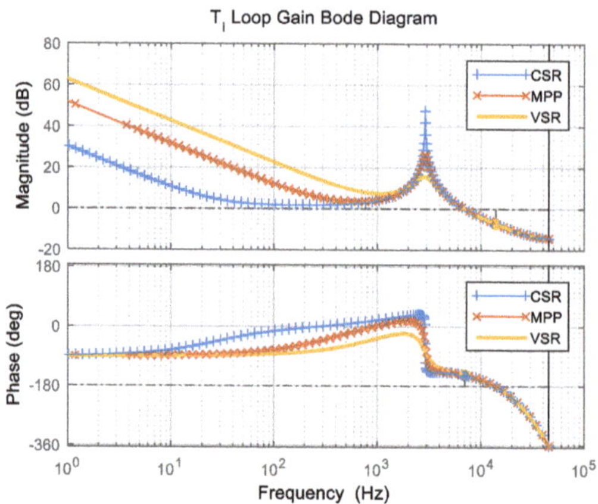

Figure 9. Small-signal block diagram for two-loop control.

In this study, a proportional and integral (PI) controller was used for current control, as shown in Equation (11), and the Ti applied with the PI current controller is shown in Equation (12). At this time, z^{-1} represents the time delay until the calculated duty is reflected, and it was considered as a one-sampling delay in this study. Figure 10 shows the Bode diagram of the Ti of the designed current controller. CSR denotes the current source region of the laser PV module, MPP denotes the region near the maximum power point, and VSR denotes the voltage source region. Since the laser PV module has nonlinear characteristics, it must be designed to ensure stability in the CSR, VSR, and MPP of the laser PV module, as shown in the Bode diagram.

$$H_i(z) = 0.075\frac{(z-0.93)}{(z-1)} \tag{11}$$

$$T_i = G_{id}(z)H_i(z)z^{-1} \tag{12}$$

Figure 10. Bode diagram of open-loop current gain.

The voltage controller of the laser PV module was designed to achieve dynamic performances (bandwidth) and stability (phase margin) for a system with a closed current loop, as shown in Equation (13). The designed PI voltage controller is expressed as shown in Equation (14); Equation (15) shows

The voltage controller of the laser PV module was designed to achieve dynamic performances (bandwidth) and stability (phase margin) for a system with a closed current loop, as shown in Equation (13). The designed PI voltage controller is expressed as shown in Equation (14); Equation (15) shows the equation of the open-loop voltage gain when the designed voltage controller is applied. As shown from the voltage loop gain of Figure 11, even when the operating point of the laser PV module has changed, the voltage controller maintained a bandwidth of 400–700 Hz, and the phase margin is designed to exceed 70°.

$$\frac{V_{sa}}{V_c} = G_{vdc}(z) = \frac{G_{vdc}(z) \cdot H(z) \cdot z^{-1}}{1 + T_i} \tag{13}$$

$$H_v(z) = 0.75\frac{(z - 0.7979)}{(z - 1)} \tag{14}$$

$$T_v = G_{vdc}(z) H_v(z) \tag{15}$$

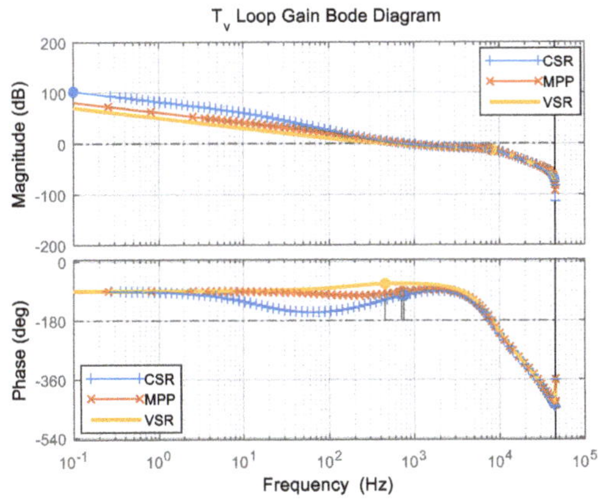

Figure 11. Board diagram of voltage loop gain of two-loop control.

2.3. MPPT Algorithm Design

In this study, the incremental conductance method with variable amplitude was applied to track the maximum power point based on the output power of the laser light source [15]. This method compares the load impedance with the impedance of the laser PV module and controls the voltage of the PV module to correspond to the maximum power point. As shown in Figure 12, when the output of the PV module is located to the left of the maximum power point, the power increases with the voltage. Conversely, when it is located to the right of the maximum power point, the power decreases with the increase in the voltage. This relationship is expressed as shown in Equations (16)–(19).

Therefore, the voltage command of the laser PV module can be positioned as the maximum power point by measuring and comparing the increment for the conductance of the laser PV module and the instantaneous resistance value. At this time, the fluctuation value of the command value was set as a variable voltage such that when the operating point was far from the maximum power point, it rapidly converged to the maximum power point, and in the vicinity of the maximum power point, the periodic vibration width reduced compared with using a fixed value. As described above, Figure 13 shows the flow chart of the MPPT algorithm with variable amplitude based on the incremental conductance method:

$$\frac{dP}{dV} = \frac{d(IV)}{dV} = I + V\frac{dI}{dV} \cong I + V\frac{\Delta I}{\Delta V} \tag{16}$$

$$\frac{\Delta I}{\Delta V} = -\frac{I}{V} \quad at\ MPP \tag{17}$$

$$\frac{\Delta I}{\Delta V} > -\frac{I}{V} \quad left\ of\ MPP \tag{18}$$

$$\frac{\Delta I}{\Delta V} > -\frac{I}{V} \quad left\ of\ MPP \tag{18}$$

$$\frac{\Delta I}{\Delta V} < -\frac{I}{V} \quad right\ of\ MPP \tag{19}$$

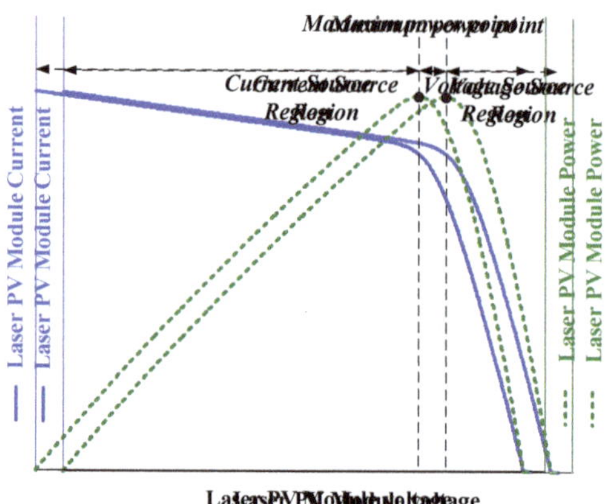

Figure 12. Characteristic curve of the laser PV Module.

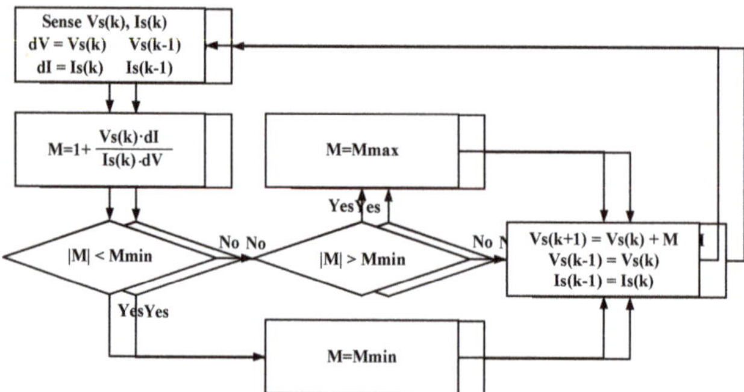

Figure 13. Algorithm flow chart of an incremental conductance method with variable amplitude.

3. Simulation and Experimental Results

3.1. Simulation Result

The method proposed herein was first verified using the simulation model shown in Figure 14. In the simulation model, the laser module was modeled with a diode model-based equation [7,16]. The laser PV module output current can be described as shown in Equations (20)–(22) in Figure 15a, where I_{sc} is the short circuit current of the laser PV module, I_o is the reverse saturation current of the diode, q is the electronic charge, v_d is the diode voltage, K is the Boltzmann constant, T is the temperature in Kelvin, n is the ideality factor, R_{sh} is the shunt resistance, R_s is the series resistance, v_{oc} is the open circuit voltage of the cell, and N_s is the serial number of cells constituting the module. Figure 15b shows the modeling of the laser PV module implemented in Matlab/Simulink using Equations (20)–(22). Illumination is a variable representing the laser power intensity irradiated to the laser PV module and has a value of 0 to 1 normalized to 2.874 W/cm².

$$I_s = I_{sc} - I_o \left(e^{\frac{qv_d}{nKT}} - 1 \right) - \frac{v_d}{R_{sh}} \tag{20}$$

$$I_o = \left(I_{sc} - \frac{v_{oc}}{R_{sh}} \right) / \left(e^{\frac{qv_{oc}}{nKT}} - 1 \right) \tag{21}$$

Table 1. Parameters of laser PV module in simulation model.

Parameters	Values	Parameters	Values
I_{sc}	1.61 A @ 2.478 W/cm^2	q	1.602×10^{-19} C
V_{oc}	0.628 V/cell	K	1.381×10^{-23} JK^{-1}
R_s	50 mΩ/cell	T	298.15 K
R_{sh}	4.5 Ω	n	1.8
Ns	16		

$$V_s = N_s(v_d - R_s I_s)^n \tag{22}$$

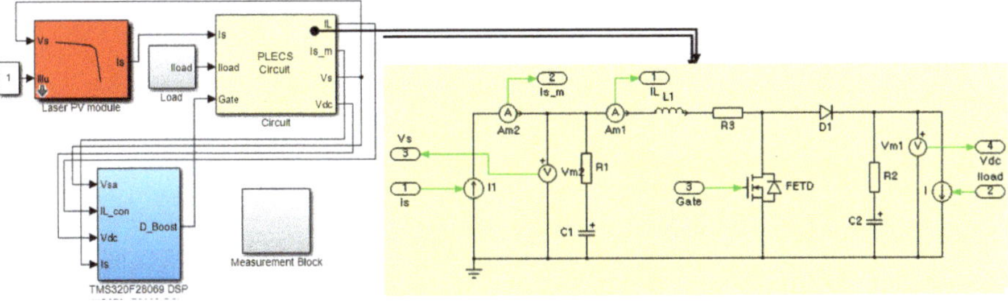

Figure 14. Simulation model using Matlab/Simulink.

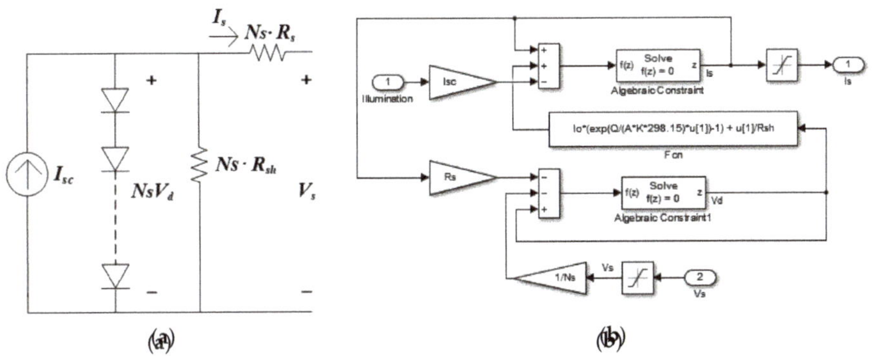

(a) **(b)**

Figure 15. Modeling of laser PV module: (a) equivalent circuit model (b) simulation model in Matlab.

The parameters used for modeling the laser PV module are shown in Table 1. In the parameters of the laser PV module, I_{sc} and V_{oc} were set as the short-circuit current and open-circuit voltage values measured in the experiment, and the R_s, R_{sh}, and n values were extracted by the trial and error method through the simulation model in Figure 15b. In order to extract more accurate parameters of the laser PV module using an optimization algorithm, the method mentioned in Sheng and Anani's articles can be applied [17,18]. The accuracy of the simulation model compared with the experimental data of the module mentioned in Section 2 is shown in Figure 16. As can be seen in Figure 16a, the root mean square error (RMSE) of the simulation and experimental results is shown as 0.0027 or less, and the simulation model reflects the experimental results well.

Table 1. Parameters of laser PV module in simulation model.

Parameters	Values	Parameters	Values
I_{sc}	1.61 A @ 2.478 W/cm^2	q	1.602×10^{-19} C
V_{oc}	0.628 V/cell	K	1.381×10^{-23} JK^{-1}
R_s	50 mΩ/cell	T	298.15 K
R_{sh}	4.5 Ω	n	1.8
Ns	16		

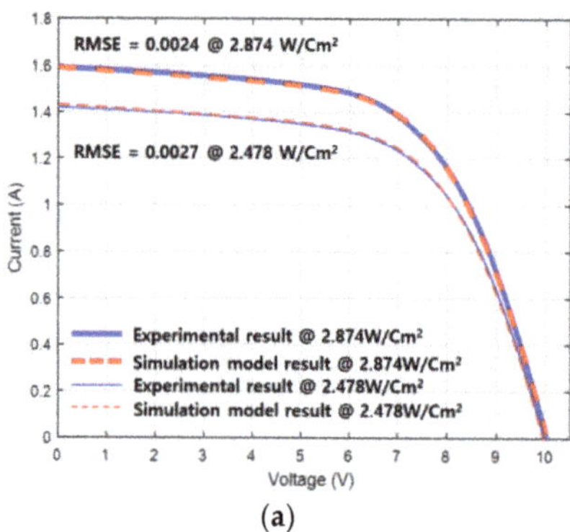
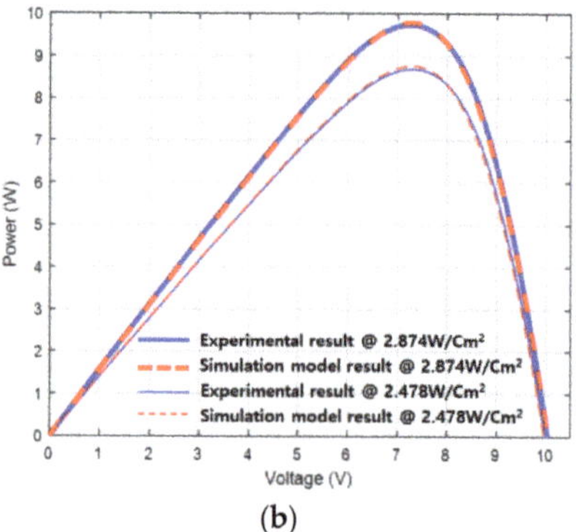

Figure 16. Characteristics of laser PV module using simulation model: (a) V-I characteristics comparison between the simulation model and experimental results; (b) V-P characteristics of the simulation model.

The power circuit of the boost converter for charging a 24 V battery was modeled using PLECS software [19]. The battery was modeled with a capacitor and ESR, which was also simulated in Figure 14 with the boost converter. A smaller capacity battery than the actual battery with a capacitor simulated ESR and the parameters were set to 2 F, which corresponded to a capacity that the actual battery in Table 2 to reduce the simulation time. The parameter values of the components applied to the simulation are summarized in Table 2.

Table 2. Experimental setup of the laser charging system.

	Name of Device/Manufacturer	Rating/Values
Laser Source	Max MFSC 200 W-300 L Air Cooling Fiber	Power range 0–140 W, Central wavelength 1080 nm
Laser Receiving Panel	Custom-made PV module	Voc: 10 V, Isc: 1.6 A @ 2.478 W/cm²
Electronic Load	ITECH LT8511	120 V/30 A
Battery	Skyholic Li-polymer battery	Nominal 22.2 V, 4.5 Ah
Battery Charging Boost Converter	Input/Output Range	Input: 3–15 V/~3.5 A, Output: 16.2–25.2 V/1 A
	Inductor	30 µH
	Input/Output Capacitor	100 µF/220 µF
	Switching Frequency	90 kHz

Figure 17 shows the battery charging control result through the MPPT operation when the laser power was 2.874 W/cm². The converter operated in the MPPT mode because the battery voltage has not reached the 24.5 V set as the full-charge voltage. The sampling period of the MPPT algorithm was set to 0.1 s to reduce the simulation time, and it was observed that the input voltage set point was generate to track the 7.5 V point in the MPP. Since the converter was well controlled by the voltage command, it can be confirmed that the battery was charged with the maximum power generated from the laser PV module.

3.2. *Experimental Results*

Figure 18 shows the experimental configuration of a battery charging system using a laser beam. The laser light source was configured to be generated by passing a multilens array at the rear end of the MFSC-200 device with a wavelength of 1080 nm. The power generated from the laser PV module was supplied to the battery after passing through the prototype boost converter, and the electronic load was connected in parallel to the battery for the load test.

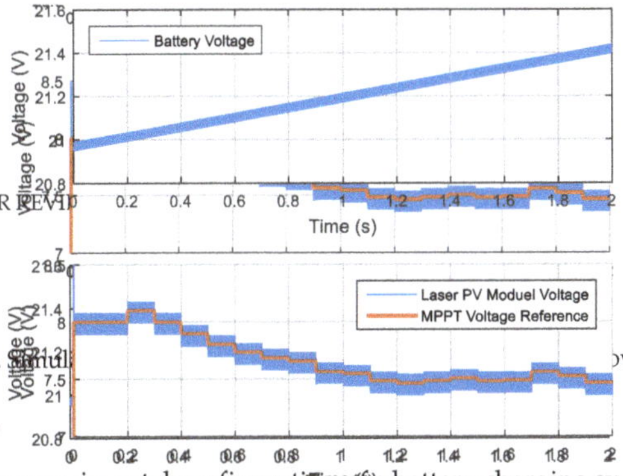

Figure 17. Simulation result of charging battery to the maximum power point.

3.2. Experimental Results

Figure 18 shows the experimental configuration of a battery charging system using a laser beam. The laser light source was configured to be generated by passing a multilens array at the rear end of the MFSC-200 device with a wavelength of 1080 nm. The power generated from the laser PV module was supplied to the battery after passing through the prototype boost converter, and the electronic load was connected in parallel to the battery for the load test.

Figure 19 shows the main configuration of the prototype boost converter system, where a TMS320f28069 MCU was applied to communicate with the UAV's host controller, perform MPPT functions, and control the converter. Table 2 shows the experimental configuration of the laser wireless power transmission system and the main specifications of the manufactured converter.

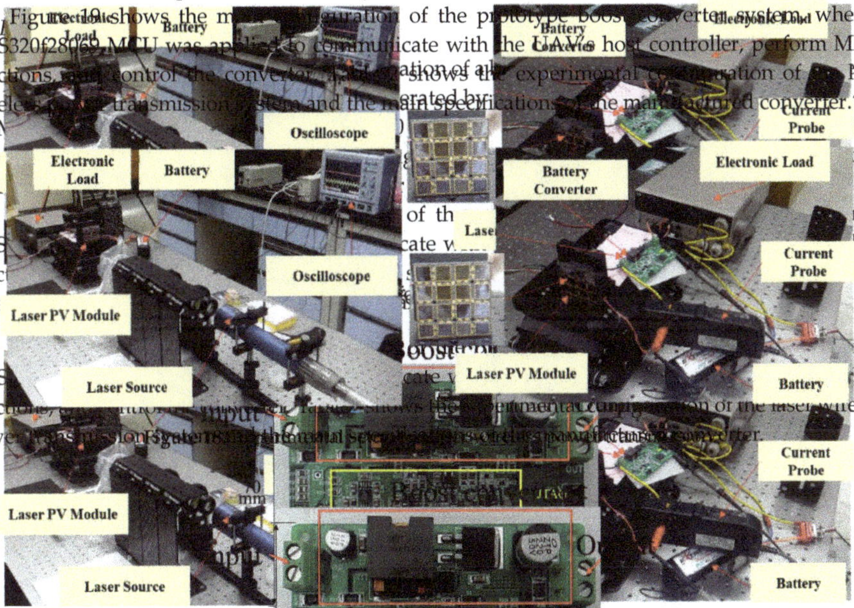

Figure 18. Experimental configuration of laser wireless power transmission.

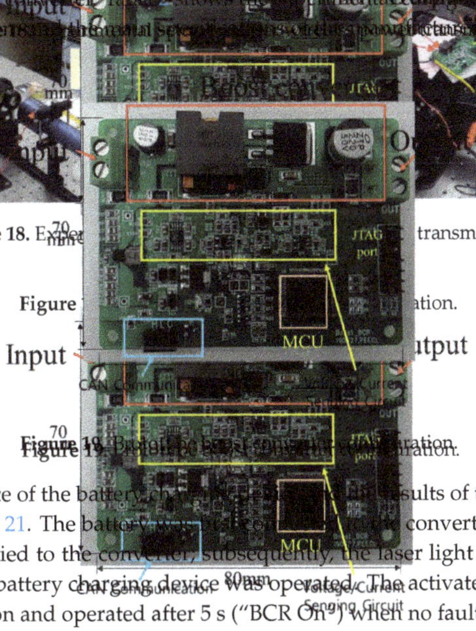

Figure 19. Prototype boost converter configuration.

The startup sequence of the battery charging device and the results of the MPPT control test are shown in Figures 20 and 21. The battery was first connected to the converter ("Battery ON") before the laser beam was applied to the converter; subsequently, the laser light source was irradiated to the PV module, and the battery charging device was operated. The activated converter performed a self-diagnostic verification and operated after 5 s ("BCR On") when no faults were encountered.

In this experiment, because the battery did not reach the full-charge voltage set at 24.5 V, it was operated in the MPPT mode; furthermore, it was confirmed that the battery was charging while moving to the maximum power point of the PV module. Figure 21 shows the enlarged waveform of the converter performing the MPPT operation. The voltage of the laser PV module, indicated by the yellow CH1 waveform, was tracking the maximum power point by changing the operating point at every MPPT sampling period with a 2 s period.

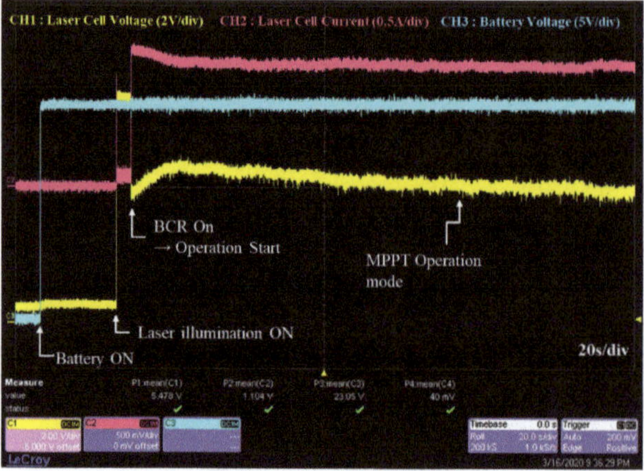

Figure 20. Experimental result of maximum power point tracking (MPPT) control of laser PV module for battery charging.

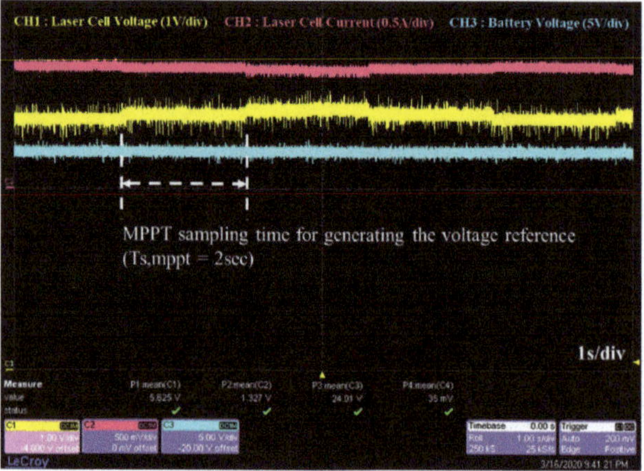

Figure 21. Enlarged waveform of MPPT control operation.

4. Conclusions

Herein, a controller design method that reflects the small-signal voltage and current characteristics of a laser PV module for a wireless power system using a laser beam was presented. The laser PV module was fabricated to generate the maximum energy from a laser light source of a specific wavelength (1080 nm in the case of the module used in this study). From the PV module experiment, it was confirmed that the voltage and current characteristics were similar to those of the solar cell module. First, the power generated by the laser PV module varies according to the power of the laser beam. Therefore, to control the operating point to the maximum power generation point that varies according to the power of the laser beam, the power conversion unit must have an MPPT control function. Secondly, the laser PV module has a characteristic of a small-signal resistance having a negative value similar to that of a solar cell module, and the small-signal resistance value increases as the operating point go to the current source region. Therefore, when designing the controller for controlling the operating point of the laser PV module, the small-signal resistance characteristics of the laser PV module must be reflected. Third, although not described in this paper, the temperature of the PV module increases when a laser beam is irradiated with the laser PV module. As the temperature of the laser PV module increases, the open-circuit voltage of the laser PV module tends to decrease rather than the magnitude of the short circuit current. Therefore, electrical characteristics analysis and control studies according to the temperature rise of the laser PV module are required as future research.

Accordingly, in this paper, a controller design method that can stably control the input voltage in the MPPT mode by inducing the transfer function of the boost converter, reflecting the small-signal characteristics of the laser PV module, was systematically presented. The method proposed herein was verified through simulation results based on Matlab/Simulink and an experiment involving a 25-W-class prototype boost converter.

Author Contributions: S.L. contributed to the main idea of this study and wrote the paper. S.L., N.L., and W.C. performed the experiments. J.B., Y.L., and J.P. revised the paper comprehensively. All authors have read and agreed to the published version of the manuscript.

Funding: This paper was supported by a Korea Institute for Advancement of Technology (KIAT) grant funded by the Korea Government (MOTIE) (P0002092, The Competency Development Program for Industry Specialist). This work was supported the Korea Institute of Energy Technology Evaluation and Planning (KETEP) grant funded by the Korea government (MOTIE) (20182410105280).

Conflicts of Interest: The authors declare no conflict of interest.

References

1. Jin, K.; Zhou, W. Wireless Laser Power Transmission: A Review of Recent Progress. *IEEE Trans. Power Electron.* **2019**, *34*, 3842–3859. [CrossRef]
2. El Rayes, M.M.; Fayoum University; Nagib, G.; Abdelaal, W.G.A. A Review on Wireless Power Transfer. *Int. J. Eng. Trends Technol.* **2016**, *40*, 272–280. [CrossRef]
3. Rhee, D.-H.; Kim, S.-M. Study on a Laser Wireless Power Charge Technology. *J. Korea Inst. Electron. Commun. Sci.* **2016**, *11*, 1219–1224. [CrossRef]
4. Nayagam, V.S.; Premalatha, L. Optimization of power losses in electric vehicle battery by wireless charging method with consideration of the laser optic effect. *Meas. Control.* **2020**, *53*, 441–453.
5. Kim, Y.; Shin, H.-B.; Lee, W.-H.; Jung, S.H.; Kim, C.Z.; Kim, H.; Lee, Y.T.; Kang, H.K. 1080 nm InGaAs laser power converters grown by MOCVD using InAlGaAs metamorphic buffer layers. *Sol. Energy Mater. Sol. Cells* **2019**, *200*, 109984. [CrossRef]
6. Helmers, H.; Bett, A.W. Photovoltaic Laser Power Converters for Wireless Optical Power Supply of Sensor Systems. In Proceedings of the 2016 IEEE International Conference on Wireless for Space and Extreme Environments (WiSEE), Aachen, Germany, 26–28 September 2016; pp. 152–154.
7. Zhang, Q.; Fang, W.; Liu, Q.; Wu, J.; Xia, P.; Yang, L. Distributed Laser Charging: A Wireless Power Transfer Approach. *IEEE Internet Things J.* **2018**, *5*, 3853–3864. [CrossRef]
8. Zhang, Y.; Chaminda, P.D.; Zhao, K.; Cheng, L.; Jiang, Y.; Peng, K. MPPT Algorithm Development for Laser Powered Surveillance Camera Power Supply Unit. In Proceedings of the IOP Conference Series: Materials Science and Engineering, Shanghai, China, 28–29 December 2017; Volume 322, pp. 1–6.
9. MFSC-200 Datasheet. Available online: http://en.maxphotonics.com/ (accessed on 20 September 2020).
10. Bae, H.; Lee, J.; Park, S.; Cho, B. Large-signal stability analysis of solar array power system. *IEEE Trans. Aerosp. Electron. Syst.* **2008**, *44*, 538–547.
11. Cho, B.H.; Lee, J.R.; Lee, F.C.Y. Large-signal stability analysis of spacecraft power processing systems. *IEEE Trans. Power Electron.* **1990**, *5*, 110–116. [CrossRef]
12. Lee, J.H.; Bae, H.S.; Cho, B.H. Resistive Control for a Photovoltaic Battery Charging System Using a Microcontroller. *IEEE Trans. Ind. Electron.* **2008**, *55*, 2767–2775. [CrossRef]

13. Polivka, W.M.; Chetty, P.R.K.; Middlebrook, R.D. State-space average modeling of converters with parasitic and storage-time modulation. In Proceedings of the IEEE Power Electronics Specialist Conference, Atlanta, GA, USA, 16–20 June 1980; pp. 119–143.

14. Kazimierczuk, M.K. *Pulse-Width Modulated DC-DC Power Converters*; Wiley: Hoboken, NJ, USA, 2008.

15. Lee, J.H.; Bae, H.; Cho, B.H. Advanced Incremental Conductance MPPT Algorithm with a Variable Step Size. In Proceedings of the 2006 12th International Power Electronics and Motion Control Conference, Portoroz, Slovenia, 30 August–1 September 2006; pp. 603–607.

16. Rathod, Y.; Hughes, L. Simulating the charging of electric vehicles by laser. *Procedia Comput. Sci.* **2019**, *155*, 527–534. [CrossRef]

17. Sheng, H.; Li, C.; Wang, H.; Yan, Z.; Xiong, Y.; Cao, Z.; Kuang, Q. Parameters Extraction of Photovoltaic Models Using an Improved Moth-Flame Optimization. *Energies* **2019**, *12*, 3527. [CrossRef]

18. Anani, N.; Ibrahim, H. Performance Evaluation of Analytical Methods for Parameters Extraction of Photovoltaic Generators. *Energies* **2020**, *13*, 4825. [CrossRef]

19. PLECS Software. Available online: https://www.plexim.com/ (accessed on 20 September 2020).

Publisher's Note: MDPI stays neutral with regard to jurisdictional claims in published maps and institutional affiliations.

Review

Wirelessly-Powered Cage Designs for Supporting Long-Term Experiments on Small Freely Behaving Animals in a Large Experimental Arena

Byunghun Lee [1] and Yaoyao Jia [2,*]

[1] Department of Electrical Engineering, Incheon National University, Incheon 22012, Korea; byunghun_lee@inu.ac.kr

[2] Department of Electrical and Computer Engineering, North Carolina State University, 890 Oval Dr, Raleigh, NC 27606, USA

* Correspondence: yjia6@ncsu.edu; Tel.: +1-919-515-7350

Received: 22 October 2020; Accepted: 18 November 2020; Published: 25 November 2020

Abstract: In modern implantable medical devices (IMDs), wireless power transmission (WPT) between inside and outside of the animal body is essential to power the IMD. Unlike conventional WPT, which transmits the wireless power only between fixed Tx and Rx coils, the wirelessly-powered cage system can wirelessly power the IMD implanted in a small animal subject while the animal freely moves inside the cage during the experiment. A few wirelessly-powered cage systems have been developed to either directly power the IMD or recharge batteries during the experiment. Since these systems adapted different power carrier frequencies, coil configurations, subject tracking techniques, and wireless powered area, it is important for designers to select suitable wirelessly-powered cage designs, considering the practical limitations in wirelessly powering the IMD, such as power transfer efficiency (PTE), power delivered to load (PDL), closed-loop power control (CLPC), scalability, spatial/angular misalignment, near-field data telemetry, and safety issues against various perturbations during the longitudinal animal experiment. In this article, we review the trend of state-of-the-art wirelessly-powered cage designs and practical considerations of relevant technologies for various IMD applications.

Keywords: wirelessly-powered cage; inductive power transmission; implantable medical device; animal experiment

---

1. Introduction

Implantable medical devices (IMDs) have been developed for behavioral neurosciences which research on small freely moving animal subjects, such as rodents [1–5]. Conventional hardwired IMD in Figure 1a, which is restricting experiments for freely behaving animal subjects [1,2], have been replaced with battery-powered IMDs. However, the battery needs be replaced after 2–4 h animal experiments [3,4], resulting in the interruption for continuous and smooth flow of the experiments, as shown in Figure 1b. When the IMD is fully implanted inside the animal body, the risk for replacing the battery dramatically increases due to the potential infection in the animal body during the surgery.

discussed with the practical design considerations that include the CLPC, coil design/optimization, scalability for wireless coverage, special/angular misalignment, near-field data telemetry, and safety issues. In Section 2, the main blocks for wirelessly-powered cage systems are introduced with the practical considerations related to the key technologies. Section 3 provides the different designs of wirelessly-powered cages with the performance comparison. Section 4 introduces the state-of-the-art wirelessly-powered cages for mm-sized IMDs, followed by a conclusion.

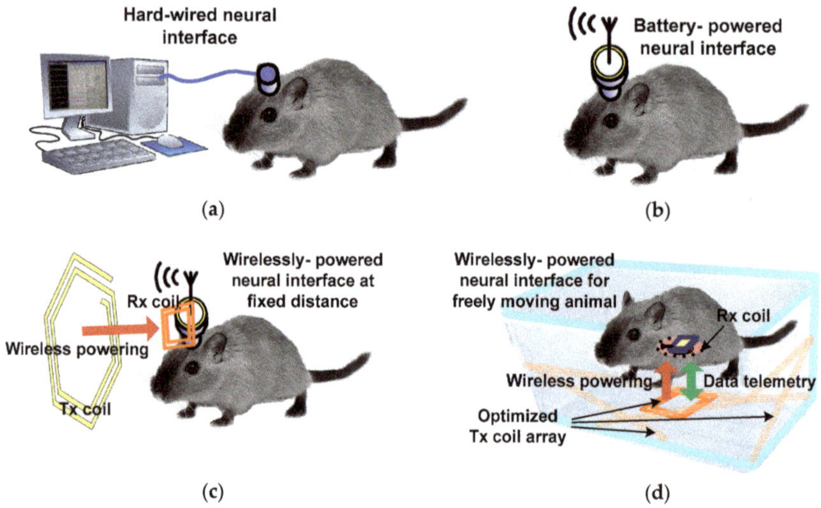

Figure 1. Neural interfaces categorized by power sources: (a) hard-wired, (b) battery-powered, (c) wirelessly-powered at a fixed distance, and (d) wirelessly-powered neural interfaces for freely moving animal.

2. Main Blocks for Wirelessly-Powered Cage System

Figure 2 shows a simplified block diagram of the wirelessly-powered cage system and the IMD attached to or implanted in the freely-moving animal subject. The wirelessly-powered cage system typically includes a main controller, a power amplifier (PA), a pair of data Tx/Rx, a Tx coil array, and

To address the problems caused by the replacement of battery in IMDs, the wirelessly-powered cage systems have been developed to recharge the batteries without detaching the IMD from the animal during the experiment. However, since these systems are designed to provide the wireless charging in the fixed distance, as shown in Figure 1c, similar to the mobile phone chargers, they are still not suitable for longitudinal animal studies over the span of several days, weeks, or months [5]. Therefore, several concepts of the wirelessly-powered cage have been proposed to extend the wireless coverage and also provide the homogenous power transfer efficiency (PTE) while the small animal is freely behaving inside the cage, as shown in Figure 1d. The key techniques for the wirelessly-powered cage include: (1) the optimized transmitter (Tx) coil design for the extended area; (2) the closed-loop power control (CLPC) for safe wireless power transmission (WPT) [6]; (3) the compensation technique for spatial/angular misalignments of receiver (Rx) coil; (4) the animal tracking technique for the scalability of wireless coverage; and (5) near-field data transmission, while other technologies can also be considered depending on the intended medical applications. Despite the exciting current art, all the requirements of an application involving high-performance or mm-scaled IMDs cannot be addressed by a current existing wirelessly-power cage design. It is important for designers to select suitable wirelessly-powered platforms considering their practical limitations with respect to the IMD design. This review article focuses on the overview of related technologies in recent wirelessly-powered cage platforms including the fundamental principles and practical considerations and provides the guidelines for designers to customize the appropriate wirelessly-powered cages with respect to their IMD applications. The optimized wirelessly-powered cage for the target application based on this article will enable the automated, high throughput, and long-term experiments in a large number of parallel standard cages or in a cage with specific shape for single or multiple animal subjects.

In this article, these key technologies for wirelessly-powered cages are categorized and discussed with the practical design considerations that include the CLPC, coil design/optimization, scalability for wireless coverage, special/angular misalignment, near-field data telemetry, and safety issues. In Section 2, the main blocks for wirelessly-powered cage systems are introduced with the practical considerations related to the key technologies. Section 3 provides the different designs of wirelessly-powered cages with the performance comparison. Section 4 introduces the state-of-the-art wirelessly-powered cages for mm-sized IMDs, followed by a conclusion.

2. Main Blocks for Wirelessly-Powered Cage System

Figure 2 shows a simplified block diagram of the wirelessly-powered cage system and the IMD attached to or implanted in the freely-moving animal subject. The wirelessly-powered cage system typically includes a main controller, a power amplifier (PA), a pair of data Tx/Rx, a Tx coil array, and a position sensor to localize the Rx in the IMD. The wireless link for the power and the data transmission is established between the Tx and the Rx array through the skin/air, while the PA/air while the Tx coil array, driving the Tx coil, tuned at the power carrier frequency, f_p. The frequency of wireless power of wireless power PA is driven from the PA, typically controlled by the main controller, namely PC or micro-controller unit (MCU), depending on the amount of received power by the Rx. Unlike the WPT system including Tx and Rx coils with fixed distance, the coupling between the Tx and Rx coils in the wirelessly-powered cage is varying due to the movements of the animal subject, resulted in the received power variations. The closed-loop power control (CLPC), composed of the received data from the IMD, the main controller, and the PA, increases the Tx power in the PA until enough power is delivered to the Rx while it reduces the Tx power when the Rx receives enough power.

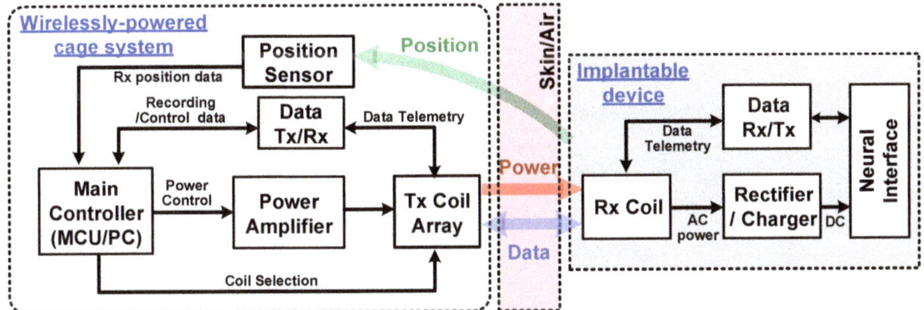

Figure 2. Simplified block diagram of the wirelessly-powered cage system as a stationary unit and the IMD as a mobile unit attached to or implanted in the freely-moving animal subject.

The Tx coil array is one of the important design considerations in the wirelessly-powered cage system to provide the high and homogeneous PTE and power delivered to the load (PDL) within the cage. The Tx coil array can be optimized for a designated arena while it also can be extended for a larger area with the modular type design. Although the achievable PTE and PDL by the inductive link are important for the Tx coil array, the homogeneity of the PTE across the cage should also be considered due to the animal subject's freely movements. In most of the wirelessly-powered cages with the Tx coil array, the position sensor selects the nearest Tx coil among the Tx coil array to the Rx coil. Since only the selected Tx coil will be activated, this position sensing mechanism helps to reduce power loss significantly. Furthermore, each single Tx coil in the modular design of Tx coil array is optimized with the Rx coil to improve the PTE within the entire wirelessly-powered arena. The near-field data telemetry between the Tx and Rx coils can be used to control the IMD, monitor the received power, and/or acquire the biomedical data from the IMD. Even though far-field communication has the advantage of longer data transmission distance, near-field communication is regarded as a more suitable method in terms of power saving in IMDs. Given that the IMD is always located within the wirelessly-powered cage or arena, the coupling between Tx and Rx coils is always enough to deliver both power and data [7]. Here are some practical considerations for different types of wirelessly-powered cage used in the freely-moving animal experiments.

2.1. Coil Design and Optimization

2.1.1. Coil Optimization for Conventional Two-/Three-/Four-Coil Inductive Links

IMDs typically have a tight limitation in their size depending on the applications, resulting in the diameter limitation of Rx coil in the body. In contrast, the Tx coil in the wirelessly-powered cage has more size relaxation in its design. In the inductive link, the coil geometries should be carefully designed to achieve the efficient inductive coupling considering the load and coil separation. Figure 3 shows the physical and electrical configurations of a two-coil inductive link with geometrical parameters used for primary (L_1) and secondary (L_2) coils, where d_{in} is inner diameter of each coil, d_{out}

2.1. Coil Design and Optimization

2.1.1. Coil Optimization for Conventional Two-/Three-/Four-Coil Inductive Links

IMDs typically have a tight limitation in their size depending on the applications, resulting in the diameter limitation of Rx coil in the body. In contrast, the Tx coil in the wirelessly-powered cage has more size relaxation in its design. In the inductive link, the coil geometries should be carefully designed to achieve the efficient inductive coupling considering the load and coil separation. Figure 3 shows the physical and electrical configurations of a two-coil inductive link with geometrical parameters used for primary (L_1) and secondary (L_2) coils, where d_{in} is inner diameter of each coil, d_{out} is outer diameter of coils, N is number of turns, and z is the coil separation between L_1 and L_2 coils. The mutual inductance between Tx and Rx coils (M_{12}) is defined by $M_{12} = k\sqrt{L_1 L_2}$, where k is the coupling coefficient of the two-coil link. M_{12} shows the ratio of magnetic flux common to both L_1 and L_2. R_1 and R_2 are the series resistance of the Tx and Rx coils, respectively. The quality factor (Q) of each coil is defined by $Q = \omega L/R$, where $\omega = 2\pi f_0$ and f_0 is the power carrier frequency.

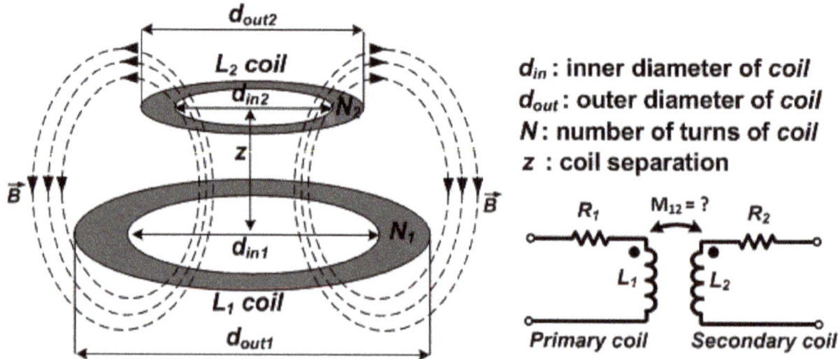

Figure 3. Physical and electrical configurations in a two-coil inductive link.

Although the two-coil inductive link has its optimized solution for highest PTE depending on a given set of Q, Q_2, and k_{12} based on [8], the optimized load resistance, $R_{L,PTE}$, is sometimes far from the nominal target load resistance R_L, which is predefined in the IMD application. Therefore, the multi-coil solution such as three- or four-coil inductive links has been widely studied to provide the designer with more degrees of freedom to convert R_L to $R_{L,PTE}$ and thereby maximize PTE, while they have a potential negative impact on the size-constrained applications. Figure 4 shows the two-, three-, and four-coil inductive links with the lumped circuit model. The PTE of two-, three-, and four-coil inductive links can be calculated based on the basic circuit theory found in [9]:

$$\eta_{2\text{-coil}} = \eta_{23} = \frac{k_{23}^2 Q_2 Q_{3L}}{1 + k_{23}^2 Q_2 Q_{3L}} \cdot \frac{Q_{3L}}{Q_L}, \tag{1}$$

$$\eta_{3\text{-coil}} = \eta_{23}\eta_{34} = \frac{(k_{23}^2 Q_2 Q_3)(k_{34}^2 Q_3 Q_{4L})}{\left[(1 + k_{23}^2 Q_2 Q_3 + k_{34}^2 Q_3 Q_{4L})(1 + k_{34}^2 Q_3 Q_{4L})\right]} \cdot \frac{Q_{4L}}{Q_L}, \tag{2}$$

$$\eta_{4\text{-coil}} = \eta_{12} \cdot \eta_{23} \cdot \eta_{34} = \frac{(k_{12}^2 Q_1 Q_2)(k_{23}^2 Q_2 Q_3)(k_{34}^2 Q_3 Q_{4L})}{\left[(1 + k_{12}^2 Q_1 Q_2) \cdot (1 + k_{23}^2 Q_2 Q_3 + k_{34}^2 Q_3 Q_{4L}) + k_{23}^2 Q_2 Q_3 + k_{34}^2 Q_3 Q_{4L}\right]} \cdot \frac{Q_{4L}}{Q_L}, \tag{3}$$

where Q_{3L} and Q_{4L} are the loaded quality factors ($Q_{3L} = Q_3 Q_L/(Q_3 + Q_L)$ and $Q_{4L} = Q_4 Q_L/(Q_4 + Q_L)$), in which the load quality factor $Q_L = R_L/\omega L_L$. Note that the source/output resistance R_S (R_Q) is included in the driver-coil resistance. The Equation (2) implies that the three-coil link gives the ideal gain with an additional degree of freedom (k_{34}) to adjust the reflected load onto L_4 to be the optimal value, $R_{L,PTE}$ when compared to the two-coil link. The PTE of the three-coil inductive link is related with k_{23}, k_{34}, Q_2, Q_3, and Q_4, for a given load condition. The four-coil link can provide an additional degree of freedom (k_{12}) from the three-coil link for the impedance matching on the source side based on (3). Since the two-, three-, and four-coil inductive links have different strengths and weaknesses in the coupling coefficient (k), PDL, and coupling variations as summarized in Table 1, the designers can select the appropriate inductive link configuration depending on the specifications of the WPT system [9].

additional degree of freedom (k_{23}) to adjust the reflected load onto L_2 to be the optimal value, $R_{L,PTE}$, compared to the two-coil link. The PTE of the three-coil inductive link is related with k_{23}, k_{34}, Q_2, Q_3, and Q_4, for a given load condition. The four-coil link can provide an additional degree of freedom (k_{12}) from the three-coil link for the impedance matching on the source side based on (3). Since the two-, three-, and four-coil inductive links have different strengths and weaknesses in the coupling coefficient (k), PDL, and coupling variations as summarized in Table 1, the designers can select the appropriate inductive link configuration depending on the specifications of the WPT system [9].

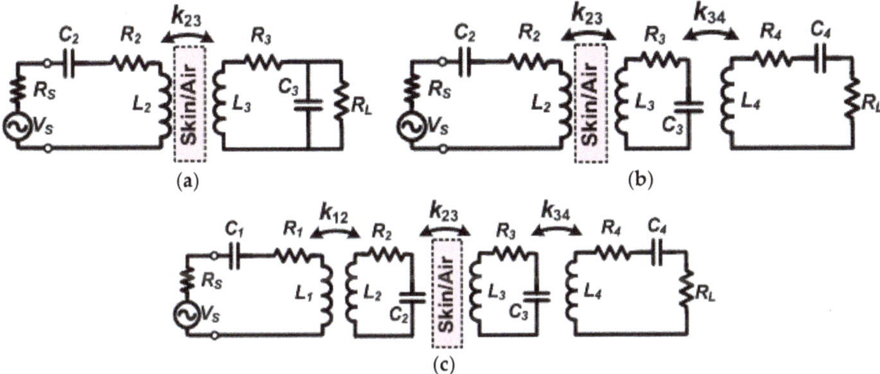

Figure 4. The lumped circuit model for (a) two-coil, (b) three-coil, and (c) four-coil inductive links.

Table 1. Comparison between two-, three-, and four-coil inductive links.

		Inductive Link Configurations		
		2-Coil	3-Coil	4-Coil
Application Conditions	Size constrain	√	×	×
	Strong coupling (k)	√	√	×
	Weak coupling (k)	×	√	√
	Large PDL (small R_s)	√	√	×
	Small PDL (large R_s)	×	×	√
	k variation w/small R_s	√	√	×
	k variation w/large R_s	×	×	√

The optimization procedures of two-, three-, and four-coil inductive links start with the design constrains imposed by the application and coil fabrication technology. The design constrains in the Rx defines the maximum outer diameter of coils in the IMD, and the coil fabrication technology indicates the minimum line width and line spacing. Depending on the application, the coil separation between L_2 and L_3 (z_{23}), the nominal load resistance (R_L), and the source resistance (R_s) are also determined. In the two-coil optimization procedure, $k_{23}Q_2Q_3$ should be maximized to achieve the maximum $\eta_{2\text{-coil}}$ based on (1). The optimum k_{23}, Q_2, and Q_3 can be derived by the proper outer and inner diameter of Tx coil, inner diameter of Rx coil, and number of turns for Tx and Rx coils at given design constraints [8]. The three-coil link optimization procedure maximizes η_{23} and Q_4 in (2), and additionally adjust k_{34} in the Rx to provide the maximum PTE, $\eta_{3\text{-coil}}$. A more detailed flow chart is discussed in [9]. In this optimization procedure, the additional L_3 coil in the Rx plays the role of an impedance-matching circuit, which can convert an arbitrary R_L to $R_{L,PTE}$ for optimal PTE compared to the conventional two-coil link. In other words, the reflected load on the Tx can be adjustable for maximizing the PTE if the designer can choose the suitable k_{23} and k_{34} in the design of a three-coil link. As shown in Figure 5 which shows the exemplar designs of two- and three-coil inductive links, the three-coil link can maintain the maximum PTE by adjusting k_{34} while the two-coil link only reaches the optimal PTE for a specific $R_L = 200\ \Omega$ [9].

Electronics **2020**, *9*, x FOR PEER REVIEW 6 of 28
Electronics **2020**, *9*, x FOR PEER REVIEW 6 of 28

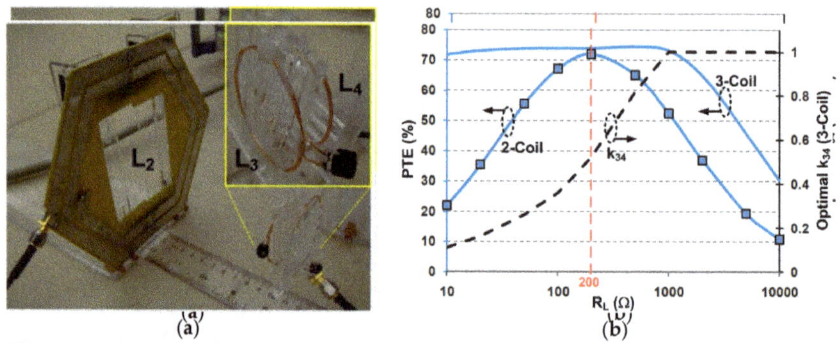

Figure 5. (a) Optimized three-coil inductive link, and (b) PTE variation in the two-coil and three-coil inductive links vs. R_L [9].

The four-coil link is sometimes very useful especially for large coil distance between the Tx and Rx and the large source impedance, R_s, since it provides the additional degree of freedom on the Tx side. Therefore, the four-coil link is widely implemented in the high carrier frequency applications because R_s in the PA is typically increased in the higher frequency. The four-coil link can tolerate the variations in k_{23} caused by the coil separation varying and maintain the high PTE by keeping k_{12} large. The four-coil link optimization maximizes the individual parameters of $k_{23}Q_3$, Q_4, Q_4/k_{12} as similar in the two- and three-coil link optimization. Then, the optimal k_{34} is chosen to provide the maximum PTE as discussed in [9]. The optimization geometries of two-, three-, or four-coil links should satisfy the specific absorption rate (SAR) limit which can be verified by a field solver. If the resulted design cannot satisfy the SAR limit, the designer needs to modify the design constraints and perform the optimization procedure again. The segmented coil design in [10] helps to reduce the average SAR while the loss of the overall link is decreased by using a segmented Tx coil. In Figure 6, the segmented coil shows a more uniform E-field distribution compared to a normal coil with the same geometry, resulting in the reduced peak E-field. Therefore, more Tx power is allowable under the same tissue environment and SAR limit.

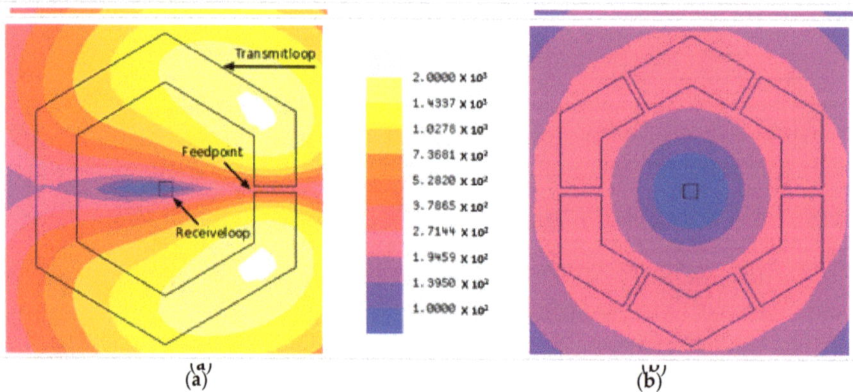

Figure 6. E-field distribution in V/m at skin surface for (a) a conventional coil and (b) a segmented coil [10].

Although previous studies provide the optimization of two-, three-, or four-coil inductive link [8,9], they only focus on the optimization procedure for fixed Tx and Rx coils that is not simply applicable for powering large arena in the cage. If the designer uses the large Tx coil around the cage, the overall PTE will significantly drop and show large variations depending on the location of the IMD. Several approaches have been studying to improve the PTE from Tx to Rx coils while maintaining the homogeneity of wireless power distribution. These approaches can be mainly

As shown in Figure 7b, the overall PTE distribution across the multi-layer Tx coil array has variations within ±24% of the average PTE. The higher PTE peaks are resulted from the Tx coils on layer 1, while the lower peaks are associated with the Tx coils in layers 2 and 3. On the one hand, the layer 1 is slightly closer to the Rx coil. On the other hand, the Tx coils in layers 2 and 3 are more overlapped and surrounded by other Tx coils. This condition leads to larger parasitic capacitance and resistance, resulting in the lower Q and PTE. Figure 7c shows the PDL distribution when the Rx is swept within the cage at the height of 70 mm. Thanks to the proposed configuration of multi-layer Tx coil array together with the GLPC mechanism, the PDL from Tx to Rx coils can be maintained at 20 mW with fluctuations of less than 2 mW. It should be noticed that one of the main disadvantages in this modular system is that one PA is required for each driving coil in the coil array, resulting in the increased complexity and cost of the Tx design.

Several approaches have been studying to improve the PTE from Tx to Rx coils while maintaining the homogeneity of wireless power distribution. These approaches can be mainly classified into two categories: modular design of coil array, as shown in Figure 7, and resonance-based multi-coil inductive link, as shown in Figure 8 in the following section.

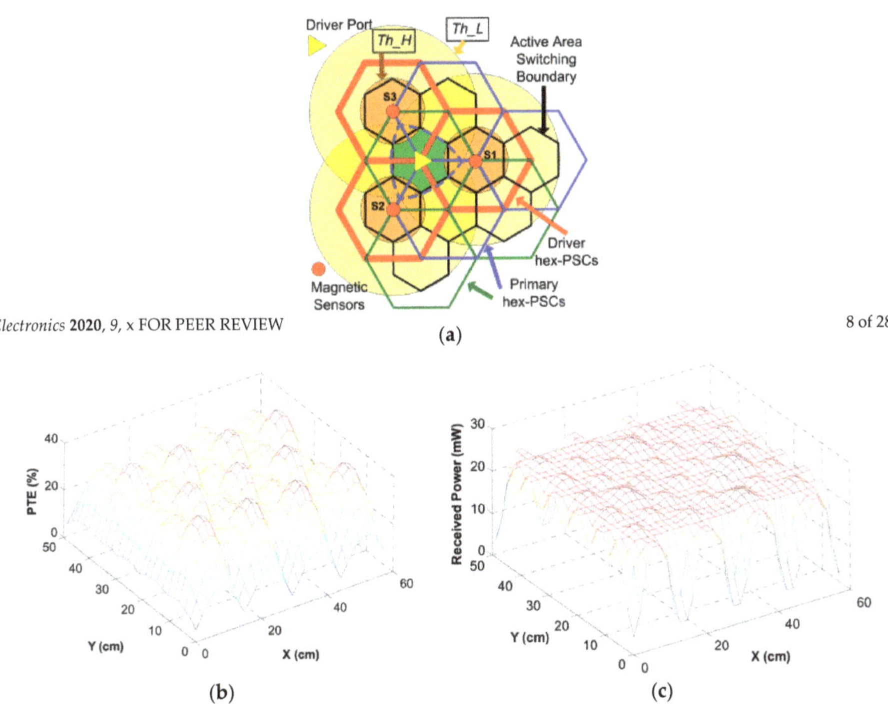

Figure 7. (a) The configuration of the scalable multi-layer Tx coil array implemented in the wirelessly powered cage [11,12]. (b) Measured PTE distribution within the cage area when the Rx is at the height of 70 mm from the bottom of the cage [11,12]. (c) Measured PDL distribution within the cage area when the Rx is swept at the height of 70 mm with GLPC set at 20 mW [11,12].

Therefore, the designer needs to choose the coil design and optimization procedure whether to adopt a modular coil design or a specific coil design dedicated to a designated area.

The [...] resonator [...] HC syste[...] HC syste[...] maximur[...] of the ca[...] wireless [...] inductive [...] keep the [...] perfectly [...]

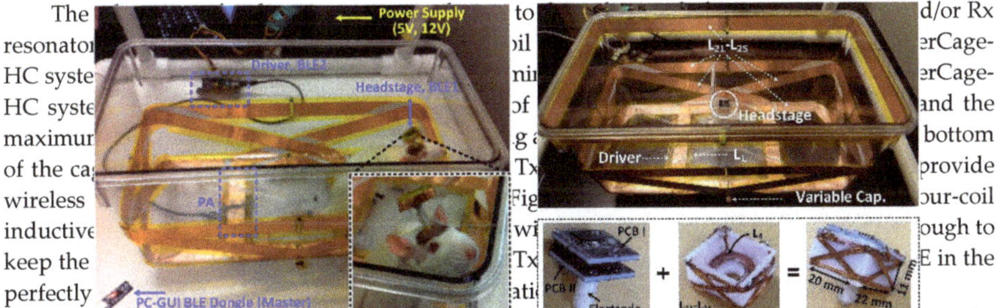

d/or Rx resonator [...] erCage-HC syste[...] erCage-HC syste[...] and the maximum [...] bottom of the ca[...] provide wireless [...] our-coil inductive [...] ough to keep the [...] E in the perfectly [...]

The coupling between loosely coupled Tx and Rx coils is the dominant factor in determining the PTE of the four-coil inductive link. Since the size of the headstage is considerably smaller than the homecage, the effectiv[...] ore Tx magnetic flux can pass through th[...] een the Tx and Rx resonators. In [15], the [...] ie area encompassed by the Rx resonator w[...] igure 8a. In [14], the largest possible area o[...] be, therefore, the Rx resonators are tilting a[...] our directions of the cubical headstage, as s[...]

Due to the large s[...] tures, the Tx and Rx resonators are loosely [...] iis system can be set, regardless of the Rx l[...] ong coupling among the Tx resonators, onl[...] resonance frequency

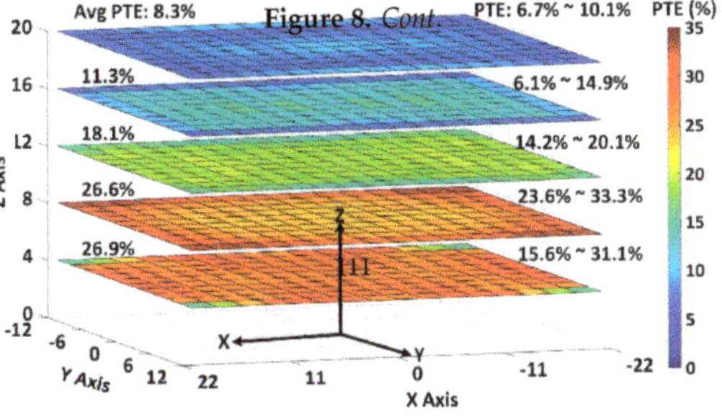

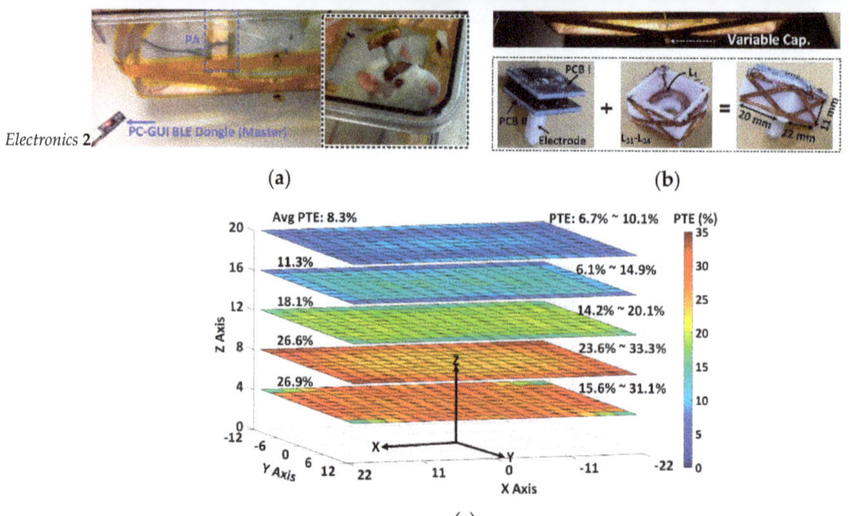

(c)

Figure 8. The configuration of the resonance-based four-coil inductive link implemented in (a) the EnerCage-HC system [11] and (b) the EnerCage-HC2 system (c) Measured PTE when the headstage is swept inside the cage at the height of 4 cm, 8 cm, 12 cm, 16 cm, and 20 cm.

2.1.2. Coil-Optimization for Inductive Links Implemented in Wirelessly-Powered Cages

For the abovementioned two categories of inductive link configurations, the optimization method for each category is different. First, we talk about the optimization method for the inductive link incorporating Tx coil array [11,12]. Instead of having a single Tx coil, identical Tx coils are repeated and tiled at the bottom of the cage so that wireless power transmission covers the entire cage arena. One of the Tx coils, which is closest to the Rx is activated, and together with the Rx coil forms a two-coil inductive link. Therefore, the optimization is required to provide enough power for the mobile device through the two-coil inductive link.

As shown in Figure 7b, the overall PTE distribution across the multi-layer Tx coil array has variations within ±24% of the average PTE. The higher PTE peaks are resulted from the Tx coils on layer 1, while the lower peaks are associated with the Tx coils in layers 2 and 3. On the one hand, the layer 1 is slightly closer to the Rx coil. On the other hand, the Tx coils in layers 2 and 3 are more overlapped and surrounded by other Tx coils. This condition leads to larger parasitic capacitance and resistance, resulting in the lower Q and PTE. Figure 7c shows the PDL distribution when the Rx is swept within the cage at the height of 70 mm. Thanks to the proposed configuration of multi-layer Tx coil array together with the CLPC mechanism, the PDL can be maintained at 20 mW with fluctuations of less than 2 mW. It should be noticed that one of the main disadvantages in this modular system is that one PA is required for each driving coil in the coil array, resulting in the increased complexity and cost of the Tx design.

The other type of coil optimization is relevant to the inductive link incorporating Tx and/or Rx resonator, for instance the resonance-based four-coil inductive link implemented in the EnerCage-HC system families [13,14]. The key factor in determining the Tx resonator geometries in EnerCage-HC system is the compatibility with dimensions of the standard-sized rodent homecage and the maximum overlap with the Tx coil. Instead of having an array of identical Tx coils tiled at the bottom of the cage, in the EnerCage-HC system, multiple Tx resonators wrap around the cage to provide wireless power coverage of the entire cage (see Figure 8). In this case, optimizing the four-coil inductive link means increasing the minimum PTE within the homecage to ensure PDL is enough to keep the headstage on when the CLPC adjusts the Tx power, as opposed to maximizing PTE in the perfectly aligned regions in traditional coil optimization.

The coupling between loosely coupled Tx and Rx coils is the dominant factor in determining the PTE of the four-coil inductive link. Since the size of the headstage is considerably smaller than the homecage, the effective area of the Rx resonators should be maximized so that more Tx magnetic flux can pass through the Rx resonator, thereby improving the coupling between the Tx and Rx resonators. In [15], the Rx resonator wrap around the headstage, maximizing the area encompassed by the Rx resonator without enlarging the size of the headstage, as shown in Figure 8a. In [14], the largest possible area of the headstage is the diagonal planes of the headstage cube, therefore, the Rx resonators are tilting an angle of 25° compared to the horizontal plane in each four directions of the cubical headstage, as shown in Figure 8b.

Due to the large separation and size difference between the Tx and Rx structures, the Tx and Rx resonators are loosely coupled. Hence, a single target resonance frequency for this system can be set, regardless of the Rx location in the homecage. Additionally, because of the strong coupling among the Tx resonators, only one Tx resonator needs to be finely tuned to match the resonance frequency of the entire Tx structure with the target carrier frequency. Such practical and convenient characteristics are also applicable on the Rx resonators, which are also strongly coupled with each other.

Figure 8c shows the PTE distribution within the 3D volume of the cage. As we can see that while the center area has a weaker magnetic flux density, mutual coupling, and thereby lower PTE, the PTE measured at each height is more uniform, with smaller variations of less than 7%. Although the PTE reduces as the height increases, the deduction of the PTE is slowed, which is mainly credited to the enhancement of EM field by the Tx resonator at the top of the cage. Although the optimization procedures in [13,14] are only dedicated to the specific geometry of the cage, which is difficult to be extended for a large arena, these techniques show high and homogeneous PTE inside the standard geometry of cage. Besides, only one PA is needed, which can significantly simplify the design of the power Tx. Therefore, the designer needs to choose the coil design and optimization procedure whether to adopt a modular coil design or a specific coil design dedicated to a designated area.

2.2. Closed-Loop Power Control (CLPC)

Compared to the wireless power transmission system between the fixed Tx and Rx coils, as shown in Figure 1c, the Rx coil attached to the animal body continuously moves inside the wirelessly-powered cage resulting in the coupling variation between Tx and Rx coils. In addition, the power consumption in the mobile device is typically not constant for recording or stimulation operation. Therefore, CLPC, which can dynamically compensate for coupling distance and load variations due to animal movements and implant functions, is required to provide enough power for the mobile device [16]. When the mobile device receives more than enough power, the CLPC reduces the Tx power automatically to minimize the power dissipation on the Tx and the EM exposure on the animal subject, resulting in the improvement of wireless link efficiency and ensuring safety.

The CLPC is typically composed of the data communication channel from the Rx to the Tx, the control unit, DC-DC converter, and the PA as shown in the exemplar design of Figure 9a. The rectifier voltage in the Rx, $V_{\rm rec}$, is monitored and sent to the Tx through the data communication channel which can be either near-field or far-field data communication. The control unit, such as a microcontroller, in the Tx collects the rectifier information through data demodulator block and determines whether the Rx receives enough power or not. If the Rx is not receiving sufficient power, the microcontroller controls the digital potentiometer to reduce the feedback voltage of DC-DC converter. Then, the DC-DC converter increases the PA supply voltage, $V_{\rm DD\_Tx}$, to increase the amount of transmitted power. Otherwise, the microcontroller adjusts the digital potentiometer to decrease

V_{DD_Tx} if the Rx is receiving surplus power. Figure 9b shows the exemplar operation of CLPC in [16]. The CLPC starts to increase the transmitted power by increasing V_{DD_Tx} when the rat moved to low PTE areas or stood up resulting in the weak coupling from the Tx to Rx coils as shown in the inset t = 14,817 s. When the Rx coil was close to the Tx coil located the bottom of the homecage or high PTE areas as shown in the inset t > 14,817 s, the Rx receives more power than necessary resulting in the increase of rectifier voltage, V_{rec}. Then, the CLPC immediately decreases the V_{DD_Tx} to reduce the transmitted power for the regulation of received Rx power. In the result, the Rx can always receive the constant power from the Tx regardless of any environmental variations during the experiment.

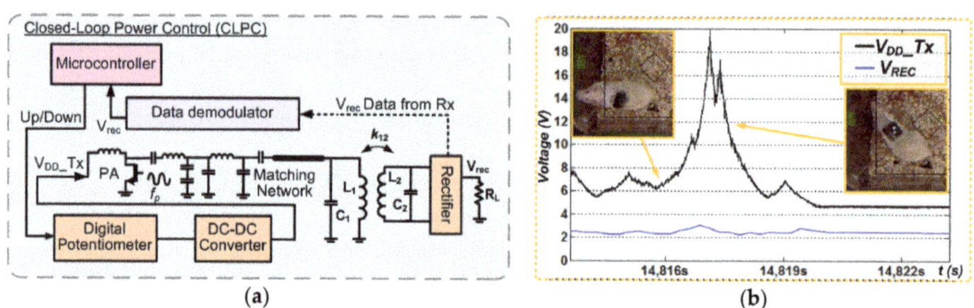

Figure 9. (**a**) Block diagram of CLPC in the two-coil inductive link, and (**b**) the experimental CLPC waveforms in the wirelessly-powered cage to compensate for the distance variation between Tx and Rx in transient caused by the animal movement [16].

2.3.3. Scalability for Wireless Coverage

The animal experimental arena can be either a square standard cage or a specific shape with a larger area depending on the experimental purposes. Some wirelessly-powered cages are dedicated to the standard cage [13,14], which have an advantage in terms of compatibility with the standard racks of rodent cages in the animal facilities compared to the modular design for a specific shape. This compatibility is beneficial for longitudinal studies on multiple animals in separate standard cages resulting in the simultaneous and massive data collection from many animal subjects [19]. The modular designs can be easily extended for large area or specific shape [11,17,18] while the optimized wireless cages dedicated to the standard cage are hard to modify the wireless coverage. One of the important considerations for the scalability using the modular coil design is to choose the suitable method for tracking the Rx coil position because the modular coil design needs to select the nearest Tx coil to the Rx coil. If the modular system does not equip the tracking method, all the Tx coils in the array will be driven simultaneously, resulting in significant power loss. In [11], a small permanent magnet is attached to the mobile device, and the wirelessly-powered cage detects the mobile device using three-axis magnetic sensors to select the nearest Tx coil to the freely moving animal subject. The permanent magnet is also utilized in [18] for the Rx coil tracking, where a single Tx coil moves mechanically on XY-rails located at the bottom of the cage. However, in the WPT system, the performance of magnetic sensors might be degraded due to the strong magnetic fields inside the cage, resulting in not sufficient tracking resolution and quality.

The optical animal tracking techniques are studied in [16,19] using an infrared range camera or a Microsoft Kinect®. The Microsoft Kinect includes infrared depth (HR-3D) and red-green-blue (RGB-2D) cameras allowing the tracking in both bright and dark conditions. Since the optical tracking method can obtain the information about the Rx coil position and the animal subject behavior at the same time, it is more beneficial than the permanent magnet or the optical cameras should be installed at the top of the cage with a few tens of centimeters. As such, the lid of the cage should not be closed during the experiment. As an alternative, the resonator-based cage design, which allows for automatic magnetic field localization, obviating the need for a tracking system or switching the coils, are introduced in [13–15,17]. In these systems, the multi-resonator coil arrangement around the cage, simply driven by a single LC-tank located at the bottom of the cage, can dynamically focus the magnetic field at the position of Rx coil. However, the parasitic resistance in multi-resonator coils still dissipates power all the time regardless of the Rx coil position, resulting in the additional power loss compared to the switchable modular Tx coil design.

2.4. Spatial and Angular Misalignment between Tx and Rx Coils

Spatial and angular misalignments and distance variation between Tx and Rx coils inside the wirelessly-powered cage happen quite frequently in practice as freely behaving animal subjects walk, sniff around, rear, and climb the walls of the cage. This will result in a significant reduction in the PTE and PDL, which might cause malfunctions in the mobile device which does not equip energy storage. The homogeneity of wireless coverage achieved by the Tx coil array alleviates the spatial misalignment as shown in Section 2.1.2, and the CLPC in Section 2.2 [11,13,16] compensates the reduced power in the Rx against some angular misalignments or distance variations. However, the Rx coil rarely receives the wireless power from the Tx coil with maximum angular misalignment of 90°. Therefore, the designer should consider the worst-case condition in the wirelessly-powered cage.

The most common technique to address the angular misalignment with omnidirectional powering is to implement 3D Rx coils in x-, y-, and z-axis, as shown in Figure 10a [20]. The 3D Rx coil is composed of three individual Rx coils followed by each rectifier as shown in Figure 10b. For the nominal condition which angular misalignment is 0°, the Rx_z coil mainly receives the power carrier from the Tx coils while Rx_x and Rx_y coils rarely receive the power. When the 3D Rx coil has large angular misalignments from the Tx coil, the received power in the Rx_x or Rx_y coil is increased depending on the misalignment axis to compensate the reduced power in the Rx_z coil. The 3D Rx coil can also be implemented by using the multiple resonator coils, L_3, instead of multiple load coils, L_4, as introduced in Figure 11 [14]. As shown in Figure 11b, the multiple resonators are strongly coupled with L_4, while the multiple resonators can receive the power against angular misalignment inside the cage. This technique only utilizes one rectifier resulting in the reduced circuit complexity compared to Figure 11b. Since these 3D Rx coils receive the inductive power from each direction, it provides better homogeneous PTE and PDL against the angular misalignments of the mobile device. However, the 3D Rx coil design in the mobile unit increases the volume compared to the conventional single planar Rx coil, resulting in the limited applications for tiny implants inside the animal body.

Instead of the 3D Rx coil, the new Tx coil configurations are studied to provide the homogeneous magnetic field along with x-, y-, and z-axis [21,22]. The proposed architecture in Figure 12a [23] includes three layers of hexagonal planar spiral coil (hex-PSCs) array for homogeneous distribution of the PTE against angular misalignments. The individual Tx coil array generates power carriers with three phases (0°, 120°, 240°) as shown in Figure 12b, and provides both vertical and lateral magnetic fluxes at the same time over the entire powered 3D volume inside the cage. When the Rx coil is aligned with the Tx coil, the Rx coil can receive the power from the vertical magnetic flux. When the Rx coil has the angular misalignment against the Tx coil by 90°, the lateral magnetic flux between the Tx coils mainly provides the power to the Rx, resulting in a PTE improvement compared to the conventional Tx coil array as shown in Figure 12b.

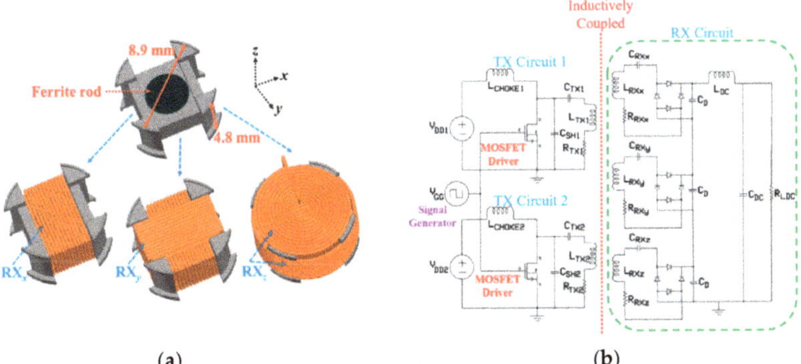

(a)   **(b)**

Figure 10. (a) The design of the 3D Rx coil structure and (b) circuit diagrams in [20] to compensate for the angular misalignment of Rx coil.

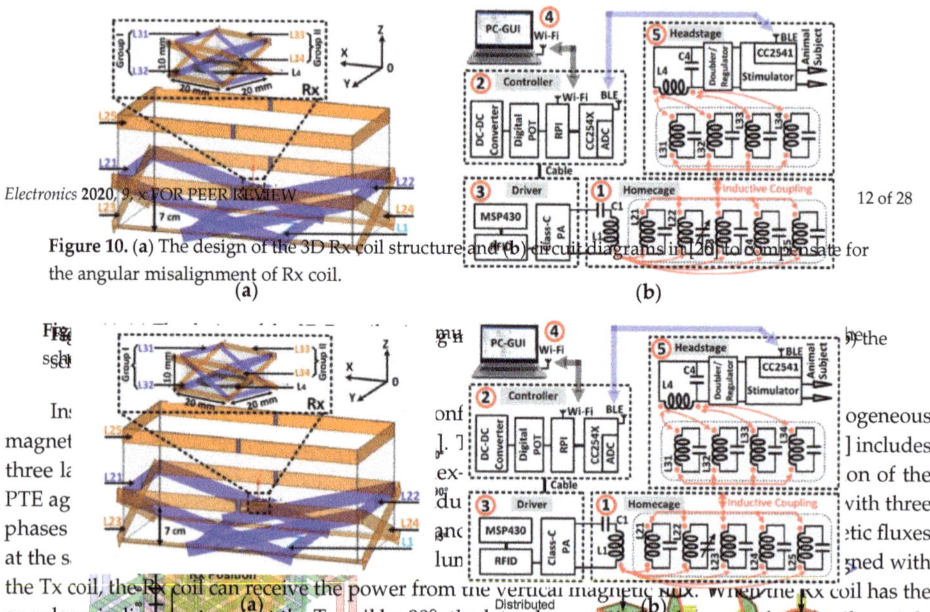

Figure 10. (**a**) The design of the 3D Rx coil structure and (**b**) circuit diagrams in [20] to compensate for the angular misalignment of Rx coil.

the Tx coil, the Rx coil can receive the power from the vertical magnetic flux. When the Rx coil has the angular misalignment against the Tx coil by 90°, the lateral magnetic flux between the Tx coils mainly provides the power to the Rx coil. The 3D Rx coil in this PTE improvement is compatible to the conventional Tx coil array as shown in Figure circuits [14].

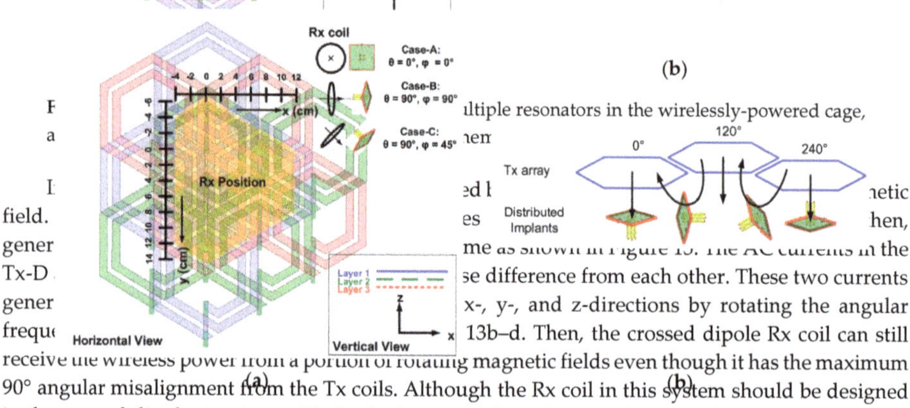

Figure 12. (**a**) The design of the 3D Rx coil using multiple resonators in the wirelessly-powered cage, and (**b**) the three-phase magnetic field excitation scheme [21].

In [22], the omnidirectional powering is achieved by two dipole coils using a rotating magnetic field. This technique utilizes the phase differences in two Tx coils, Tx-D and Tx-Q, and then generates the rotating magnetic field in the 3D volume as shown in Figure 13. The AC currents in the Tx-D and Tx-Q have same magnitude with 90° phase difference from each other. These two currents generate the magnetic field along with x-, y-, and z-directions by rotating the angular frequency from 0 to 2π as partly shown in Figure 13b–d. Then, the crossed dipole Rx coil can still receive the wireless power from a portion of rotating magnetic fields even though it has the maximum 90° angular misalignment from the Tx coils. Although the Rx coil in this system should be designed in the crossed dipole structure with the ferrite material resulting in the limited applications for tiny biomedical implants, the rotating magnetic field technique can be potentially applicable for external portable biomedical devices attached to the animal body.

the rotating magnetic field technique can be potentially applicable for external portable biomedical devices attached to the animal body.

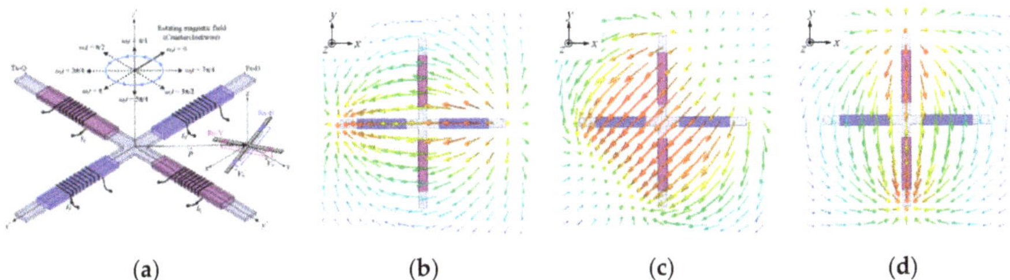

Figure 13. (a) The design of two crossed-dipole coils for Tx-D (Tx1, x-axis) and Tx-Q (Tx2, y-axis) coils for generating omnidirectional magnetic fields, and the simulated rotating magnetic flux at relative angular frequency of (b) 0 (c) $\pi/4$, and (d) $\pi/2$ [22].

2.5. Near-Field Data Telemetry

Since the IMD in the wirelessly-powered cage typically monitors different types of biomedical signals recorded from the electrodes and performs stimulation to tissue for the treatment, the data transmission between the mobile device and the wirelessly-powered cage is required to collect the data and to control the device. In terms of the average power dissipation in the recent IMDs composed of amplifiers, signal conditioning, digitization, processing, and reading frequency (RF) transmission blocks for stimulating application, the far-field data or near-field data blocks dedicate a relatively large portion of the IMD power budget compared to the other functional blocks [7]. Therefore, the near-field communication can be regarded as a suitable method in terms of power saving in IMDs considering that the IMDs also subjecting within the wirelessly powered cage and the Tx/Rx coils were already implemented. While there are many wireless data telemetry techniques are introduced, it is important for the designers to select suitable data telemetry methods, considering the practical limitations of IMD design related to key aspects, such as power budget, silicon area, IMD dimensions, temperature elevation, data bandwidth, sensitivity, reliability (Bit-Error Rate, BER), and robustness against various perturbations [7].

2.6. Specific Absorption Ratio (SAR) and Safety Issues

The transmitted power from the Tx coil should be less than allowable specific absorption rate (SAR) limit, determined by the induced electric field intensity, E, in tissue when exposed RF EMF field. There are mainly two official international safety standard establishment organizations which have deeply impacted each country for their safety standards. The International Commission on Non-Ionizing Radiation Protection (ICNIRP) has the series exposure guideline for each E-field and H-field. The recent ICNIRP (2010) used the induced E-Field instead of induced current density metric. Meanwhile, the guideline also provided the exposure limits in central nervous system (CNS) tissues of the head and other tissues of the body. Institute of Electrical and Electronics Engineers (IEEE) also provided [23] for exposure standard. Compared with ICNIRP guideline, IEEE standard is widely used in US, Canada, Japan, Korea, etc. It also has basic restriction and reference restriction. When setting protective limits for localized tissue heating due to the difficulty of SAR measurement, the point SAR is mass averaged, in recognition of the thermal diffusion properties of tissue. The averaging mass in ICNIRP is any 10 g of contiguous tissue. IEEE standard specifies a smaller 1 g averaging mass in the shape of a cube. The IEEE limits differ from the ICNIRP guidelines in both the maximum level specified which are 100 W/mm² for IEEE and 50 W/m² for ICNIRP, and the frequency at which these peak values reach are 2000 MHz for ICNIRP and 3000 MHz for IEEE [24].

SAR can be calculated by $\sigma|E|^2/\rho$, where σ and ρ are tissue conductivity and density, respectively. Since the SAR limit allows the less amount of RF exposure for the higher carrier frequency, the designer also needs to select a suitable power carrier frequency for the wirelessly-powered cage system based on the expected maximum power used in the Tx and SAR limit for safety, prior to the PTE or PDL optimization. RF exposure can be evaluated by the simulation using a "phantom", which emulates the electrical characteristics of the human head, body, or tissue. In [25], the SAR can be calculated by $SAR = \sigma E^2/\rho$, where σ and ρ are the tissue conductivity and density respectively. Since the SAR limit allows the least amount of RF exposure for the higher carrier frequency, the designer also needs to set a suitable power carrier frequency for the wireless link, so that the expected maximum power reached the Tx and Rx band for safety in Figure 14a. [...] Under different settings of tissue layers, the SAR values with the WPT [...]

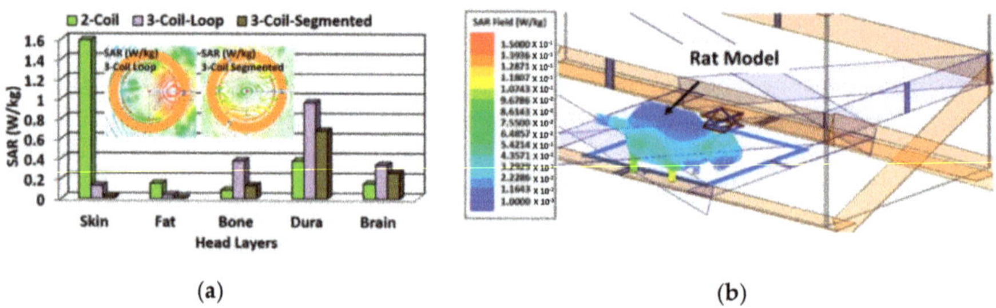

Figure 14. Examples of EM simulation of SAR limits for (a) lamb's head layers [25] and (b) a rat model made entirely out of brain tissue [14].

3. Designs of Wirelessly-Powered Cage Systems

In this section, we will introduce different designs of wirelessly-powered cage systems, which can be classified into four categories. Based on the design considerations discussed in Section 2, we will discuss the main features of each wirelessly-powered cage design.

3.1. Wirelessly-Powered Cage Systems with Single Tx Coil Configuration

The single Tx coil configuration can simplify the design of the driver control in the wirelessly-powered cage systems, significantly improving the system robustness. However, the following examples of the wirelessly-powered cage systems with such configuration also suffer from other limitations, which we will discuss in detail [18,28–30].

Figure 15a,b show two wirelessly-powered cage designs, each of which a large Tx coil wraps around the cage to enhance the magnetic field within the volume covered by the Tx coil [28,29]. In [...] A π-capacitor matching circuit, which matches the Rx coil to the load [...]

Compared to the abovementioned wirelessly-powered cage designs which suffer from the size mismatch between the Tx and Rx coils, the wirelessly-powered cage in Figure 15d does not have this issue [18]. The coils in the inductive link are geomecally optimized. Besides, the optimum alignment between the Tx and Rx coils is also achieved. In Figure 15d, the IRPower system uses a servo-controlled Tx coil moving under the cage on the X and Y rails. A permanent magnet in the mobile device and an array of magnetic sensors placed around the Tx coil form the animal tracking system. Given the real-time optimized size and optimized alignment of the Tx and Rx coils, the PTE of this inductive link design is high. The IRPower system in Figure 15d also features a CLPC, which compensates the PDL variations by adjusting the amount of power transmitted into the Tx coil. However, the mechanical setup of and X and Y rails enlarges the size of the IRPower system, making it incompatible with the standard racks of rodent cages in animal facilities. For high throughput experiments, a large number of cages can be located in standard racks in animal research facilities without occupying precious laboratory space. In addition, with the Tx coil underneath the cage, the systems in Figure 15c,d will have the dramatic reduction in the magnetic field strength and then the PTE, as the distance between the Tx and Rx coil increases. Table 2 compares the features of the abovementioned systems.

In Figure 15b, the Tx coil completely encompasses the area in which the animals can roam freely, resulting in an effective space of 40 × 40 × 20 (height) cm³ [29]. These two designs can improve the WPT resilience to lateral and vertical misalignments between the Tx and Rx coils. However, the angular misalignments still can negatively impact the PTE of the inductive link. Moreover, due to huge size difference, the pair of Tx and Rx coils is away from their optimal size matching, resulting in the low PTE of these two designs. For a larger experimental arena, the size of the Tx coil will be enlarged accordingly, which will aggravate the coil size mismatch and then the PTE reduction. For the designs in Figure 15a,b, the Tx coil covers the 3D volume of the experimental arena, which can ensure the strength of the magnetic field within the covered volume.

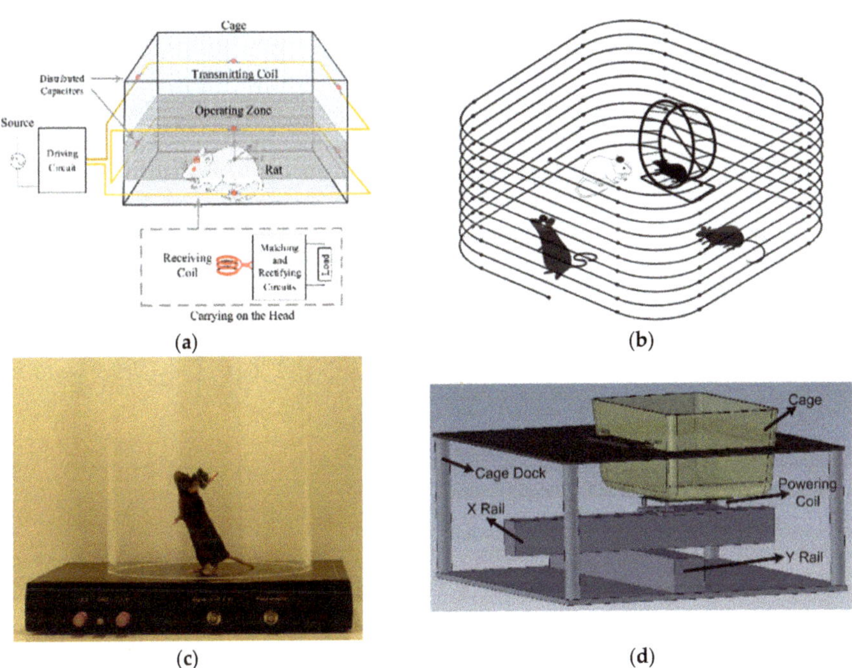

Figure 15. Examples of wirelessly-powered cage designs with a single Tx coil configuration presented in (a) [28], (b) [29], (c) [30], and (d) [18].

For the wirelessly-powered cage design shown in Figure 15c, the Tx coil is located underneath the circular mouse cage with a diameter of 8 inches, and the power Rx is a 15 mm-long, 2 mm-diameter solenoid coil wrapping around a copper ferrite core [30]. The coil size mismatch is eased in this design, at the cost of smaller wirelessly-powered arena, compared to the designs in Figure 15a,b. The Rx coil continuously receivers a large amount of power of 2 W over distances of several centimeters within the cage. Given that the inductive link is non-optimal, it is challenging to ensure the wireless power transmission safety by not overcoming the SAR limitation in the case of strong power delivery. In this wirelessly-powered cage system, the Tx coil also needs to cover the bottom area of the circular mouse cage. Such a configuration is also not conductive for the system scalability.

Compared to the abovementioned wirelessly-powered cage designs which suffer from the size mismatch between the Tx and Rx coils, the wirelessly-powered cage in Figure 15d does not have this issue [18]. The coils in the inductive link are geomecally optimized. Besides, the optimum alignment between the Tx and Rx coils is also achieved. In Figure 15d, the IRPower system uses a servo-controlled Tx coil moving under the cage on the X and Y rails. A permanent magnet in the mobile device and an array of magnetic sensors placed around the Tx coil form the animal tracking system. Given the real-time optimized size and optimized alignment of the Tx and Rx coils, the PTE of this inductive link design is high. The IRPower system in Figure 15d also features a CLPC, which compensates the PDL variations by adjusting the amount of power transmitted into the Tx coil. However, the mechanical setup of and X and Y rails enlarges the size of the IRPower system, making it incompatible with the standard racks of rodent cages in animal facilities. For high throughput experiments, a large number of cages can be located in standard racks in animal research facilities without occupying precious laboratory space. In addition, with the Tx coil underneath the cage, the systems in Figure 15c,d will have the dramatic reduction in the magnetic field strength and then the PTE, as the distance between the Tx and Rx coil increases. Table 2 compares the features of the abovementioned systems.

Table 2. Benchmarking of the wirelessly-powered cage systems with a single Tx coil configuration.

Publications	[28]	[29]	[30]	[18]
Inductive link	2-coil	2-coil	2-coil	2-coil
Frequency	6.78 MHz	13.56 MHz	120 kHz	13.56 MHz
WPT coverage, length × width × height	$40 \times 24 \times 4$ cm^3	$40 \times 40 \times 20$ cm^3	diameter of 8 inch	34×18 cm^2
Mobile device size, length × width × height	$24.5 \times 13 \times 16$ mm^3	$10 \times 10 \times 10$ mm^3	<1 cm^3	12×12 mm^2
PTE	3.8–7.5%	>0.56%	N/A	17%
PDL	100 mW	8.5 mW	2 W	1.7 mW
CLPC	×	×	×	√
Scalability	×	×	×	√
Compatible to racks	√	√	√	×
Coil size matching	×	×	×	√
Vertical misalignment Resilience	√	√	×	×

3.2. Wirelessly-Powered Cage Systems with a Scalable Tx Coil Array

With the configuration of the scalable Tx coil array, the wirelessly-powered cage systems introduced in this section have scalable designs [11,17,31]. Each Tx coil in the array has its optimum diameter matched with the Rx coil at a given Rx-Tx separation distance. Within the array, the Tx coil closest to the animal, which is often in the best position to power the mobile unit, is activated. Several different methods have been used to dynamically and automatically activate the Tx coil that is in the optimum alignment with the Rx coil.

The inductive link configuration in Figure 16a includes a single Rx coil embedded in the mobile device and two overlapping arrays of planar Tx coils placed at the bottom of the cage [31]. Every four coils are connected to a single driver module. The 50% overlap in both X and Y directions eliminates dead spots on the power transmission side. Rx tracking is performed by the impedance tracking technique. Particularly, the proximity of the Rx coil will induce a reflected impedance adding onto the Tx coil, causing a reduction in the current drawn from the driver. The Tx power controller detects the impedance variation of each Tx coil by sensing its supply current variation so that the Tx coil closest to the Rx can be identified as the one with supply current reduction.

In Figure 16b, the wirelessly-power cage system includes a resonance-based inductive link with its configuration adjustable [11]. The stationary unit includes a three-layer array of overlapping hex-PSCs that tile the floor of the experimental arena, Tx coils on the third layer driven by PAs and Tx resonators on the first and second layers. The Tx coil can directly link to the mobile unit by forming a three-coil inductive link if the nearest to the Rx is a Tx coil. Otherwise, a four-coil inductive link will be formed if a Tx resonator is closest to the Rx. The animal location tracking is performed via an array of magnetic sensors under the stationary unit and a magnetic tracer in the mobile device. This system is equipped with CLPC, which can enhance the robustness of the wireless power delivery against the Tx-Rx distance variation.

recording for analyzing animal behavior.

In sum, the abovementioned wirelessly-powered cage systems feature the modular and scalable architecture, which allows for experimental arenas with arbitrary shapes and dimensions. While the Tx includes multiple coils, these systems can still minimize the power interferences as only one Tx coil is activated at a time. Table 3 compares the features of the abovementioned systems.

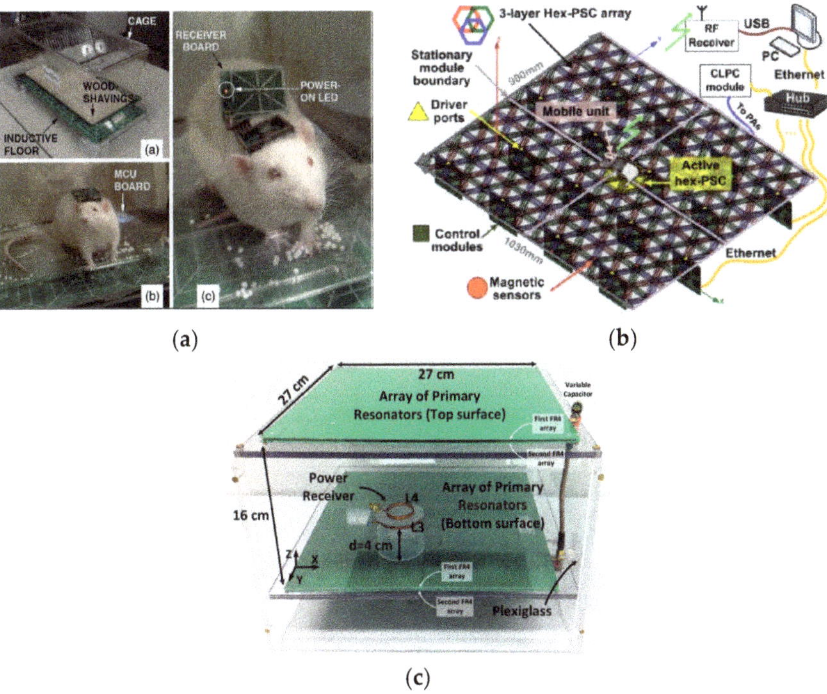

(a) (b)

(c)

Figure 16. Examples of wirelessly-powered cage system designs with the Tx coil array configuration presented in (a) [31], (b) [11], and (c) [17].

Figure 16c shows another wirelessly-powered cage design based on a 13.56 MHz resonance-based four-coil inductive link [17]. The power Tx includes a Tx coil powered by a PA and two arrays of Tx resonators connected in parallel and tiled at the top and bottom surfaces of the chamber to form uniform power distribution in 3D. The top and bottom resonator arrays have more or less contribution to the power transmission, depending on the location of the mobile device in the z-direction. The distance between the surfaces is set for ensuring almost constant power delivery to the mobile device in the z-direction. Such an arrangement also can improve the system tolerance to angular misalignment as the Tx resonators at two surfaces will contribute to directing the EM field towards the mobile device. The Tx resonator configuration features the natural localization mechanism that the Tx resonators located directly below and above the mobile device will be automatically activated. Without the requirement of Rx location detection and Tx coil switching, the system implementation can be significantly simplified. However, placing a resonator array on the top surface of the cage limits access to the animals and blocks the field of view of video recording for analyzing animal behaviors.

Table 3. Benchmarking of the wirelessly-powered cage systems with scalable Tx coil array configuration.

Publications	[31]	[11]	[17]
Inductive link	2-coil	3/4-coil	4-coil
Frequency	1.5 MHz	13.56 MHz	13.56 MHz
WPT coverage, length × width × height	42 × 18 × (8–11) cm³	35 × 38 × 12 cm³	27 × 27 × 16 cm³
Mobile device size, length × width × height	40 × 40 mm²	20 × 20 mm²	42 mm diameter
PTE	13–39%	1.6–12.6%	50%
PDL	21–235 mW	20 mW	100 mW
Control complexity	[illegible]	[illegible]	[illegible]
Compatible to animal behavior tracks	√	×	√

In sum, the abovementioned wirelessly-powered cage systems feature the modular and scalable architecture, which allows for experimental arenas with arbitrary shapes and dimensions. While the Tx includes multiple coils, these systems can still minimize the power interferences as only one Tx coil is activated at a time. Table 3 compares the features of the abovementioned systems.

3.5. Wirelessly-Powered Cage Systems with Slanted Tx Resonators

The wirelessly-powered cage systems introduced in this section have multiple slanted Tx resonators wrapping around the cage [13,16,32,33]. The slanted Tx resonators help bend and direct the magnetic flux towards the Rx coil. Such a configuration can improve the system tolerance to angular misalignments. Additionally, the geometry of the Tx and Rx coils are optimized and matched, which is beneficial to improve PTE.

The wirelessly-powered cage system, shown in Figure 17a, is built based on a resonance-based inductive link whose configuration is adjustable [16]. The stationary unit includes one square-shaped central Tx coil at the bottom of the cage and four triangular-shaped slanted Tx resonators on the corners of the cage. The Tx coil can directly form a three-coil inductive link with the coils on the mobile device when the animal moves to the center of the cage, or form a four-coil inductive link with one Tx resonator when the mobile device is at a corner of the cage. The position of the mobile device is tracked in real-time by a Microsoft Kinect installed above the cage. Depending on the recorded position information, the stationary unit will either deactivate all Tx resonators or activate the Tx resonator closest to the mobile device by shorting this Tx resonator with a resonant capacitor for forming an LC-tank with a high quality-factor. However, installing the Microsoft Kinect will enlarge the experimental volume needed for the entire system setup, making the system not compatible to the racks of standard-sized rodent cages.

In Figure 8a, the wirelessly-powered cage system, named EnerCage-HC, includes four-coil inductive link built around a standard-sized rodent cage [13]. The inductive link has a square-shaped Tx coil attached at the center and bottom of the cage and four Tx resonators wrapping around the cage at different heights/orientations, and planar Rx resonator and Rx coil in the mobile device. The slanted Tx resonators, which direct the magnetic field to the mobile device, together with the CLPC can support sufficient and constant PDL up to 80-degree rotation. In Figure 17b,c, two similar wireless cage designs are presented [32,33]. In the wirelessly-powered cage system, shown in Figure 17b, two more Tx resonators, wrapping around the bottom and top rim of the cage, are added to enhance the homogenous distribution of the transmitted power in the z-direction [32]. The Tx coil consists of multiple identical spiral coils connected in series. However, the Tx coil located at the top of the cage may block the line-of-view of the camera. The wireless cage design, shown in Figure 17c, has a similar Tx resonator configuration [33]. The Tx coil is formed by three identical coils connected in parallel and tiled at the bottom of the cage to make the transmitted power more evenly distributed in 3D.

Thanks to the slanted Tx resonators, the abovementioned wirelessly-powered cage systems show improved robustness to the angular misalignment. Additionally, the Tx coil configuration makes the magnetic field more evenly distributed in the 3D volume of the cage, and the required number of PAs is reduced to only one. For the abovementioned four designs, there is no need for Tx resonator switching, significantly reducing the design complexity, and they feature the compatibility to the standard racks. However, the abovementioned wirelessly-powered cage designs are not scalable. The size of the Tx resonators increases as the cage side increases, which will weaken the advantage of promoting EM uniform distribution. Table 4 compares the features of the abovementioned systems.

Table 3. Benchmarking of the wirelessly-powered cage systems with scalable Tx coil array configuration.

Publications	[31]	[11]	[17]
Inductive link	2-coil	3/4-coil	4-coil
WPT coverage, length × width × height	42 × 18 × (8–19) cm	58 × 58 × 12 cm	27 × 27 × 16 cm
Mobile device size, length × width × height	40 × 40 mm	7 × 20 × 20 mm	42 mm diameter
PTE	13%	5.4%	5%
PDL	21–225 mW	20 mW	100 mW
CLPC	×	√	×
Compatible to racks	×	×	√
Angular misalignment resilience	√	×	√
No view-blocking	×	×	√

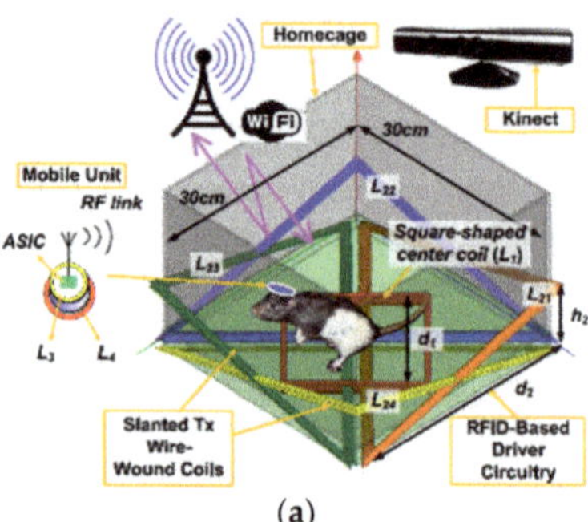

(a)

Figure 17. *Cont.*

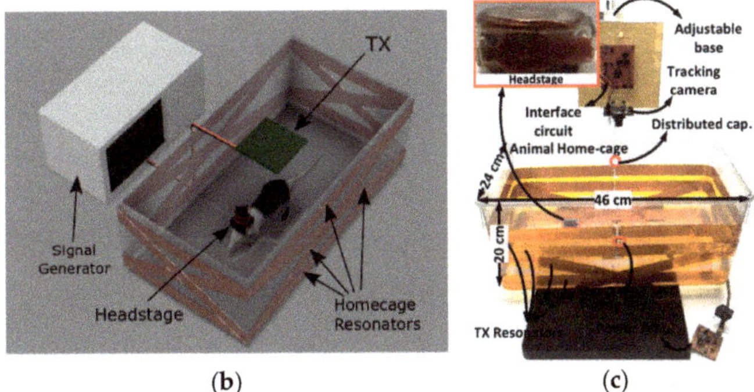

Figure 17. Examples of wirelessly-powered cage system designs with the slanted Tx resonator configuration presented in (a) [16], (b) [32], and (c) [33].

In Figure 4b, the wirelessly-powered cage system a system EnerCage-HC includes four-coil inductive link built around a standard-sized rodent cage [13]. The inductive link has a square-shaped Tx coil attached at the center and bottom of the cage and four Tx resonators wrapping around the cage at different heights/orientations, and planar Rx resonator and Rx coil in the mobile device. The slanted Tx resonators, which direct the magnetic field to the mobile device, together with the CLPC can support sufficient and constant PDL up to 80-degree rotation. In Figure 17b,c, two similar wireless cage designs are presented [32,33]. In the wirelessly-powered cage system, shown in Figure 17b, two more Tx resonators, wrapping around the bottom and top rim of the cage, are added to enhance the homogenous distribution of the transmitted power in the z-direction [32]. The Tx coil consists of multiple identical spiral coils connected in series. However, the Tx coil located at the top of the cage may block the line of view of the camera. The wireless cage design, shown in Figure 17c, has a similar Tx resonator configuration [33]. The Tx coil is formed by three identical coils connected in parallel and tiled at the bottom of the cage to make the transmitted power more evenly distributed in 3D.

Table 3. Benchmarking of the wirelessly-powered cage systems with the slanted Tx resonator configuration.

Publications	[16]	[13]	[32]	[33]
Inductive link	3/4-coil	4-coil	4-coil	42-coil
Frequency	13.56 MHz	13.56 MHz	13.56 MHz	13.56 MHz
WPT coverage, length × width × height	$30 \times 30 \times 17\ cm^3$	$46 \times 24 \times 20\ cm^3$	$28.5 \times 18 \times 13\ cm^3$	$46 \times 24 \times 20\ cm^3$
Mobile device size, length × width × height	$40 \times 40 \times 20\ mm^3$	$20 \times 22 \times 5\ mm^3$	$26 \times 26 \times 35\ mm^3$	$15 \times 15 \times 10\ mm^2$
PTE	16.1–36.3%	14%	34–42%	17%
PDL	24 mW	42 mW	13 mW	62 mW
CEPC	✓	✓	×	✓
Tx control simplicity	×	✓	✓	✓
Compatible to racks	×	✓	×	✓
No view blocking	✓	✓	×	✓

Thanks to the slanted Tx resonators, the abovementioned wirelessly-powered cage systems show improved robustness to the angular misalignment. Additionally, the Tx coil configuration makes the magnetic field more evenly distributed in the 3D volume of the cage, and the required number of PAs is reduced to only one. For the abovementioned four designs, there is no need for Tx resonator switching, significantly reducing the design complexity, and they feature the compatibility to the standard racks. However, the abovementioned wirelessly-powered cage designs are not scalable. The size of the Tx resonators increases as the cage side increases, which will weaken the advantage of promoting EM uniform distribution. Table 4 compares the features of the abovementioned systems.

3.4. Wirelessly-Powered Cage Systems for Omnidirectional Power Transmission

In the experiments involving small freely-moving animals such as rodents, the horizontal distance and/or angular misalignments between the power Tx and Rx components may frequently happen due to the animal's free movements. The misalignments will result in a significant reduction in the PTE and the PDL. In this section, we will introduce several omnidirectional power transmission solutions that can address the issue of wirelessly powered cage systems with the presences of arbitrary position and/or orientation variations of the Rx device.

Table 4. Benchmarking of robust wirelessly powered cage systems with the presences of arbitrary position and/or orientation variations of the Rx device.

Publications	[16]	[13]	[32]	[33]
Inductive link	3/4-coil	4-coil	4-coil	42-coil
Frequency	13.56 MHz	13.56 MHz	13.56 MHz	13.56 MHz
WPT coverage, length × width × height	$30 \times 30 \times 17\ cm^3$	$46 \times 24 \times 20\ cm^3$	$28.5 \times 18 \times 13\ cm^3$	$46 \times 24 \times 20\ cm^3$
Mobile device size, length × width × height	$40 \times 40 \times 20\ mm^3$	$20 \times 22 \times 5\ mm^3$	$26 \times 26 \times 35\ mm^3$	$15 \times 15 \times 10\ mm^2$
PTE	16.1–36.3%	14%	34–42%	17%
PDL	24 mW	42 mW	13 mW	62 mW
CEPC	✓	✓	×	✓
Tx control simplicity	×	✓	✓	✓
Compatible to racks	×	✓	×	✓
No view blocking	✓	✓	×	✓

One omnidirectional power transmission solution is a cavity resonator-based wireless cage design, as shown in Figure 18a [34]. On the power Tx side, the H-field produced is rotational around the center of the cavity. The amplitude of the H-field increases radially from the center to the periphery and is not a function of height. On the Rx side, the implantable device has two Rx coils oriented perpendicular to each other. This placement of the two Rx coils can address the device orientation insensitivity within two planes. To achieve truly omnidirectional powering, a third Rx coil oriented perpendicular to the first two Rx coils should be added. Even then, there are still five WPT blind spots, namely the center and four corners of the cavity, where the H-field is very weak.

For the wirelessly-power cage system, shown in Figure 11, we call it EnerCage-HC2 system [14]. Based on the EnerCage-HC system shown in Figure 8a, the fifth Tx resonator is added at the top rim of the homecage to enhance the magnetic flux density at top of the cage. More importantly, the Rx resonators in the EnerCage-HC2 system encompass four diagonal planes of the mobile device. Such a novel Rx coil arrangement directs the transmitted EM field, which is already homogenized by the Tx resonators, towards the Rx coil at the bottom of the mobile device at any arbitrary orientations.

3.4. Wirelessly-Powered Cage Systems for Omnidirectional Power Transmission

In the experiments involving small freely-moving animals, such as rodents, the horizontal, distance, and/or angular misalignments between the power Tx and Rx components may frequently happen due to the animal's free movements. The misalignments will result in a significant reduction in the PTE and the PDL. In this section, we will introduce several omnidirectional power transmission solutions that can address robust wireless power delivery in the presence of any position and/or orientation variations of the Rx device.

One omnidirectional power transmission solution is a cavity resonator-based wireless cage design, as shown in Figure 18a [34]. On the power Tx side, the H-field produced is rotational around the center of the cavity. The amplitude of the H-field increases radially from the center to the periphery and is not a function of height. On the Rx side, the implantable device has two Rx coils oriented perpendicular to each other. This placement of the two Rx coils can address the device orientation insensitivity within two planes. To achieve truly omnidirectional powering, a third Rx coil oriented perpendicular to the first two Rx coils should be added. Even then, there are still five WPT blind spots, namely the center and four corners of the cavity, where the H-field is very weak.

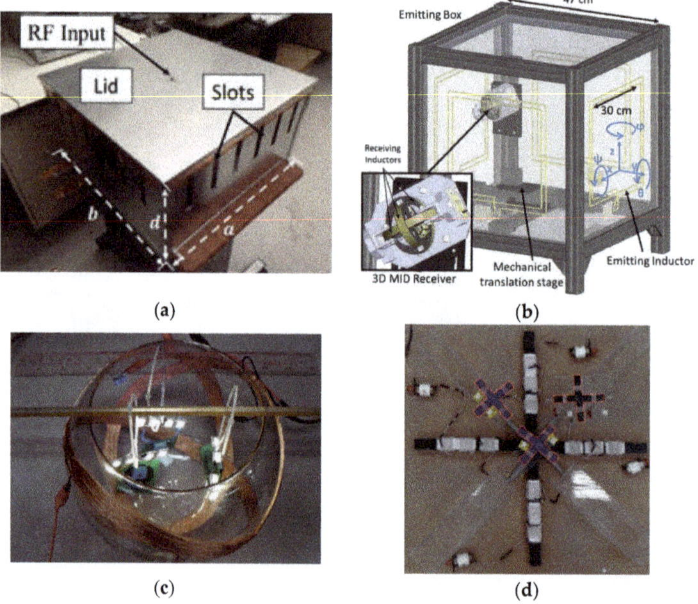

(a)   (b)

(c)   (d)

Figure 18. Examples of wirelessly-powered cage systems offering omnidirectional power transmission, presented in (a) [34], (b) [35], (c) [36], and (d) [22].

For the wirelessly-power cage system, shown in Figure 11, we call it EnerCage-HC2 system [14]. Based on the EnerCage-HC system shown in Figure 8a, the fifth Tx resonator is added at the top rim of the homecage to enhance the magnetic flux density at top of the cage. More importantly, the Rx resonators in the EnerCage-HC2 system encompass four diagonal planes of the mobile device. Such a novel Rx coil arrangement directs the transmitted EM field, which is already homogenized by the Tx resonators, towards the Rx coil at the bottom of the mobile device at any arbitrary orientations. Together with CLPC, the inductive link achieves omnidirectional power delivery to the mobile device, irrespective of the subject's location or head orientation. When the mobile device is 90° rotated while locating along the center lines of the homecage, the magnetic fields passing through the two Rx resonators are bent in the opposite direction and cancel with each other. In this case, little magnetic

Table 5. Benchmarking of the wirelessly-powered cage systems offering omnidirectional power transmission.

Publications	[34]	[14]	[35]	[36]	[22]
Inductive link	Cavity	4 coil	2 coil	2 coil	2 coil
Frequency	346.6 MHz	13.56 MHz	6.78 MHz	530 kHz	280 kHz
WPT coverage, length × width × height	61 × 61 × 30 cm³	46 × 24 × 20 cm³	30 × 30 × 30 cm³	N/A	N/A
Mobile device size, length × width × height	π × (7)² × 25 mm³	20 × 22 × 11 mm³	~64 × 64 × 64 mm³	N/A	20 × 20 × 0.5 cm³
PTE	14.32%	23.6–33.3%	7.9%	N/A	33.6%
PDL	6.1–13 mW	42 mW	1.4 W	N/A	10 W
CLPC	×	√	×	×	×
Scalability	×	×	×	×	×
Tx control simplicity	√	√	√	×	×
Compatible to racks	√	√	√	√	×

field can pass through the Rx coil, resulting in PTE drop. It should be noted that the probability of this case happening is quite low in practice. Therefore, we add a small supercapacitor following the rectifier in the mobile device to prevent any sudden drop in the supply voltage.

In addition to the EnerCage-HC2 system, there are three other inductive link configurations that have been proposed to address omnidirectional wireless powering [22,35,36]. One possible coil configuration is a set of triple orthogonal Rx coils together with a single Tx coil. For achieving higher uniformity of magnetic flux, a pair of Tx coils facing each other can be used instead of a single Tx coil. On the Rx side, each Rx coil is followed by an AC-DC converter, such as a rectifier. The three rectifiers can be connected in series or in parallel to drive their common load. In the series connection scenario, the three DC output voltages of the rectifiers are stacked for providing supply voltage for the load. With the parallel connection scheme, the highest Rx coil voltage is automatically delivered to the load. In Figure 18b, two pairs of Tx coils facing each other are attached to the vertical sides of the cage and driven by their own class-D PA simultaneously [35]. On the Rx side, three coils are placed at 90° from each other, which can make the system insensitive to angular misalignments. However, the Tx coil configuration causes two "cold corners" formed by opposite magnetic fields and two "hot corners" formed by constructive magnetic fields in the cage. In the "cold corners", the triple orthogonal Rx coils still cannot ensure omnidirectional powering. Besides, the Rx coils block the access to the electronics inside the mobile device, making it difficult to repair or to have electrode feed-through.

The triple orthogonal Tx coils together with a single Rx coil is another coil configuration offering the omnidirectional powering. If the three Tx coils are driven by the same current, a constant magnetic field in a fixed direction will be generated. For omnidirectional powering, the three Tx coil currents should be driven appropriately different from each other in order to generate a rotating magnetic field. To achieve this objective, there are three possible modulation methods in general: (1) phase-domain modulation (PDM); (2) amplitude-domain modulation (ADM); and (3) frequency-domain modulation (FDM). Given the simplicity of implementing system control, the PDM method is widely used. In Figure 18c, the PDM method is used. More specifically, the three Tx coils excited with three-phase currents with a peak magnitude of 250 mA and 120° displacement are wound on the outer surface of a glass bowl [36]. The magnetic field vectors rotate periodically in all directions. Three loads hung inside the glass bowl with the Rx coil facing in different directions. Each load consists of a planar Rx coil, a rectifier, and four LEDs. It can be seen that the LEDs in all three loads are on, confirming the omnidirectional nature of the wireless power transmission.

The third viable coil configuration offering omnidirectional powering is the double orthogonal Tx coils together with double orthogonal Rx coils. The two orthogonal Tx coils should be driven by two nonidentical currents to form a rotating magnetic field. Figure 18d shows an example of such a coil configuration [22]. Instead of using loop coils, both the Tx and Rx coil, operating at 280 kHz, have crossed-dipole coil structure. The crossed-dipole Tx and Rx coils are composed of a dual set of two orthogonal coils, wire-wounded around ferromagnetic cores. The currents of the two Tx coils have the same magnitude with a 90° phase difference from each other in order to generate a rotating magnetic field. The magnetic field produced can always pass through the crossed-dipole Rx coils at any orientations, to allow for the omnidirectional power receiving. Additionally, the physical dimensions of Tx and Rx are reduced from volume to plane, which is crucial for practical applications, where volumetric coil structure is highly prohibited. Table 5 compares the features of the abovementioned systems for omnidirectional power transmission.

Table 5. Benchmarking of the wirelessly-powered cage systems offering omnidirectional power transmission.

Publications	[34]	[14]	[35]	[36]	[22]
Inductive link	Cavity	4-coil	2-coil	2-coil	2-coil
Frequency	346.6 MHz	13.56 MHz	6.78 MHz	530 kHz	280 kHz
WPT coverage, length × width × height	$61 \times 61 \times 30$ cm^3	$46 \times 24 \times 20$ cm^3	$30 \times 30 \times 30$ cm^3	N/A	1 m^3

Table 5. *Cont.*

Publications	[34]	[14]	[35]	[36]	[22]
Mobile device size, length × width × height	$\pi \times (7)^2 \times 25$ mm^3	$20 \times 22 \times 11$ mm^3	$\sim\!64 \times 64 \times 64$ mm^3	N/A	$20 \times 20 \times 0.5$ cm^3
PTE	14.32%	23.6–33.3%	7.9%	N/A	33.6%
PDL	6.1–13 mW	42 mW	1.4 W	N/A	10 W
CLPC	×	√	×	×	×
Scalability	×	×	×	×	×
Tx amplifier	√	√	√	×	×
Compatible to racks	√	√	√	√	×
Blind spots	Center + 4 corners	Center lines	2 corners	None	None
No Rx access blocking	√	√	√	×	√

wirelessly operate the tiny implants in a cage. As a result of the significant size difference, the coupling between the Tx and Rx coils is quite weak. The optimal operating frequency for a mm-sized coil to maximize its quality factor is within several tens to hundreds of megahertz range. At such high operating frequencies, it also increases the risk of unsafe exposure to EM radiation and interference, which, in turn, imposes new challenges on SAR compliance. The misalignments caused by free movements of the animal subject and/or the arbitrary placement of the tiny implants, which combine with the other challenges, make it challenging to maintain sufficient power delivery to the tiny implants within a cage. Thus, it is important to develop WPT systems that can overcome these challenges.

4. Designs of Wirelessly-Powered Cage for mm-Sized Implants

Instead of a single, large, and centralized mobile device [37,38], distributed neural interfaces made of a large number of tiny implants are poised to play a key role in future brain-computer interfaces (BCI) because of their less damage to the surrounding tissue. However, it is challenging to wirelessly operate the tiny implants within a cage. As a result of the significant size difference, the coupling between the Tx and Rx coils is quite weak. The optimal operating frequency for a mm-sized coil to maximize its quality factor is within several tens to hundreds of megahertz range. At such high operating frequencies, it also increases the risk of unsafe exposure to EM radiation and interference, which, in turn, imposes new challenges on SAR compliance. The misalignments caused by free movements of the animal subject and/or the arbitrary placement of the tiny implants, which combine with the other challenges, make it challenging to maintain sufficient power delivery to the tiny implants within a cage. Thus, it is important to develop WPT systems that can overcome these challenges.

A wireless cage based on the resonant cavity was introduced in [39] to power and control tiny implanted devices. An aluminum resonant cavity, which has 21 cm diameter and 15 cm height with a surface lattice of hexagons (2.5 cm diameter), is utilized to couple EM energy with 1.5 GHz carrier frequency to the tissue of a mouse as shown in Figure 19a. Compared to the conventional inductive power transfer system through direct coupling between two coils, the energy in this system is confined to the mouse placed on the grid due to the resonance excitation of the limited EM field pattern. This system can provide self-tracking wireless power to the tiny implant within the mouse resulting in no needs of tracking algorithms over the experimental area. Figure 19b shows a schematic of wireless implant for optogenetic stimulation, designed in the size of 10~25 mm^3. This light-emitting implants can receive 5.6~15.7 mW with 3.2 W of averaged input power. Although the wireless operation in RF bands are typically susceptible to signal reflection, interference, and absorption by obstructions, the proposed system in [39] has been successfully demonstrated by behavioral testing environments and has shown the possibility of using RF bands for wireless cage applications.

EM WPT at high frequency in the GHz range facilitate the Rx size reduction due to shorter wavelength, but they suffer from difficulties in creating homogeneous WPT in a large experimental arena [40]. The SAR of EM field in the tissue, which mainly consists of water, at high frequencies is rather high [23]. At high frequency, it also increases the risk of unsafe personnel exposure to EM radiation and interference with nearby instruments [40]. The microwave chamber delivers power at low efficiency through the animal body. Its extension to larger animals and eventually humans without surpassing SAR limits at this frequency does not seem to be feasible, even though it might be a reasonable solution for ultra-low-power applications.

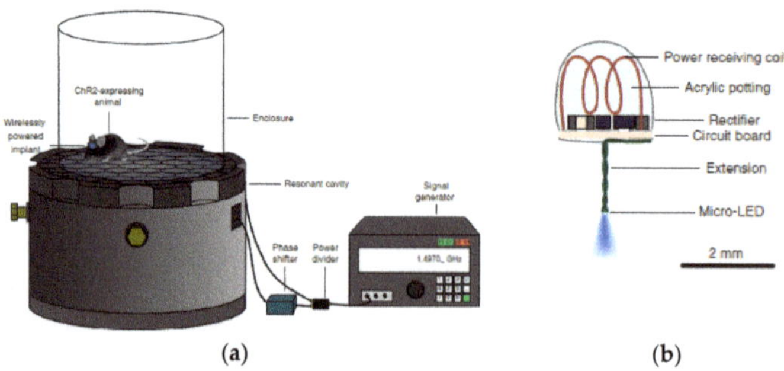

Figure 19. (a) A diagram of wirelessly-powered cage system, and (b) a schematic of wireless implant customized for the brain in [39].

To address this issue, Figure 20 shows the wireless cage for powering tiny implants using inductive coupling at 13.56 MHz frequency with near-field communication hardware [41]. The wireless cage equips a double-loop coil with turns at heights of 4 and 11 cm from the bottom of the animal enclosure as shown in Figure 20a. This technique is similar with the design in Figure 15a, which improves the inductive coupling at further distance from the bottom of the cage resulting in

Electronics **2020**, *9*, x FOR PEER REVIEW 23 of 28

EM WPT at high frequency in the GHz range facilitate the Rx size reduction due to shorter wavelength, but they suffer from difficulties in creating homogeneous WPT in a large experimental arena [40]. The SAR of EM field in the tissue, which mainly consists of water, at high frequencies is rather high [23]. At high frequency, it also increases the risk of unsafe personnel exposure to EM radiation and interference with nearby instruments [40]. The microwave chamber delivers power at low efficiency through the animal body. Its extension to larger animals and eventually humans without surpassing SAR limits at this frequency does not seem to be feasible, even though it might be a reasonable solution to get a bit more uniform coverage. The height for wireless coverage and its uniformity in coverage can be optimized for a given experimental area based on the separation of the coil.

While the wireless cage with high frequency is typically suffer from the SAR in the tissue, the wireless cage in the near-field regime has a difficulty in the reduction of Rx size due to the poor EM field focusing. In Figure 20b, the two-coil inductive link configuration results in the Rx coil with a diameter of 9.8 mm, which becomes the main limiting factor in the device miniaturization. In addition, the single Tx coil configuration, which can simplify the design of the power control circuitry, suffers from low PTE at the center of the cage. Therefore, they transmitted a large amount of power to energize the entire cage. This will result in excessive heat dissipation and large EM interference between the cage and other lab instruments.

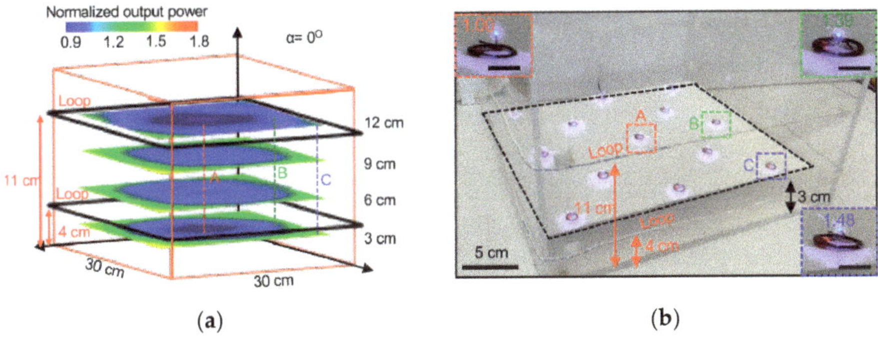

(a) (b)

Figure 20. (a) Simulated normalized output power inside the wireless cage without angular misalignment and (b) wireless operation of 13 implants placed at height of 3 cm from the bottom [41].

In Figure 21a, the dual-band EnerCage-HC system, built upon the EnerCage-HC2 system (Figure 11), uses a two-stage inductive power transmission technique to wirelessly deliver sufficient PDL to the tiny implant within the cage [42]. Given the mismatch between the optimal operating frequencies of the Tx and Rx coils due to their size difference, the dual-band EnerCage-HC system includes two inductive links operating simultaneously at their optimal frequencies. To do so, this system also includes an intermediate unit (a headstage device) in addition to the power Tx and Rx. This intermediate unit acts as a converter, which receives power from the cage via a four-coil inductive link at 13.56 MHz and delivers it to the tiny implant via a three-coil inductive link at 60 MHz. This intermediate unit is also used as a power buffer that uses its stored energy to support the continuous operation of the tiny implants in the presence of misalignments in the four-coil inductive link. A dual-loop sequential CLPC mechanism, which includes two loops, further stabilizes the amount of power delivered to the tiny implant despite any misalignments by adjusting PDLs of the four-coil inductive link and the three-coil inductive link at designated levels, respectively. Figure 21b shows the final in vitro measurement setup for enabling optical stimulation in the dual-band EnerCage-HC system. The experiment was performed on 6 × 6 × 7 cm cube from a fresh sheep head, including brain, skull, fat, and skin, and placed the cube in the center of the cage. The headstage was mounted on top of the tissue cube to receive 13.56 MHz carrier from the EnerCage-HC coils while the mm-sized implant was placed ~2 mm under the skull. It shows both the orange LED that is an indicator of the headstage receiving sufficient power at 13.56 MHz, and the selected µLED (blue a power buffer that uses its stored energy to support the continuous operation of the tiny implants in the presence of misalignments in the four-coil inductive link. A dual-loop sequential CLPC mechanism, which includes two loops, further stabilizes the amount of power delivered to the tiny implant despite any misalignments by adjusting PDLs of the four-coil inductive link and the three-coil inductive link at designated levels, respectively. Figure 21b shows the final in vitro measurement setup for enabling optical stimulation in the dual-band EnerCage-HC system. The experiment was performed on $6 \times 6 \times 7$ cm cube from a fresh sheep head, including brain, skull, fat, and skin, and placed the cube in the center of the cage. The headstage was mounted on top of the tissue cube to receive 13.56 MHz carrier from the EnerCage-HC coils while the mm-sized implant was placed ~2 mm under the skull. It shows both the orange LED that is an indicator of the headstage receiving sufficient power at 13.56 MHz, and the selected μLED (blue color) for optogenetic stimulation that is powered at 60 MHz.

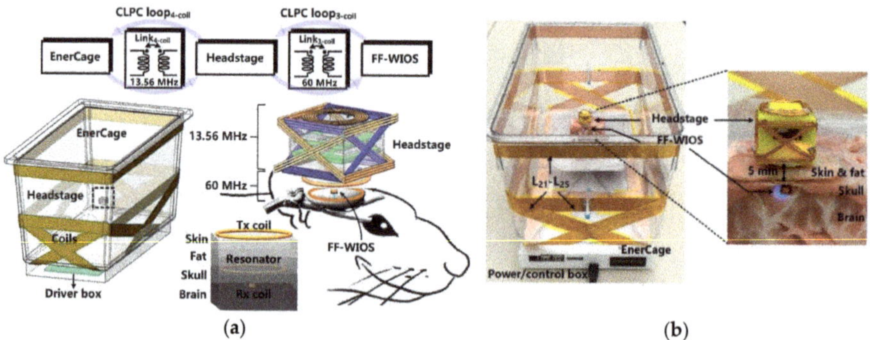

Figure 21. A prototype of inductive power transfer systems for tiny implants, proposed in [42] for (a) conceptual system overview and (b) in vitro measurement setup.

Figure 22a presents a new wireless power transfer system that can wirelessly power a large number of tiny implants arbitrarily distributed over a large tissue area. This is achieved by a power Tx platform which can generate a vertical and lateral magnetic field alternately at any places within the experimental arena [21]. This power Tx platform includes three overlapping layers of hex-PSCs tiled at the bottom of the cage. The three layers are horizontally shifted in a way that the centers of every three adjacent hex-PSCs on three different layers are on the corners of an equilateral triangle. In each layer, every three adjacent hex-PSCs are driven at carrier signals with the same magnitude and phase of 0°, 120°, and 240°, respectively, as shown in Figure 22b. To avoid any counteracts among the three layers, they are controlled by time-division multiplexing (TDM). The appropriate TDM period, T, is decided by the carrier frequency, time constant of the capacitance following the Rx coil and rectifier, Rx loading, and the acceptable level of ripple on the regulated supply voltage of the electronics that are powered by the Rx coil as shown in Figure 22c. To this end, this power Tx platform in Figure 22d, which takes advantage of the overlapping placements of the three layers, the three-phase power carrier signals, and the TDM control mechanism, eventually achieves omnidirectional WPT towards a large number of tiny implants, which prohibits any method that [21]/precise spatial misalignment distribution. But the measured result, the proposed method for 90° achieves 5% of PTE distributions and 5 mW of PDL with PTE for a 90° angular misalignment of Rx coil compared to 0.8% of PTE for the conventional WPT system.

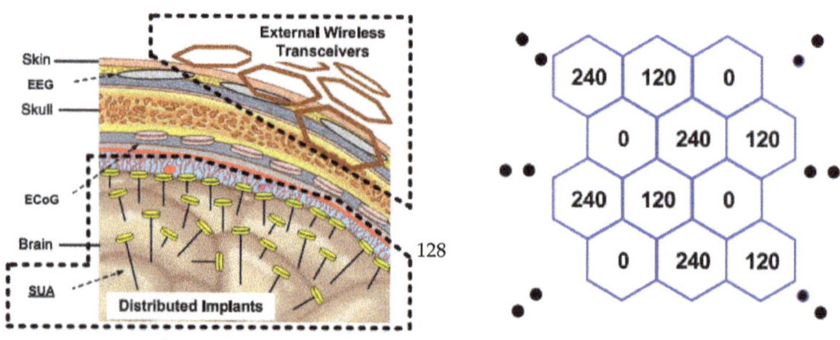

In Figure 22d, which takes advantage of the overlapping placements of the three layers, the three-phase power carrier signals, and the TDM control mechanism, eventually achieves omnidirectional WPT towards a large number of tiny implants with arbitrary angular and/or spatial misalignments. In the measured result, the proposed method in [21] achieves 5% of PTE distributions and 5 mW of PDL within the area for 90° angular misalignment of Rx coil compared to 0.8% of PTE for the conventional WPT system.

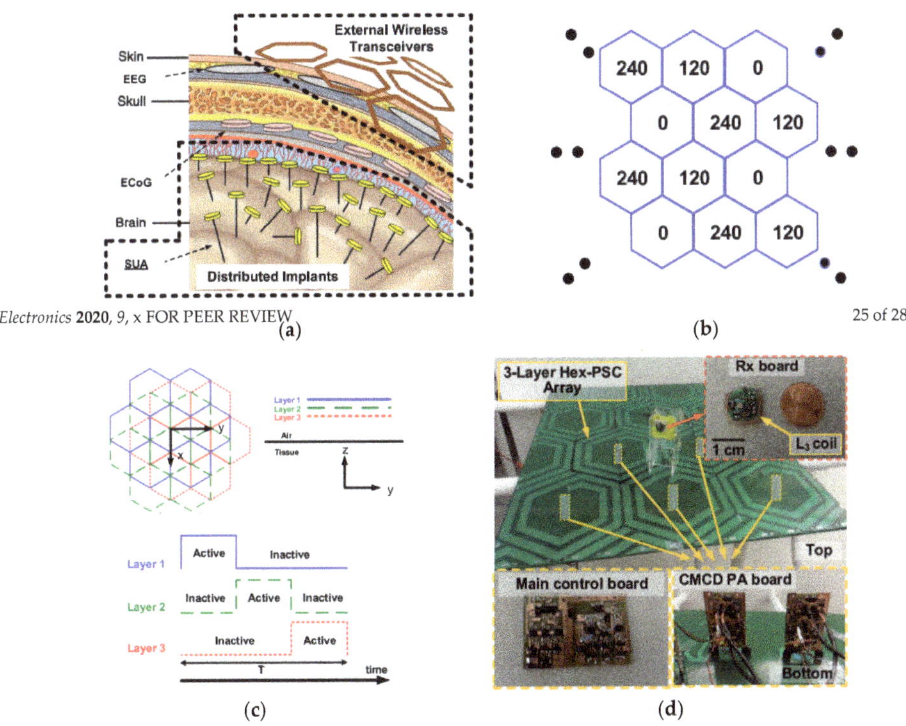

(a)   (b)   (c)   (d)

Figure 22. (a) Rendered view of distributed BCI concept, (b) phase distribution among the hex-PSCs in one conductive layer, (c) TDM driving technique to avoid the magnetic interference between three layers, and (d) a prototype measurement setup for the three-phase TDM overlapping hex-PSC array in [21].

5. Conclusions

Wirelessly-powered cages for IMD applications are categorized and discussed with their key techniques, such as closed-loop IMD operation, coil design/optimization, scalability for wireless coverage, spatial/angular IMD misalignments, near-field data recording, and stimulations. Since the limits for performance IMDs herein the focus systems on long-term freely-behaving animal subjects [37,38], it is important to design a suitable wireless power platform having an a large subjects, there are important different techniques a suitable wireless power platform that can bring a large experimental area considering various practical issues wirelessly power this article. Although benchmarked in Tables 2–5 consider PTE and PDL for specific experimental conditions, the physical constraints of Rx coils, coil separations, and magnetic field homogeneity over the experimental area should be compared for a more suitable choice. As the most recent class of IMDs are millimeter-sized and distributed over a large area in the brain or the rest of the body [43–45], the importance of wirelessly-powered cages for tiny multiple implants is elevated as introduced in Section 4. Hence, it is important for designers to select suitable wirelessly-powered cages, considering the practical limitations of IMD design related to key aspects, such as PTE, PDL, closed-loop power control, scalability, spatial/angular misalignment, near-field data telemetry, and safety issues against various perturbations during the longitudinal animal experiment. With respect to various IMD applications from conventional implantable devices to recent mm-sized implants, this article contributes the design strategy of the appropriate wirelessly-powered cage designs, which have the ability to create an automated enriched environment inside the experimental arena for long-term electrophysiology experiments.

Author Contributions: Conceptualization: B.L. and Y.J.; methodology: B.L. and Y.J.; validation: B.L. and Y.J.; formal analysis: B.L. and Y.J.; investigation: B.L. and Y.J.; data curation: B.L. and Y.J.; writing—original-draft preparation: B.L. and Y.J.; writing—review and editing: B.L. and Y.J.; visualization: B.L. and Y.J.; supervision, project administration, and funding acquisition: B.L. and Y.J. All authors have read and agreed to the published version of the manuscript.

Author Contributions: Conceptualization: B.L. and Y.J.; methodology: B.L. and Y.J.; validation: B.L. and Y.J.; formal analysis: B.L. and Y.J.; investigation: B.L. and Y.J.; data curation: B.L. and Y.J.; writing—original-draft preparation: B.L. and Y.J.; writing—review and editing: B.L. and Y.J.; visualization: B.L. and Y.J.; supervision, project administration, and funding acquisition: B.L. and Y.J. All authors have read and agreed to the published version of the manuscript.

Funding: This work was supported by the Incheon National University (#2019-0299) research grant in 2019.

Conflicts of Interest: The authors declare no conflict of interest.

References

1. Manns, J.R.; Eichenbaum, H. A congnitive map for object memory in the hippocampus. *Learn. Mem.* **2009**, *16*, 616–624. [CrossRef]
2. Choi, Y.; Park, S.; Chung, Y.; Gore, R.K.; English, A.W. PDMS microchannel scaffolds for neural interfaces with the peripheral nervous system. In Proceedings of the 27th International Conference on Micro Electro Mechanical Systems (MEMS), San Francisco, CA, USA, 26–30 January 2014; pp. 873–876.
3. Lee, S.B.; Yin, M.; Manns, J.R.; Ghovanloo, M. A wideband dual antenna receiver for wireless recording from animals behaving in large arenas. *IEEE Trans. Biomed. Eng.* **2013**, *60*, 1993–2004.
4. Fan, D.; Rich, D.; Holtzman, T.; Ruther, P.; Dalley, J.W.; Lopez, A.; Rossi, M.A.; Barter, J.W.; Meza, D.S.; Herwik, S.; et al. A wireless multi-channel recording system for freely behaving mice and rats. *PLoS ONE* **2011**, *6*, e22033. [CrossRef]
5. Lee, S.B.; Lee, H.M.; Kiani, M.; Jow, U.; Ghovanloo, M. An inductively powered scalable 32-channel wireless neural recording system-ona-chip for neuroscience applications. *IEEE Trans. Biomed. Circuits Syst.* **2010**, *4*, 360–371. [CrossRef] [PubMed]
6. Lee, B.; Kiani, M.; Ghovanloo, M. A triple-loop inductive power transmission system for biomedical applications. *IEEE Trans Biomed Circuits Syst.* **2015**, *10*, 138–148. [CrossRef] [PubMed]
7. Lee, B.; Ghovanloo, M. An overview of data telemetry in inductively powered implantable biomedical devices design and implementation of devices. *IEEE Commun. Mag.* **2019**, *57*, 74–80. [CrossRef]
8. Harrison, R.R. Designing efficient inductive power links for implantable devices. In Proceedings of the 2007 International Symposium on Circuits and Systems, New Orleans, LA, USA, 27–30 May 2007; pp. 2080–2083.
9. Kiani, M.; Jow, U.; Ghovanloo, M. Design and optimization of a 3-coil inductive link for efficient wireless power transmission. *IEEE Trans. Biomed. Circuits Syst.* **2011**, *5*, 579–591. [CrossRef]
10. Mark, M.; Björninen, T.; Ukkonen, L.; Sydänheimo, L.; Rabaey, J.M. SAR reduction and link optimization for mm-size remotely powered wireless implants using segmented loop antennas. In Proceedings of the 2011 IEEE Topical Conference on Biomedical Wireless Technologies, Networks, and Sensing Systems, Phoenix, AZ, USA, 16–19 January 2011; pp. 7–10.
11. Jow, U.E.; McMenamin, P.; Kiani, M.; Manns, J.R.; Ghovanloo, M. EnerCage: A smart experimental arena with scalable architecture for behavioral experiments. *IEEE Trans. Biomed. Eng.* **2014**, *61*, 139–148. [CrossRef] [PubMed]
12. Jow, U.E.; Kiani, M.; Huo, X.; Ghovanloo, M. Towards a Smart Experimental Arena for Long-Term Electrophysiology Experiments. *IEEE Trans. Biomed. Circuits Syst.* **2012**, *6*, 414–423. [CrossRef] [PubMed]
13. Mirbozorgi, S.A.; Jia, Y.; Canales, D.; Ghovanloo, M. A Wirelessly-Powered Homecage With Segmented Copper Foils and Closed-Loop Power Control. *IEEE Trans. Biomed. Circuits Syst.* **2016**, *10*, 979–989. [CrossRef] [PubMed]
14. Jia, Y.; Mirbozorgi, S.A.; Wang, Z.; Hsu, C.H.; Madsen, T.E.; Rainnie, D.; Ghovanloo, M. Position and Orientation Insensitive Wireless Power Transmission for EnerCage-Homecage System. *IEEE Trans. Biomed. Eng.* **2017**, *64*, 2439–2449. [CrossRef] [PubMed]
15. Mirbozorgi, S.A.; Jia, Y.; Zhang, P.; Ghovanloo, M. Towards a High-Throughput Wireless Smart Arena for Behavioral Experiments on Small Animals. *IEEE Trans. Biomed. Circuits Syst.* **2020**, *67*, 2359–2369. [CrossRef] [PubMed]
16. Lee, B.; Kiani, M.; Ghovanloo, M. A Smart Wirelessly Powered Homecage for Long-Term High-Throughput Behavioral Experiments. *IEEE Sens. J.* **2015**, *15*, 4905–4916. [CrossRef] [PubMed]
17. Mirbozorgi, S.A.; Bahrami, H.; Sawan, M.; Gosselin, B. A Smart Cage with Uniform Wireless Power Distribution in 3D for Enabling Long-Term Experiments With Freely Moving Animals. *IEEE Trans. Biomed. Circuits Syst.* **2016**, *10*, 424–434. [CrossRef] [PubMed]

18. Kilinc, E.G.; Conus, G.; Weber, C.; Kawkabani, B.; Maloberti, F.; Dehollain, C. A System for Wireless Power Transfer of Micro-Systems In-Vivo Implantable in Freely Moving Animals. *IEEE Sens. J.* **2014**, *14*, 522–531. [CrossRef]

19. Ou-Yang, T.H.; Tsai, M.L.; Yen, C.T.; Lin, T.T. An infrared range camera-based approach for three-dimensional locomotion tracking and pose reconstruction in a rodent. *J. Neurosci. Methods* **2011**, *201*, 116–123. [CrossRef]

20. Khan, S.R.; Desmulliez, M.P.Y. Towards a Miniaturized 3D Receiver WPT System for Capsule Endoscopy. *Micromachines* **2019**, *10*, 545. [CrossRef]

21. Lee, B.; Ahn, D.; Ghovanloo, M. Three-phase time-multiplexed planar power transmission to distributed implants. *IEEE J. Emerg. Sel. Top. Power Electron.* **2015**, *4*, 263–272. [CrossRef]

22. Choi, B.H.; Lee, E.S.; Sohn, Y.H.; Jang, G.C.; Rim, C.T. Six Degrees of Freedom Mobile Inductive Power Transfer by Crossed Dipole Tx and Rx Coils. *IEEE Trans. Power Electron.* **2016**, *31*, 3252–3272. [CrossRef]

23. *IEEE Standard for Safety Levels with Respect to Human Exposure to Radio Frequency Electromagnetic Fields, 3 kHz to 300 GHz*; IEEE Standard C95.1-2005; IEEE Standard: Piscataway, NJ, USA, 2005.

24. McIntosh, R.L.; Anderson, V.; McKenzie, R.J. A numerical evaluation of SAR distribution and temperature changes around a metallic plate in the head of a RF exposed worker. *Bioelectromagnetics* **2005**, *26*, 377–388. [CrossRef]

25. Mirbozorgi, S.A.; Yeon, P.; Ghovanloo, M. Robust wireless power transmission to mm-sized free-floating distributed implants. *IEEE Trans. Biomed. Circuits Syst.* **2017**, *11*, 692–702. [CrossRef] [PubMed]

26. Karipott, S.S.; Veetil, P.M.; Nelson, B.D.; Guldberg, R.E.; Ong, K.G. An Embedded Wireless Temperature Sensor for Orthopedic Implants. *IEEE Sens. J.* **2018**, *18*, 1265–1272. [CrossRef]

27. Lu, D.; Yan, Y.; Avila, R.; Kandel, I.; Stepien, I.; Seo, M.H.; Bai, W.; Yang, Q.; Li, C.; Haney, C.R.; et al. Bioresorbable, wireless, passive sensors as temporary implants for monitoring regional body temperature. *Adv. Healthc. Mater.* **2020**, *9*, 2000942. [CrossRef] [PubMed]

28. Chow, J.P.-W.; Chung, H.S.-H.; Chan, L.L.-H.; Shen, R.; Tang, S.C. Optimal Design and Experimental Assessment of a Wireless Power Transfer System for Home-Cage Monitoring. *IEEE Trans. Power Electron.* **2019**, *34*, 9779–9793. [CrossRef]

29. Nassirinia, F.; Straver, W.; Hoebeek, F.E.; Serdijn, W.A. Wireless power transfer and optogenetic stimulation of freely moving rodents. In Proceedings of the 2017 8th International IEEE/EMBS Conference on Neural Engineering (NER), Shanghai, China, 25–28 May 2017; pp. 456–460.

30. Bernstein, J.G.; Boyden, E.S. Optogenetic tools for analyzing the neural circuits of behavior. *Trends Cogn. Sci.* **2011**, *15*, 592–600. [CrossRef]

31. Soltani, N.; Aliroteh, M.S.; Salam, M.T.; Velazquez, J.L.P.; Genov, R. Low-Radiation Cellular Inductive Powering of Rodent Wireless Brain Interfaces: Methodology and Design Guide. *IEEE Trans. Biomed. Circuits Syst.* **2016**, *10*, 920–932. [CrossRef]

32. Biswas, D.K.; Martinez, J.H.A.; Daniels, J.; Bendapudi, A.; Mahbub, I. A Novel 3-D Printed Headstage and Homecage based WPT System for Long-term Behavior Study of Freely Moving Animals. In Proceedings of the 2020 IEEE Radio and Wireless Symposium (RWS), San Antonio, TX, USA, 26–29 January 2020; pp. 268–271.

33. Maghsoudloo, E.; Gagnon-Turcotte, G.; Rezaei, Z.; Gosselin, B. A Smart Neuroscience Platform with Wireless Power Transmission for Simultaneous Optogenetics and Electrophysiological Recording. In Proceedings of the 2018 IEEE International Symposium on Circuits and Systems (ISCAS), Florence, Italy, 27–30 May 2018; pp. 1–5.

34. Mei, H.; Thackston, K.A.; Bercich, R.A.; Jefferys, J.G.R.; Irazoqui, P.P. Cavity resonator wireless power transfer system for freely moving animal experiments. *IEEE Trans. Biomed. Eng.* **2017**, *64*, 775–785. [CrossRef]

35. Sergkei, K.; Lombard, P.; Semet, V.; Allard, B.; Moguedet, M.; Cabrera, M. The Potential of 3D-MID Technology for Omnidirectional Inductive Wireless Power Transfer. In Proceedings of the 2018 13th International Congress Molded Interconnect Devices (MID), Würzburg, Germany, 25–26 September 2018; pp. 1–6.

36. Ng, W.M.; Zhang, C.; Lin, D.; Hui, S.Y.R. Two-and three-dimensional omnidirectional wireless power transfer. *IEEE Trans. Power Electron.* **2014**, *29*, 4470–4474. [CrossRef]

37. Lee, B.; Koripalli, M.K.; Jia, Y.; Acosta, J.; Sendi, M.S.E.; Choi, Y.; Ghovanloo, M. An implantable peripheral nerve recording and stimulation system for experiments on freely moving animal subjects. *Sci. Rep.* **2018**, *8*, 1–12. [CrossRef]

38. Lee, B.; Jia, Y.; Mirbozorgi, S.A.; Connolly, M.; Tong, X.; Zeng, Z.; Mahmoudi, B.; Ghovanloo, M. An inductively-powered wireless neural recording and stimulation system for freely-behaving animals. *IEEE Trans. Biomed. Circuits Syst.* **2019**, *13*, 413–424. [CrossRef]

39. Ho, J.S. Fully internal, wirelessly powered systems for optogenetics. In Proceedings of the 2016 International Conference on Optical MEMS and Nanophotonics (OMN), Singapore, 31 July–4 August 2016; pp. 1–2.

40. Amar, A.B.; Kouki, A.B.; Cao, H. Power approaches for implantable medical devices. *Sensors* **2015**, *15*, 28889–28914. [CrossRef] [PubMed]

41. Shin, G.; Gomez, A.M.; Hasani, R.A.; Jeong, Y.R.; Kim, J.; Xie, Z.; Banks, A.; Lee, S.M.; Han, S.Y.; Yoo, C.J.; et al. Flexible Near-Field Wireless Optoelectronics as Subdermal Implants for Broad Applications in Optogenetics. *Neuron* **2017**, *93*, 509–521. [CrossRef] [PubMed]

42. Jia, Y.; Mirbozorgi, S.A.; Zhang, P.; Inan, O.T.; Li, W.; Ghovanloo, M. A dual-band wireless power transmission system for evaluating mm-sized implants. *IEEE Trans. Biomed. Circuits Syst.* **2019**, *13*, 595–607. [CrossRef] [PubMed]

43. Park, S.I.; Brenner, D.S.; Shin, G.; Morgan, C.D.; Copits, B.A.; Chung, H.U.; Pullen, M.Y.; Noh, K.N.; Davidson, S.; Oh, S.J.; et al. Soft, Stretchable, Fully Implantable Miniaturized Optoelectronic Systems for Wireless Optogenetics. *Nat. Biotechnol.* **2015**, *33*, 1280–1286. [CrossRef]

44. Piech, D.K.; Johnson, B.C.; Shen, K.; Ghanbari, M.M.; Li, K.Y.; Neely, R.M.; Kay, J.E.; Carmena, J.M.; Maharbiz, M.M.; Muller, R. A wireless millimeter-scale implantable neural stimulator with ultrasonically powered bidirectional communication. *Nat. Biomed. Eng.* **2020**, *4*, 207–222. [CrossRef]

45. Liu, S.; Moncion, C.; Zhang, J.; Balachandar, L.; Kwaku, D.; Riera, J.J.; Volakis, J.L.; Chae, J. Fully Passive Flexible Wireless Neural Recorder for the Acquisition of Neuropotentials from a Rat Model. *ACS Sens.* **2019**, *4*, 3175–3185. [CrossRef]

46. Yang, K.W.; Oh, K.; Ha, S. Challenges in Scaling Down of Free-Floating Implantable Neural Interfaces to Millimeter Scale. *IEEE Access* **2020**, *8*, 133295. [CrossRef]

Publisher's Note: MDPI stays neutral with regard to jurisdictional claims in published maps and institutional affiliations.

Article

A Study on Precise Positioning for an Electric Vehicle Wireless Power Transfer System Using a Ferrite Antenna

Jae Yong Seong and Sang-Sun Lee *

Department of Electronics and Computer Engineering, Hanyang University, Seoul 04763, Korea; jyseong@hyundai.com
* Correspondence: ssnlee@hanyang.ac.kr; Tel.: +82-2-2299-0372

Received: 7 June 2020; Accepted: 7 August 2020; Published: 11 August 2020

Abstract: In the last decade, engineers from automotive manufacturers and charging infrastructure suppliers have widely studied the application of wireless power transfer (WPT) technology to electric vehicles. Since this time, engineers from automotive manufacturers have studied precise positioning methods suitable for WPT using methods such as mechanical, communication-based or video-based. However, due to high costs, electromagnetic interference and environmental factors, the experts of the SAE J2954 was focused on the WPT's precise positioning method by ferrite antennas and low power excitation. In this study, we present how to use the ferrite antennas to find a central alignment point between the primary and secondary units within the alignment tolerance area that requires the minimum power transfer efficiency of the EV WPT system. First, we analyze the ferrite antenna already applied in the automotive and verifies whether it is suitable for the precise positioning of the WPT system for EV. We use modeling and simulation to show that it is necessary to calculate all induced loop voltages in the relationship between incident magnetic field signal strength and induced loop voltage because of the short distance between the transmitter and receiver of the ferrite antenna in WPT. In addition, we also suggest a sequence to find the fitting location of the ferrite antenna, the number of antennas used and the center alignment point. After the simulation is performed on the suggestions, component-level and vehicle-level tests were conducted to verify the validity of the simulation results. As a result, it is shown that a ferrite antenna is suitable as a method for the secondary device to find the center alignment point of the primary device.

Keywords: electric vehicle; wireless power transfer; center alignment point; ferrite antenna

---

1. Introduction

Faraday's experiment on electromagnetic induction and energy transfer was the first experiment to transfer electrical energy wirelessly [1]. Since then, researchers have been interested in wireless power transfer and radio-frequency communication technologies. The concept of wireless power transfer (WPT) was demonstrated by Nikola Tesla in the early 1900 s [2]. However, engineers have experienced difficulties in commercialization due to specific problems with WPT, viz., low efficiency and difficulty in long-distance transmission when compared with conductive power transfer. Therefore, researchers have focused on contact wireless power transfer, which has been commercialized and used in many electronic and electrical devices [3–5]. Long-range WPT technology regained attention in 1964 owing to William C. Brown [6], who successfully supplied power to fuel-free helicopters using 2.45 GHz microwaves. In 2007, owing to the continued research and development by engineers, Professor Marin Soljacic at Massachusetts Institute of Technology (MIT) succeeded in verifying 40% efficiency of WPT technology at a distance of two meters using a coil with a diameter of 60 cm [7].

The application of WPT technology to electric vehicles (EVs) is widely studied. In 2010, the United States (U.S.) Department of Energy (DOE) and the Society of Automotive Engineers (SAE) led the wireless power transfer and alignment task force and began research and standardization of WPT technology [8]. Researchers from major automotive manufacturers and charging infrastructure suppliers [9] wished to develop WPT technology suitable for EVs while participating in the EV WPT standardization. The primary goal of the development of WPT for the EVs in its early stages was to ensure the safety of users during the implementation of WPT and maximizing the charging efficiency of wireless charging. Therefore, the coil shape or electric power circuit was studied [10].

Further, engineers from automotive manufacturers began to study fine and precise positioning methods suitable for WPT in EVs and various indoor and outdoor positioning technologies. The contents of the discussion are as follows [11]: (i) As a mechanical method, stopping a vehicle in the center of a primary device using a curbside block or parking block was considered. Further, a method of using a robotic arm to move the primary device to the center of the secondary device after the vehicle is parked was considered. (ii) As a communication-based method [12–16], technologies such as global positioning system (GPS), Bluetooth low energy (BLE), radio frequency identification (RFID), Wi-Fi and ultra-wideband (UWB) were discussed. (iii) As a video-based method [17,18], a parking assistant system (PAS) applied to a vehicle, 2D/3D marker notified to a user by a camera installed in a parking lot and optical character recognition (OCR) was mentioned. However, the mechanical methods were excluded from the discussion owing to an increase in the production cost of the WPT manufacturer, and the communication-based methods were excluded because it is difficult to satisfy the vehicle electromagnetic compatibility (EMC) standards regulated by the International Telecommunication Union (ITU) [19]. The video-based method was found to be difficult to apply to external public parking lots due to weather or environmental factors. Therefore, the experts of the SAE J2954 task force is focused on technologies that could easily be mounted on an EV, was inexpensive, did not interfere with an electronic component in the vehicle, and, satisfied the conditions for positioning within the alignment tolerance range for the WPT [20]. It also focuses on technology that satisfies the fine positioning condition, where the central alignment distance between the primary and secondary devices is approximately 1.5 m or more and the precise positioning condition, where the primary and secondary devices begin to overlap. Among them, low power excitation (LPE) is a technology whose primary device, i.e., the power transfer device, performs fine and precise positioning by transmitting a minute quantity of power to the secondary device. Another method is to mount a ferrite antenna using a low frequency (LF) in primary device or a secondary device and perform fine and precise positioning using the magnetic field change value of the ferrite antenna. Therefore, LPE and ferrite antennas were applied to the SAE J2954 standard as a method for fine and precise positioning in an EV WPT system [20]. In addition, the International Electrotechnical Commission (IEC), an international standards and conformity assessment body, also addresses LPE and ferrite antennas as a method to fine and precise positioning in the 61980-2 document [21].

This article describes how to find the central alignment point between the primary device and secondary device within the alignment tolerance area that requires the minimum power transfer efficiency of the EV WPT system using the ferrite antenna. This method suggests that it is necessary to calculate all induced loop voltages in the relationship between the incident magnetic field signal strength and the induced loop voltage because of the distance between the transmitter and receiver of the ferrite antenna in EV WPT precise positioning is short—to within 250 mm. It also suggests a sequence to find the fitting location of the ferrite antenna, the number of antennas used and the center alignment point. After the simulation is performed on the suggestions, unit-level, component-level and vehicle-level tests are performed to validate the simulation results. Therefore, we propose that the ferrite antenna was suitable for the precise positioning of EV WPT.

The content is organized as follows: In Section 2, we review the SAE J2954 document, the magnetic flux density of ferrite antenna and open-circuit voltage of a ferrite antenna. In Section 3, we verify that ferrite antennas already applied in the automotive field are suitable for use in the precise positioning of WPT systems for EVs. Then, we simulate and validate the performance of the ferrite antenna as a transmitter and receiver as a method of finding the central alignment point within the alignment tolerance area of the EV WPT system. In Section 4, we extend the magnetic flux density and open-circuit voltage models of a ferrite antenna to the EV WPT system and conducts simulations. In Section 5, we present the results of performed experiments at the component-level and vehicle-level to verify the validity of modeling and simulation results and compare them with simulation results. We present conclusions in Section 6. In Section 7, we describe registered patents. Finally, in Appendix A, we describe the magnetic flux density, the open-circuit voltage applied to the geometric dimensions of the WPT system for EV and the power received by the ferrite antenna of the primary device.

2. Background

2.1. SAE J2954 Standard

SAE J2954 covers WPT for light-duty electric and plug-in EVs [8,20]. The scope of SAE J2954 is the requirements of the WPT system, such as interoperability, electromagnetic compatibility (EMC), electromagnetic force (EMF), minimum performance, safety, alignment and testing. Of particular importance among these requirements, the minimum target efficiency for each class of wireless power transfer should exceed 85% at the center alignment point and 80% at the alignment tolerance area (e.g., WPT1 power class is 3.7 kW, WPT2 power class is 7.7 kW and WPT3 power class is 11.1 kW) [20]. The maximum wireless power transfer capacity is determined by the lower of the primary device and secondary device power class ratings. The WPT power classes for light-duty EVs defined in SAE J2954 are described in Table 1 [20].

Table 1. Wireless power transfer (WPT) power classification with target efficiency for light-duty electric vehicles [20].

WPT Power Class	Input Power of the Primary Device	Output Power of the Secondary Device	Target Efficiency	
			Center Point	Tolerance Area
WPT 1	≤3.7 kW	≤3.7 kW	≤85%	≤80%
WPT 2	≤7.7 kW	>3.7 kW and ≤7.7 kW	≤85%	≤80%
WPT 3	≤11.1 kW	>7.7 kW and ≤11.1 kW	≤85%	≤80%

The center alignment point is the point at which the geometry centers of the primary and secondary devices are correctly aligned with each other. The alignment tolerance area is the offset required between the primary device and secondary device geometric centers to achieve the center position for the WPT, and the maximum positioning deviation is ±75 mm for the x-direction, ±100 mm for the y-direction. The x-direction of the alignment tolerance is positive in the rearward vehicle direction, the y-direction of the alignment tolerance is positive toward the right-hand-side of the vehicle.

The first step of positioning is to find the WPT charger for vehicles with WPT or the WPT charger for vehicles with WPT. Here, although a few communication problems indoors and outdoors are observed, Wi-Fi is widely used. This is called WPT charging-spot discovery. The second step is to position the vehicle in the parking spot for the primary and secondary devices to overlap. This is called fine positioning. The last step is for the secondary device to find the center alignment point of the primary device after the primary and the secondary devices overlap. This is called precise positioning. It is important for the positioning device not to affect the charging efficiency and not interfere with existing installed electrical components during the wireless charging of the vehicle. Therefore, IEC61980-2 and SAE J2954 selected two types of alignment methodologies [20,21] that can maximize the wireless power transfer efficiency between the EV device and the supply device. First, magnetic field alignment using the existing coil is that the primary device provides a small magnetic field, which can be detected by the secondary device and used as a method of aligning the EV. This is called low power excitation (LPE). Second, magnetic field alignment using an auxiliary coil is that the magnetic field should be transmitted or received by a separate magnetic coil system which is generally not the same magnetic assembly used to WPT. Since the separate magnetic coil system should not use the resonant frequency used in the WPT system and should not affect the WPT system, an auxiliary means such as a ferrite antenna is used.

2.2. Magnetic Flux Density

In order to determine the magnetic flux density of a ferrite antenna, it is necessary to consider the steady magnetic field in free space. When a current I is flowing in a small circular loop of radius a, to define the magnetic dipole at observation point $r(x, y, z)$, consider that an infinitesimal current creates a magnetic dipole magnetic field [22]. The magnetic entities are magnetized because there are many infinitesimal currents inside the material. The model of the small circular loop with a current I is shown in Figure 1 where $r'(x', y', 0)$ is a point with an electric current that is the source of the magnetic field and dl is the infinitesimal current at r'. The center of the small circular loop is selected to be the origin of the spherical coordinates. Here, it is necessary to consider the condition that the distance R of the observation point at the center of the circular loop is always greater than the radius a of the circular loop [22].

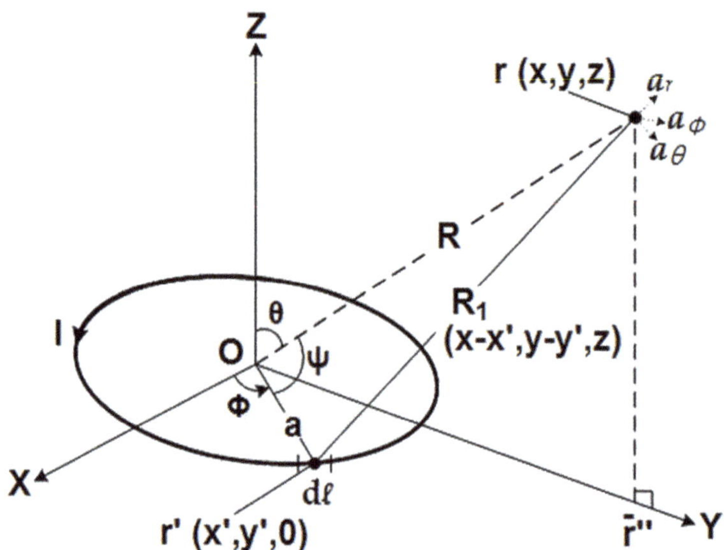

Figure 1. Model of the small circular loop with current I.

Equation (1) describes the magnetic vector potential.

$$\mathbf{A} = \frac{\mu_0 I}{4\pi} \oint_{C'} \frac{d\mathbf{l}'}{R_1} \tag{1}$$

where μ_0 is the permeability of free space, I is the current and path C' represents the line integral.

From Equation (1), spherical coordinates of the magnetic vector potential for observation r in Figure 1 is given by:

$$\mathbf{A} = \frac{\mu_0 I a^2 \sin\theta}{4R^2} a_\phi \tag{2}$$

where a_ϕ is the unit vector in spherical coordinates at observation point r.

Using the concept of the moment as an indicator of the tendency of the magnetic dipole to change and substituting Equation (2) is given by [22]:

$$\mathbf{A} = \frac{\mu_0 \mathbf{m} \times \mathbf{a}_r}{4\pi R^2} \tag{3}$$

where **m** is the magnetic dipole moment and $\mathbf{a}_r$ is the unit vector in spherical coordinates at observation point r. The magnetic dipole moment is a vector and its magnitude is the product of the current and the area of the loop. Therefore, the magnetic flux density produced by the magnetic dipole is given by [22]:

$$\mathbf{B} = \nabla \times \mathbf{A} = \frac{\mu_0 m}{4\pi R^3}(2cos\theta\mathbf{a}_r + sin\theta\mathbf{a}_\theta) \tag{4}$$

where $\mathbf{a}_\theta$ is the unit vector in spherical coordinates at observation point r.

Therefore, the magnetic flux density of a uniformly magnetized ferrite antenna is determined as follows: A model of the ferrite antenna is shown in Figure 2, where L is the length of the antenna, a is the radius and $\mathbf{M} = \mathbf{a}_z M_0$ is the uniform magnetization along z-axis [22].

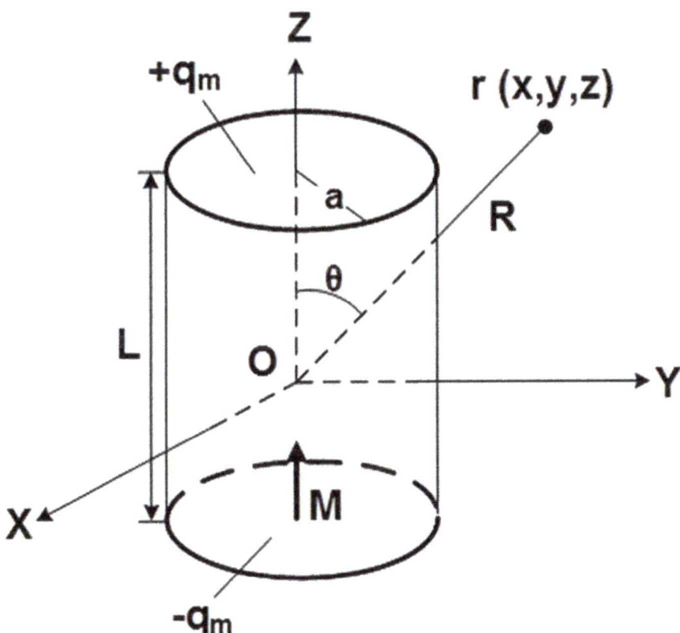

Figure 2. Model of the ferrite antenna.

Using the concept of equivalent magnetic charge density to determine the magnetic flux density of the ferrite antenna, the magnetic flux density of the ferrite antenna obtains the same value as Equation (4) [22]. Converting the magnetic flux density expressed in the spherical coordinate system of Equation (4) into the Cartesian coordinate system gives:

$$\begin{bmatrix} a_x \\ a_y \\ a_z \end{bmatrix} = \begin{bmatrix} sin\theta cos\varnothing & cos\theta cos\varnothing & -sin\varnothing \\ sin\theta sin\varnothing & cos\theta sin\varnothing & cos\varnothing \\ cos\theta & -sin\theta & 0 \end{bmatrix} \begin{bmatrix} a_r \\ a_\theta \\ a_\varnothing \end{bmatrix} \tag{5}$$

Substituting Equation (5) into Equation (4), the magnetic flux density is given by:

$$\mathbf{B} = \frac{\mu_0\mu_r M_T}{4\pi R^3}\left\{(2cos\theta sin\theta cos\varnothing + sin\theta cos\theta cos\varnothing)\boldsymbol{a_x} + (2cos\theta sin\theta sin\varnothing + sin\theta cos\theta sin\varnothing)\boldsymbol{a_y} + \left(2cos\theta^2 - sin\theta^2\right)\boldsymbol{a_z}\right\} \quad (6)$$

where μ_r is the relative permeability of the ferrite antenna and $M_T = \pi a^2 L M_0$ is the total dipole moment of the ferrite antenna.

Here, the ferrite antenna mounted in the EV WPT system only considers the a_z component. Therefore, Equation (6) can be summarized as:

$$\mathbf{B} = \frac{\mu_0\mu_r M_T}{4\pi R^3}\left(2cos\theta^2 - sin\theta^2\right)\boldsymbol{a_z} \quad (7)$$

2.3. Open-Circuit Voltage

By Faraday's law, the voltage induced in an electrically small loop antenna is equal to the rate of change of the magnetic flux. In addition, when the loop antenna is composed of multiple turns, the induced voltage of each turn is in series with all other turns [23]. Therefore, the open-circuit voltage is given by:

$$V_{oc} = N\frac{d\Phi}{dt} \quad (8)$$

where V_{oc} is the open-circuit voltage, N is the number of turns, Φ is magnetic flux.

If the loop is small compared with a wavelength, magnetic flux may be assumed to be constant throughout the loop area at any instant of time. The relationship between magnetic flux and magnetic flux density is given by [23]:

$$\Phi = \mathbf{B}Acos\theta \quad (9)$$

where A is the loop area, $\mathbf{B}$ is magnetic flux density and θ is the angle between the plane of the loop axis and the incoming flux.

An electromagnetic radiation field contains electric and magnetic field component. The field is given by [22]:

$$\mathbf{E} = c\mathbf{B} \quad (10)$$

where $\mathbf{E}$ is the electric field and c is the speed of light.

Solving Equation (10) for magnetic flux density and substituting it into Equation (9) allows us to express the open-circuit voltage induced in a loop as a function of the electric field strength of the incoming electromagnetic signal [23].

$$V_{oc} = \frac{NA}{c}\frac{dE}{dt}cos\theta \quad (11)$$

The electric and magnetic field strength varies with time in a sinusoidal form. This study considers the peak magnitude of V_{oc}. Therefore, dE/dt is simply $\mathbf{E}2\pi f$. Equation (11) can be summarized as:

$$V_{oc} = \frac{NAE2\pi f}{c}cos\theta = j\omega NA\mathbf{B}cos\theta \quad (12)$$

where ω is the angular frequency of θ. When the loop plane is 90° to the magnetic field, $cos\theta$ is 0 and V_{oc} is zero.

The basic theory of the ferrite antenna is based on an electrically small loop antenna [24,25]. The ferrite antenna uses a ferrite rod in the loop to increase the radiation resistance, which results in better antenna efficiency, without increasing the physical size compared to a loop antenna with an air core.

From Equation (12), the open-circuit voltage of the ferrite antenna is given by:

$$V_{oc} = j\omega N A \mu_{eff} B_z^i \tag{13}$$

where μ_{eff} is the effective permeability of the ferrite antenna and B_z^i is the z-component of the incident magnetic flux density of the ferrite antenna.

The permeability of the ferrite antenna is characterized by a combination of ferrite material permeability, the shape of the ferrite antenna and the dimensions of the ferrite antenna [26]. The relative permeability of the ferrite antenna is given by:

$$\mu_r = \mu_{eff} \sqrt[3]{\frac{l_r}{l_c}} \tag{14}$$

where μ_r is the relative permeability of the ferrite antenna, μ_{eff} is the effective permeability of the ferrite antenna, l_r is the length of the ferrite antenna and l_c is the length of the ferrite antenna's coil.

The open-circuit voltage of a single-turn loop in the middle of the ferrite antenna is increased by the factor μ_r for the value of the same loop in free space.

3. Analysis and Modeling to Use Ferrite Antenna

3.1. Analysis of Ferrite Antenna Used in the Automotive

As mentioned in SAE J2954, the ferrite antenna used for the precise positioning of the WPT system for EV should not electromagnetically interfere with EV during precise positioning and should not affect the charging efficiency during the WPT. In addition, for the precise positioning of the EV WPT system, the ferrite antenna transmitter should transmit a strong magnetic field within a range that satisfies the ITU regulation suitable for automotive [19] and the ferrite antenna receiver should accurately detect the magnetic field received by the ferrite antenna transmitter. Therefore, it was examined that the ferrite antenna, which was verified and used in an automotive, was suitable for the precise positioning of the WPT for an EV.

Figure 3 shows a ferrite antenna that was applied and used in the automotive. In the automotive, the ferrite antenna is used in two systems. One is a keyless entry system (see Figure 3a) and the other is a smart key system (see Figure 3b). The ferrite antenna used in the keyless entry system had a length of 2.5 mm, a width of 10 mm and a thickness of 10 mm. Two hundred turns of copper wire were wound around the width and thickness, and 91 turns were wound around the length. The ferrite antenna used in the smart key system had a length of 90 mm, a width of 7 mm, a thickness of 4 mm and 70 turns of copper wire. Both of the ferrite antennas demonstrated a relative magnetic permeability of 150, an operating frequency of 125 kHz, and did not electromagnetically interference with the EV. Therefore, to confirm that the ferrite antennas shown in Figure 3 are suitable for the precise positioning of the WPT system for EV, the magnetic flux density in the near-field was verified. We used FEKO from the Altair software package [27], simulation software widely used in industry and academia.

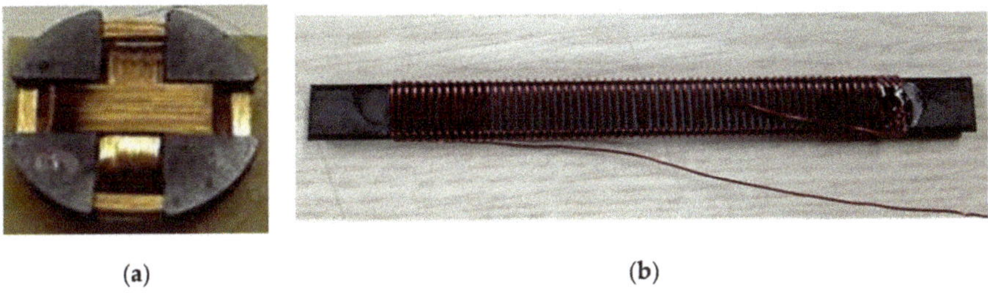

(a)   (b)

Figure 3. Ferrite antennas used in the automotive field. (a) Keyless entry system; (b) smart key system.

Figure 4 shows the design for simulation with FEKO software. All of the constants applied in the simulation were applied with all of the above-mentioned values. The result of the simulation of the ferrite antenna is shown in Figure 5. The ferrite antenna used in the keyless entry system has a magnetic flux density distance of 1 m and the ferrite antenna used in the smart key system has a magnetic flux density distance of 1.6 m. Precise positioning in the WPT system for EV requires that the secondary device mounted on the EV locates and aligns the primary device because the primary device is installed to the parking spot. Therefore, the ferrite antenna used in the smart key system is more suitable for precise positioning of the EV WPT systems than the ferrite antenna used in the keyless entry systems.

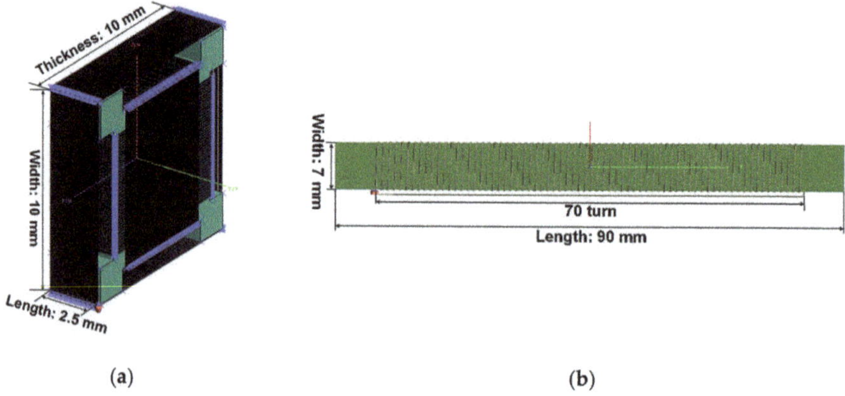

(a)　　　　　　　　　　(b)

Figure 4. Geometric dimension of ferrite antenna used in automotive field. (**a**) Keyless entry system; (**b**) smart key system.

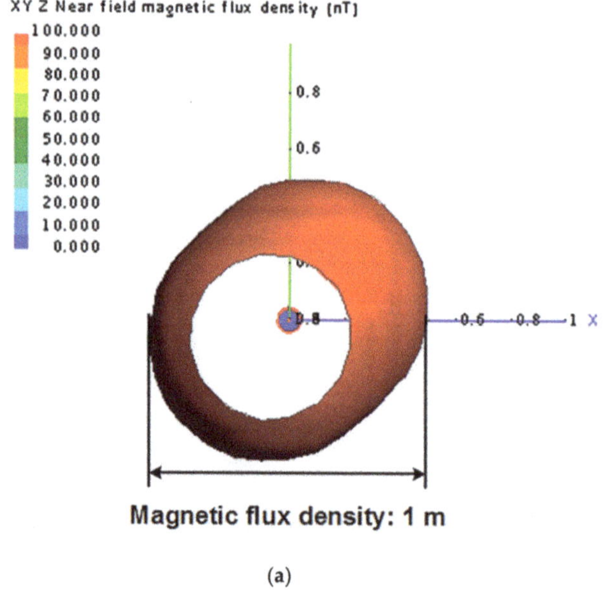

(a)

Figure 5. *Cont.*

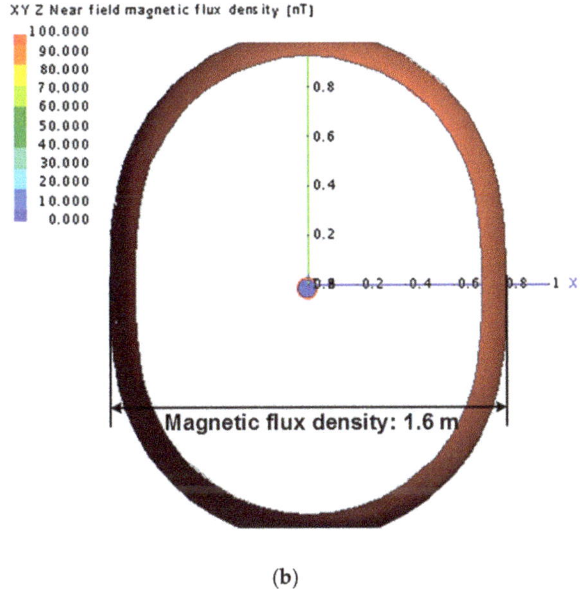

(**b**)

Figure 5. Near-field magnetic flux density of ferrite antenna used in automotive field. (**a**) Keyless entry system; (**b**) smart key system.

The ferrite antenna shown in Figure 6 was simulated using FEKO software to verify that the charging efficiency of the WPT system for EV was affected. The geometric dimensions of the WPT system for EV were based on the WPT1 power class of SAE J2954 [20]. Where the ferrite antenna can be mounted in the WPT system for EV is the corner of the primary and secondary devices or the center of each side. Figure 6 shows the ferrite antenna arranged on the WPT system for EV.

Figure 7 shows the simulation results when the ferrite antenna used in the keyless entry system and smart key system are mounted on each corner of the WPT system for EV. In the ferrite antenna used in the keyless entry system, the coils wound in three directions can be observed to affect the WPT system of EV (see Figure 7a). It was confirmed that the ferrite antenna with the coil wound on the ferrite rod in one direction did not affect the EV WPT system (see Figure 7b).

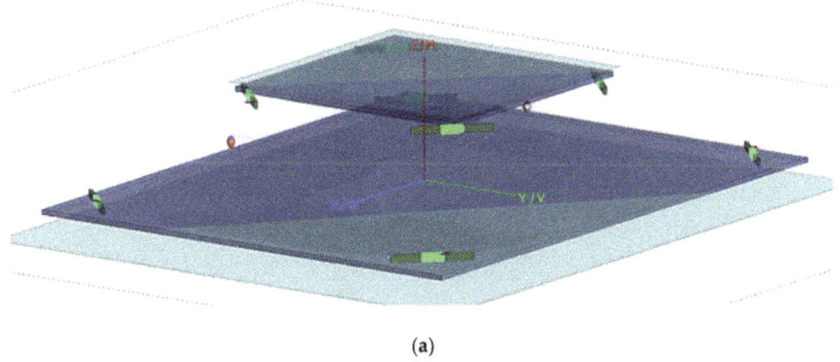

(**a**)

Figure 6. *Cont.*

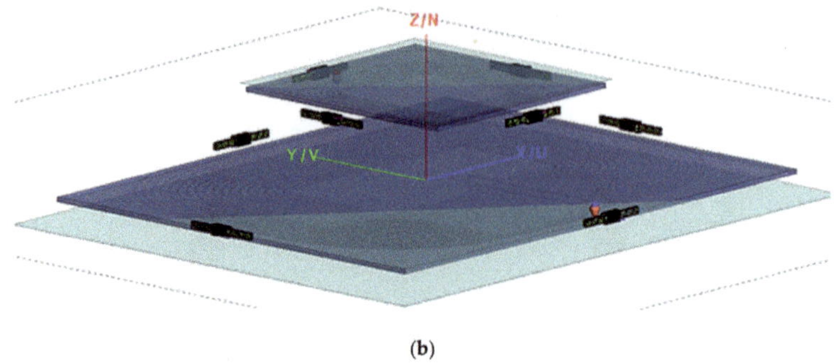

(b)

Figure 6. Ferrite antenna mounted on WPT system for EV. (**a**) Mounted on each corner; (**b**) mounted on the center of each side.

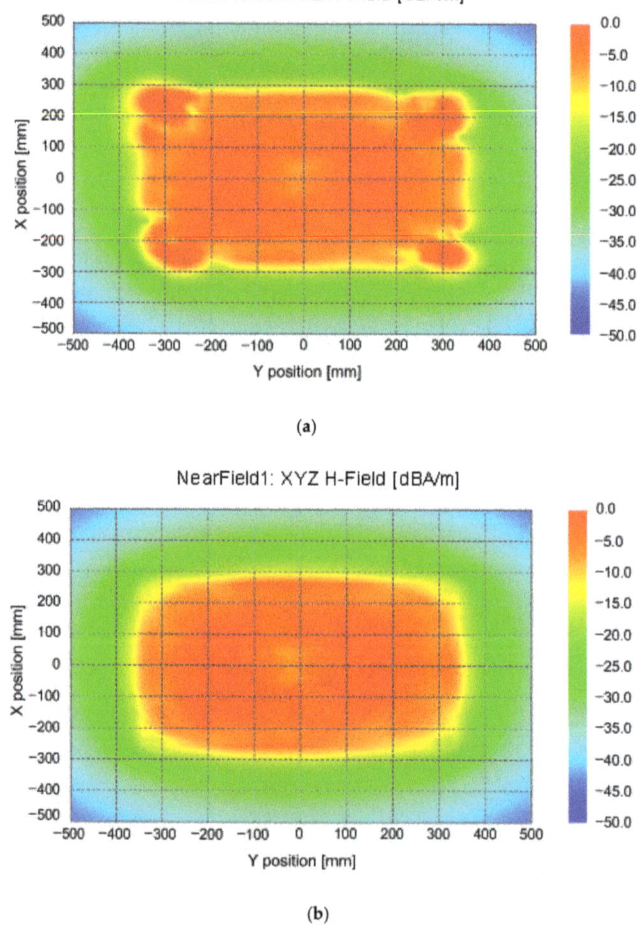

(a)

(b)

Figure 7. Simulation results when the ferrite antenna is mounted on each corner of the WPT system for EV. (**a**) Keyless entry system; (**b**) smart key system.

Figure 8 shows the simulation results when the ferrite antenna used in the keyless entry system and smart key system are mounted on the center of each side of the WPT system for EV. In the ferrite antenna applied in the keyless entry systems, the coils wound in three directions can be observed to affect the WPT system for EV (see Figure 8a). It was confirmed that the ferrite antenna with the coil wound on the ferrite rod in one direction did not affect the WPT system for EVs (see Figure 8b).

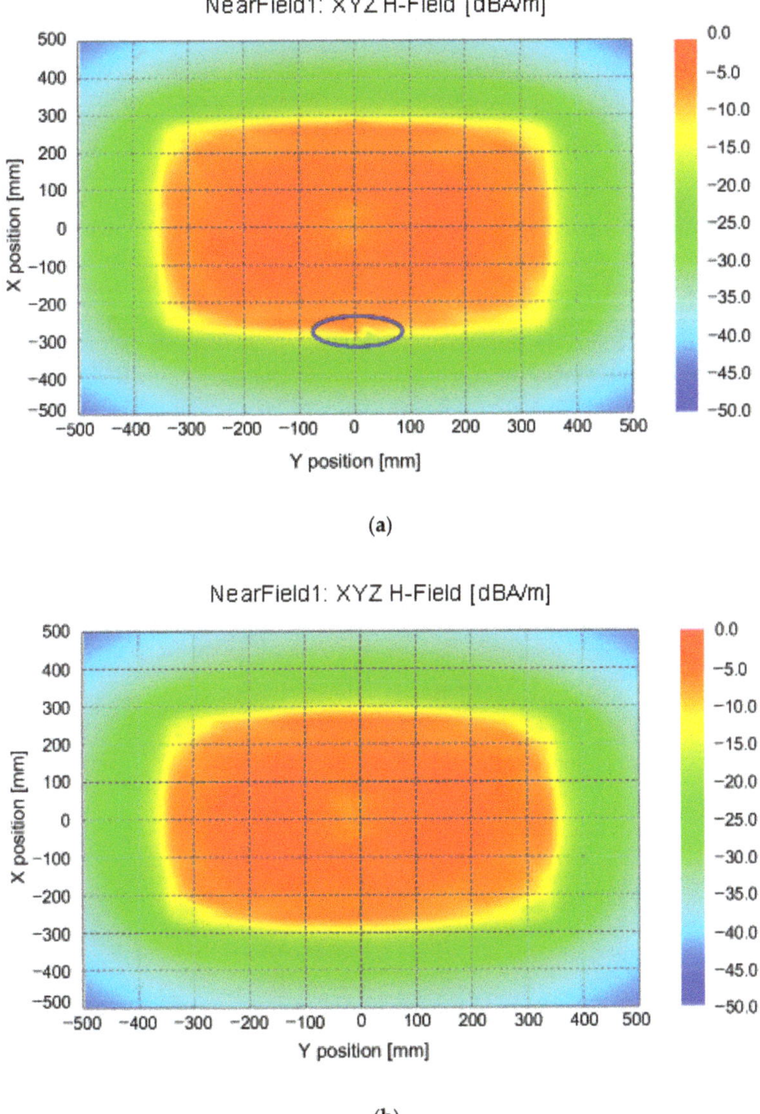

(a)

(b)

Figure 8. Simulation results when the ferrite antenna is mounted on the center of each side of the WPT system for EV. (**a**) Keyless entry system; (**b**) smart key system.

Based on the results shown in Figures 7 and 8, the ferrite antenna, which affects the charging efficiency when the WPT system for the EV is wireless charging, is a coil wound to the ferrite rod in three directions. The magnetic vector when the primary device supplies power to the secondary device that of the ferrite antenna wound in one direction on the ferrite rod is in the same direction. Hence, the ferrite antenna wound in one direction on the ferrite rod does not affect the wireless charging efficiency. However, the magnetic vector when the primary device supplies power to the secondary device and that of the ferrite antenna wound in three directions on the ferrite rod is in a different direction. Hence, the ferrite antenna wound in three directions on the ferrite rod is observed to affect the wireless charging efficiency.

3.2. Ferrite Antenna Modeling for Application to the Precise Positioning of the EV WPT System

In order to use the ferrite antenna for the precise positioning of the WPT system for EV, it is necessary to verify whether it is suitable as a transmitter and receiver even at a short distance. The reason is that the ferrite antenna, generally used to a vehicle, operates as a transmitter and receiver at a distance of 1 m or more. Equations (7) and (13) are the result of one ferrite antenna. In the WPT system for EV, the ferrite antenna serves as both a transmitter and a receiver and consists of a pair. Therefore, we constructed modeling for simulation and conducted unit-level tests. Figure 9 shows the geometric dimensions of the ferrite antenna used for verification and the relationship between the transmitter and receiver of the ferrite antenna. Figure 10 is an yz-plane view for calculating the magnetic flux density and open-circuit voltage of the ferrite antenna.

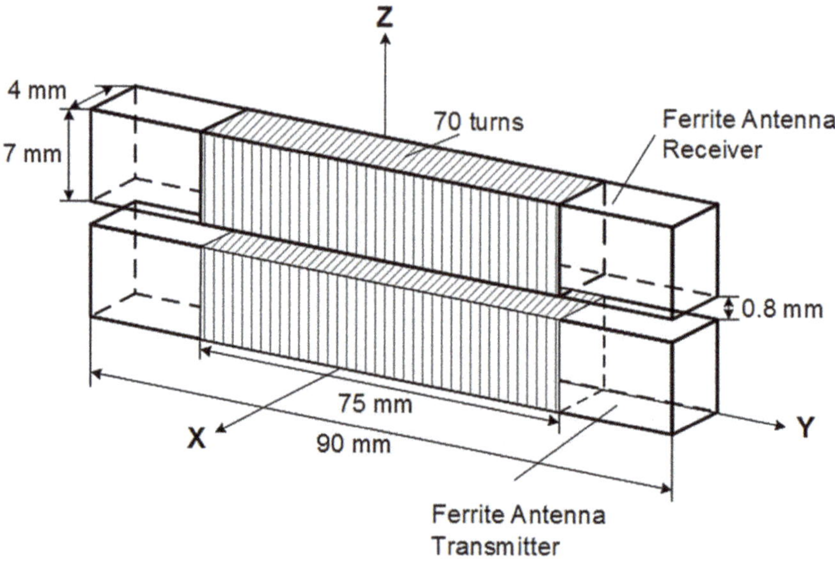

Figure 9. Geometric dimensions of the transmitter and receiver of the ferrite antenna.

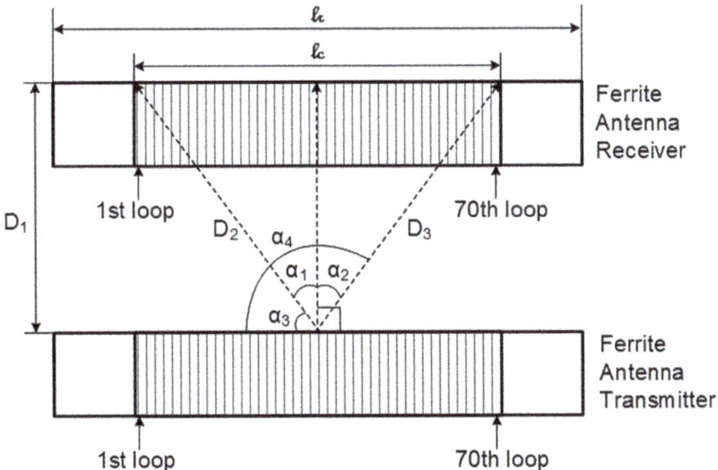

Figure 10. YZ-plane view of the transmitter and receiver of the ferrite antenna.

In Figure 10, α_1 is the angle between the Nth loop of the ferrite antenna transmitter and the 1st loop of the ferrite antenna receiver, α_2 is the angle between the Nth loop of the ferrite antenna transmitter and the 2nd to 70th loops of the ferrite antenna receiver, α_3 is the 90 degrees minus α_1, α_4 is the sum of α_1, α_2 and α_3, D_1 is the vertical distance between the transmitter and receiver of the ferrite antenna, D_2 is the distance between the Nth loop of the ferrite antenna transmitter and the 1st loop of the ferrite antenna receiver, D_3 is the distance between the Nth loop of the ferrite antenna transmitter and the 2nd to 70th loops of the ferrite antenna receiver, l_c is the length of the ferrite antenna's coil and l_r is the length of the ferrite antenna.

As mentioned previously, when the loop antenna consists of multiple turns, the induced voltage of each turn is in series with all other turns [22]. Therefore, the open-circuit voltage of a ferrite antenna with 70 turns is the sum of all the open-circuit voltages all the 70 turns. If the magnetic flux density of the ferrite antenna shown in Figure 10 is expressed in the form of Equation (7), it is given by:

$$\mathbf{B}_{tot} = \frac{\mu_0 \mu_r m}{4\pi r^3}\left(2\cos\theta^2 - \sin\theta^2\right)a_z$$

$$\theta = \sum_{k=\alpha_3}^{\alpha_4}\left(\frac{\alpha_1+\alpha_2}{69}\right)k + \sum_{N=1}^{70}\frac{\alpha_1+\alpha_2}{69}(1-N)$$

$$\alpha_1 = cos^{-1}\frac{D_1}{\sum_{N=1}^{70}\sqrt{((1-N)D_1)^2+D_1^2}},$$

$$\alpha_2 = cos^{-1}\frac{D_1}{\sum_{N=1}^{70}\sqrt{(l_c+(1-N)D_1)^2+D_1^2}},$$

$$r = \sqrt{\left(\sum_{N=1}^{70}(1-N)D_1 - \frac{69D_1}{\alpha_1+\alpha_2}\left(\sum_{k=\alpha_3}^{\alpha_4}\frac{\alpha_1+\alpha_2}{69} - \alpha_3\right)\right)^2 + D_1^2}$$

$$\tag{15}$$

where $\mathbf{B}_{tot}$ is the sum of the magnetic flux density transmitted by the 70-turn loop of the ferrite antenna transmitter to the 70-turn loop of the receiver. The magnetic dipole moment of the ferrite antenna (m) is 4.76×10^{-13} A·m^2, the relative permeability of the ferrite antenna (μ_r) is 150, the operating frequency of the ferrite antenna is 125 kHz.

In addition, the open-circuit voltage that the ferrite antenna receiver receives from the transmitter is given by:

$$V_{octot} = j\omega\mu_r A\mathbf{B}_{tot}a_z \tag{16}$$

where V_{octot} is the sum of the open-circuit voltage that the ferrite antenna receiver receives from the transmitter.

As the value measured in the measurement instrument is expressed in dBm, Equation (16) can be written as:

$$Power_{ferrite\ antenna} = 10log_{10}\left(1000\frac{V_{octot}^2}{R}\right) \tag{17}$$

where R is the resistance of the ferrite antenna and the value of resistance is 1.

A Ferrite antenna model results were calculated using MATLAB software from MathWorks [28]. A ROHDE and SCHWARZ SMC100A signal generator [29] was used as the instrument for generating a magnetic flux density in the ferrite antenna of the transmitter. The instrument for measuring the magnetic flux density of the ferrite antenna of the receiver was an RTE 1104 oscilloscope [30] from ROHDE and SCHWARZ. The characteristic impedance of the port connected to the signal generator to perform the test is 50 ohms. Figure 11 shows the ferrite antenna used for the unit-level measurement and the measurement results. The ferrite antenna (see Figure 11a) used for unit-level measurement has the same geometric dimensions and constants used for modeling. The measurement results when x-direction is 0 mm, y-direction is 0 mm and z-direction is 8 mm are shown in Figure 11b. When the ferrite antenna for the transmitter and the ferrite antenna for the receiver are placed in parallel in the z-direction, the measured value of the induced open-circuit voltage received by the ferrite antenna for the receiver is −13.61 dBm.

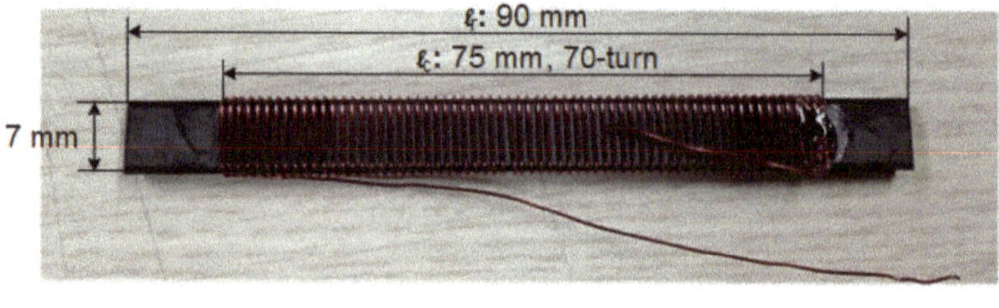

(a)

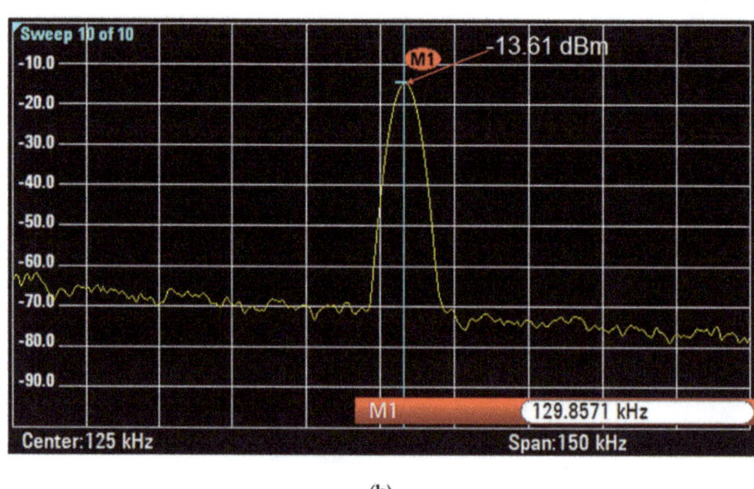

(b)

Figure 11. Experimental test of unit-level. (**a**) Ferrite antenna used for unit-level test; (**b**) experimental result is that x-direction is 0 mm, y-direction is 0 mm and z-direction is 8 mm.

Table 2 shows the comparison between the simulation and experimental results calculated by applying Equation (17) at the boundary of the alignment tolerance range of x-, y- and z-direction. It can be seen that the simulation and the unit-level experimental results are similar even when the transmitter and receiver of the ferrite antenna are matched, in addition to each alignment tolerance range boundary line.

Table 2. Comparison of modeling and unit-level experimental results in the alignment tolerance area.

Distance			Unit Vector z-Direction [2]	Simulation Result	Test Result
x-Direction	y-Direction	z-Direction			
0 mm	0 mm	8 mm [1]	Parallel	−13.21 dBm	−13.61 dBm
0 mm	0 mm	8 mm [1]	Parallel	−13.21 dBm	−13.61 dBm
0 mm	0 mm	8 mm [1]	Parallel	−13.21 dBm	−13.61 dBm
0 mm	0 mm	8 mm [1]	Parallel	−13.21 dBm	−13.61 dBm
0 mm	0 mm	8 mm [1]	Parallel	−13.21 dBm	−13.61 dBm
0 mm	0 mm	8 mm [1]	Parallel	−13.21 dBm	−13.61 dBm
0 mm	0 mm	8 mm [1]	Parallel	−13.21 dBm	−13.61 dBm
0 mm	0 mm	8 mm [1]	Parallel	−13.21 dBm	−13.61 dBm

[1] enclosure thickness of each transmitter and receiver of the ferrite antenna is 4 mm, [2] magnetic flux density unit vector z-direction relationship between the ferrite antenna transmitter and receiver.

4. Simulation and Analysis for Component-Level of EV WPT System

In this section, we discuss extending the magnetic flux density and open-circuit voltage models of the ferrite antenna to the EV WPT system.

In the SAE J2954, when an EV mounted with WPT system parks in a parking spot for charging, front parking is recommended [20]. The geometric center of the primary device shall be installed at the center of the width of the parking spot and 2 m from the inner of the line in front of the parking spot, as shown in Figure 12. Therefore, considering the location of the primary device installed at the parking point, the mounting of the ferrite antenna is optimal in the direction ±y of the primary device and the secondary device. In addition, the minimum number of ferrite antennas to be mounted on the primary device and the secondary device is two for accurate direction recognition [31].

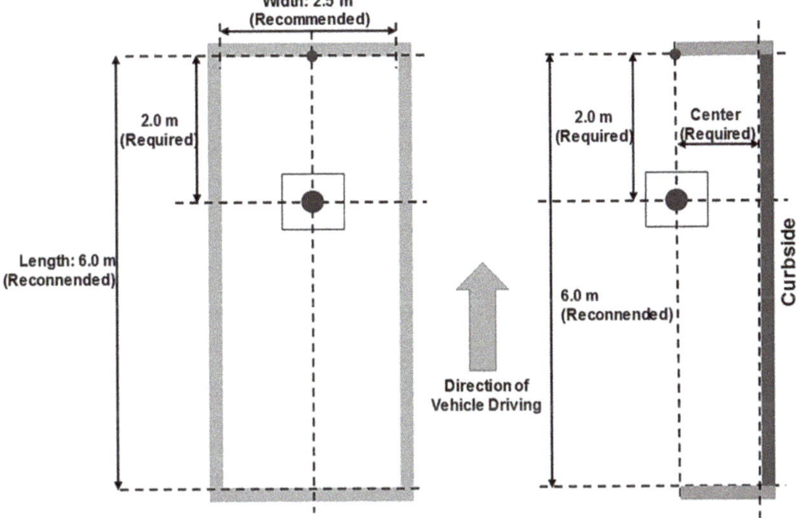

Figure 12. Location of the geometric center of the primary device at the parking spot [20].

Figure 13 shows the secondary device to find the center alignment point of the primary device within the alignment tolerance area of the EV WPT system when the ferrite antenna arrangement is optimized for the primary device and the secondary device. Figure 14 shows the geometric dimensions and the xy-plane view for applying the simulation model.

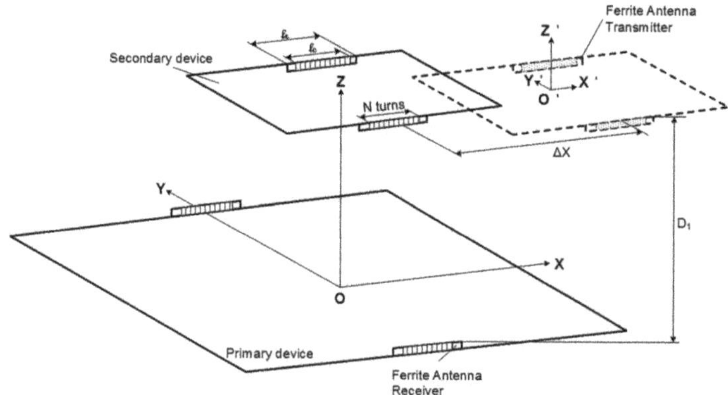

Figure 13. Diagram for finding center alignment point in the electric vehicle (EV) wireless power transfer (WPT) system.

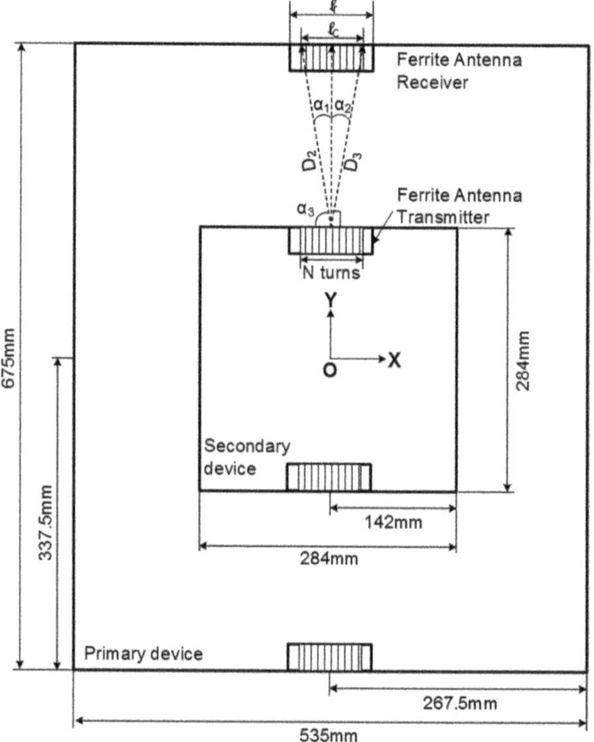

Figure 14. XY-plane view of the ferrite antenna transmitter in the ±y-direction mounted on the secondary device and the ferrite antenna receiver in the ±y-direction mounted on the primary device.

In Figure 14, α_1 is the angle between the Nth loop of the ferrite antenna transmitter in the ±y-direction mounted on the secondary device and the 1st loop of the ferrite antenna receiver in the ±y-direction mounted on the primary device, α_2 is the angle between the Nth loop of the ferrite antenna transmitter in the ±y-direction mounted on the secondary device and the 2nd to 70th loops of the ferrite antenna receiver in the ±y-direction mounted on the primary device, D_2 is the angle between the Nth loop of the ferrite antenna transmitter in the ±y-direction mounted on the secondary device and the 1st loop of the ferrite antenna receiver in the ±y-direction mounted on the primary device, D_3 is the angle between the Nth loop of the ferrite antenna transmitter in the ±y-direction mounted on the secondary device and the 2nd to 70th loops of the ferrite antenna receiver in the ±y-direction mounted on the primary device, and D_1 is the vertical distance between the transmitter and receiver of the ferrite antenna. The application of the geometric dimensions of Figure 14 to Equations (15) and (16) are detailed in Appendix A.

In order to find the center alignment point of the primary device within the alignment tolerance area, the secondary device should check the WPT efficiency at a point 20 mm from the ±x- and ±y-direction. Figure 15 shows 99 test points to find the center alignment point of the primary device within the alignment tolerance area. In Figure 15, the gray area is the primary device. The size of the primary device is 675 mm × 535 mm. The white area is the secondary device. The size of the secondary device is 284 mm × 284 mm. The distance between each test point is 20 mm.

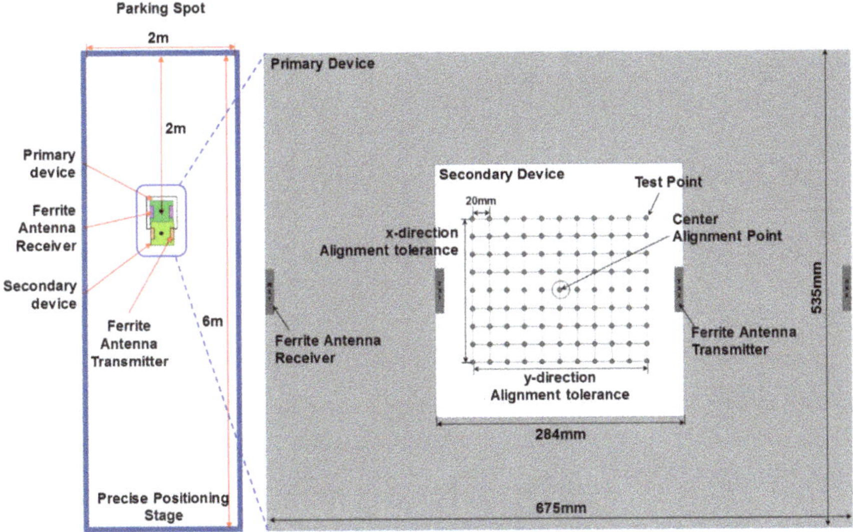

Figure 15. Test point for checking minimum wireless power transfer efficiency in precise positioning.

The sequence for finding the center alignment point of the primary device and secondary device using the ferrite antenna within the alignment tolerance area of the EV WPT system is proposed as follows: (i) The primary device's ferrite antenna receiver is received magnetic flux density from the secondary device's ferrite antenna transmitter; (ii) The primary device is compared to the received magnetic flux density in the +y- and −y-directions; (iii) If the magnetic flux density value in the +y-direction is greater than in the −y-direction, the secondary device is moved to the left. If the magnetic flux density value in the +y-direction is smaller than in the −y-direction, the secondary device is moved to the right. (iv) If the magnetic flux density value in the +y- and −y-direction was equaled, the secondary device is at the center of the width of the parking spot. (v) If the magnetic flux density value was equaled and has decreased after an increase, the secondary device is moved to the backward because it has already passed the center alignment point of the primary device. (vi) If the magnetic flux density value was equaled, but has not decreased after an increase, the secondary device is moved to the forward because the secondary device has not reached the center alignment point of the primary device. (vii) If the magnetic flux density value in the +y- and −y-direction is the maximum, the secondary device is stopped because it has reached the center alignment point of the primary device. The sequence for finding the center alignment point within the alignment tolerance area is shown in Figure 16.

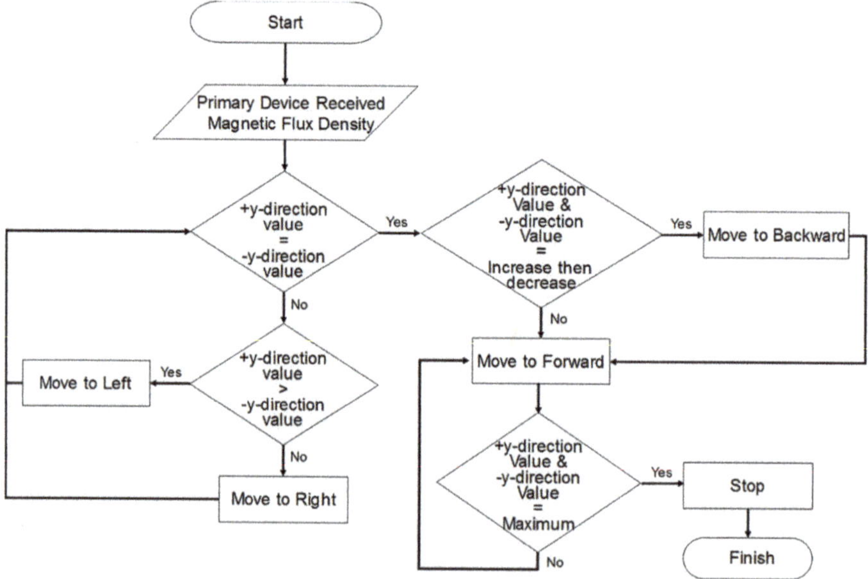

Figure 16. Sequence for finding the center alignment point.

Using MATLAB software, model results of 99 points were calculated by the secondary device to find the center alignment point of the primary device within the alignment tolerance area of the EV WPT system. As component-level and vehicle-level verification are required, the EV WPT system data are reflected as much as possible. We used an EV WPT system with an output power level of up to 7.7 kVA for the supply device and an input power level of up to 3.7 kVA for the EV device. The height of the primary device and the secondary device was 180 mm [32].

Figure 17 shows the results of simulation by applying Equation from (1) to (3) at 99 test points. In Figure 17, P_1 and P'_1 have a secondary device located at 0 mm in the x-direction and −100 mm in the y-direction, respectively, among the alignment tolerance areas of the primary device. Here, the ferrite antenna received power of the primary device is −52.72 dBm for ferrite antenna (Rx1) located at −y-direction and −65.99 dBm for ferrite antenna (Rx2) located at +y-direction. It shows that the secondary device is located to the left of the center alignment point of the primary device. Therefore, the secondary device is moved to the right. Similar results were observed for in P_2 and P'_2. As P_2 and P'_2 mean that the secondary device is located to the right, the secondary device is moved to the left. In P_3, the secondary device is located in the alignment tolerance area of the primary device. Here, by comparing the ferrite antenna received power of the primary device, it can be seen that the values of Rx1 located in −y-direction and Rx2 located in +y-direction are the same as −59.31 dBm.

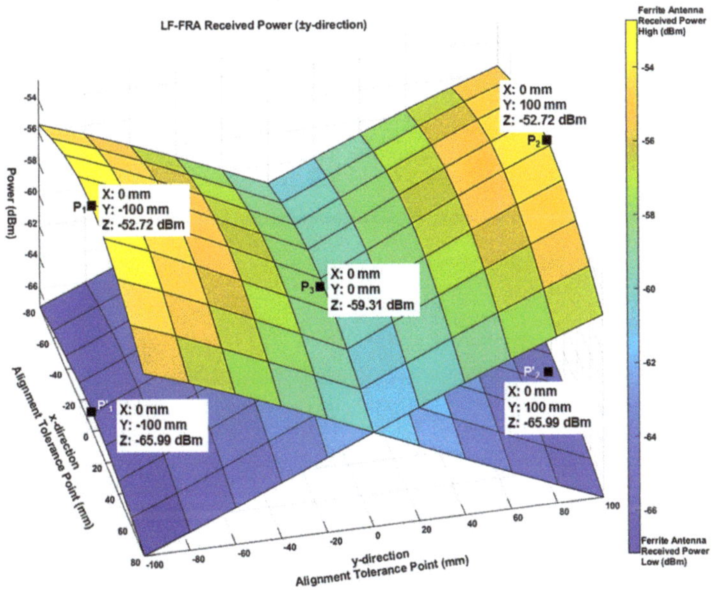

Figure 17. Simulation results for power received by the ferrite antenna receiver of the +y-direction and −y-direction.

5. Experimental Results

In the EV WPT system, using the ferrite antenna mounted on the secondary device, an actual test was conducted to ensure that the secondary device could correctly locate the center alignment point of the primary device within the primary device's alignment tolerance area. In the actual test, the WPT system used a primary device with a maximum input power of 7.7 kW and a secondary device with a maximum output power of 3.7 kW. Parameter of the primary and secondary devices are shown in Table 3.

Table 3. Parameter of the primary and secondary devices.

Parameter	Primary Device	Secondary Device
Coil size	650 mm × 500 mm	270 mm × 270 mm
Coil material	Litz wire φ 7 mm	Litz wire φ 4 mm
Number of turns	15 turns of single layer	8 turns of single layer
Ferrite tile size	675 mm × 535 mm	284 mm × 284 mm
Shield-plate size	Aluminum	Aluminum
Shield-plate material	750 mm × 600 mm	800 mm × 800 mm

To verify the validity of the modeling and simulation results, a component-level and vehicle-level experimental benches were built. Figure 18a shows a component-level experimental bench to find the center alignment point within the alignment tolerance area of the EV WPT system. Figure 18b shows a vehicle-level test bench to find the center alignment point within the alignment tolerance area by installing the EV WPT system on a KIA SOUL EV. A ROHDE and SCHWARZ SMC100A signal generator was used to generate magnetic flux density in the ferrite antenna transmitter of the secondary device. The characteristic impedance of the connected to the signal generator to perform the test is 50 ohms. A ROHDE and SCHWARZ RTE 1104 oscilloscope was used to measure the open-circuit voltage in the ferrite antenna receiver of the primary device. The WPT Testing Platform from Chroma [33] is a device that can move 1 mm steps for ±x-, ±y and ±z-direction. As it is difficult to move the vehicle by 20 mm in a vehicle-level test, the primary device was moved by 20 mm to conduct the test. In order to match the experimental conditions of the component-level and vehicle-level, the primary device was tested by moving it by 20 mm. This was also performed at the component-level. The simulation and actual experimental results measured at 99 points were compared and analyzed for each point. In addition, in order to increase the reliability of the measured data, after measuring 99 points at both the component-level test and the vehicle-level test, the same test was repeated 10 times to check the difference in measured values for each number of times.

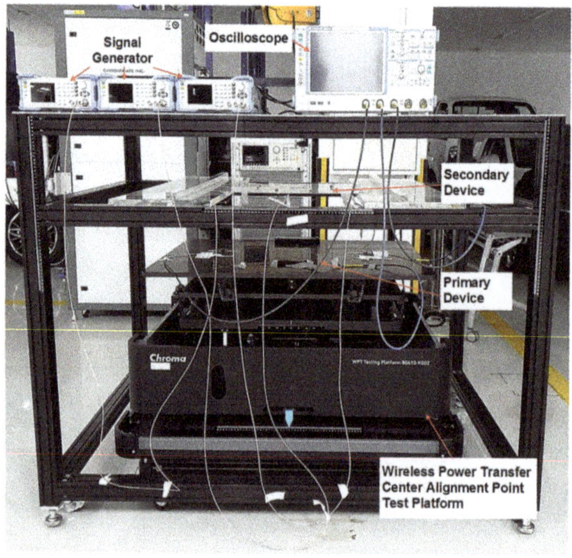

(a)

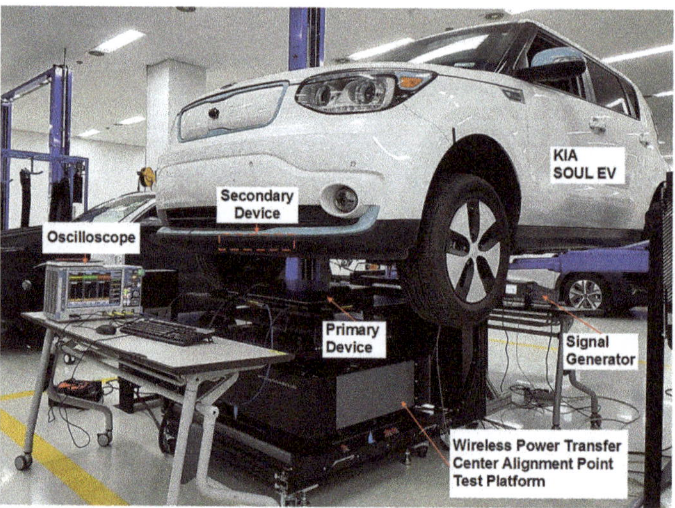

(b)

Figure 18. Experimental test conditions. (**a**) Component-level test; (**b**) vehicle-level test.

Figures 19 and 20 shows the results of the oscilloscope measurement waveform when the secondary device is positioned at the center alignment point of the primary device. In each Figure, cursor result 1 is the result measured by the −y-direction of the ferrite antenna receiver of the primary device. In addition, cursor result 3 is the result measured by the +y-direction of the ferrite antenna receiver of the primary device.

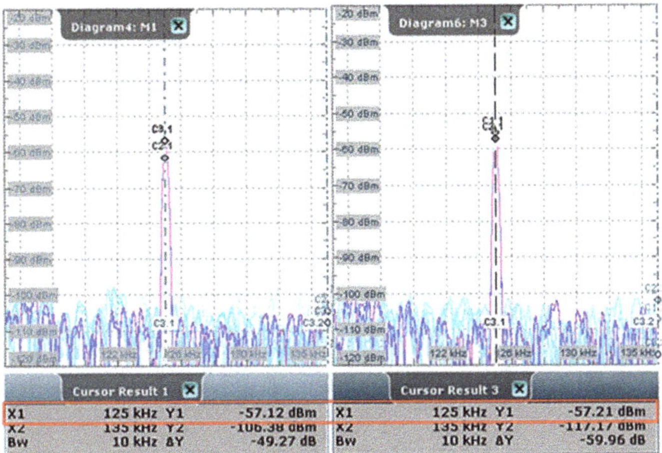

Figure 19. Oscilloscope measurement waveform results for component-level experiments at the center alignment point of the electric vehicle (EV) wireless power transfer (WPT) system.

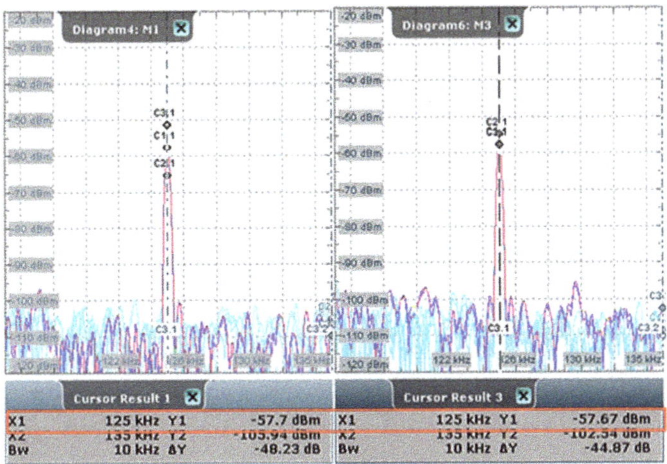

Figure 20. Oscilloscope measurement waveform results for vehicle-level experiments at the center alignment point of the EV WPT system.

As mentioned in Section 4, according to the simulation results, if the secondary device is located at the center alignment point of the primary device, the output of the ±y-direction ferrite antenna receiver mounted on the primary device should be the same. In Figure 19, the value of cursor results 1 is −57.12 dBm and that of cursor result 3 is −57.21 dBm. The measurement error guaranteed by the signal generator and oscilloscope is 2% and the difference between the value of cursor result 1 and the value of cursor result 2 is 0.09 dBm. Therefore, in the component-level test, it can be observed that the output of the ±y-direction ferrite antenna receiver mounted on the primary device is the same. Similarly, in the vehicle-level test (see Figure 20), the difference between the value of cursor result 1 and that of cursor result 3 is 0.03 dBm. Hence, it can be seen that the output of the ±y-direction ferrite antenna receiver mounted on the primary device is the same. Therefore, it can be confirmed that the secondary device is located at the center alignment point of the primary device in the component-level and the vehicle-level test.

Tables 4 and 5 compare the results of simulation, component-level and vehicle-level tests at important points among the 99 test points. The reason for selecting the five important points mentioned in Tables 4 and 5, is that the results at each endpoint of the primary device and its center alignment point are of paramount importance. Analysis of the results in Tables 4 and 5 shows that the component-level and vehicle-level measurement results showed a difference of less than 1 dBm. This is the difference less than 2%, which is the measurement error guaranteed by the signal generator and oscilloscope, the measurement equipment. Hence, the measurement results of component-level and vehicle-level are the same. However, the difference between the simulation and the actual experimental result is less than 3 dBm at maximum. This is more than the measurement error guaranteed by the measurement equipment. Therefore, in order to reduce the difference between the simulation value and the actual experimental result, it is necessary to review the environmental conditions of the EV WPT system as well as the environmental conditions of the EV.

Table 4. Comparison of experimental and simulated results in the −y-direction of the ferrite antenna receiver of the primary device.

Alignment Tolerance		Simulation Results	Component Results	Vehicle Results
x-Direction	y-Direction			
0 mm	0 mm	−59.31 dBm	−57.12 dBm	−57.70 dBm
80 mm	100 mm	−55.58 dBm	−57.48 dBm	−57.03 dBm
80 mm	−100 mm	−67.16 dBm	−68.27 dBm	−68.54 dBm
−80 mm	100 mm	−55.58 dBm	−57.59 dBm	−56.81 dBm
−80 mm	−100 mm	−67.16 dBm	−68.31 dBm	−67.75 dBm

Table 5. Comparison of experimental and simulated results in the +y-direction of the ferrite antenna receiver of the primary device.

Alignment Tolerance		Simulation Results	Component Results	Vehicle Results
x-Direction	y-Direction			
0 mm	0 mm	−59.31 dBm	−57.21 dBm	−57.67 dBm
80 mm	100 mm	−67.16 dBm	−68.61 dBm	−68.31 dBm
80 mm	−100 mm	−55.58 dBm	−57.73 dBm	−58.18 dBm
−80 mm	100 mm	−67.16 dBm	−68.56 dBm	−67.65 dBm
−80 mm	−100 mm	−55.58 dBm	−57.78 dBm	−58.18 dBm

Figure 21 show the component-level and vehicle-level results for 99 points. Figure 21 shows that the results of the component-level and vehicle-level show the same tendency when the secondary device is positioned at 0 mm in the x-direction and −100 mm in the y-direction among the alignment tolerance areas of the primary device. In component-level (see Figure 21a), the ferrite antenna received power of the primary device is −50.91 dBm for ferrite antenna (Rx1) located at the −y-direction and −64.31 dBm for ferrite antenna (Rx2) located at the +y-direction. This indicates that the secondary device is on the left-side of the center alignment point of the primary device. Therefore, the secondary device is moved to the right. Similarly, in vehicle-level (see Figure 21b), the ferrite antenna received power of the primary device is −52.35 dBm for ferrite antenna (Rx1) located at the −y-direction and −65.25 dBm for ferrite antenna (Rx2) located at the +y-direction. This also indicates that the secondary device is on the right-side of the center alignment point of the primary device. Therefore, in vehicle-level testing, the secondary device is moved to the left. It can be confirmed that these results show the same tendencies as those of the simulation result, as shown in Figure 17.

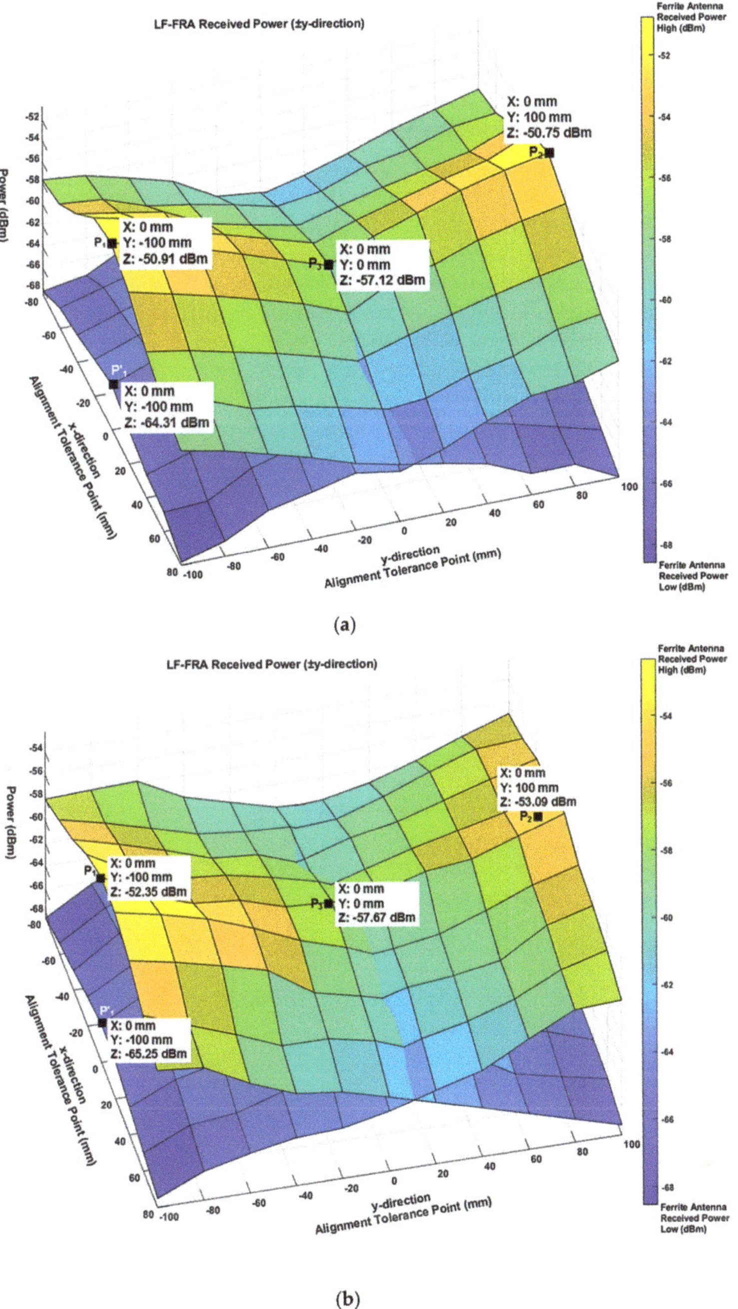

Figure 21. Experimental result for power received by the ferrite antenna receiver of the +y-direction and −y-direction. (**a**) Component level; (**b**) vehicle level.

6. Conclusions

In order to maximize the charging efficiency of the EV MF-WPT system, it is important to match the center alignment points of the primary and secondary devices. In this article, we proposed how to use the ferrite antennas to find a central alignment point between the primary and secondary units within the alignment tolerance area that requires the minimum power transfer efficiency of the EV WPT system. First, the ferrite antenna used for precise positioning of the EV WPT should not affect the charging efficiency when the primary device transmits power to the secondary device. Therefore, it was analyzed that the mounting position of the ferrite antenna should be mounted at the center of each side with ±y-direction of the primary device and the secondary device using electromagnetic simulation. Second, the method suggested that it is necessary to calculate all induced loop voltages in the relationship between incident magnetic field signal strength and induced loop voltage because of the distance between the transmitter and receiver of the ferrite antenna in EV WPT precise positioning is short to within 250 mm. After the simulation was performed on the suggestions, the unit-level test was performed to validate the simulation results. Since the difference between the simulation result and the experimental result was within one decibel-milliwatt, it could be confirmed that the proposed induced loop voltage formula was applied correctly. The reason is that the error guarantee range of the measurement equipment is two decibel-milliwatts. Third, even if two ferrite antennas are mounted on each of the primary device and the secondary device, a method that can match the central alignment points of the primary and the secondary devices were proposed, and a sequence was also proposed. The difference between the simulation results and the experimental results was within two decibel-milliwatts for component-level and within three decibel-milliwatts for vehicle-level. Consequently, we confirmed that the ferrite antenna is a suitable method to find the center alignment points of the primary and secondary devices within the alignment tolerance area of the EV WPT system.

In the future, it is necessary to review the environmental conditions of EV and EV WPT systems to reduce the error of less than one decibel-milliwatt between simulation results and vehicle-level experimental results. In addition, the EV driver must make considerable effort to directly align the vehicle to identify the central alignment point of the base unit. Therefore, it is necessary to review the research that combines the precise positioning of the autonomous driving parking system and the electric vehicle wireless transmission system.

7. Patents

Seong, J.Y. Method and apparatus for position alignment using low-frequency antennas in wireless power transfer system. EP.3364522.B1. US.10622846.B2. 2020, http://kpat.kipris.or.kr/kpat/biblioa.do?method=biblioFrame (accessed on 20 May 2020).

Author Contributions: Conceptualization, J.Y.S. and S.-S.L.; methodology, J.Y.S.; software, J.Y.S.; validation, J.Y.S.; investigation, J.Y.S.; data curation, J.Y.S.; writing—original draft preparation, J.Y.S.; writing—review and editing, J.Y.S. and S.-S.L.; visualization, J.Y.S.; supervision, S.-S.L. All authors have read and agreed to the published version of the manuscript.

Funding: This research was supported by the 2016–2017 R&D Technical Fund of Hyundai Motor Company.

Conflicts of Interest: The authors declare no conflicts of interest.

Appendix A

From Equation (15), the magnetic flux density applying the geometric dimensions of Figure 14 is given by:

$$\mathbf{B}_{WPTtot} = \frac{\mu_0 \mu_r m}{4\pi r^3}\left(2\cos\theta^2 - \sin\theta^2\right)\mathbf{a}_z$$

$$\theta = \sum_{k=\min(90+\alpha_1,90-\alpha_2)}^{\max(90+\alpha_1,90-\alpha_2)} \left(\frac{\alpha_1+\alpha_2}{69}\right)k,$$

$$\alpha_1 = \cos^{-1}\frac{\sqrt{\left(0.3375-SD_{\Delta y}\right)^2+D_1^2}}{\sqrt{\left(0.3375-SD_{\Delta y}\right)^2+\left(0.2675-SD_{\Delta x}+\sum_{N=1}^{70}(1-N)D_1\right)^2+D_1^2}},$$

$$\alpha_2 = \cos^{-1}\frac{\sqrt{\left(0.3375-SD_{\Delta y}\right)^2+D_1^2}}{\sqrt{\left(0.3375-SD_{\Delta y}\right)^2+\left(0.2675-SD_{\Delta x}+l_c+\sum_{N=1}^{70}(1-N)D_1\right)^2+D_1^2}},$$

$$r = \sqrt{\left(0.3375-SD_{\Delta y}\right)^2 + \left(0.2675-SD_{\Delta x}+l_c+\sum_{N=1}^{70}(1-N)D_1+\left(\frac{69\cdot\min(90+\alpha_1,90-\alpha_2)-\theta}{\alpha_1+\alpha_2}\right)D_1\right)^2+D_1^2}$$

(A1)

where $\mathbf{B}_{WPTtot}$ is the sum of the magnetic flux density transmitted by the 70-turn loop of the ferrite antenna transmitter mounted on the secondary device to the 70-turn loops of the ferrite antenna receiver mounted on the primary device, μ_0 is the permeability of free space, μ_r is the relative permeability of the ferrite antenna, m is the magnetic dipole moment, α_1 is the angle between the Nth loop of the ferrite antenna transmitter in the ±y-direction mounted on the secondary device and the 1st loop of the ferrite antenna receiver in the ±y-direction mounted on the primary device, α_2 is the angle between the Nth loop of the ferrite antenna transmitter in the ±y-direction mounted on the secondary device and the 2nd to 70th loops of the ferrite antenna receiver in the ±y-direction mounted on the primary device, $SD_{\Delta x}$ is the distance the secondary device moves in the ±x-direction, $SD_{\Delta y}$ is the distance the secondary device moves in the ±y-direction and r is the distance between the ferrite antenna in the ±y-direction mounted on the secondary device and ferrite antenna in the ±y-direction mounted on the primary device.

The Equation (16) of the open-circuit voltage received by the receiver of the ferrite antenna mounted on the primary device of the EV WPT system is replaced by:

$$V_{ocWPTtot} = j\omega\mu_r A\mathbf{B}_{WPTtot}\mathbf{a}_z \tag{A2}$$

Substituting Equation (17) into (A2),

$$Power_{WPT\,ferrite\,antenna} = 10log_{10}\left(1000\frac{V_{ocWPTtot}^2}{R}\right) \tag{A3}$$

References

1. Faraday, M. Experimental researches in electricity. *Philos. Trans. Roy. Soc. Lond.* **1832**, *122*, 125–162.
2. Tesla, N. The transmission of electrical energy without wires as a means for furthering peace. *Elect. World Eng.* **1905**, *1*, 21–24.
3. Covic, G.A.; Boys, J.T. Inductive power transfer. *Proc. IEEE* **2013**, *101*, 1276–1289. [CrossRef]
4. RamRakhyani, A.K.; Mirabbasi, S.; Chiao, M. Design and optimization of resonance-based efficient wireless power delivery systems for biomedical implants. *IEEE Trans. Biomed. Circuits Syst.* **2011**, *5*, 48–63. [CrossRef] [PubMed]
5. Ahson, S.; Ilyas, M. *RFID Handbook: Applications, Technology, Security, and Privacy*; CRC Press/Tayler & Francis Group: Boca Raton, FL, USA, 2008; pp. 3–16.
6. McSpadden, J.O.; Mankins, J.C. Space Solar Power Programs and Microwave Wireless Power Transmission Technology. *IEEE Microw. Mag.* **2002**, *3*, 46–57. [CrossRef]
7. Kurs, A.; Karalis, A.; Moffatt, R.; Joannopoulos, J.D.; Fisher, P.; Soljačić, M. Wireless power transfer via strongly coupled magnetic resonances. *Science* **2007**, *317*, 83–86. [CrossRef] [PubMed]

8.   Society of Automotive Engineers Home Page. Available online: https://www.sae.org/servlets/works/committeeHome.do?comtID=TEVHYB10 (accessed on 29 April 2020).

9.   Xia, J.; Yuan, X.; Li, J.; Lu, S.; Cui, X.; Li, S.; Fernández-Ramírez, L.M. Foreign Object Detection for Electric Vehicle Wireless Charging. *Electronics* **2020**, *9*, 805. [CrossRef]

10.   Abou Houran, M.; Yang, X.; Chen, W. Magnetically Coupled Resonance WPT: Review of Compensation Topologies, Resonator Structures with Misalignment, and EMI Diagnostics. *Electronics* **2018**, *7*, 296. [CrossRef]

11.   Society of Automotive Engineers International. *Wireless Power Transfer for Light-Duty Plug-In/Electric Vehicles and Alignment Methodology*; SAE Information Report J2954 (issued. 201605); Society of Automotive Engineers International: Troy, MI, USA, 2016.

12.   Manandhar, D.; Okano, K.; Ishii, M.; Torimoto, H.; Kogure, S.; Maeda, H. Development of ultimate seamless positioning system based on QZSS IMES. In Proceedings of the ION GNSS, Savannah, GA, USA, 16–19 September 2008.

13.   Ji, M.; Kim, J.; Jeon, J.; Cho, Y. Analysis of positioning accuracy corresponding to the number of BLE beacons in indoor positioning system. In Proceedings of the 17th International Conference on Advanced Communication Technology, Seoul, Korea, 1–3 July 2015.

14.   Koutsou, A.D.; Seco, F.; Jimenez, A.R.; Roa, J.O.; Ealo, J.L.; Prieto, C.; Guevara, J. Preliminary Localization Results With An RFID Based Indoor Guiding System. In Proceedings of the 2007 IEEE International Symposium on Intelligent Signal Processing, Alcala de Henares, Spain, 3–5 October 2007.

15.   Sathyan, T.; Humphrey, D.; Hedley, M. WASP: A System and Algorithms for Accurate Radio Localization Using Low- Cost Hardware. *IEEE Trans. Syst. Man Cyber. Part C* **2011**, *41*, 211–222. [CrossRef]

16.   Alarifi, A.; Al-Salman, A.; Alsaleh, M.; Alnafessah, A.; Al-Hadhrami, S.; Al-Ammar, M.A.; Al-Khalifa, H.S. Ultra Wideband Indoor Positioning Technologies: Analysis and Recent Advances. *Sensors* **2016**, *16*, 707. [CrossRef]

17.   Lin, C.C.; Wang, M.S. A Vision Based Top-View Transformation Model for a Vehicle Parking Assistant. *Sensors* **2012**, *12*, 4431–4446. [CrossRef]

18.   Zhang, L.; Li, X.; Huang, J.; Shen, Y.; Wang, D. Vision-Based Parking-Slot Detection: A Benchmark and A Learning-Based Approach. *Symmetry* **2018**, *10*, 64. [CrossRef]

19.   International Telecommunication Union. *Technical and Operating Parameters and Spectrum Use for Short-Range Radiocommunication Devices*; ITU-R SM.2153-7; International Telecommunication Union: Geneva, Switzerland, 2019.

20.   Society of Automotive Engineers International. *Wireless Power Transfer for Light-Duty Plug-In/Electric Vehicles and Alignment Methodology*; SAE Recommended Practice J2954 (rev. 201904); Society of Automotive Engineers International: Troy, MI, USA, 2019.

21.   International Electrotechnical Commission. *Electric Vehicle Wireless Power Transfer (WPT) Systems—Part. 2: Specific Requirements for Communication between Electric Road Vehicle (EV) and Infrastructure*; IEC TS 61980-2:2019; International Electrotechnical Commission: Geneva, Switzerland, 2019.

22.   Cheng, D.K. *Field and Wave Electromagnetics*, 2nd ed.; Addison-Wesley Publishing Company, Inc.: Boston, MA, USA, 1989; pp. 225–262.

23.   Guru, B.S. *Electromagnetic Field Theory Fundamentals*; PWS Publishing Company: Boston, MA, USA, 1997; pp. 245–260.

24.   Straw, R.D. *The ARRL Antenna Book*; The American Radio Relay League, Inc.: Newington, CT, USA, 1997; Volume 5, pp. 1–11.

25.   DeVore, R.; Bohley, P. The electrically small magnetically loaded multiturn loop antenna. *IEEE Trans. Antennas Propag.* **1977**, *25*, 496–505. [CrossRef]

26.   Johnson, R.C. *Antenna Engineering Handbook*, 3rd ed.; McGraw-Hill, Inc.: New York, NY, USA, 1993; Volume 5, pp. 1–9.

27.   Altair HyperWorks Home Page. Available online: https://altairhyperworks.com (accessed on 12 February 2020).

28.   MathWorks Home Page. Available online: https://www.mathworks.com (accessed on 4 March 2020).

29.   Rohde and Schwarz Home Page: R&S®SMC100A Signal Generator. Available online: https://www.rohde-schwarz.com/product/smc100a-productstartpage_63493-10181.html (accessed on 4 March 2020).

30.   Rohde and Schwarz Home Page: R&S®RTE1000 Oscilloscopes. Available online: https://www.rohde-schwarz.com/product/rte-productstartpage_63493-54848.html (accessed on 4 March 2020).

31. Alan, B. *Wireless Positioning Technologies and Applications*; Artech House: Norwood, MA, USA, 2016; pp. 147–167.
32. Seong, J.; Jang, J.; Koh, Y. A study of electric vehicle wireless charging system integration and vehicle alignment optimization. In Proceedings of the 30th International Electric Vehicle Symposium & Exhibition, Stuttgart, Germany, 9–11 October 2017.
33. Chroma ATE Home Page. Available online: http://www.chromaate.com/product/EV_wireless_power_transfer_ATS.htm (accessed on 21 March 2020).

Article

The Influence of Substrate Size Changes on the Coil Resistance of the Wireless Power Transfer System

Jiacheng Li [1,2], Linlin Tan [1,2,*], Xueliang Huang [1,2], Ruoyin Wang [1,2] and Ming Zhang [1,2]

1   Department of Electrical Engineering, Southeast University, No. 2 Sipailou, Nanjing 210096, China; wldfy@foxmail.com (J.L.); xlhuang@seu.edu.cn (X.H.); ruoy_wang@seu.edu.cn (R.W.); zhangming1212@seu.edu.cn (M.Z.)
2   Key Laboratory of Smart Grid Technology and Equipment in Jiangsu Province, Nanjing 210096, China
*   Correspondence: tanlinlin@seu.edu.cn; Tel.: +86-25-8379-4691 (ext. 815)

Received: 29 May 2020; Accepted: 18 June 2020; Published: 21 June 2020

Abstract: Wireless power transfer (WPT) technology has been widely used in many fields. Nevertheless, in the field of high power transmission, such as the WPT system of electric vehicles, the power transmission efficiency of WPT system lags behind that of wired charging due to losses brought by substrate shielding materials. In this regard, the conduction resistance of Litz-wire coils without substrate is analyzed first in this paper. Secondly, the induction resistance of the coil with single-layer and double-layer substrate materials is modeled. Then, through the establishment of a coil simulation and experimental platform with a single-layer substrate, a contrastive analysis of the variation trend of coil equivalent series resistance (coil ESR) at changing thickness and area and constant volume of the substrate is carried out in combination with the theory. The variation law of coil ESR at changing thickness and area and constant volume of double-layer substrate is also explored at the end of this paper. This is expected to contribute to the reduction of coil losses in the WPT system through a systematic study of the influence of substrate size changes on the coil resistance of the WPT system.

Keywords: wireless power transfer; coil resistance; substrate size

---

1. Introduction

The wireless power transfer (WPT) technology uses energy carriers (e.g., magnetic field, electric field, electromagnetic wave, etc.) in the physical space and realizes the transfer of electric energy from the power supply side to the load side based on a no-wire contacting mode [1,2]. This technology provides effective solutions and measures for reliable power supply of inspection robots [3], electric vehicles (EVs) [4], mobile phones [5], unmanned aerial vehicles [6], on-line EVs [7] and industrial application in high-voltage transmission networks [8], and avoids contact sparks, plug wear [9,10], and other problems.

As the core index of the technology's advantages over traditional wired charging technology, the power transmission efficiency (PTE) of WPT system plays a decisive role in the large-scale application of the technology. Despite the WPT system works on the principle of high-frequency magnetic field propagating in the air, the large reluctance of the air (a relative permeability of only 1) greatly affects the PTE of the system. The main methods to improve the PTE of the system include: increasing the working frequency of the system, reducing the loss of the system itself, and increasing the coupling between the coils. As a rule, high operating frequency corresponds to the smaller limit index in the safety standard of electromagnetic environment [11], which may cause the electromagnetic environment around the system to fail to meet the limit index and thus fail to be normally applied. Also, a high frequency puts forward higher requirements for the design and stable operation of the inverter. In addition, restricted by the size, space and weight of the system, as well as the processing cost of magnetic materials, it is challenging to improve the coupling strength between coils, which justifies the feasibility of minimizing the resistance loss of the system itself. In general, the loss of the system comes from that of high-frequency inverter, the transceiver coil and the rectifier converter. In view of the large number of components involved in high-frequency inverters and rectifier conversion devices, apart from the optimal control of soft switches, it is expected to conduct optimal selection of devices in terms of hardware to reduce loss. Thus, the minimization of coil losses is believed to be feasible.

Since the Massachusetts Institute of Technology (MIT) proposed magnetic coupling resonance WPT technology in 2007 [1], academia and industry have done a lot of research for solving the problem of WPT system transmission distance and PTE improvement [12–16], and proposed frequency tracking [13], impedance matching [14], changing the coil structure, and adding magnetic circuit optimization structure [15,16]. These methods have improved the PTE of the system to a certain extent, but more focus on the principle of increasing the coupling between the coils. There is relatively little research on the calculation of the coil self-resistance in the field of WPT.

For the calculation of self- and mutual impedances in planar inductors, Hurley et al. [17] studied the variation of the mutual impedance of two coaxial circular filaments with frequency in the presence of a semi-infinite substrate. On this basis, the authors proposed a theoretical model for calculating the self-inductance and equivalent resistance of a circular filament on a substrate with a limited thickness [18], and studied the variation of the mutual impedance of two coaxial circular filaments with frequency.

Based on the above research, the research on the change law of the impedance of the heating coil on a substrate is more in-depth in the field of induction heating [19–23]. A set of analytical expressions was derived for the equivalent impedance in a planar circular induction heating system [19]. This method was extended to the case of multilayer substrates with finite thickness and infinite area [20]. A theory of inductance calculation of planar spiral windings was extended to determine the inductance of planar spiral windings shielded by a double-layer planar EM shield which consists of a layer of soft magnetic material and a layer of conductive material [21]. The variation trend of the equivalent impedance of the induction heating coil wound with Litz-wire on the substrate of finite thickness and infinite area was also analyzed [22,23].

However, previous studies have mostly focused on the variation rule of coil effective series resistance (ESR) with frequency when the substrate contains ferrite or aluminum plate, and few studies have focused on the influence of the substrate size changes on coil ESR. Although the change rule of coil ESR when there are multiple ferrite cores based on the method of simulation and theory is mentioned in [24], it is still not systematically studied with respect to the change rule of coil ESR when the thickness and area of substrate change. So, here, the influence of substrate size on the coil ESR, for single-layer and double-layer substrates, is studied. The purpose is to get the variation law of coil ESR at changing area and thickness and constant weight of the substrate and help the researchers for efficiently selecting the substrate and further improving the transmission efficiency of the system.

For the above purpose, the resistance of circular disk coils without substrate is firstly analyzed theoretically in this paper. Based on existing research, the theoretical analysis of the coil resistance of single-layer and double-layer substrate with finite thickness and infinite area is studied respectively, followed by a discussion on the shifts of calculation methods when the substrate thickness is close to the skin depth. Through the establishment of the actual model and the introduction of COMSOL Multiphysics finite element simulation and experimental measurement, this is compared with the theoretical calculation and summarized the variation rule of coil ESR at changing thickness of the substrate. Finally, the variation law of coil ESR at changing area and thickness and constant weight of the substrate is discussed to explore ways to reduce coil ESR.

The structure of this paper is as follows. In Section 2, the theoretical calculation method of the coil's conduction resistance and induction resistance with the single-layer and double-layer substrates is modeled and analyzed. In Section 3, the change rule of coil ESR when the size of the single-layer substrate changes is analyzed by theory, simulation, and experiment. When the size of the double-layer substrate changes, the change rule of coil ESR is also analyzed in Section 4. Section 5 concludes this paper.

2. Modeling of Coil ESR

The loss of the coil mainly includes the conduction resistance of the Litz-wire bundle itself and the induction resistance caused by the substrate material around the coil [25]. That is to say, coil ESR R_{ESR} with substrate material consists of two parts: conduction resistance R_{AC} and substrate induction resistance R_{Ind}, which is defined as $R_{ESR} = R_{AC} + R_{Ind}$. In order to reduce the coil loss as much as possible, it is not only necessary to realize the accurate calculation of the conducting resistance of the coil itself, but also to grasp the influence law of the substrate material on the change of the coil resistance. Firstly, the conduction resistance of the coil is modeled and analyzed in this section.

2.1. Coil Conduction Resistance

2.1.1. Calculation of Litz-Wire Bundle Resistance

In order to reduce the conduction resistance of the coil, multiple film-insulated round copper wires are often used to wind the coil by twisting into a Litz-wire bundle. The coil of WPT system works under the condition of high frequency current, and there is skin effect in the flow of high frequency current in film-insulated round copper wire. Therefore, after the working frequency of WPT system is determined, the diameter of single wire should be selected based on the skin depth of copper wire to ensure the optimization of cost and loss. Based on the recommendation of Litz-wire manufacturer, the model of the copper wire corresponding to the commonly used working frequency band is shown in Table 1. The specifications of the copper wire in Table 1 meet the NEMAMW1000 (Revision2008) standard.

Table 1. The relationship between the strand type and working frequency of Litz-wire.

Working Frequency Band	AWG Size	Bare Wire Diameter d_s (mm)	Single Paint Film Diameter $d_{s'}$ (mm)	Unit Resistance R_{S_Unit} (Ω/m)
60–1000 Hz	28	0.320	0.356	0.2123
1–10 kHz	30	0.254	0.284	0.3371
10–20 kHz	33	0.180	0.206	0.6689
20–50 kHz	36	0.127	0.147	1.350
50–100 kHz	38	0.102	0.119	2.111
100–200 kHz	40	0.079	0.094	3.519
200–350 kHz	42	0.064	0.076	5.421
350–850 kHz	44	0.051	0.061	8.495
850k–1.4 MHz	46	0.0399	0.0508	13.802
1.4–2.8 MHz	48	0.0315	0.0381	22.129

For the resistance value of the Litz-wire bundle, the DC resistance R_{DC} is to be calculated firstly. Then the AC resistance R_{AC} is calculated in high frequency. According to [26], the accurate theoretical calculation of the DC resistance of the Litz-wire bundle is

$$R_{DC} = \frac{4\rho l}{n_{Litz}\pi d_s^2}\left(1 + \frac{n_{Litz}\pi^2 d_s^2}{4K_a P^2}\right) \tag{1}$$

where ρ is the resistivity of the Litz-wire strand, l is the actual length of the Litz-wire strand, P is the pitch of the Litz-wire bundle, d_s is the bare wire diameter of the Litz-wire strand, n_{Litz} is the strand number of the Litz-wire bundle, and K_a is the packing factor of the Litz-wire bundle and

$$K_a = \frac{A_{cu}}{A_b} = \frac{n_{Litz}d_s^2}{d_b^2} \tag{2}$$

$$K_a = \frac{A_{cu}}{A_b} = \frac{n_{\text{Litz}} d_s^2}{d_b^2} \tag{2}$$

$$l = l_{tot}\sqrt{1 + (\pi d_b / P)^2} \tag{3}$$

where A_b is the overall bundle area ($\pi d_b^2 / 4$), A_{cu} is the sum of the cross sectional areas of all the strands($n_{\text{Litz}} \pi d_s^2 / 4$), d_b is the overall bundle diameter of the Litz-wire bundle and $d_b \approx d_{s'} \times \sqrt{n_{\text{Litz}} / 0.75}$, $d_{s'}$ is the paint film diameter of the Litz-wire strand, and l_{tot} is the length of the Litz-wire bundle for coil.

Furthermore, according to [27], the accurate calculation equation of AC resistance of the Litz-wire bundle can be obtained by

$$R_{AC} = R_{DC}\left(1 + \frac{K_a n_{\text{Litz}} \pi^4 d_s^4 f^2 \times 10^{-14}}{8\rho^2}\right) \tag{4}$$

where f is the frequency of current in Litz-wire bundle.

Of course, in the case of low error requirements, it might be estimated by the ratio of the unit resistance of a copper strand in the corresponding working frequency band in Table 1 to the number of strands of the Litz-wire bundle as

$$R_{AC} \approx R_{DC} = l_{tot}\frac{R_{S\_Unit}}{n_{\text{Litz}}}. \tag{5}$$

2.1.2. Circumference Calculation of Disk Coil

As can be seen from the Equation (3), the accurate calculation of the coil conduction resistance requires the total length of the wire bundle winding the coil, that is, the coil conductor resistance requires the total length of the disk coil. This paper relates to the length of the disk coil as shown in Figure 1. Their number of turns is N, the minimum radius of the coil ring is $r_{\min}$ and the maximum radius of the coil ring is $r_{\max}$ in the disk coil. The turn spacing of the coil is ΔD.

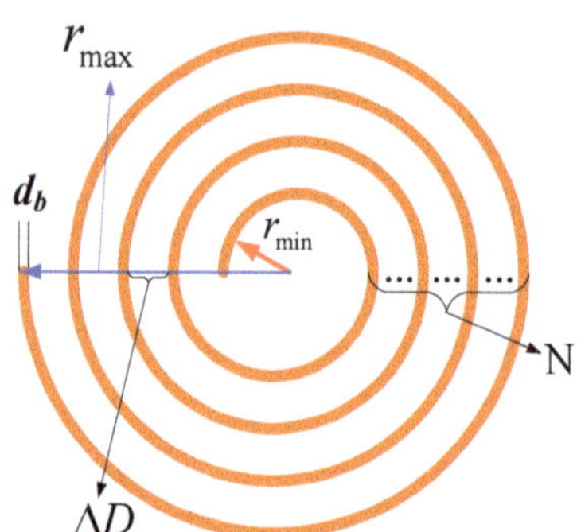

Figure 1. The Structure diagram of disk coil.

For a more accurate simulation of the actual winding of the coil, it is assumed that the disk coil is tangentially connected by a semi-arc with a gradually increasing diameter, and the diameter increases by $\Delta D + d_b$ with each additional semi-arc. Let the diameter of the first semi-arc of the first turn be $2r_{\min}$, the diameter of the second semi-arc of the first turn is $2r_{\min} + \Delta D + d_b$, the diameter of the first semi-arc of the second turn is $2r_{\min} + \Delta D + d_b + \Delta D + d_b$, and the diameter of the second semi-arc of the second turn is $2r_{\min} + \Delta D + d_b + \Delta D + d_b + \Delta D + d_b$. By analogy, the diameter of the second semi-arc of the Nth turn is $2r_{\min} + (2N - 1)(\Delta D + d_b)$. In summary, the total length of the Litz-wire bundle winding the coil shall be

$$l_{tot} = \frac{\pi}{2}[4Nr_{\min} + N(2N - 1)(\Delta D + d_b)]. \tag{6}$$

$$l_{tot} = \frac{\pi}{2}[4Nr_{min} + N(2N-1)(\Delta D + d_b)]. \tag{6}$$

Therefore, the conduction resistance of the disk coil without substrate can be calculated by simultaneous Equations (1)–(4) and (6).

2.2. Coil Induction Resistance with Substrate

In order to meet the requirements of electromagnetic environment limits and improve the coupling between coils, substrate materials are often used around coils to increase coupling and shield electromagnetic fields. Due to the eddy current effect, the substrate material shields the magnetic field, but it also affects the resistance of the coil, i.e., induction resistance R_{Ind}. In this paper, the theoretical modeling calculation of induction resistance is carried out next.

2.2.1. Coil Induction Resistance with Single-Layer Substrate

The coil model with single-layer substrate is as shown in Figure 2. The diameter of the coil wire in it is the overall bundle diameter d_b of the Litz-wire bundle. The outer radius of the outermost turn of the coil is r_{1o}, and the inner radius is r_{1i}. The secondary outer turns are r_{2o} and r_{2i} respectively. The innermost turns are r_{No} and r_{Ni} respectively, and so on. μ_{r1} and σ_1 are the relative permeability and conductivity of the substrate material, respectively. t_1 is the thickness of the substrate material. d_c is the vertical distance from the center point of the cross section of the coil to the origin O.

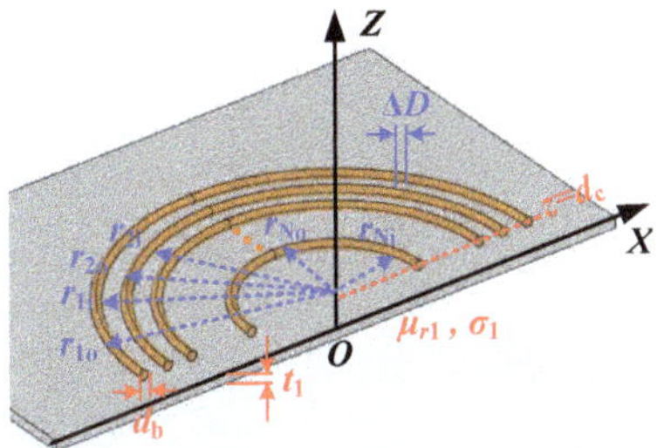

Figure 2. The coil model with single-layer substrate.

Based on the calculation equation of the mutual impedance of the coil with a finite thickness of substrate material in [18], the detailed calculation equation of the coil induction resistance with single-layer substrate in Figure 2 can be obtained by

$$R_{\text{Ind-1}} = real\left[\sum_{m=1}^{N}\sum_{m=1}^{N}\frac{j\omega\mu_0\pi}{d_b^2\ln(\frac{r_{mo}}{r_{mi}})\ln(\frac{r_{no}}{r_{ni}})}\int_0^{\infty}S(kr_{mo},kr_{mi})\cdot S(kr_{no},kr_{ni})Q(kd_b)\lambda(t_1)e^{-2kd_c}dk\right] \tag{7}$$

where

$$S(kr_{mo}, kr_{mi}) = \frac{J_0(kr_{mo}) - J_0(kr_{mi})}{k} \tag{8}$$

$$Q(kd_b) = \frac{2}{k}(d_b + \frac{e^{-kd_b} - 1}{k}) \tag{9}$$

$$\lambda(t_1) = \phi(k)\frac{1 - e^{-2\eta t_1}}{1 - \phi(k)^2 e^{-2\eta t_1}} \tag{10}$$

$$\eta = \sqrt{k^2 + j\omega\mu_0\mu_{r1}\sigma_1} \tag{11}$$

$$\phi(k) = \frac{\mu_{r1}k - \eta}{\mu_{r1}k + \eta} \tag{12}$$

μ_0 is the permeability of free space ($4\pi \times 10^{-7}$ H/m). $J_0(kr)$ is the first kind of 0-order Bessel function.

$$\phi(k) = \frac{\mu_{r1}k - \eta}{\mu_{r1}k + \eta} \tag{12}$$

μ_0 is the permeability of free space ($4\pi \times 10^{-7}$ H/m). $J_0(kr)$ is the first kind of 0-order Bessel function.

Electronics 2020, 9, 1025

The application condition of Equation (7) is that the thickness of substrate shall be less than 5 times of the skin depth of material. In case that the thickness of substrate fails to meet the condition, it might be equivalent to a semi-infinite plane and calculated by

$$R_{Ind-1} = real\left[\sum_{m=1}^{N}\sum_{n=1}^{N}\frac{j\omega\mu_0\pi}{d_b^2 \ln(\frac{r_{mo}}{r_{mi}})\ln(\frac{r_{no}}{r_{ni}})}\int_0^{\infty} S(kr_{mo}, kr_{mi})\cdot S(kr_{no}, kr_{ni})Q(kd_b)\phi(k)e^{-2kd_c}dk\right] \tag{13}$$

2.2.2. Coil Induction Resistance with Double-Layer Substrate

The electromagnetic shielding around the coil of WPT system usually adopts a double-layer substrate composed of ferrite and aluminum plate. The first layer of substrate adjacent to the coil is ferrite, and the second layer is aluminum plate, as shown in Figure 3.

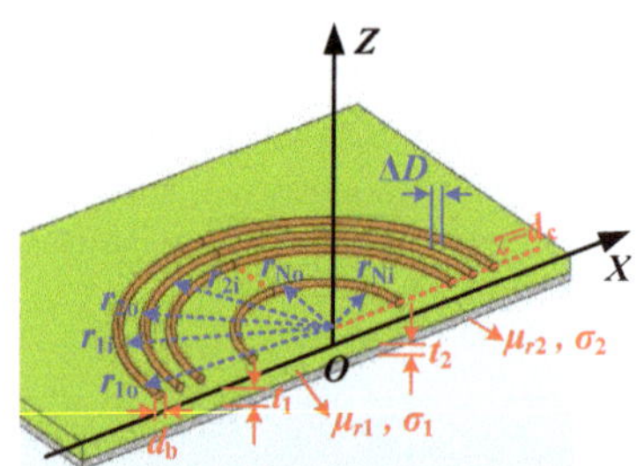

Figure 3. The coil model with double -layer substrate.

Compared with Figure 2, a layer of substrate material with thickness t_2, relative permeability μ_{r2} and conductivity σ_2 is added on the substrate material with thickness t_1, relative permeability μ_{r1} and conductivity σ_1. Based on the calculation equation of the coil induction resistance with a double-layer substrate material given in [21], the calculation equation of the coil induction resistance with a double-layer substrate material can be obtained as follows

$$R_{Ind-2} = real\left[\sum_{m=1}^{N}\sum_{n=1}^{N}\frac{j\omega\mu_0\pi}{d_b^2 \ln(\frac{r_{mo}}{r_{mi}})\ln(\frac{r_{no}}{r_{ni}})}\int_0^{\infty} S(kr_{mo}, kr_{mi})\cdot S(kr_{no}, kr_{ni})Q(kd_b)\lambda(t_1,t_2)e^{-2kd_c}dk\right] \tag{14}$$

where

$$\lambda(t_1,t_2) = \frac{\phi_1(k) + \frac{\theta(t_2)-m}{\theta(t_2)+m}e^{-2\eta_1 t_1}}{1 + \phi_1(k)\frac{\theta(t_2)-m}{\theta(t_2)+m}e^{-2\eta_1 t_1}} \tag{15}$$

$$\theta(t_2) = \frac{1 - \phi_2(k)e^{-2\eta_2 t_2}}{1 + \phi_2(k)e^{-2\eta_2 t_2}} \tag{16}$$

$$m = \frac{\mu_{r1}\eta_2}{\mu_{r2}\eta_1} \tag{17}$$

$$\eta_1 = \sqrt{k^2 + j\omega\mu_0\mu_{r1}\sigma_1} \tag{18}$$

$$\eta_2 = \sqrt{k^2 + j\omega\mu_0\mu_{r2}\sigma_2} \tag{19}$$

$$\phi_1(k) = \frac{\mu_{r1}k - \eta_1}{\mu_{r1}k + \eta_1} \tag{20}$$

$$\phi_2(k) = \frac{\mu_{r2}k - \eta_2}{\mu_{r2}k + \eta_2} \tag{21}$$

$$\phi_1(k) = \frac{\mu_{r1}k - \eta_1}{\mu_{r1}k + \eta_1} \tag{20}$$

$$\phi_2(k) = \frac{\mu_{r2}k - \eta_2}{\mu_{r2}k + \eta_2} \tag{21}$$

All of the above parameters are in SI units. To sum up, this paper has modeled the calculation method of the equivalent resistance of the coil with single-layer and double-layer substrate materials. Next, the above method is verified by some examples, and the influence of substrate material size change on coil ESR is discussed.

3. Analysis of Coil ESR with Single-Layer Substrate

The coil as shown in Figure 4 is used for analysis in this paper, and its detailed parameters are shown in Table 2. The Litz wire of the winding coil is insulated with its own paint film layer, and external woolen material. The substrate around the coil to shield the electromagnetic field is usually made of ferrite, aluminum, and other materials [15], whose conductivity and permeability are shown in Table 3. Among them, the conductivity of Ferrite 1 is 0 S/m, which is the ideal ferrite. The conductivity of Ferrite 2 is 10 S/m, which is closer to reality. Due to the common charging frequency of EVs wireless charging system is 85 kHz. Therefore, this paper focuses on the study of the influence of substrate material on coil ESR at 85 kHz. The skin depth of substrate materials at 85 kHz is also listed in Table 3.

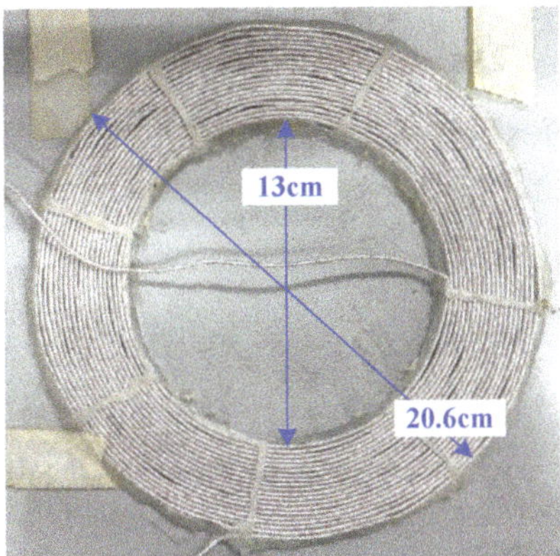

Figure 4. The prototype of coil.

Table 2. The mechanical parameters of the coil.

Parameters	Value	Parameters	Value
the strand diameter d_s	0.1 mm	strands number n_{Litz}	120
the overall bundle diameter d_b	1.5 mm	coil turn spacing ΔD	0.5 mm
coil inner diameter $2r_{min}$	13 cm	coil outer diameter $2r_{max}$	20.6 cm
coil turns N	19	current flow area	1 mm^2

Table 3. The substrate material parameters.

Parameters	Ferrite 1	Ferrite 2	Aluminum
Relative permeability μ_r	3300	$3300 \pm 25\%$	1
Conductivity σ (S/m)	0	10	3.82×10^7
Skin depth @ 85 kHz (mm)	/	545.9	0.2793

Then, the change law of coil ESR is compared and analyzed by the methods of theory, finite element simulation software (Figure 5a) and experimental test (Figure 5b). The condition when the side length of substrate is far larger than the diameter of the outer ring of the coil is first analyzed.

Conductivity σ (S/m)	0	10	3.82×10^7
Skin depth @ 85 kHz (mm)	/	545.9	0.2793

Then, the change law of coil ESR is compared and analyzed by the methods of theory, finite element simulation software (Figure 5a) and experimental test (Figure 5b). The condition when the side length of substrate is far larger than the diameter of the outer ring of the coil is first analyzed.

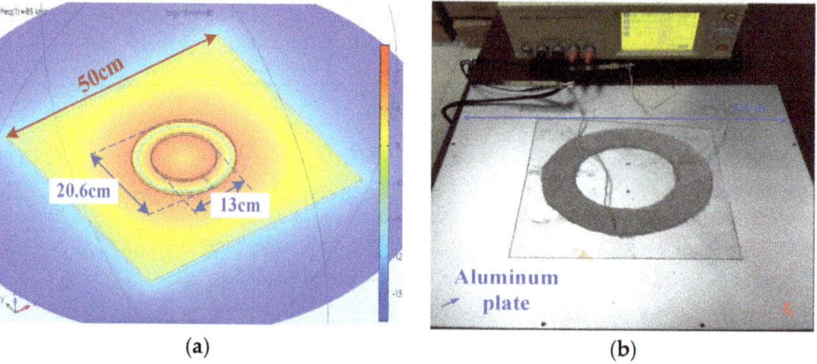

(a)   (b)

Figure 5. The model of coil with single-layer substrate (a) Simulation (COMSOL Multiphysics); (b) Experimental test platform.

3.1. When the Thickness of Substrate Material Changes

The theoretical calculation curve of coil ESR variation shall be obtained by substituting the data in Table 3 into Equations (4) and (7), i.e., the dot curve in Figure 6. Please note that a thickness of 1 mm of the aluminum plate, it should be calculated by Equation (7) because it is less than five times of skin depth. For the thickness of 2–10 mm, Equation (13) shall be adopted because it is greater than five times of skin depth.

The simulation model as shown in Figure 5a is established by setting the current flow area of a single conductor of 1 mm² in COMSOL Multiphysics Coil Domain node and the substrate area of 50 cm × 50 cm (i.e., the side length of 50 cm, which meets the requirement of far larger than the diameter of the outer ring of the coil). The finite element calculation curve of coil ESR, i.e., the trigonometric curve in Figure 6, is obtained by parameterized scanning of the thickness of substrate. HIOKI 3532-50 LCR HiTESTER (HIOKI, Shanghai, China) (LCR analyzer) is introduced to build the experimental measurement platform in Figure 5b, and the measurement results of the variation of coil ESR including single-layer substrate are shown in the dotted line in Figure 6. Although Ferrite 1 with conductivity of 0 in Table 3 is an ideal material, it does not exist in reality, the experimental measurement curve of which is therefore not included in Figure 6. In addition, due to the complex

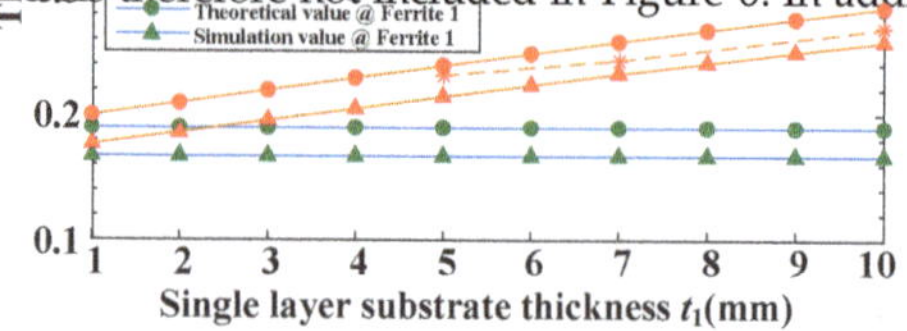

Figure 6. Coil effective series resistance (ESR) change curves when the thickness of single-layer substrate material changes.

As the curve in Figure 6 implies, the theoretical calculation, finite element simulation calculation and experimental measurement results of ferrite are highly consistent. The change of thickness of Ferrite 1 has no effect on coil ESR. Coil ESR increases with the increase of the thickness of Ferrite 2, and the resistance increases by 31.73% at the thickness of 10 mm compared with at the thickness of 1 mm, suggesting that the conductivity of substrate contributes to the main growth of coil ESR.

Also, according to Figure 6, coil ESR first decreases and then increases with the increase of the aluminum plate thickness. Coil ESR decreases to the lowest at the thickness of 5 mm of the aluminum plate substrate, which is a decrease of 38.66% compared with that at the thickness of 1 mm and a decrease of 19.25% compared with that at the thickness of 10 mm. The variation might be attributed to the combination of eddy current and skin effect in aluminum plate. In spite of the high consistency of the results of finite element simulation and experimental measurement of coil ESR of single-layer aluminum plate, there are significant deviations between the theoretical calculation results and the simulation and experimental results. This may be due to the assumptions of the theoretical calculation model are in the transition stage when the thickness of the aluminum plate is ferrite, only the ferrite with thickness of 5 mm, 7 mm, and 10 mm is available, which is thus measured in this paper.

As the curve in Figure 6 implies, the theoretical calculation, finite element simulation calculation and experimental measurement results of ferrite are highly consistent. The change of thickness of Ferrite 1 has no effect on coil ESR. Coil ESR increases with the increase of the thickness of Ferrite 2, and the resistance increases by 31.73% at the thickness of 10 mm compared with at the thickness of 1 mm, suggesting that the conductivity of substrate contributes to the main growth of coil ESR.

Also, according to Figure 6, coil ESR first decreases and then increases with the increase of the aluminum plate thickness. Coil ESR decreases to the lowest at the thickness of 5 mm of the aluminum plate substrate, which is a decrease of 38.66% compared with that at the thickness of 1 mm and a decrease of 19.25% compared with that at the thickness of 10 mm. The variation might be attributed to the combination of eddy current and skin effect in aluminum plate. In spite of the high consistency of the results of finite element simulation and experimental measurement of coil ESR of single-layer aluminum plate, there are significant deviations between the theoretical calculation results and the simulation and experimental results. This may be due to the assumptions of the theoretical calculation model are in the transition stage when the thickness of the aluminum plate is 5–50 times of the skin depth (i.e., 1–7 mm). Therefore, in the coil design and optimization analysis, a simple assessment of resistance loss might be made by theoretical calculation before a more accurate evaluation of resistance by the Coil Domain calculation of COMSOL Multiphysics, thus improving the complex calculation in the evaluation of equivalent resistance in coil design.

From the above, when the size of single-layer substrate is far larger than that of coil, shielding the magnetic field with materials with the smallest conductivity is one of the effective ways to reduce coil ESR. Due to the extremely high conductivity of substrate materials (such as aluminum), the increase of thickness can be resorted to in case of high requirement for the shielding effect, whereas there is an optimal value of thickness. The thickness of substrate shall be appropriately selected based on a comprehensive consideration of coil weight and working performance.

According to the analysis above, for aluminum plate with finite thickness and infinite area, the theoretical calculation results are low consistency with experimental measurement. The experimental measurement results are high consistency with COMSOL Multiphysics simulation calculation. Therefore, the methods of simulation and experimental measurement shall be adopted to explore the variation rule of coil ESR at changing area and volume of substrate.

3.2. When the Area of Substrate Material Changes

Based on the three materials in Table 3, the change of coil ESR when the area of different substrate materials changes from 5 cm × 5 cm to 26 cm × 26 cm, i.e., the side length changes from 5 cm to 26 cm, is calculated by simulation, as shown in Figure 7. In Figure 7a, the red curve is the change curve of coil ESR when the substrate material is Ferrite 1, and the blue curve is the change curve of coil ESR when the substrate material is Ferrite 2. Figure 7b is the change curve of coil ESR when the substrate material is aluminum plate.

Electronics 2020, 9, 1025

substrate materials changes from 5 cm × 5 cm to 26 cm × 26 cm, i.e., the side length changes from 5 cm to 26 cm, is calculated by simulation, as shown in Figure 7. In Figure 7a, the red curve is the change curve of coil ESR when the substrate material is Ferrite 1, and the blue curve is the change curve of coil ESR when the substrate material is Ferrite 2. Figure 7b is the change curve of coil ESR when the substrate material is aluminum plate.

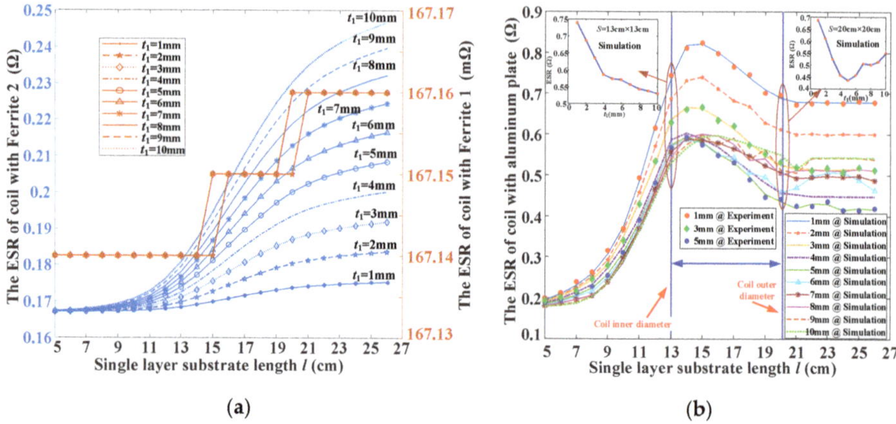

Figure 7. Coil ESR change curves when the area of single-layer substrate material changes: (**a**) With Ferrite 1 and Ferrite 2; (**b**) With aluminum plate.

As shown in Figure 7a, at changes of the thickness and area of Ferrite 1, coil ESR changes only within 0.01 mΩ, that is negligible, suggesting that the change of Ferrite 1 in area and thickness shall not affect coil ESR. At the constant thickness of Ferrite 2, coil ESR increases with the increase of area. At the thickness of 1 mm, coil ESR of that with the area of 26 cm × 26 cm increases by 4.92% compared with the area of 5 cm × 5 cm. At the thickness of 10 mm, coil ESR of that with the area of 26 cm × 26 cm increases by 47.49% compared with the area of 5 cm × 5 cm. At the constant area of Ferrite 2, the change of coil ESR is positively correlated with the increase of the thickness of Ferrite 2. At the area of 1 cm × 1 cm, coil ESR of that with the thickness of 10 mm increases by 3.15% compared with the thickness of 1 mm. At the area of 26 cm × 26 cm, coil ESR of that with the thickness of 10 mm increases by 40.77% compared with the thickness of 1 mm. Therefore, in terms of practical application on the premise of achieving mutual inductance between coils, coil ESR should be reduced by minimizing the thickness and area of Ferrite 2.

From the curve in Figure 7b, it can be concluded that the effect of aluminum plate substrate on coil ESR is different from that of ferrite. Coil ESR decreases with the increase of the thickness of the aluminum plate when the side length of the aluminum plate is less than 13 cm (the diameter of the inner ring of the coil), and increases with the increase of the area of the aluminum plate at the same thickness. At the area of 9 cm × 9 cm, Coil ESR of that at the thickness of 10 mm is 21.99% lower than that of 1 mm thickness. At the thickness of 5 mm, coil ESR of that at the area of 13 cm × 13 cm is 32.6% higher than that of 5 cm × 5 cm.

As the side length of the aluminum plate continues to increase from 13 cm, coil ESR shall gradually increase to the highest before starting to decrease and finally tend to be stable. At the stable thickness, coil ESR of 1 mm is much greater than that of other aluminum plates with the same area. At the area of 20 cm × 20 cm, coil ESR of 1 mm is 57.57% larger than that of 5 mm thickness and 25.02% larger than that of 10 mm thickness. Coil ESR of the plate with the thickness of 5 mm is smaller than that of the plate with the same area. When the side length of the aluminum plate with the diameter of 20 cm is close to the outer ring of the coil. Thereafter, with the increase of aluminum plate area, there is basically no change to coil ESR, and it remains the smallest at the thickness of 5 mm.

According to the small diagram in the upper right corner of Figure 7b, at the area of 20 cm × 20 cm of the aluminum plate, coil ESR first decreases and then increases with the increase of the aluminum plate thickness, and reaches the minimum at the thickness of 5 mm. This implies that, when the side length of the aluminum plate is not less than the diameter of the outer ring of the coil, the value of coil ESR is the smallest at the thickness of 5 mm of the aluminum plate, which however is unavailable in

According to the small diagram in the upper right corner of Figure 7b, at the area of 20 cm × 20 cm of the aluminum plate, coil ESR first decreases and then increases with the increase of the aluminum plate thickness, and reaches the minimum at the thickness of 5 mm. This implies that, when the side length of the aluminum plate is not less than the diameter of the outer ring of the coil, the value of coil ESR is the smallest at the thickness of 5 mm of the aluminum plate, which however is unavailable in reality due to the limitations of the application environment. This is suggested to take into account the appropriate thickness and area of aluminum plate substrate based on the system performance, coil size and the application requirements.

Since the weight of the whole packaged coil is one of the major concerns in the industrial application of WPT system, the variation rule of coil ESR in case of constant weight and changing thickness and area of single-layer substrates shall be studied.

3.3. When the Weight of Substrate Material is Constant

At the constant substrate density, the constancy of weight might be equivalent to the constancy of volume. According to Section 3.1, in the area of 20 cm × 20 cm, the volume of 5 mm, 7 mm, and 10 mm thickness are taken as the reference respectively, that is, to maintain a constant volume of $V = 200\,\text{cm}^3$, $V = 280\,\text{cm}^3$, and $V = 400\,\text{cm}^3$. The simulation software is adopted to calculate the variation of coil ESR at the thickness of 1–10 mm of the three materials in Table 3, as shown in Figure 8.

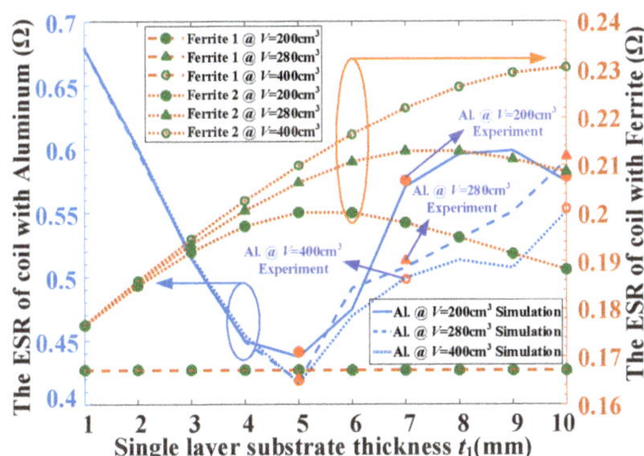

Figure 8. Coil ESR change curves when the weight of single-layer substrate material is constant.

As shown in Figure 8, Ferrite 1 with the same volume and different thickness has basically no effect on coil ESR. In the case of constant volume of Ferrite 2, coil ESR increases first and then decreases with the increase of thickness, and there is a maximum value. The maximum value of coil ESR of each volume appears at the area of 20 cm × 20 cm, i.e., at the thickness of 5 mm, 7 mm, and 10 mm, respectively. However, for a single-layer aluminum plate substrate, with the increase of thickness, coil ESR decreases first and then increases at the constant volume. Coil ESR of the three reference volumes is the smallest at the thickness 5 mm and the area of 20 cm × 20 cm of the aluminum plate.

Therefore, at the constant volume, the closer the side length of Ferrite 2 is to the diameter of the coil outer ring, the larger coil ESR is and the larger the variation is. The minimum value of coil ESR of aluminum plate substrate appears at the thickness of 5 mm aluminum plate. To sum up, when the coil transmission performance and weight requirements shall be met, the smaller the thickness and the larger the area of Ferrite 2 are, and the closer the thickness of aluminum plate is to 5 mm, the better for coil transmission performance and weight requirements shall be met, the smaller the thickness and the larger the area of Ferrite 2 are, and the closer the thickness of aluminum plate is to 5 mm, the better for single-layer substrates.

4. Analysis of Coil ESR with Double-Layer Substrate

In this section, the effect of double-layer substrate composed of different materials on coil ESR shall be identified. There are two types of combinations of double-layer substrate, namely Aluminum plate + Ferrite 1 and Aluminum plate + Ferrite 2. Based on the coil shown in Figure 4, the simulation and experiment platform, as shown in Figure 9, is established to calculate coil ESR and the results are compared with the theoretical calculation results.

In this section, the effect of double-layer substrate composed of different materials on coil ESR shall be identified. There are two types of combinations of double-layer substrate, namely Aluminum plate + Ferrite 1 and Aluminum plate + Ferrite 2. Based on the coil shown in Figure 4, the simulation and experiment platform, as shown in Figure 9, is established to calculate coil ESR and the results are compared with the theoretical calculation results.

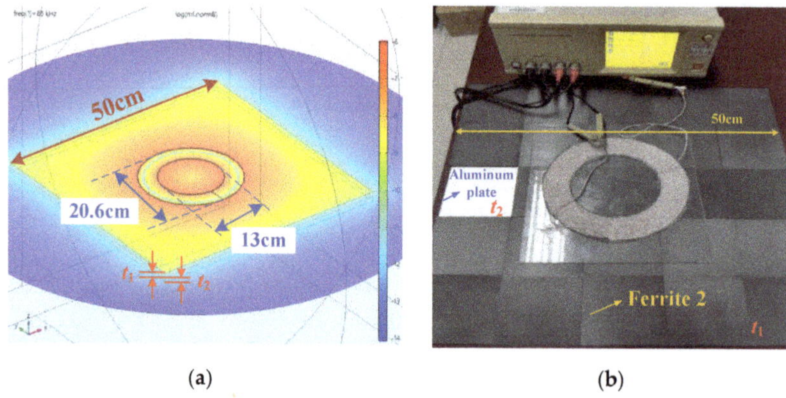

(a)   (b)

Figure 9. The model of coil with double-layer substrate (a) Simulation in COMSOL, (b) Experimental test platform.

4.1. When the Thickness of Double-Layer Substrate Material Changes

Similar to Section 3.1, we first analyze the case when the side length of the double-layer substrate is far larger than the diameter of the outer ring of the coil. On account of the data in Tables 2 and 3, when the thickness of each layer in the double-layer substrate varies between 1 mm and 10 mm, Equation (14) is adopted to carry out the theoretical calculation of coil ESR to obtain the theoretical value curve of coil ESR, i.e., the orange curve in Figure 10. COMSOL Multiphysics is introduced to build the simulation model in Figure 9a, which is used to conduct parametric scanning of the substrate's thickness and obtain the simulation value curve of the variation of coil ESR. At the same time, LCR analyzer is introduced to build the experimental platform in Figure 9b. Similar to the ferrite in single-layer substrates, only the variation curve of coil ESR in the first layer of Ferrite 2 in double-layer substrate is measured at the thickness of 5 mm, 7 mm, and 10 mm. To conduct a more intuitive comparison with the simulation value and the measured value, only the case of "Aluminum plate + Ferrite 2" is shown in Figure 10.

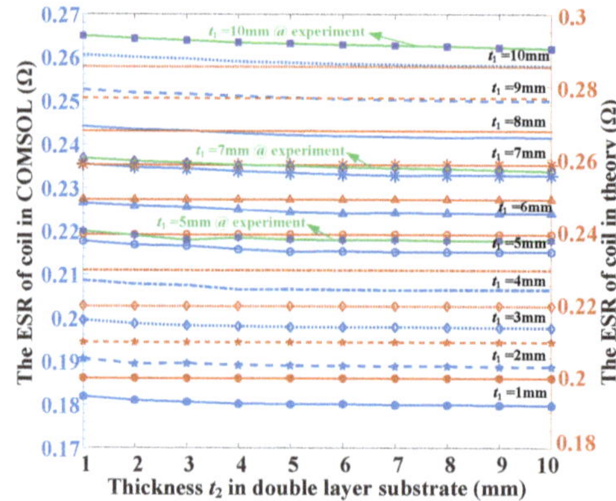

Figure 10. Coil ESR change curves when the thickness of double-layer substrate material changes.

As indicated by the abscissa in Figure 10, at the constant thickness of ferrite in the first layer of double-layer substrate, the change in the thickness of aluminum plate in the second layer of substrate has little effect on coil ESR. That is to say, when the double-layer substrate is large enough and the thickness of ferrite substrate in the first layer is determined, the change in the thickness of aluminum plate substrate in the second layer has no effect on coil ESR. According to the ordinate in Figure 10, coil ESR increases with the increase of the thickness of Ferrite 2 when the thickness of aluminum plate in the second layer of double-layer substrate is constant. At the thickness of $t_2 = 5$ mm of aluminum plate in the second layer of double-layer substrate, coil ESR at $t_1 = 10$ mm thickness of substrate in the first layer is 43.68% larger than that at $t_1 = 1$mm thickness. Therefore, it plate substrate in the second layer has no effect on coil ESR. According to the ordinate in Figure 10, coil ESR increases with the increase of the thickness of Ferrite 2 when the thickness of aluminum plate in the second layer of double-layer substrate is constant. At the thickness of $t_2 = 5$ mm of aluminum plate in the second layer of double-layer substrate, coil ESR at $t_1 = 10$ mm thickness of substrate in the first layer is 43.68% larger than that at $t_1 = 1$mm thickness. Therefore, it is suggested to minimize the thickness of ferrite substrate in the first layer on the premise of achieving the mutual inductance and shielding requirements between coils.

It can also be seen from Figure 10 that the coil ESR is almost unaffected by the change of thickness t_2. So, the coil ESR change law of theoretical, simulation and experimental results are a high consistency, when the size of the double-layer substrate is infinite compared with the coil. Nevertheless, the theoretical calculation is only applicable to the cases where the side length of the substrate is far larger than the diameter of the outer ring of the coil. When the side length of the substrate is close to the diameter of the outer ring of the coil, the next step of this paper is to analyze and study by combining simulation and measurement.

4.2. When the Area of Double-Layer Substrate Material Changes

In the case that the substrate size is close to that of the coil, the impact of the change in the area of double-layer substrate on coil ESR at the ferrite thickness of 5 mm, 7 mm, and 10 mm shall be studied in this section based on the actual situation. The six types of combinations of double-layer substrate to be studied in this section are as shown in Table 4.

Table 4. The combinations parameters of double layer substrate.

Combinations	1st Layer Material	1st Layer Thickness	2nd Layer Material	2nd Layer Thickness
1	Ferrite 2	5 mm	Aluminum	$1\,\text{mm} \leq t_2 \leq 10\,\text{mm}$
2	Ferrite 1	5 mm	Aluminum	$1\,\text{mm} \leq t_2 \leq 10\,\text{mm}$
3	Ferrite 2	7 mm	Aluminum	$1\,\text{mm} \leq t_2 \leq 10\,\text{mm}$
4	Ferrite 1	7 mm	Aluminum	$1\,\text{mm} \leq t_2 \leq 10\,\text{mm}$
5	Ferrite 2	10 mm	Aluminum	$1\,\text{mm} \leq t_2 \leq 10\,\text{mm}$
6	Ferrite 1	10 mm	Aluminum	$1\,\text{mm} \leq t_2 \leq 10\,\text{mm}$

The results in Figure 11 are obtained by introducing finite element simulation and experimental measurements. The abscissa represents the change of area with the change of side length of double-layer substrate in Figure 11.

It can be concluded from Figure 11 that, when the side length of the substrate is smaller than the diameter of the inner ring of the coil, coil ESR gradually reaches a peak with the increase of the side length of the substrate. Coil ESR decreases and tends to be stable when the side length of substrate continues to increase to be larger than the diameter of outer coil ring. In Figure 11a,c,e, when the thickness of the aluminum plate is 5 mm and the first layer thickness of double-layer substrate is 5 mm, 7 mm, and 10 mm respectively, coil ESR of that with the area of 26 cm × 26 cm decreases by 53.34%, 39.52%, and 31.46% compared that with the area of 14 cm × 14 cm. That is to say, with the constant thickness of double-layer substrate, coil ESR increases first and then decrease with the increase of the substrate area. Therefore, when the thickness of double-layer substrate is constant and the side length is larger than the diameter of the inner ring of the coil, the continuous increase of the side length of double-layer substrate shall effectively reduce coil ESR.

At the constant side length of double-layer substrate, coil ESR decreases with the increase of the thickness at the thickness of 1–5 mm of the aluminum plate. When the thickness of aluminum plate is more than 5 mm, coil ESR hardly changes with the thickness. Therefore, increase of the thickness of aluminum plate in double-layer substrate shall reduce coil ESR when the side length of the substrate is constant.

The results in Figure 11 are obtained by introducing finite element simulation and experimental measurements. The abscissa represents the change of area with the change of side length of double-layer substrate in Figure 11.

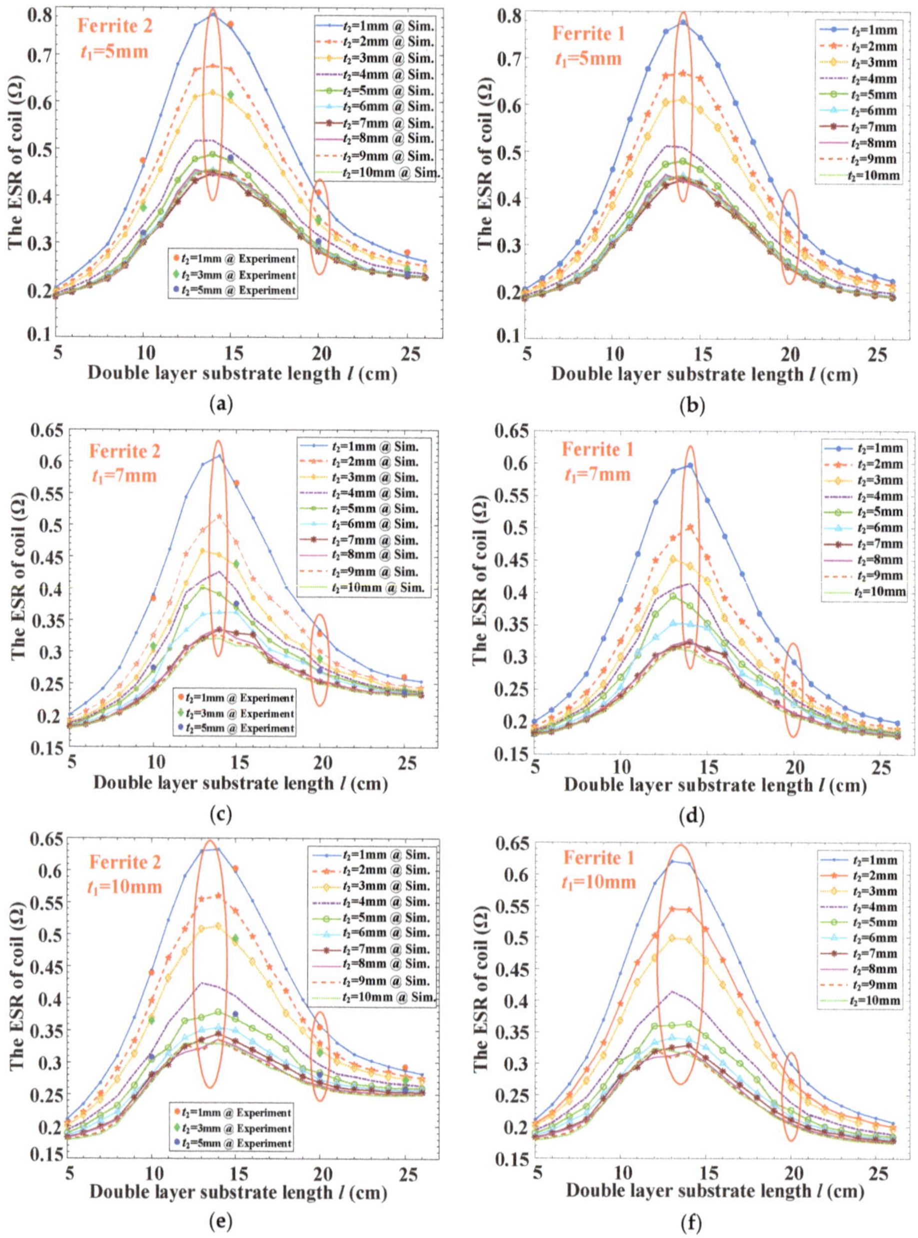

Figure 11. Coil ESR change curves when the area of double-layer substrate material changes: (a) Combination 1; (b) Combination 2; (c) Combination 3; (d) Combination 4; (e) Combination 5; (f) Combination 6.

It can be concluded from Figure 11 that, when the side length of the substrate is smaller than the diameter of the inner ring of the coil, coil ESR gradually reaches a peak with the increase of the side length. When the side length of double-layer substrate is larger than the diameter of the coil outer ring and the thickness of the aluminum plate is constant, the double-layer of aluminum plate in double-layer substrate shall reduce coil ESR when the side length of the substrate is constant.

Compared with Ferrite 1, with the same thickness of ferrite in the first layer of double-layer substrate, there is almost no change to coil ESR, suggesting that coil ESR is mainly affected by the aluminum plate in double-layer substrate. When the side length of double-layer substrate is larger than the diameter of the coil outer ring and the thickness of the aluminum plate is constant, the double-layer substrate with Ferrite 1 assumes smaller impact on coil ESR. When the thickness of the aluminum plate is 5 mm, the first layer of double-layer substrate is respectively 5 mm, 7 mm and 10 mm thick and has an area of 26 cm × 26 cm, coil ESR of Ferrite 1 is 16.86%, 22.68%, and 28.91% smaller than that of Ferrite 2. That is, ferrite with lower conductivity might be employed to reduce coil ESR when the size of double-layer substrate is larger than the diameter of the outer ring of coil.

In summary, for a double-layer substrate, coil ESR shall be reduced by increasing the side length of the substrate, increasing the thickness of aluminum plate substrate in the second layer and reducing the conductivity of ferrite in the first layer. Similar to the analysis of a single-layer substrate, research on a double-layer substrate at constant weight shall be carried out as follows.

4.3. When the Weight of Double-Layer Substrate Material is Constant

For double-layer substrate materials, the constant weight can also be equivalent to the constant material volume. In the area of 20 cm × 20 cm, the volume of 5 mm and 7 mm thickness are also taken as the reference respectively, that is, to maintain a constant volume of $V = 200\ \text{cm}^3$, $V = 280\ \text{cm}^3$ and $V = 400\ \text{cm}^3$. The variation trend of coil ESR is analyzed by changing the substrate thickness of the two layers at the same time. When the second layers of double-layer substrates are both aluminum plate, the value of coil ESR is calculated by COMSOL Multiphysics software when the first layer of substrate is Ferrite 1 and Ferrite 2 respectively, as shown in Figure 12.

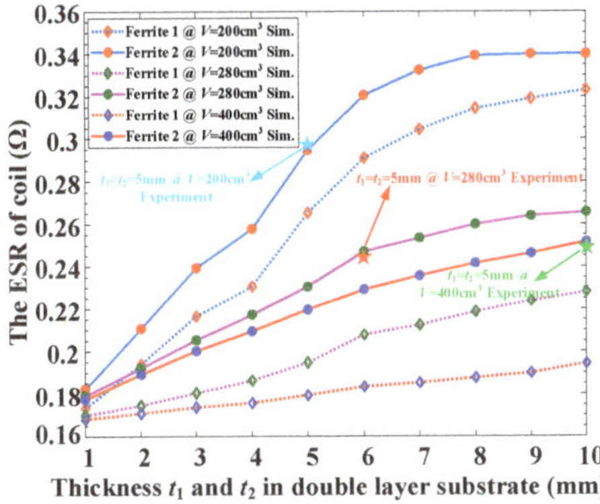

Figure 12. Coil ESR change curves when the weight of double-layer substrate material is constant.

As shown in Figure 12, at the same thickness of each layer of double-layer substrate, the larger the total volume of the substrate, the smaller coil ESR is. At the constant volume of double-layer substrate, coil ESR shall gradually increase with the increase of the thickness of each layer of substrate. When $V = 200\ \text{cm}^3$, the thickness of each layer of substrate greater than 5 mm shall be equivalent to the side length of each layer of substrate less than 20 cm, that is, smaller than the diameter of the outer ring of the coil. According to the theory in Section 4.2, coil ESR shall demonstrate a sharp growth, which is consistent with the results in Figure 11. Hence, in the case of coil weight limitation, the thinner the thickness of double-layer substrate, the better. Of course, the conclusion is based on the volume reference value when $V = 200\ \text{cm}^3$, $V = 280\ \text{cm}^3$ and $V = 400\ \text{cm}^3$, and in the case of larger volume, the variation law of coil ESR with substrate thickness might be different. For WPT system, however, it is of significance to only discuss the case when the substrate is close to the coil size.

5. Conclusions

In view of the increase of coil ESR caused by substrate, based on theoretical modeling, simulation calculation, and experimental measurement, a systematic study on the variation rule of coil ESR with a single-layer substrate and double-layer substrate is carried out in this paper. According to the results, when the size of single-layer substrate is much larger than the coil size, the material with the smallest conductivity should be selected as the substrate. The thickness should be increased in case of large conductivity of the substrate, with an optimal value whereas. When the size of single-layer substrate is close to the diameter of the coil outer ring, coil ESR is proportional to the area of ferrite. However, with the increase of the area of aluminum plate, coil ESR increases first and then decreases. At the constant volume, the closer the side length of ferrite substrate is to the diameter of coil outer ring, the larger coil ESR is. Coil ESR value with aluminum plate reaches the minimum at the thickness of 5 mm. For the double-layer substrate, when its size is far larger than the coil size, coil ESR is only subject to the thickness of ferrite substrate in the first layer, rather than that of aluminum plate in the second layer. When the size of substrate is close to the diameter of coil outer ring, coil ESR increases first and then decreases with the increase of area and finally tends to be constant. The peak value appears when the size is close to the diameter of coil inner ring. At the constant volume, coil ESR increases with the increase of the thickness of double-layer substrate. It is believed that the foresaid rules shall assist researchers' efficient selection of the thickness, area and volume of substrate, thus reducing the coil loss and realizing the optimal design of WPT system. Although this paper summarizes the influence of substrate with different size on the coil ESR, the variation law of the coil ESR needs further study when the excitation frequency of the coil, the distance between the substrate and the coil, and the shape of the substrate change.

Author Contributions: Conceptualization, J.L.; methodology, J.L., R.W., and M.Z.; software, J.L.; investigation, M.Z.; writing—original draft preparation, J.L.; writing—review and editing, L.T. and X.H. All authors have read and agreed to the published version of the manuscript.

Funding: This research was funded by the National Key Research and Development Project, grant number 2018YFB0106300.

Conflicts of Interest: The authors declare no conflict of interest.

References

1. Kurs, A.; Karalis, A.; Moffatt, R.; Joannopoulos, J.D.; Fisher, P.; Soljacic, M. Wireless power transfer via strongly coupled magnetic resonances. *Science* **2007**, *317*, 83–86. [CrossRef] [PubMed]
2. Huang, X.; Wang, W.; Tan, L. Technical progress and application development of magnetic coupling resonant wireless power transfer. *Autom. Electr. Power Syst.* **2017**, *41*, 2–14.
3. Liu, H.; Huang, X.; Tan, L.; Guo, J.; Wang, W.; Yan, C.; Xu, C. Dynamic Wireless Charging for Inspection Robots Based on Decentralized Energy Pickup Structure. *IEEE Trans. Ind. Inform.* **2018**, *14*, 1786–1797. [CrossRef]
4. Wen, F.; Chu, X.; Li, Q.; Gu, W. Compensation Parameters Optimization of Wireless Power Transfer for Electric Vehicles. *Electronics* **2020**, *9*, 789. [CrossRef]
5. Zhu, J.; Ban, Y.; Zhang, Y.; Yan, Z.; Xu, R.; Mi, C.C. Three-Coil Wireless Charging System for Metal-Cover Smartphone Applications. *IEEE Trans. Power Electr.* **2020**, *35*, 4847–4858. [CrossRef]
6. Nguyen, M.T.; Nguyen, C.V.; Truong, L.H.; Le, A.M.; Quyen, T.V.; Masaracchia, A.; Teague, K.A. Electromagnetic Field Based WPT Technologies for UAVs: A Comprehensive Survey. *Electronics* **2020**, *9*, 461. [CrossRef]
7. Huh, J.; Lee, W.; Cho, G.; Lee, B.; Rim, C. Characterization of Novel Inductive Power Transfer Systems for On-Line Electric Vehicles. In *Proceedings of the 2011 Twenty-Sixth Annual IEEE Applied Power Electronics Conference and Exposition (APEC), Fort Worth, TX, USA, 6–11 March 2011*; IEEE: Piscataway, NJ, USA, 2011; pp. 1975–1979.
8. Zhang, C.; Lin, D.; Tang, N.; Hui, S.Y.R. A Novel Electric Insulation String Structure With High-Voltage Insulation and Wireless Power Transfer Capabilities. *IEEE Trans. Power Electr.* **2018**, *33*, 87–96. [CrossRef]
9. Abou Houran, M.; Yang, X.; Chen, W. Magnetically Coupled Resonance WPT: Review of Compensation Topologies, Resonator Structures with Misalignment, and EMI Diagnostics. *Electronics* **2018**, *7*, 296. [CrossRef]
10. Joo, J.; Kwak, S.I.; Kwon, J.H.; Song, E. Simulation-Based System-Level Conducted Susceptibility Testing Method and Application to the Evaluation of Conducted-Noise Filters. *Electronics* **2019**, *8*, 908. [CrossRef]

11. International Commission on Non-Ionizing Radiation Protection. Guidelines for limiting exposure to time-varying electric and magnetic fields (1 Hz to 100 kHz). *Health Phys.* **2010**, *99*, 818–836.

12. Yan, Z.; Li, Y.; Zhang, C.; Yang, Q. Influence Factors Analysis and Improvement Method on Efficiency of Wireless Power Transfer Via Coupled Magnetic Resonance. *IEEE Trans. Magn.* **2014**, *50*, 1–4.

13. Moon, J.; Shin, H.; Jeong, H.; Kim, S. Analysis of Receiver Circuit Efficiency for Wireless Power Transfer System According to Operating Frequency Variation. In *Proceedings of the IEEE International Symposium on Consumer Electronics, Madrid, Spain, 24–26 June 2015*; IEEE: Piscataway, NJ, USA, 2015.

14. Jeong, S.; Lin, T.; Tentzeris, M.M. Range-adaptive Impedance Matching of Wireless Power Transfer System Using a Machine Learning Strategy Based on Neural Networks. In *Proceedings of the 2019 IEEE MTT-S International Microwave Symposium (IMS) 2019, Boston, MA, USA, 2–7 June 2019*; IEEE: Piscataway, NJ, USA, 2019; pp. 1423–1425.

15. Zhao, L.; Thrimawithana, D.J.; Madawala, U.K.; Hu, A.P.; Mi, C.C. A Misalignment-Tolerant Series-Hybrid Wireless EV Charging System With Integrated Magnetics. *IEEE Trans. Power Electr.* **2019**, *34*, 1276–1285. [CrossRef]

16. Li, H.; Liu, Y.; Zhou, K.; He, Z.; Li, W.; Mai, R. Uniform Power IPT System With Three-Phase Transmitter and Bipolar Receiver for Dynamic Charging. *IEEE Trans. Power Electr.* **2019**, *34*, 2013–2017. [CrossRef]

17. Hurley, W.G.; Duffy, M.C. Calculation of self and mutual impedances in planar magnetic structures. *IEEE Trans. Magn.* **1995**, *31*, 2416–2422. [CrossRef]

18. Hurley, W.G.; Duffy, M.C. Calculation of self- and mutual impedances in planar sandwich inductors. *IEEE Trans. Magn.* **1997**, *33*, 2282–2290. [CrossRef]

19. Acero, J.; Alonso, R.; Burdio, J.M.; Barragan, L.A.; Puyal, D. Analytical equivalent impedance for a planar circular induction heating system. *IEEE Trans. Magn.* **2006**, *42*, 84–86. [CrossRef]

20. Acero, J.; Alonso, R.; Barragan, L.A.; Burdio, J.M. Modeling of Planar Spiral Inductors Between Two Multilayer Media for Induction Heating Applications. *IEEE Trans. Magn.* **2006**, *42*, 3719–3729. [CrossRef]

21. Su, Y.P.; Liu, X.; Hui, S.Y.R. Extended theory on the inductance calculation of planar spiral windings including the effect of double-layer electromagnetic shield. *IEEE Trans. Power Electr.* **2008**, *23*, 2052. [CrossRef]

22. Acero, J.; Alonso, R.; Burdio, J.M.; Barragan, L.A.; Puyal, D. Frequency-Dependent Resistance in Litz-Wire Planar Windings for Domestic Induction Heating Appliances. *IEEE Trans. Power Electr.* **2006**, *21*, 856–866. [CrossRef]

23. Lope, I.; Acero, J.; Carretero, C. Analysis and Optimization of the Efficiency of Induction Heating Applications with Litz-Wire Planar and Solenoidal Coils. *IEEE Trans. Power Electr.* **2016**, *31*, 5089–5101. [CrossRef]

24. Liu, J.; Deng, Q.; Czarkowski, D.; Kazimierczuk, M.K.; Zhou, H.; Hu, W. Frequency Optimization for Inductive Power Transfer Based on AC Resistance Evaluation in Litz-Wire Coil. *IEEE Trans. Power Electr.* **2019**, *34*, 2355–2363. [CrossRef]

25. Nguyen, M.H.; Fortin Blanchette, H. Optimizing AC Resistance of Solid PCB Winding. *Electronics* **2020**, *9*, 875. [CrossRef]

26. Tang, X.; Sullivan, C.R. Stranded wire with uninsulated strands as a low-cost alternative to litz wire. In *Proceedings of the IEEE 34th Annual Conference on Power Electronics Specialist, 2003, PESC '03, Acapulco, Mexico, 15–19 June 2003*; IEEE: Piscataway, NJ, USA, 2003; pp. 289–295.

27. Howe, G.W.O. The High-Frequency Resistance of Multiply-Stranded Insulated Wire. *Proc. R. Soc. Lond. Ser. A Contain. Pap. Math. Phys. Character* **1917**, *93*, 468–492.

Article

Compensation Parameters Optimization of Wireless Power Transfer for Electric Vehicles

Feng Wen [1,2,*], Xiaohu Chu [1], Qiang Li [1] and Wei Gu [1]

[1] School of Automation, Nanjing University of Science and Technology, Nanjing 210094, China; chu@njust.edu.cn (X.C.); chnliqiang@njust.edu.cn (Q.L.); Jingly@njust.edu.cn (W.G.)
[2] Jiangsu Provincial Key Laboratory of Smart Grid Technology and Equipment, Nanjing 210096, China
* Correspondence: wen@njust.edu.cn; Tel.: +86-159-9620-0950

Received: 16 March 2020; Accepted: 7 May 2020; Published: 11 May 2020

Abstract: For wireless charging of electric vehicles (EVs), increasing the output power level is particularly important. In this paper, the purpose of improving the output power while maintaining optimal transmission efficiency is achieved by optimizing the parameters of the compensation topology under the premise that the coupled coils of the system does not need to be redesigned. The series-series (SS) and hybrid-series-parallel (LCC, composed by an inductor and two capacitors) compensation topology are studied. The influence factors of load resistance to achieve optimal efficiency, the influence of LCC compensation parameters on the power output level, and the influence of parameter changes on system safety are analyzed. Theorical results show that by rationally designing the LCC compensation parameters, larger output power and optimal transfer efficiency can be achieved under different load resistance by adjusting the inductances of the primary and secondary compensation circuits. The output power of the optimized system with adjusted LCC compensation topology is increased by 64.2% with 89.8% transfer efficiency under 50 ohms load in experiments. The correctness and feasibility of this parameter design method are verified by both theorical and experimental results.

Keywords: Wireless Power Transfer (WPT); compensation topology; optimal load; output power level; electric vehicle (EV)

---

1. Introduction

Since the concept of magnetic coupled resonant wireless power transfer (MCR-WPT) was proposed by MIT in 2007, the technology has developed rapidly and has been widely applied in implantable medical devices, home appliances, mobile devices and electric vehicles [1–8]. Increasing transmission power is particularly important for wireless charging of electric vehicles. The main ways to improve power levels are: (1) power supply parallel topology, which uses multiple parallel power supplies for power distribution, but the parallel topology is too redundant [9]; (2) multi-phase parallel, there are shortcomings of current imbalance between phases and serious coil loss. In addition, the compensation topology also has a great impact on the output power level.

At present, the basic compensation topologies are series-series (SS), series-parallel (SP), parallel-series (PS), and parallel-parallel (PP) [10,11]. The most widely used is the SS topology, but this topology has inherent disadvantages: once coil parameters are determined, the rated output power can be regulated only by the input voltage. In addition to the four basic resonance topologies described above, hybrid-series-parallel (LCC, composed by an inductor and two capacitors) compensation topology has been extensively studied for its excellent performance. It can implement zero phase angle (ZPA) and zero voltage switching (ZVS), and its output current is independent of the load [12–17].

Most of the current studies on WPT compensation topology focus on improving the output performance by adopting complex control strategies. However, the load resistance to achieve optimal efficiency of SS topology WPT system is unchangeable after the coil-parameters are fixed whereas the one of LCC topology can be adjusted according to the parameters of compensation topology. In this paper, the WPT system is modeled and the transfer performance is analyzed. Methods are proposed to increase output power level while maintaining optimal transmission efficiency without redesign of the coupled coils by optimizing the parameters of LCC compensation topology. Both theorical and experimental results indicate that the proposed parameter optimization strategy is effective in improving transfer efficiency and adjusting output power under different load resistances. It avoids the disadvantages of redesigning the coil due to power level changes to maintain high transmission efficiency. At the same time, it has obvious advantages in safety under the condition of output short circuit.

2. Theoretical Analysis

2.1. System Analysis

Figure 1 shows the schematic diagram of the magnetic coupled wireless power transfer system, where AC power on the grid side is rectified to form DC power, the high-frequency inverter is composed of H-bridge, and the direct current is converted by inverter circuit into alternating current with a system rated frequency, which flows through the transmitter end and transmits the power to the receiver coil. The primary and secondary compensation circuit is composed of SS or LCC compensation topology. The transmitter and receiver coils are planar spiral coils designed in Section 3, and the receiver end is connected to electric vehicle battery through a rectifier circuit.

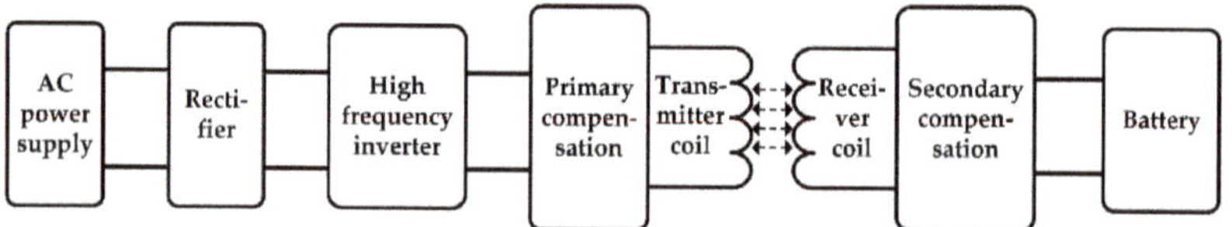

Figure 1. The structure diagram of wireless power transfer system.

Because the resonance part of wireless power transfer system has the characteristics of band-pass filtering, only the fundamental wave component is considered later, and the square wave generated by the full-bridge inverter can be equivalent to an AC voltage source for theoretical analysis. The square wave is expanded according to the Fourier series, the duty cycle is set to D, and the fundamental wave of the square wave whose amplitude is U_d is as follow:

$$f_t = \frac{4U_d}{\pi} \sin(\pi D) \cos(\omega t - \pi D) \tag{1}$$

The battery is a typical non-linear load. With the use of battery, the battery charging current and the state of charge are different, and different external load characteristics are displayed. In order to simplify the difficulty of system analysis, the battery is generally equivalent to a resistive load according to the ratio of charging voltage and current of battery and then analyzed. Therefore, the battery and its management system are equivalent to a load in this paper.

2.2. Theoretical Analysis of SS and LCC Compensation Topology

Modeling and analysis of SS-type and LCC-type compensation topologies are shown below. Figures 2 and 3 show simplified topologies of SS-type and LCC-type resonant circuits, respectively. The parasitic resistance of inductors and capacitors are much smaller than the internal resistance of coil, so it can be ignored. The subsequent experimental results show that this approximate analysis does not affect the accuracy of experiment.

2.2.1. Analysis of SS-type Resonance Circuit

The SS topology is the most basic resonance structure. Both primary and secondary side use series resonance. The equivalent circuit is shown in Figure 2, where U_S is a voltage-stabilized source, the source internal resistance is ignored, M is the mutual inductance between two coils.

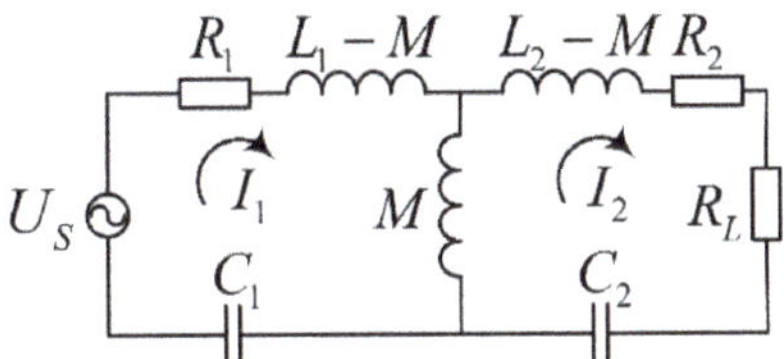

Figure 2. The equivalent circuit model of SS-type.

According to Figure 2 and Kirchhoff voltage law (KVL), we can obtain:

$$\begin{cases} (R_1 + j\omega L_1 + \dfrac{1}{j\omega C_1})I_1 - j\omega M I_2 = U_s \\ (R_2 + R_L + j\omega L_2 + \dfrac{1}{j\omega C_2})I_2 - j\omega M I_1 = 0 \end{cases} \tag{2}$$

When the system meets the resonance conditions:

$$f_0 = \frac{1}{2\pi \sqrt{L_1 C_1}} = \frac{1}{2\pi \sqrt{L_2 C_2}} \tag{3}$$

When R_L is much greater than R_1 and R_2, R_1 and R_2 can be ignored, the currents are as follows:

$$\begin{cases} I_1 = \dfrac{U_s R_L}{(\omega M)^2} \\ I_2 = \dfrac{U_s}{\omega M} \end{cases} \tag{4}$$

The transmission power and efficiency of the system are:

$$P_{ss} = \frac{U_s^2 R_L}{(\omega M)^2} \tag{5}$$

$$\eta_{ss} = 1 - \frac{R_L R_1}{(\omega M)^2} - \frac{R_2}{R_L} \tag{6}$$

The maximum efficiency and corresponding load value are:

$$\eta_{ss\_max} = 1 - \frac{2 \sqrt{R_1 R_2}}{\omega M} \tag{7}$$

$$R_{ss} = \sqrt{\frac{R_2}{R_1}(\omega M)^2} \tag{8}$$

Due to the inverse relationship between distance and mutual inductance, in practical applications, it is common for charging object to be far away from primary coil. It can be seen from (4) and (5) that when the secondary side is open, the primary current will increase sharply; when a short circuit occurs on the secondary side, the primary current and transmission power will decrease. On the other hand, in the case of the SS structure with constant coil parameters, the load resistance to achieve optimal efficiency is unchangeable.

2.2.2. Analysis of LCC-type Resonance Circuit

The LCC structure is a new type of composite resonant structure. As shown in Figure 3, L_{f1}, C_1, C_{p1} and L_{f2}, C_2, C_{p2} are the corresponding resonant circuit units of primary and secondary coils, respectively.

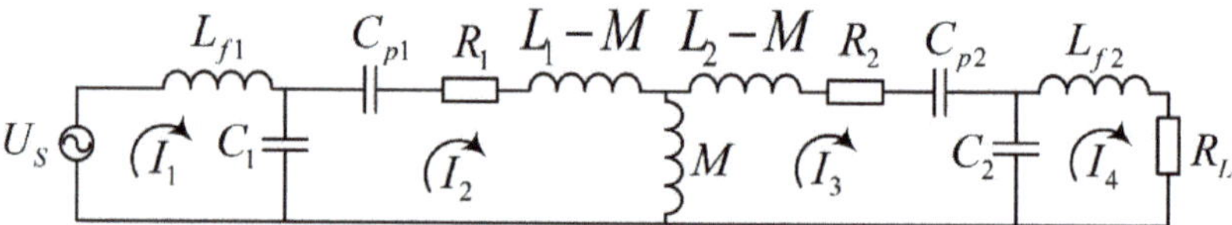

Figure 3. The equivalent circuit model of LCC-type.

Similarly, we can formulate KVL equation according to Figure 3:

$$
\begin{cases}
(j\omega L_{f1} + \dfrac{1}{j\omega C_1})I_1 - \dfrac{1}{j\omega C_1}I_2 = U_s \\[2mm]
(R_1 + j\omega L_1 + \dfrac{1}{j\omega C_1} + \dfrac{1}{j\omega C_{p1}})I_2 - \dfrac{1}{j\omega C_1}I_1 - j\omega M I_3 = 0 \\[2mm]
(R_2 + j\omega L_2 + \dfrac{1}{j\omega C_2} + \dfrac{1}{j\omega C_{p2}})I_3 - \dfrac{1}{j\omega C_2}I_4 - j\omega M I_2 = 0 \\[2mm]
(R_L + j\omega L_{f2} + \dfrac{1}{j\omega C_2})I_4 - \dfrac{1}{j\omega C_2}I_3 = 0
\end{cases}
\tag{9}
$$

The resonance conditions are as follows:

$$
\begin{aligned}
C_1 &= \frac{1}{\omega^2 L_{f1}}, \quad C_{p1} = \frac{1}{\omega^2 L_1 - \dfrac{1}{C_1}}, \\[3mm]
C_2 &= \frac{1}{\omega^2 L_{f2}}, \quad C_{p2} = \frac{1}{\omega^2 L_2 - \dfrac{1}{C_2}}
\end{aligned}
\tag{10}
$$

The currents of the loops are:

$$
\begin{cases}
I_1 = \omega^6 C_1{}^2 C_2{}^2 M^2 R_L U_s \\
I_2 = -\omega C_1 U_s \\
I_3 = \omega^4 C_1 C_2{}^2 M R_L U_s \\
I_4 = -\omega^3 C_1 C_2 M U_s
\end{cases}
\tag{11}
$$

The transmission power and efficiency of the system are:

$$
P_{LCC} = \frac{M^2 U_s{}^2 R_L}{\omega^2 L_{f1}{}^2 L_{f2}{}^2}
\tag{12}
$$

$$
\eta_{LCC} = 1 - \frac{R_1 L_{f2}{}^2}{R_L M^2} - \frac{R_2 R_L}{\omega^2 L_{f2}{}^2}
\tag{13}
$$

The maximum efficiency and corresponding load values are:

$$
\eta_{LCC\_max} = 1 - \frac{2\sqrt{R_1 R_2}}{\omega M}
\tag{14}
$$

$$
R_{LCC} = \sqrt{\frac{R_1 \omega^2 L_{f2}{}^4}{R_2 M^2}} = \frac{R_1 L_{f2}{}^2}{R_2 M^2} R_{SS}
\tag{15}
$$

It can be known from (11) that when the mutual inductance decreases, the currents of primary and secondary coils decrease, which is a safe working condition. And when the system coil-parameters are fixed, the load resistance to achieve optimal efficiency can be adjusted by the inductance of the secondary resonance circuit L_{f2}. C_2 also needs to be adjusted to keep resonant according to (10). Compared with the SS structure, it avoids the disadvantages of redesigning the coil parameters to maintain efficient operation when the system's rated parameters are changed, which makes the system design more flexible and reduces the manufacturing cost.

3. Parameter Design

3.1. Effect of Coil Design on Optimal Efficiency Load

When a WPT system is actually designed, some parameters are fixed due to the limitation of frequency or coil size. At this time, choosing a specific resonance mode to obtain better transmission efficiency according to the existing parameters becomes the key to the design.

This paper presents a design scheme of planar spiral coil [18] that optimizes the system transmission efficiency. The core idea of designing a high-efficiency MCR-WPT system is to maintain the system's high-efficiency operating state by adjusting the other coil parameters (mainly turn number, wire diameter, turn pitch, etc.) when the rated load is the optimal load to achieve maximum efficiency.

Figure 4 is a schematic diagram of planar spiral coil. The internal resistance R, self-inductance L and mutual inductance M between the coils [19–21] can be obtained by:

$$\begin{cases} R = \sqrt{\dfrac{\rho\mu_0\omega}{2}}\dfrac{Nr_{avg}}{a} \\ L = \mu_0 N^2 r_{avg} c_1\left[\ln(c_2/\lambda) + c_3\lambda + c_4\lambda^2\right] \\ M = \dfrac{\mu_0\pi N_1 N_2\left(\dfrac{r_{1\_avg}}{2}\right)^2\left(\dfrac{r_{2\_avg}}{2}\right)^2}{2\left[h^2 + \left(\dfrac{r_{1\_avg}}{2}\right)^2\right]^{1.5}} \end{cases} \tag{16}$$

where ρ is the conductor resistivity, μ_0 is the vacuum permeability, ω is the current angular frequency, N is the number of coil turns, r_{min} is the inner radius of coil, d is the turn spacing of coil, $r_{avg} = r_{min} + (N-1)d/2$ is the average radius of coil, a is the conductor radius, c_1, c_2, c_3 and c_4 are fitting coefficients, for circular coils they are $1, 2.46, 0$ and 0.2, respectively, $\lambda = (N-1)d/(2^*r_{avg})$, h is the transmission distance.

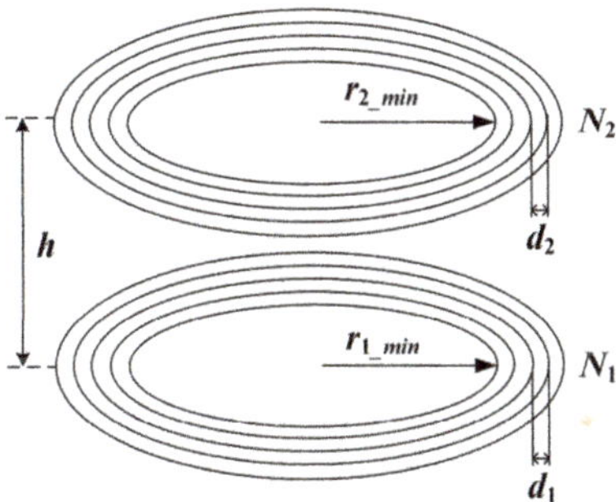

Figure 4. Schematic diagram of planar spiral coil.

According to the actual application scenario, some parameters of the system can be determined. In this paper, the transmission distance $h = 20$cm, the inner radius of the transmitter and receiver coil $r_{1\_min} = r_{2\_min} = 2$ cm. In the case of tightly wound, the conductor diameter and turn spacing are not considered, the influence of coil turns on transmission efficiency occupies the main factor. Where the load resistance R_L is optimal to realize maximum transmission efficiency, the coupling coefficient $k = M/\sqrt{L_1 * L_2}$, self-inductance L_1 and L_2, the coil internal resistance R_1 and R_2, and the optimal load R_L are all related to the turn number of primary coil N_1 and secondary coil N_2:

$$\begin{cases} P_L = \varsigma(k(N_1,N_2), L_1(N_1), L_2(N_2), R_1(N_1), R_2(N_2), R_L(N_1,N_2)) \\ \eta = \xi(k(N_1,N_2), L_1(N_1), L_2(N_2), R_1(N_1), R_2(N_2), R_L(N_1,N_2)) \end{cases} \tag{17}$$

(17) shows that the transmission power P_L and efficiency η are quantities related to the coupling coefficient, load value, self-inductance, and coil internal resistance. Under the tightly wounding condition, the independent variable is mainly affected by the turn number of the coils. Therefore, the ultimate optimization goal of this paper can be expressed as finding the optimal coil turn number while meeting the rated power value of the system design, that is, to find $\max(\eta(N_1,N_2))$.

Figure 5 can be made by Equation (16). It can be seen from Figure 5a that the transmission power decreases with increasing turns; from Figure 5b, the efficiency increases quickly first, then slowly increases as the turn number increases, only when the turn number of primary and secondary coils reaches a critical point can the efficiency be maintained at a higher level. Finally, the iterative method is used to obtain the number of turns, and the coil parameters are reasonably designed to achieve a high-efficiency transmission system that meets the power requirements. The specific design process is shown in Figure 6.

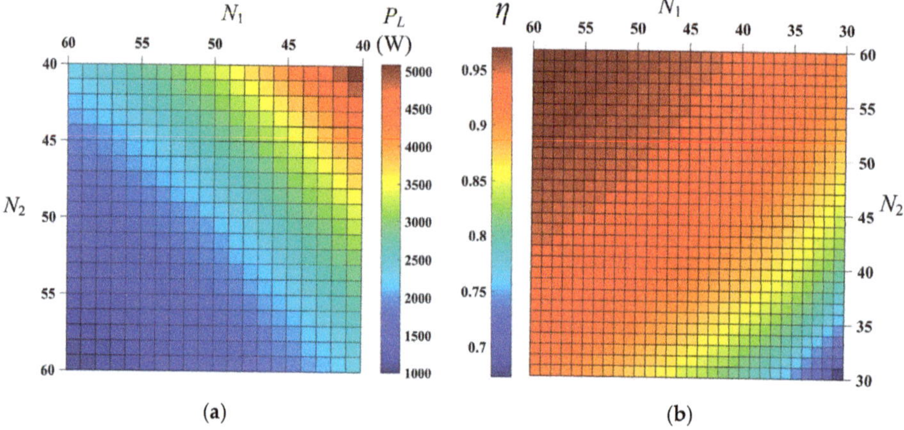

Figure 5. (**a**) Transmission power and (**b**) efficiency with respect to the turn number of coil.

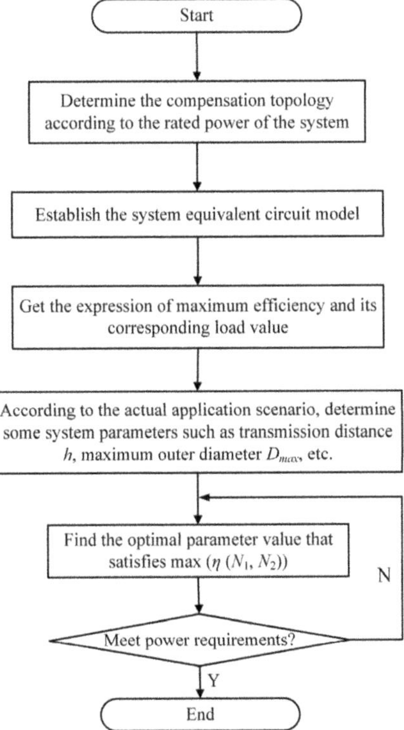

Figure 6. Flow chart of system design.

The following Table 1 shows the optimized resonant parameters of SS compensation topology designed to the load (50 Ω) corresponding to optimal efficiency point.

Table 1. The resonant circuit parameters of SS-type.

Parameter	Value	Unit
resonance inductance L_1	477.4	μH
resonance inductance L_2	426	μH
primary coil turns N_1	50	-
secondary coil turns N_2	48	-
resonance capacitor C_1	7.34	nF
resonance capacitor C_2	8.23	nF
input voltage	300	V
output power	2	kW

3.2. Selection of LCC Resonance Parameters

By optimizing the coil design, for the SS resonance structure, the system is in an optimal efficiency condition when the load is 50 ohms. According to (15), after the coil parameters of LCC resonance structure are determined, the optimal efficiency load value is only related to the parameter L_{f2}. The optimal efficiency load of LCC resonance structure can be determined by designing the inductance L_{f2}. From (5) and (12), we have:

$$P_{LCC} = \frac{M^4}{L_{f1}^2 L_{f2}^2} P_{SS} \tag{18}$$

For the LCC resonance structure, since the load and coil parameters have been determined, it can be known that by adjusting the value of L_{f1}, the level of rated output power can be changed. The parameter configuration method is as follows: (1) Firstly, according to the design parameters of SS compensation structure system, the value of L_{f2} is obtained by Equation (15); (2) Secondly, according to the rated power of system, determine the value L_{f1} through Equation (18); (3) Finally, determine the remaining parameter values according to Equation (10).

As the rated output power of SS resonant structure is 2 kW, a margin of 0.85 is taken, and the rated output power of LCC resonant structure is 2.45 kW, so the values of two inductors L_{f1} and L_{f2} can be determined. The specific parameter values are as follows in Table 2.

Table 2. The resonant circuit parameters of LCC-type.

Parameter	Value	Unit
resonance inductance L_{f1}	64.9	μH
resonance inductance L_{f2}	93.6	μH
resonance capacitor C_1	54.0	nF
resonance capacitor C_2	37.5	nF
resonance capacitor C_{p1}	9.1	nF
resonance capacitor C_{p2}	10.5	nF
output power	2.45	kW

4. Simulation and Experimental Verification

4.1. Magnetic Simulation

For the coil designed in Section 3, the finite element simulation software ANSYS EM was used for simulation, and the vertical distance between two coils was 200 mm, the remaining parameters are shown in Tables 1 and 2. The 3D finite element model is shown in Figure 7a, the boundary condition is radiation, the excitation currents $I_p = 7$A and $I_s = 7$A, the maximum mesh element lengths of solution region, transmitter and receiver coil are 75 mm, 40 mm and 40 mm, respectively. Figure 7b is the distribution diagram of magnetic field strength around the system. It can be seen that the magnetic field distribution around the system is evenly distributed, mainly concentrated near the coils, and has little effect on the surrounding environment.

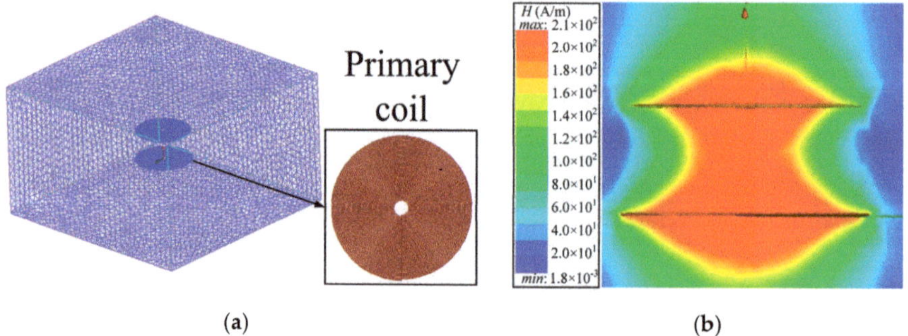

(a) (b)

Figure 7. (**a**) Simulation setup and (**b**) magnetic field strength distribution.

Another purpose of simulation is to obtain the self-inductance, mutual inductance and coupling coefficient of the coil. There are many theoretical calculation methods for calculating these parameters, which will not be repeated here. They can also be obtained by actual measurement. According to the SAEJ2954 standard [22], the actual measured coupling coefficient formula is:

$$k = \sqrt{V_{oc}I_{sc}/V_1 I_1} \tag{19}$$

where, V_1 and I_1 are the voltage and current of primary coil respectively, V_{oc} and I_{sc} are the open circuit voltage and short circuit current of secondary coil, respectively.

The actual mutual inductance value can be obtained by "open circuit and short circuit test". This method needs to measure three quantities, L_{p1} is the measured primary inductance when the secondary circuit is open, L_{s1} is the measured secondary inductance when the primary circuit is open, and L_{p2} is the measured primary inductance when the secondary circuit is shorted, as follow:

$$M = \sqrt{(L_{p1} - L_{p2})L_{s1}} \tag{20}$$

Table 3 shows the theoretical, simulated and actual measured parameter values of the system under rated parameters. It can be seen from Table 3 that the deviation of the theoretical, simulated and actual value of the primary and secondary coil self-inductance can be ignored.

Table 3. System parameter values.

Parameter	Theoretical Value	Simulation Value	Measured Value
primary coil self-inductance (μH)	477.4	481.5	475.2
secondary coil self-inductance (μH)	426.1	428.3	423.1
mutual inductance (μH)	82.0	82.6	81.2
coupling coefficient	0.18	0.18	0.18

4.2. Experiment

The experimental setup is shown in Figure 8, the experimental parameters of compensation circuit are generally consistent with Tables 1 and 2. The coils are planner spiral coil wound by Litz wire, and tightly wound with the turn spacing $d = 2a$, the conductor radius $a = 1.8$ mm, the inner radius of the transmitter and receiver coil $r_{1\_min} = r_{2\_min} = 2$ cm, the out radius of the transmitter $r_{1\_max} = 20$ cm, the out radius of the receiver $r_{2\_max} = 19.28$ cm. The quality factor of the transmitter and receiver coil are 612 and 623, respectively. The gate drive signal of full-bridge inverter circuit can achieve frequency adjustment, and its duty cycle is 0.5. In the resonance state, the input power can be obtained from the input voltage value and input current value of rectifier bridge. The output power is obtained by measuring the load voltage with a differential probe and measuring the output current with a current probe. Considering that the voltage on the capacitor at resonance is Q (quality factor) times the voltage on the circuit, the tuning capacitor is composed of CBB capacitor series and parallel.

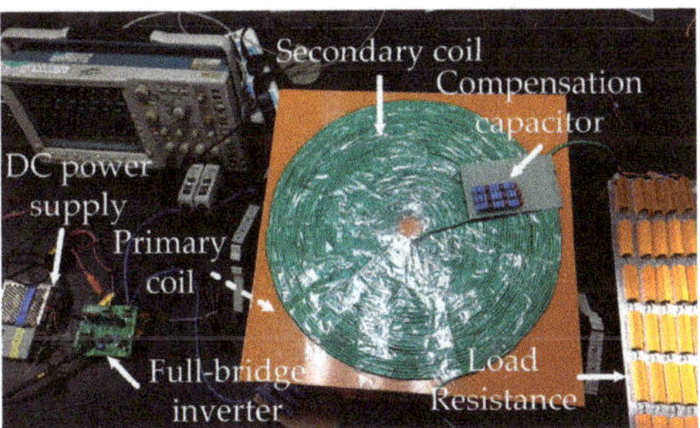

Figure 8. Experimental set up.

A comparative experiment is performed for SS and LCC compensation topologies. Figure 9 shows the experimental waveforms of SS and LCC under the system's optimal efficiency load. In Figure 9b, the primary current shows wonky parts. This is because LCC compensation topology is more complicated than SS, and the coil winding and capacitance matching errors are larger, which results in the failure to realize ideal resonance at the rated frequency. Therefore, the I_P waveform of LCC structure is not as stable as the SS structure.

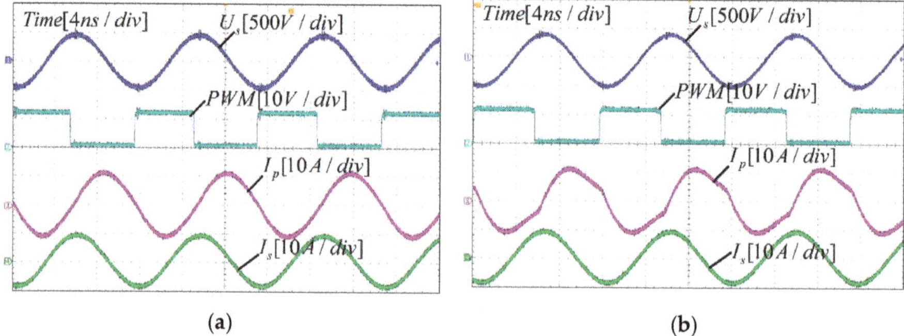

Figure 9. The input and output waveform of optimal state, (**a**) SS-type; (**b**) LCC-type.

As can be seen from Table 4, the actual system operating efficiency and output power are basically consistent with the design goals. It illustrates the correctness of the above design method. The error between experimental results and simulation values is due to the fact that the actual system cannot fully work in the ideal state, the coil and resonant capacitor cannot be perfectly matched at the rated frequency in experiment. It is verified through experiments that the output power is increased by 13.5% under the same input voltage without a significant decrease in efficiency.

Table 5 shows the experimental data of the power and efficiency of the SS and LCC compensation topology under different loads. It can be seen from Table 5 that both SS and fixed LCC compensated WPT system achieve maximum efficiency at 50 ohms load, which is consistent with theoretical analysis. By adjusting the parameters of LCC compensation topology according to the load resistances, the output power and transfer efficiency of the adjusted LCC compensated WPT system are improved and more stable under different loads.

Table 4. System experimental results and simulation values.

Parameter	SS Simulation	SS Experiment	LCC Simulation	LCC Experiment
input voltage	300 V	300 V	300 V	300 V
input current	7.2 A	7.56 A	8.857 A	8.52 A
input power	2.161 kW	2.268 kW	2.657 kW	2.556 kW
load voltage	316 V	319 V	352.1 V	339.6 V
output power	1.994 kW	2.011 kW	2.475 kW	2.282 kW
efficiency	92.26%	88.67%	93.14%	89.28%

Table 5. System power and efficiency under different load resistances.

Load	SS Power	SS Efficiency	LCC with Fixed Parameters Power	LCC with Fixed Parameters Efficiency	LCC with Adjusted Parameters Power	LCC with Adjusted Parameters Efficiency	Adjusted L_{f1}	Adjusted L_{f2}
10 Ω	455.2 W	68.6%	478.7 W	73.7%	3295.1 W	89.0%	1.4 µH	18.7 µH
33 Ω	1331.8 W	82.8%	1559.5 W	87.7%	3323.5 W	89.3%	66.3 µH	61.8 µH
50 Ω	2011.0 W	88.7%	2282.1 W	89.3%	3302.1 W	89.8%	54.3 µH	93.6 µH
83 Ω	3170.0 W	85.6%	3210.7 W	86.7%	3313.0 W	89.5%	41.2 µH	155.4 µH

Based on the experiment of transmission distance change under SS and LCC structures, Figure 10 was produced. When the transmission distance changes, the system resonance frequency remains unchanged, so as to observe the vertical offset stability of the system under SS and LCC structures. It can be seen from Figure 10 that for the SS resonant topology, the input current and output current increase with distance; for the LCC structure, the input and output current decrease with distance. In practical applications, this positive correlation of the SS structure is very dangerous.

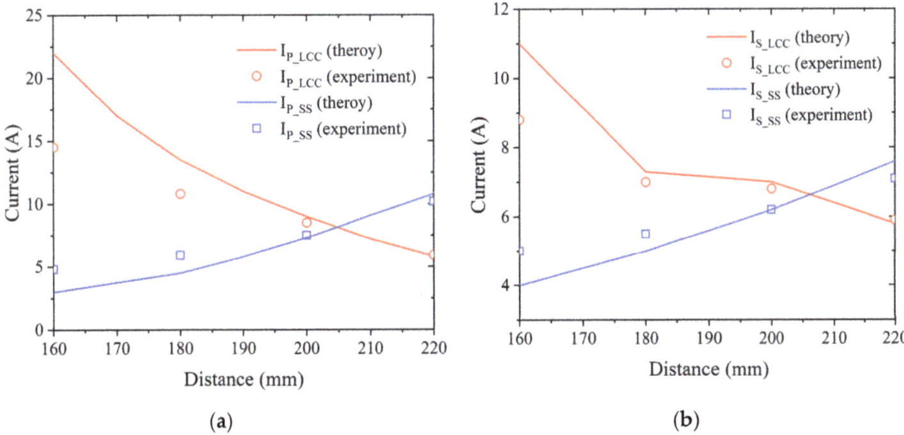

Figure 10. (**a**) Input current and (**b**) output current with respect to distance.

There is a deviation between the theoretical and experimental current values, the main reason is that the mutual inductance measured by the actual wound coil is smaller than the theoretical value. However, the current trend is basically the same.

5. Conclusions

In this paper, the optimal transfer efficiency of the WPT system under certain load resistance is achieved by optimizing the coil-turns number. Once the coupled coil-parameters are fixed, the load resistance to make the optimal efficiency can be adjusted and the output power can be increased by optimization design of the parameters of LCC compensation topology. The major contributions of this paper are as follows:

(1) The method is proposed to keep the optimal transfer efficiency of the WPT system under different load resistances by adjusting the inductance value of the LCC compensate topology L_{f1}.
(2) The method is proposed to increase the output power of the WPT system to satisfy more power requirements by adjusting the inductance value of the LCC compensate topology L_{f2}.
(3) The optimized LCC compensated WPT system has significant advantages over the SS type. There's no need to redesign the coupled coils to maintain transmission power and efficiency under different power levels and transfer distance. When the load resistance equals 50 Ω, the output power of the optimized WPT system is increased by 64.2% with 89.8% transfer efficiency.

Author Contributions: F.W. conceived and designed the study, and this work was performed under the advice of and regular feedback from him. X.C. and Q.L. was responsible for the models, simulations and X.C. wrote the article. W.G. was responsible for the experiments and data analysis. All authors have read and agreed to the published version of the manuscript.

Funding: This work was supported by the Basic Research Program of Jiangsu Province (No. BK20180485), Jiangsu Provincial Key Laboratory of Smart Grid Technology and Equipment Project, and the Fundamental Research Funds for the Central Universities (No. 30919011241).

References

1. Kurs, A.; Karalis, A.; Moffatt, R. Wireless power transfer via strongly coupled magnetic resonances. *Science* **2007**, *317*, 83–86. [CrossRef] [PubMed]

2.	Abid, A.; O'Brien, J.; Bensel, T. Wireless Power Transfer to Millimeter-Sized Gastrointestinal Electronics Validated in a Swine Model. *Sci. Rep.* **2017**, *7*, 46745. [CrossRef] [PubMed]

3.	Roshan, Y.M.; Park, E.J. Design Approach for a Wireless Power Transfer System for Wristband Wearable Devices. *IET Power Electron.* **2017**, *10*, 931–937. [CrossRef]

4.	Patil, D.; McDonough, M.K.; Miller, J.M. Wireless Power Transfer for Vehicular Applications: Overview and Challenges. *IEEE Trans. Transp. Electrif.* **2018**, *4*, 3–37. [CrossRef]

5.	Yang, W.; Gao, Y.; Farley, K.B. EV usage and city planning of charging station installations. In Proceedings of the 2015 IEEE Wireless Power Transfer Conference (WPTC), Boulder, CO, USA, 13–15 May 2015; pp. 1–4.

6.	Bu, Y.; Endo, S.; Mizuno, T. Improvement in the Transmission Efficiency of EV Wireless Power Transfer System Using a Magnetoplated Aluminum Pipe. *IEEE Trans. Magn.* **2018**, *54*, 1–5. [CrossRef]

7.	Elnait, K.E.I.; Huang, L.; Tan, L. Resonant Reactive Current Shield Design in WPT Systems for Charging EVs. In Proceedings of the 2018 IEEE PES Asia-Pacific Power and Energy Engineering Conference (APPEEC), Kota Kinabalu, Malaysia, 7–10 October 2018; pp. 56–59.

8.	Li, W.; Wang, Q.; Wang, Y. Three-dimensional rotatable omnidirectional MCR WPT systems. *IET Power Electron.* **2020**, *13*, 256–265. [CrossRef]

9.	Hao, H.; Covic, G.A.; Boys, J.T. A Parallel Topology for Inductive Power Transfer Power Supplies. *IEEE Trans. Power Electron.* **2014**, *29*, 1140–1151. [CrossRef]

10.	Liu, F.; Zhang, Y.; Chen, K. A comparative study of load characteristics of resonance types in wireless transmission systems. In Proceedings of the 2016 Asia-Pacific International Symposium on Electromagnetic Compatibility, APEMC 2016, Shenzhen, China, 17–21 May 2016; Volume 01, pp. 203–206.

11.	Villa, J.L.; Sallan, J.; Osorio, J.F.S. High-Misalignment Tolerant Compensation Topology for ICPT Systems. *IEEE Trans. Ind. Electron.* **2011**, *59*, 945–995. [CrossRef]

12.	Shen, D.; Du, G.; Zeng, W. Research on Optimization of Compensation Topology Parameters for a Wireless Power Transmission System with Wide Coupling Coefficient Fluctuation. *IEEE Access.* **2020**, *8*, 59648–59658. [CrossRef]

13.	Chen, Y.; Yang, N.; Liu, L. Two/Three-Coil Hybrid Topology for WPT Systems Charging Electric Bicycles. In Proceedings of the 2019 IEEE Applied Power Electronics Conference and Exposition (APEC), Anaheim, CA, USA, 17–21 March 2019; pp. 3084–3087.

14.	Jenson, J.; Therattil, J.P.; Johnson, J.A. A Novel LCC-LCL Compensation WPT System for Better Performance. In Proceedings of the 2019 IEEE International Conference on Electrical, Computer and Communication Technologies (ICECCT), Coimbatore, India, 20–22 February 2019; pp. 1–6.

15.	Mai, R.; Chen, Y.; Zhang, Y. Optimization of the Passive Components for an S-LCC Topology-Based WPT System for Charging Massive Electric Bicycles. *IEEE Trans. Ind. Electron.* **2018**, *65*, 5497–5508. [CrossRef]

16.	Zhang, H.; Lu, F. An improved design methodology of the double-sided LC-Compensated CPT system considering the inductance detuning. *IEEE Trans. Power Electron.* **2019**, *34*, 11396–11406. [CrossRef]

17.	Li, S.; Li, W.; Deng, J. A Double-Sided LCC Compensation Network and Its Tuning Method for Wireless Power Transfer. *IEEE Trans. Veh. Technol.* **2015**, *64*, 2261–2273. [CrossRef]

18.	Kiani, M.; Ghovanloo, M. A Figure-of-Merit for Designing High-Performance Inductive Power Transmission Links. *IEEE Trans. Ind. Electron.* **2013**, *60*, 5292–5305. [CrossRef] [PubMed]

19.	Balanis, C.A. *Antenna Theory: Analysis and Design*, 3rd ed.; Wiley-Interscience: Hoboken, NJ, USA, 2005.

20.	Mohan, S.S.; Del Mar Hershenson, M.; Boyd, S.P. Simple Accurate Expressions for Planar Spiral Inductances. *IEEE J. Solid-State Circuits* **1999**, *34*, 1419–1424. [CrossRef]

21.	Wei, W.; Kawahara, Y.; Kobayashi, N. Characteristic Analysis of Double Spiral Resonator for Wireless Power Transmission. *IEEE Trans. Antennas Propag.* **2014**, *62*, 411–419. [CrossRef]

22.	*SAE J2954: Wireless Power Transfer for Light-Duty Plug-in/Electric Vehicles and Alignment Methodology*; SAE International: Warrendale, PA, USA, 2019.

Article

Design of a Cylindrical Winding Structure for Wireless Power Transfer Used in Rotatory Applications

Mohamad Abou Houran, Xu Yang and Wenjie Chen *

School of Electrical Engineering, Xi'an Jiaotong University, Xi'an 710049, China;
eng.horan@yahoo.com (M.A.H.); yangxu@mail.xjtu.edu.cn (X.Y.)
* Correspondence: cwj@xjtu.edu.cn; Tel.: +86-29-82665223

Received: 26 February 2020; Accepted: 19 March 2020; Published: 23 March 2020

Abstract: A cylindrical joint structure for wireless power transfer (WPT) systems is proposed. The transmitter (Tx) and receiver (Rx) coils were wound on hemicylindrical and cylindrical structures, respectively. The Rx coil rotates freely around the axial direction of the Tx coil. Different methods of winding the Tx and Rx coils are given and discussed. Electromagnetic fields (EMFs) around the WPT windings should be lower than the limits set by WPT standards. Therefore, the WPT windings were designed to reduce EMF level and maintain constant power-transfer efficiency (PTE). The design procedures of the windings are discussed in detail. EMF analysis was done under different rotation angles (α). The selected design reduced the variation of the mutual inductance (M). As a result, it maintained a constant PTE while rotating the Rx coil between 0° and 85°. Moreover, leakage magnetic fields (LMFs) near the WPT coils of the chosen design were reduced by 63.6% compared with other winding methods that have the same efficiency. Finally, a prototype was built to validate the proposed idea. Experiment results were in good agreement with the simulation results. The WPT system maintained constant efficiency in spite of the rotation of Rx coil, where efficiency dropped by only 2.15% when the Rx coil rotated between 0° and 85°.

Keywords: cylindrical joint; electromagnetic fields; rotation-free structure; wireless power transfer

---

1. Introduction

Wireless power transfer (WPT) systems have proven their reliability and have become a widely used technique. A WPT system transfers power for many applications, such as implantable medical devices (IMDs) [1–3], the charging of electric vehicles (EVs) [4–9], autonomous underwater vehicles (AUVs) [10], unmanned aerial vehicles (UAV) [11], robotic systems [12], light detection and ranging equipment [13], and the Internet of Things (IoT) [14–16]. In addition, it is used in some appliances, for instance, smartwatch straps [17], smartphones [18], battery-powering systems [19], and electrical drones [20,21]. Earlier, many research works investigated different types of structures, such as pancake coils, square coils, and circular coils [22,23]. Recently, in order to extend the transfer area, three-dimensional (3D) geometries have been proposed. For example, a rectangle-shaped resonant cavity was presented [1]. It charged multiple IMDs in a freely behaving animal. A WPT system made of a bowl-shaped transmitter (Tx) coil and a box-shaped receiver (Rx) coil was investigated and could be embedded in an in-ear hearing aid [24]. Hou, et al. [25] fabricated 3D windings for the WPT system. Moreover, Ha-Van et al. [26] studied an omnidirectional WPT system with a cube-shaped Tx coil that could be a possible way of charging portable devices. Many other structures were reported in [27]. However, those structures only considered fixed coils without built-in rotatory parts.

Therefore, recent studies have considered the rotation of the coils to improve WPT performance and maintain constant output power. Yan et al. [10] presented a rotation-free WPT system for AUV charging. Zhang et al. [28] designed a ball-joint WPT system. The joint consisted of a small ball that was rotated inside a socket structure. Houran et al. [29] investigated a spherical-joint structure that was made of a small ball that rotated inside a hemisphere structure. Han et al. [30] discussed a rotatory WPT system for multiload applications. Sugino et al. [12] fabricated a linear-free motion WPT used in robotic applications. These references provided good contributions to the development of the WPT winding study. Further development is given in this paper, where the coils are optimized for low electromagnetic fields (EMFs) and high power-transfer efficiency (PTE) regardless of angular rotation.

On the other hand, compliance with EMF safety regulations and standards is a main concern regarding the design of the WPT system [31]. If the coupling coefficient (k) has low values, it will create high levels of leakage magnetic fields (LMFs) in the windings' vicinities. Thus, there is an exposure to EMFs for anyone who approaches the application during charging [32]. In addition, using the WPT charging systems next to other electronic devices could create electromagnetic field interference (EMI). To comply with standards and regulations, such as the International Commission on Nonionizing Radiation Protection (ICNIRP) 2010 [33], many EMF- and EMI-reduction methods have been presented. For example, using ferrite [31], metamaterials [34–36], and reducing LMFs by three-phase power [37]. Moreover, in [38], the authors presented three active shielding methods: three-dB dominant EMF cancellation (3DEC), independent self-EMF cancellation (ISEC), and linkage-free EMF cancellation (LFEC). In [18], the authors presented resonant reactive shields for a planar WPT system. The above-mentioned methods require additional components, such as coils, power supplies, and capacitors. In addition, choosing an applicable EMF mitigation method depends on several important factors, for example, application type, available space, weight, and cost. However, in order to reduce LMFs, it is better to reduce the source of LMFs and optimize the coils.

In this paper, a new joint structure is proposed. The main contributions are as follows. (1) A moveable WPT winding structure is proposed that combines a hemicylindrical-structure Tx coil and a cylindrical-structure Rx coil. By means of the proposed structure, the Rx coil could rotate inside the Tx coil within angles of up to 85°. (2) A detailed design procedure of the proposed structure is introduced. It includes the design of the windings, electromagnetic-field assessment, and necessary simulations. (3) The Tx and Rx coils were designed in order to minimize the leakage of magnetic fields around the coils' vicinities and maintain high power-transfer efficiency (PTE). Furthermore, EMF analysis was done under different rotation angles. (4) A prototype of the WPT coils was fabricated and measurements were done. Efficiency was almost constant under different rotation angles.

The rest of this paper is organized as follows. In Section 2, a detailed description of the proposed structure is given. In Section 3, design procedures of the WPT system for high efficiency are presented. Design procedures of the WPT system for low-leakage magnetic fields are given in Section 4. Experiment results are shown in Section 5 in order to validate the proposed structure. Section 6 concludes this paper.

2. Design of Proposed WPT System

The design of the cylindrical joint of the WPT system is shown in Figure 1. In this design, the Rx coil can rotate within the Tx coil up to 85°. The Tx coil (blue) was wound on the hemicylindrical structure, and the Rx coil (brown) was wound on the cylindrical structure.

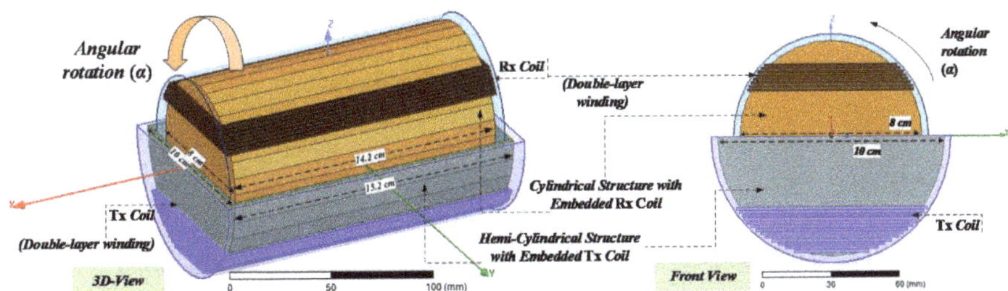

Figure 1. Cylindrical-joint structure of wireless-power-transfer (WPT) system (rotation-free structure).

The transferred power (P) across the gap, given in Equation (1), is proportional to frequency (f), mutual inductance (M), and the square of the Tx current (I_2) [28]. The design of the proposed WPT system can be optimized by maximizing the mutual inductance and reducing its fluctuation at different rotation angles (α). Mutual inductance is given by Equation (2).

$$P \propto f M I_2^{\,2} \tag{1}$$

$$M = k \sqrt{L_{Tx} L_{Rx}} \tag{2}$$

where k is the coupling coefficient, L_{Tx} is the self-inductance of the Tx coil, and L_{Rx} is the self-inductance of the Rx coil. L_{Tx} and L_{Rx} depend on the resonators' geometries. Several variables were considered to parametrize the coils, as follows. Turn numbers are given as N_{Tx}, N_{Rx}, where N_{Tx} is the number of turns of Tx coil and N_{Rx} is the number of turns of Rx coil. Number of winding layers are given as single-layer (SL) winding and double-layer (DL) winding. In addition, the space between turns and variation in the z-axis position, which affects the value of the coupling coefficient, was considered. Therefore, there are many possibilities for winding Tx and Rx coils on a joint structure.

Some winding models are illustrated in Figure 2. Figure 2a is a hemicylindrical winding method of Tx and Rx coils that can be written as DL (64, 50), where 64 is the number of turns of the Tx coil, and 50 is the number of turns of the Rx coil. Figure 2a–c shows very high coupling coefficients (close to 0.4). In such models, the fluctuation of mutual inductance with rotation is very high. For example, Figure 4 displays the variations of mutual inductance and coupling coefficient of the winding structures that are given in Figure 2c. M variation reached 95% when the Rx coil rotated between 0° and 85°. Therefore, efficiency drops a lot in this case. In addition, a high coupling could result in a frequency-splitting issue, thus reducing output power. For choosing the right model, low variations of mutual inductance while rotating the Rx coil should be considered. Thus, a WPT system maintains constant efficiency. The coupling coefficient between the studied models ranged between 0.08 and 0.5. In further steps, the chosen model should consider low-leakage magnetic fields, which is explained in the next section.

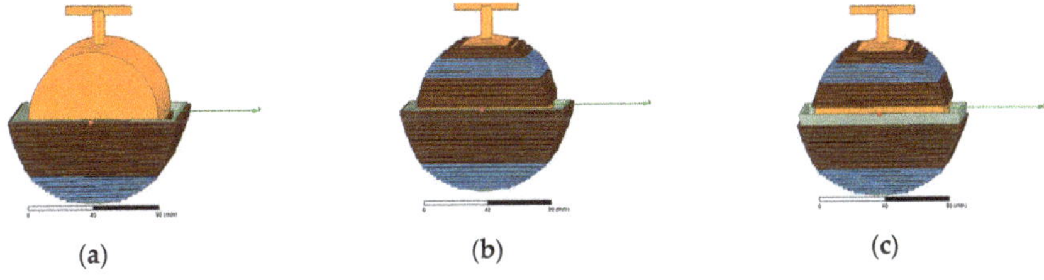

(a)   (b)   (c)

Figure 2. *Cont.*

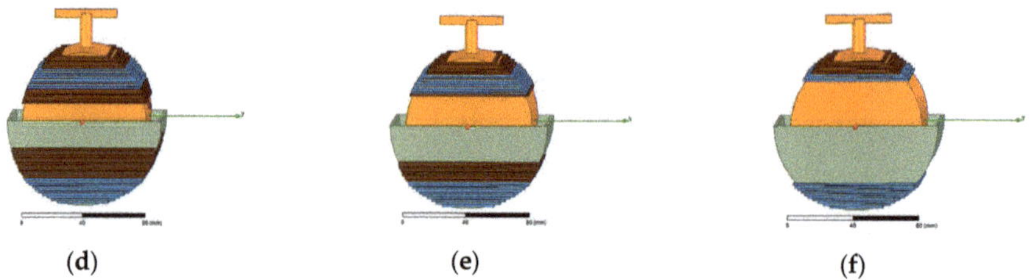

(d)   (e)   (f)

Figure 2. Winding methods of transceiver (Tx) and receiver (Rx) coils. Blue coils represent limits of design scenario (or part of it). (**a**) Hemicylindrical windings of Tx and Rx coils: DL (64, 50). (**b**) Opposite hemicylindrical windings of Tx and Rx coils (Case 1). (**c**) Opposite hemicylindrical windings (Case 2). (**d**–**f**) Other models with fewer number of turns (conducted with Ansys Maxwell 3D, USA).

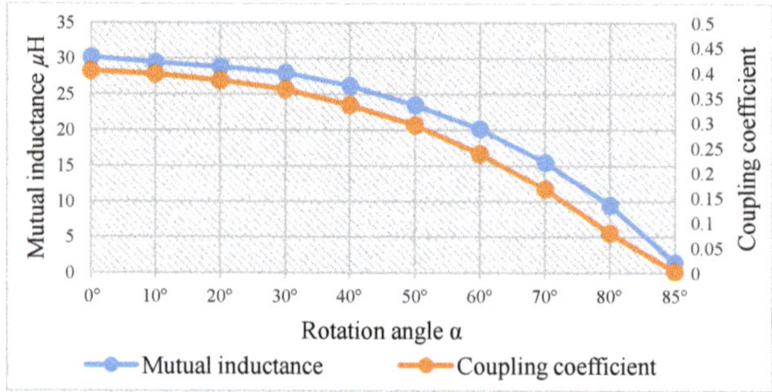

Figure 3. Mutual inductance and coupling coefficient according to rotation angle of hemicylindrical winding structure.

3. Design Procedures for High Efficiency

Different methods of winding the Tx and Rx coils were obtained. In Figure 4, twelve different coil designs are displayed. In Figure 4a, the Tx and Rx coils took the same shape of hemicylindrical structures, and could be denoted as DL (64, 50). In Figure 4h, the Tx and Rx coils took the same shape of hemicylindrical structures and could be denoted as SL (32, 25). As mentioned before, a constant PTE during the rotation of the Rx coil depends on mutual inductance and coupling coefficient, which follows the winding structure. Different methods of winding the Tx and Rx coils resulted in different values of the coupling coefficients and mutual inductances, as illustrated in Figure 5. DL (64, 50) and SL (32, 25) designs had very high values of M and k. However, cost, volume, and weight were higher than those of other models. In addition, for the magnetically coupled resonant (MCR) WPT design, the k value could not be very high. Therefore, the design could be selected as one of the models that are marked in black in Figure 5.

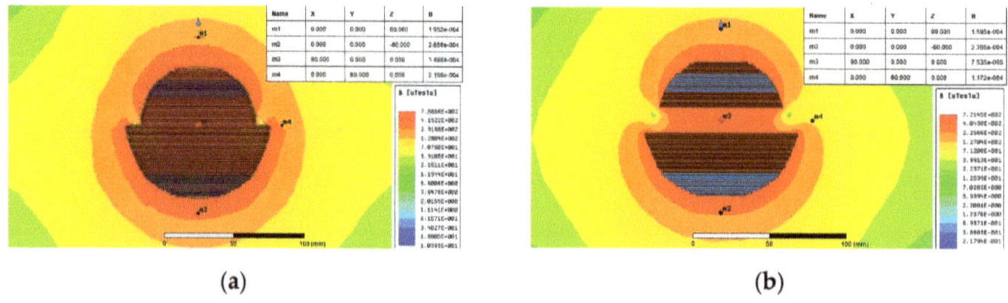

(a)   (b)

Figure 4. *Cont.*

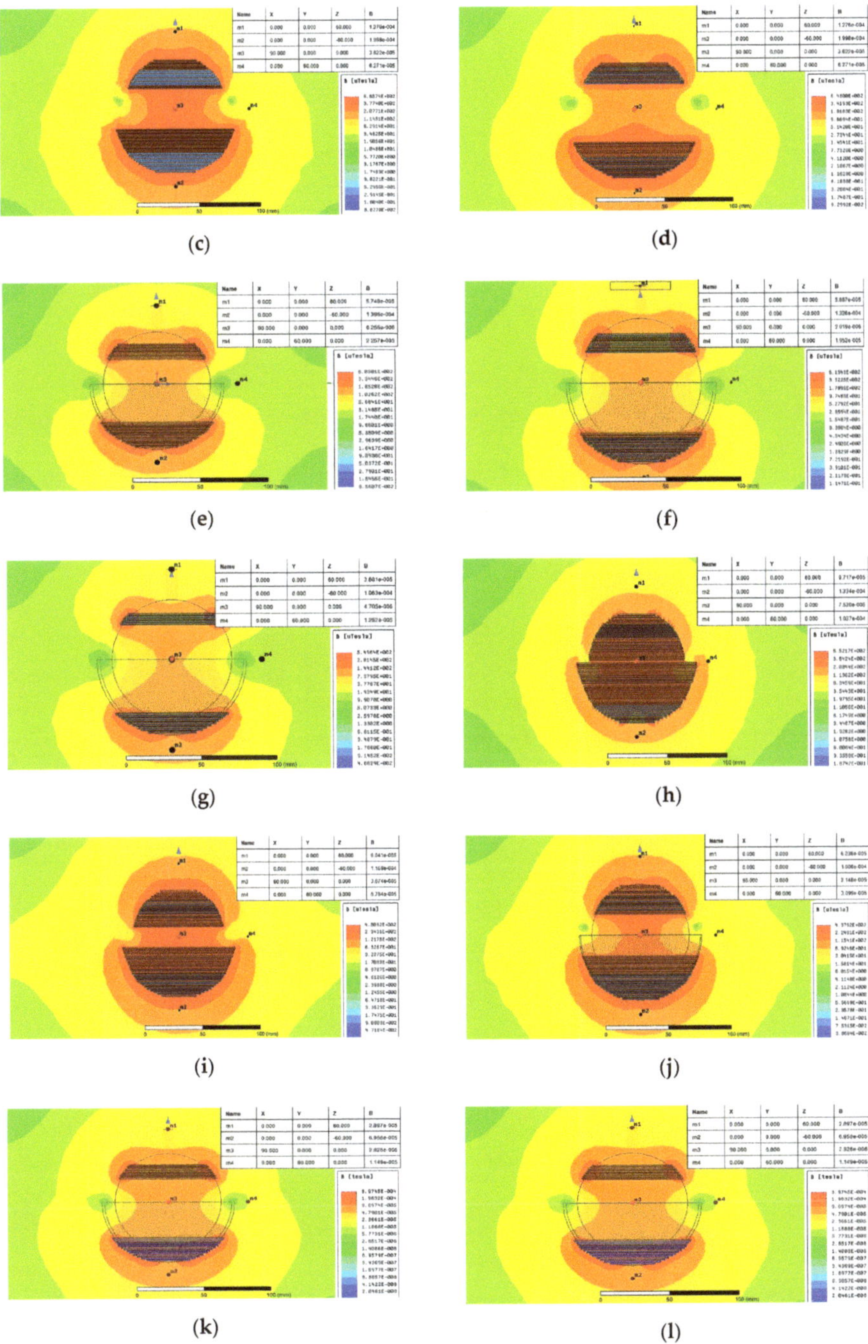

Figure 4. Winding methods (N_{Tx}, N_{Rx}) and magnetic-field densities taken at four positions. (**a**) Double layer (DL) (64, 50). (**b**) DL (54, 40). (**c**) DL (44, 30). (**d**) DL (34, 20). (**e**) DL (26, 16). (**f**) DL (24, 16). (**g**) DL (18, 10). (**h**) Single layer (SL) (32, 25). (**i**) SL (27, 20). (**j**) SL (22, 15). (**k**) SL (22, 15). (**l**) SL (13, 8) (conducted with Ansys Maxwell 3D, PA, USA).

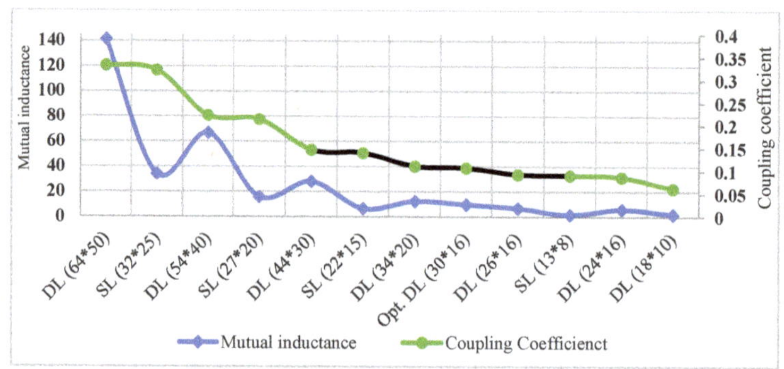

Figure 5. Mutual inductance and coupling coefficient of different winding methods.

The simulations of the WPT system were conducted with Ansys Maxwell 3D and Ansys Simplorer for cosimulation. Series–series (SS) compensation topology was considered. Resonant frequency was 950 kHz. Figure 6 illustrates the efficiency of each SS-compensated WPT system at a resonant frequency of 950 kHz. On the basis of the value of the coupling coefficient (shown in Figure 5) and the efficiency values in Figure 6, the chosen design was DL (30, 16). Figure 6b displays the DL efficiency (30, 16).

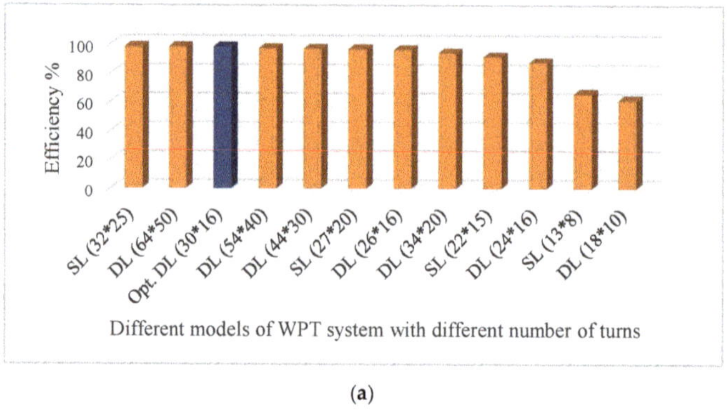

(**a**)

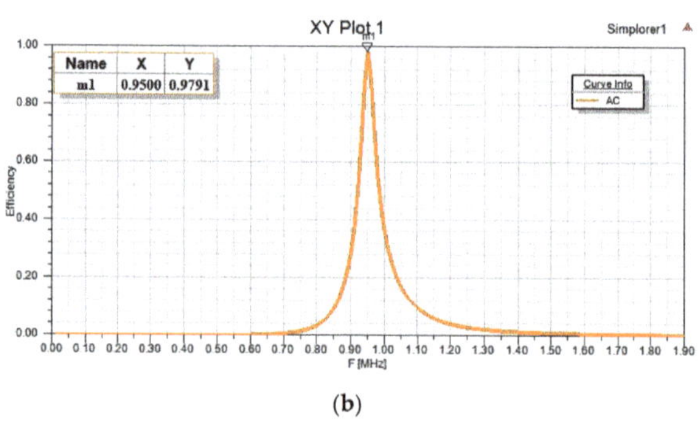

(**b**)

Figure 6. Efficiency values of different series–series (SS)-compensated WPT systems, where the chosen model is marked in blue. (**a**) Different models of WPT system with different numbers of turns. (**b**) Efficiency of chosen design DL (30, 16).

DL (30, 16) self-inductances were L_{Tx1} = 120.68 µH and L_{Rx1} = 52.068 µH. Figure 7 presents the relation between frequency, load, and efficiency, showing that the system was steady at low and high loads.

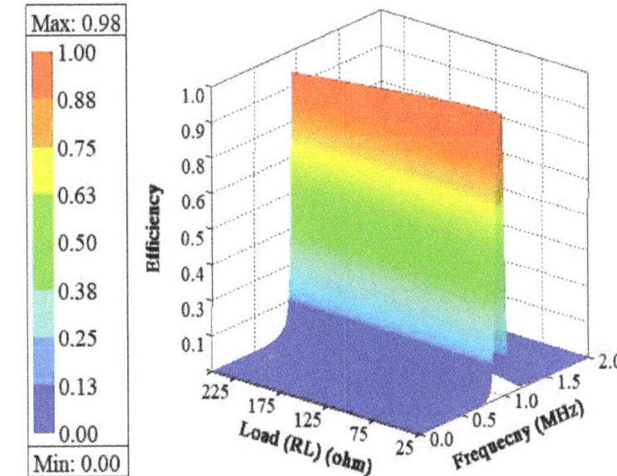

Figure 7. Three-dimensional representation of relation between frequency, load, and efficiency.

4. Design Procedures for Low-Leakage Magnetic Fields

Maintaining low-leakage magnetic fields (LMFs) around Tx and Rx coils is another key point of WPT design. Figure 8 illustrates a comparison of the magnetic-field density (*B*) of different WPT systems (given in Figure 4). *B* was calculated around the coils' vicinities. The worst winding scenario was DL (64, 50), which had very-high-leakage magnetic fields (LMFs) of 74.67 µT. In addition, DL (64, 50) had a very high value of coupling coefficient at $\alpha = 0°$, and this value dropped close to zero at $\alpha = 85°$. On the basis of the efficiency value given in Figure 6, and the magnetic-field-density value presented in Figure 8, the chosen design was the same as DL (30, 16) (dark green), which had a low LMF level of 27.1 µT. This value is almost the same as the exposure limit that was set by ICNIRP-2010. Therefore, the chosen design decreased LMFs by 63.6% compared with DL (64, 50), which has similar efficiency. Moreover, the magnetic-field density of the chosen design at different rotation angles (0°–85°) is illustrated in Figure 9. With the rotation of the Rx coil, *B* was reduced. Furthermore, Figure 10 displays the *B* of SL (32, 25). Compared with the best selected design, the level of LMFs was decreased by 22.5%.

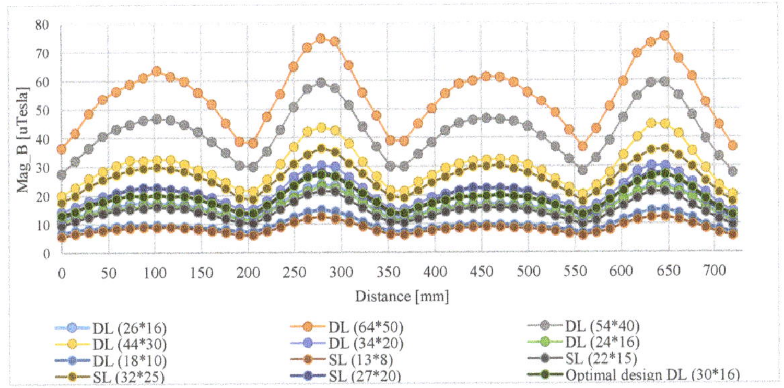

Figure 8. Magnetic-field density (*B*): comparison of different WPT systems.

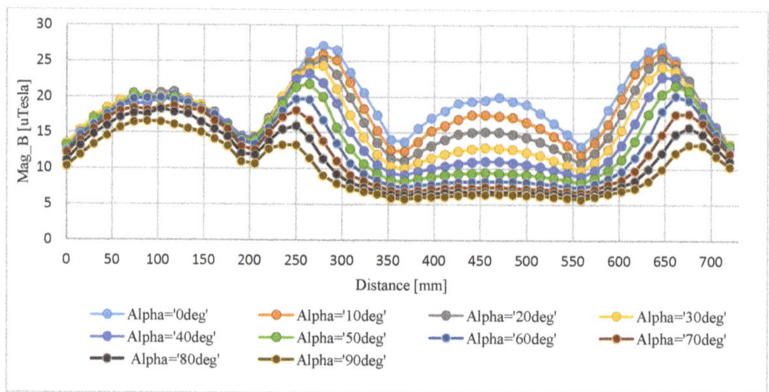

Figure 9. Magnetic-field density of chosen design of DL (30, 16) at different rotation angles of Rx coil.

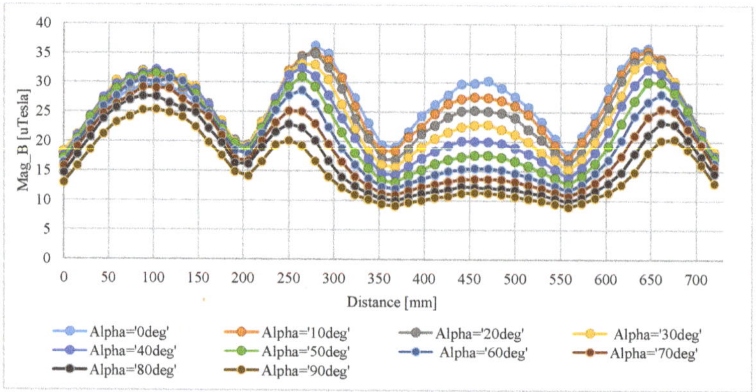

Figure 10. Single-layer winding model SL (32, 25) at different rotation angles of Rx coil.

There are some models that have almost the same level of LMFs as those of the chosen design, such as DL (24, 16), as shown in Figure 11 at different rotation angles. Nevertheless, on the basis of efficiency values, DL (30, 16) had efficiency of 97.9%, whereas DL (24, 16) had efficiency of 87.1%.

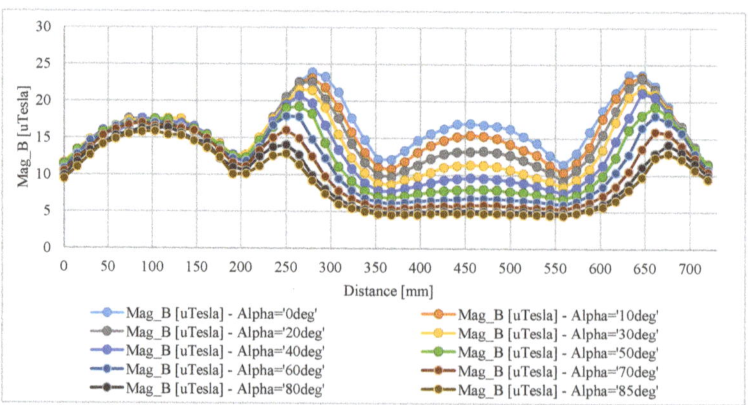

Figure 11. Double-layer winding model DL (24, 16) at different rotation angles of Rx coil.

5. Experiment Results and Validation

To validate the selected design (DL (30, 16)), a prototype was built. Series–series (SS) topology was chosen. The experiment setup is given in Figure 12. Multistrand Litz wire was used to wind the coils. This reduced the skin effect and power losses at high frequency. In addition, radio-frequency (RF) mica capacitors were used for better performance of the WPT system. The measured parameters are given in Table 1, where R_1 and R_2 (Ω) are the resistances of Tx and Rx windings, respectively; and C_{Tx} and C_{Rx} (nF) are the compensation capacitors of Tx and Rx coils, respectively.

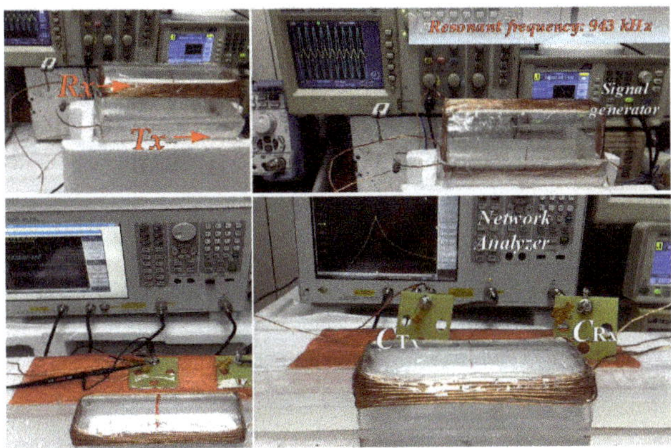

Figure 12. Experiment setup of WPT system.

Table 1. Measured parameters.

WPT Prototype	N_1/N_2	f_0 (kHz)	L_{Tx}, L_{Rx} (μH)	R_1, R_2 (Ω)	M μH, k at $\alpha = 0°$	C_{Tx} C_{Rx}, (nF)
Tx	23		139	0.64	9.94	0.2
Rx	16	943	58.8	0.25	0.11	0.48

The input and output voltages are displayed in Figure 13. The output voltage slightly changed when Rx coil rotated. The voltages are presented at four angles between 0° and 85°.

The chosen design had low fluctuations of mutual inductance while rotating the Rx coil. The measured and simulated mutual inductances are presented in Figure 14. Simulated mutual inductance is always larger than measured mutual inductance. The simulations gave ideal values of the coils' inductances and mutual inductance. However, measurements take into consideration some factors such as losses. In addition, in the fabricated prototype, the distance between the Tx and Rx coils was slightly different than that in the simulated model, so the coupling coefficient in the simulation was 0.13, whereas the measured one was 0.11. Therefore, simulated mutual inductance was larger than the measured mutual inductance. The measured M varied between 9.94 μH at $\alpha = 0°$, 11.56 μH at $\alpha = 60°$, and 10.033 μH at $\alpha = 85°$. This affected the measured efficiency. A network analyzer (E5061B) was used for measuring the S-parameters at the resonant frequency of 943 kHz. Ports 1 and 2 were connected to the Tx and Rx coils, respectively. Power-transfer efficiency (PTE) could be obtained in terms of the linear magnitude of the S-parameter ($|S_{21}|$) [26]. In Figure 14, PTE was given according to rotation angle. The measurements indicated that the given WPT system could maintain almost constant PTE in spite of rotation. At $\alpha = 0°$, PTE = 83.50%; at $\alpha = 30°$, PTE = 84.24%; at $\alpha = 60°$, PTE = 85.01%; and at $\alpha = 85°$, PTE = 81.35%. Thus, PTE was increased by 1.51% when the Rx coil rotated from 0° to 60°, and dropped by only 2.15% when the Rx coil rotated between 0° and 85°.

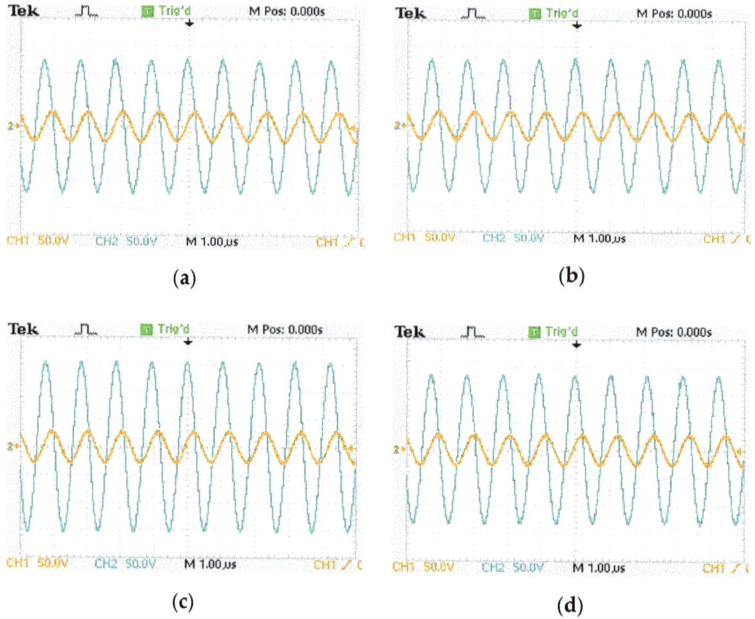

Figure 13. Input voltage (CH1) and output voltage (CH2) of first design: (**a**) $\alpha = 0°$; (**b**) $\alpha = 30°$; (**c**) $\alpha = 60°$; (**d**) $\alpha = 85°$.

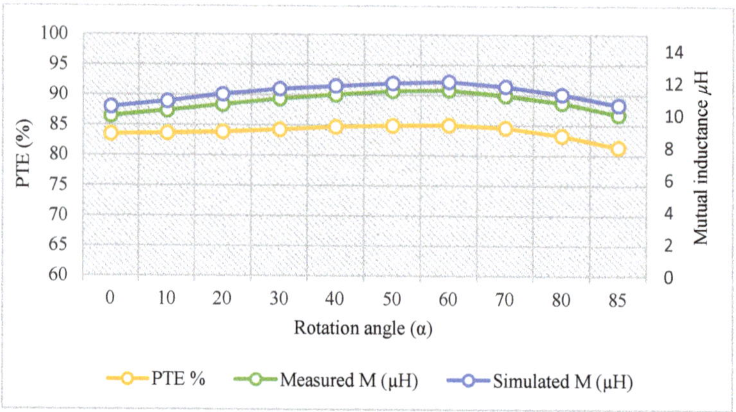

Figure 14. Simulated and measured mutual inductance, and efficiency according to rotation angle.

6. Conclusions

In this paper, a new cylindrical winding structure for WPT systems was proposed. In the proposed design, the Tx coil was wound on a hemicylindrical structure, and the Rx coil was wound on a cylindrical structure. Therefore, the Rx coil could freely rotate within the Tx coil. Different winding methods were presented and compared. The best winding method reduced the leakage of magnetic fields (LMFs) and maintained constant power-transfer efficiency (PTE), while rotating the Rx coil. Design procedures were discussed in detail. Moreover, EMF analysis was done under different rotation angles between 0° and 85°. The leakage of magnetic fields in the coils' vicinities of the chosen design was reduced by 63.6% compared with other winding methods with the same efficiency. Thus, the given design reduces the cost, weight, shielding (if required), and volume of designed WPT coils. A prototype was built to validate the chosen design. Measurements confirmed that variations of mutual inductance were very low, where M varied between 9.94 µH at $\alpha = 0°$ and 10.03 µH at $\alpha = 85°$. As a result, the WPT system maintained a constant PTE. Efficiency was decreased by only 2.15% when the Rx coil rotated from $0°$ to $85°$. The proposed WPT system is a good choice for power transfer for applications that require angular movements.

Author Contributions: Conceptualization, M.A.H. and W.C.; data curation, M.A.H.; formal analysis, M.A.H.; Investigation, M.A.H.; methodology, M.A.H. and W.C.; resources, X.Y. and W.C.; software, M.A.H.; validation, M.A.H.; writing, original draft, M.A.H.; review and editing, X.Y. and W.C.; supervision, X.Y. All authors have read and agreed to the published version of the manuscript.

Funding: This research received no external funding.

Acknowledgments: The authors would like to thank the anonymous reviewers for their constructive comments.

Conflicts of Interest: The authors declare no conflict of interest.

References

1. Mei, H.; Thackston, K.A.; Bercich, R.A.; Jefferys, J.G.R.; Irazoqui, P.P. Cavity Resonator Wireless Power Transfer System for Freely Moving Animal Experiments. *IEEE Trans. Biomed. Eng.* **2017**, *64*, 775–785. [CrossRef] [PubMed]

2. Abdulfattah, A.N.; Tsimenidis, C.C.; Al-Jewad, B.Z.; Yakovlev, A. Performance Analysis of MICS-based RF Wireless Power Transfer System for Implantable Medical Devices. *IEEE Access* **2019**, *7*, 11775–11784. [CrossRef]

3. Ben Fadhel, Y.; Ktata, S.; Sedraoui, K.; Rahmani, S.; Al-Haddad, K. A Modified Wireless Power Transfer System for Medical Implants. *Energies* **2019**, *12*, 1890. [CrossRef]

4. Kim, M.; Joo, D.M.; Lee, B.K. Design and control of inductive power transfer system for electric vehicles considering wide variation of output voltage and coupling coefficient. *IEEE Trans. Power Electron.* **2019**, *34*, 1197–1208. [CrossRef]

5. Luo, Z.; Wei, X. Analysis of Square and Circular Planar Spiral Coils in Wireless Power Transfer System for Electric Vehicles. *IEEE Trans. Ind. Electron.* **2018**, *65*, 331–341. [CrossRef]

6. Rozman, M.; Ikpehai, A.; Adebisi, B.; Rabie, K.M.; Gacanin, H.; Ji, H.; Fernando, M. Smart wireless power transmission system for autonomous EV charging. *IEEE Access* **2019**, *7*, 112240–112248. [CrossRef]

7. Mou, X.; Groling, O.; Sun, H. Energy-efficient and adaptive design for wireless power transfer in electric vehicles. *IEEE Trans. Ind. Electron.* **2017**, *64*, 7250–7260. [CrossRef]

8. Shah, I.A.; Yoo, H. Assessing Human Exposure with Medical Implants to Electromagnetic Fields from a Wireless Power Transmission System in an Electric Vehicle. *IEEE Trans. Electromagn. Compat.* **2019**, 1–8. [CrossRef]

9. Campi, T.; Cruciani, S.; Maradei, F.; Feliziani, M. Pacemaker Lead Coupling with an Automotive Wireless Power Transfer System. *IEEE Trans. Electromagn. Compat.* **2019**, *61*, 1935–1943. [CrossRef]

10. Yan, Z.; Song, B.; Zhang, Y.; Zhang, K.H.; Mao, Z.; Hu, Y. A Rotation-Free Wireless Power Transfer System with Stable Output Power and Efficiency for Autonomous Underwater Vehicles. *IEEE Trans. Power Electron.* **2018**, *34*, 4005–4008. [CrossRef]

11. Xu, J.; Zeng, Y.; Zhang, R. UAV-enabled wireless power transfer: Trajectory design and energy optimization. *IEEE Trans. Wirel. Commun.* **2018**, *17*, 5092–5106. [CrossRef]

12. Sugino, M.; Kondo, H.; Takeda, S. Linear motion type transfer robot using the wireless power transfer system. In Proceedings of the IEEE International Symposium on Antennas and Propagation (ISAP), Okinawa, Japan, 24–28 October 2016; pp. 508–509.

13. Cardoso, L.A.L.; Monteiro, V.; Pinto, J.G.; Nogueira, M.; Abreu, A.; Afonso, J.A.; Afonso, J.L. Design of an Intrinsically Safe Series-Series Compensation WPT System for Automotive LiDAR. *Electronics* **2020**, *9*, 86. [CrossRef]

14. Choi, K.W.; Aziz, A.A.; Setiawan, D.; Tran, N.M.; Ginting, L.; Kim, D.I. Distributed Wireless Power Transfer System for Internet of Things Devices. *IEEE Internet Things J.* **2018**, *5*, 2657–2671. [CrossRef]

15. Wang, W.; Zhang, Q.; Lin, H.; Liu, M.; Liang, X.; Liu, Q. Wireless Energy Transmission Channel Modeling in Resonant Beam Charging for IoT Devices. *IEEE Internet Things J.* **2019**, *6*, 3976–3986. [CrossRef]

16. Liu, B.; Xu, H.; Zhou, X. Resource Allocation in Wireless-Powered Mobile Edge Computing Systems for Internet of Things Applications. *Electronics* **2019**, *8*, 206. [CrossRef]

17. Jeong, S.; Kim, D.H.; Song, J.; Kim, H.; Lee, S.; Song, C.; Lee, J.; Song, J.; Kim, J. Smartwatch Strap Wireless Power Transfer System with Flexible PCB Coil and Shielding Material. *IEEE Trans. Ind. Electron.* **2019**, *66*, 4054–4064. [CrossRef]

18. Park, J.; Kim, D.; Hwang, K.; Park, H.H.; Kwak, S.I.; Kwon, J.H.; Ahn, S. A Resonant Reactive Shielding for Planar Wireless Power Transfer System in Smartphone Application. *IEEE Trans. Electromagn. Compat.* **2017**, *59*, 695–703. [CrossRef]

19. Wang, T.; Liu, X.; Jin, N.; Tang, H.; Yang, X.; Ali, M. Wireless Power Transfer for Battery Powering System. *Electronics* **2018**, *7*, 178. [CrossRef]

20. Song, C.; Kim, H.; Kim, Y.; Kim, D.; Jeong, S.; Cho, Y.; Lee, S.; Ahn, S.; Kim, J. EMI reduction methods in wireless power transfer system for drone electrical charger using tightly coupled three-phase resonant magnetic field. *IEEE Trans. Ind. Electron.* **2018**, *65*, 6839–6849. [CrossRef]

21. Campi, T.; Cruciani, S.; Maradei, F.; Feliziani, M. Innovative Design of Drone Landing Gear Used as a Receiving Coil in Wireless Charging Application. *Energies* **2019**, *12*, 3483. [CrossRef]

22. Yang, L.; Li, X.; Liu, S.; Xu, Z.; Cai, C.; Guo, P. Analysis and Design of Three-Coil Structure WPT System with Constant Output Current and Voltage for Battery Charging Applications. *IEEE Access* **2019**, *7*, 87334–87344. [CrossRef]

23. Shi, Z.H.; Chen, X.Y.; Qiu, Z.C. Modeling of mutual inductance between superconducting pancake coils used in wireless power transfer (WPT) systems. *IEEE Trans. Appl. Supercond.* **2019**, *29*, 1–4. [CrossRef]

24. Kim, J.; Kim, D.H.; Choi, J.; Kim, K.H.; Park, Y.J. Free-positioning wireless charging system for small electronic devices using a bowl-shaped transmitting coil. *IEEE Trans. Microw. Theory Tech.* **2015**, *63*, 791–800. [CrossRef]

25. Hou, T.; Xu, J.; Elkhuizen, W.S.; Wang, C.C.; Jiang, J.; Geraedts, J.M.; Song, Y. Design of 3D Wireless Power Transfer System Based on 3D Printed Electronics. *IEEE Access* **2019**, *7*, 94793–94805. [CrossRef]

26. Ha-Van, N.; Seo, C. Analytical and Experimental Investigations of Omnidirectional Wireless Power Transfer Using a Cubic Transmitter. *IEEE Trans. Ind. Electron.* **2018**, *65*, 1358–1366. [CrossRef]

27. Abou Houran, M.; Yang, X.; Chen, W. Magnetically Coupled Resonance WPT: Review of Compensation Topologies, Resonator Structures with Misalignment, and EMI Diagnostics. *Electronics* **2018**, *7*, 296. [CrossRef]

28. Zhang, C.; Lin, D.; Hui, S.R. Ball-Joint Wireless Power Transfer Systems. *IEEE Trans. Power Electron.* **2018**, *33*, 65–72. [CrossRef]

29. Abou Houran, M.; Yang, X.; Chen, W. Free Angular-Positioning Wireless Power Transfer Using a Spherical Joint. *Energies* **2018**, *11*, 3488. [CrossRef]

30. Han, H.; Mao, Z.; Zhu, Q.; Su, M.; Hu, A.P. A 3D Wireless Charging Cylinder with Stable Rotating Magnetic Field for Multi-Load Application. *IEEE Access* **2019**, *7*, 35981–35997. [CrossRef]

31. Campi, T.; Cruciani, S.; Maradei, F.; Feliziani, M. Near-Field Reduction in a Wireless Power Transfer System Using LCC Compensation. *IEEE Trans. Electromagn. Compat.* **2017**, *59*, 686–694. [CrossRef]

32. Cirimele, V.; Freschi, F.; Giaccone, L.; Pichon, L.; Repetto, M. Human exposure assessment in dynamic inductive power transfer for automotive applications. *IEEE Trans. Magn.* **2017**, *53*, 1–4. [CrossRef]

33. ICNIRP. Guidelines for limiting exposure to time-varying electric and magnetic fields (1 Hz to 10 MHz). *Health Phys.* **2010**, *99*, 818–836.

34. Besnoff, J.; Chabalko, M.; Ricketts, D.S. A Frequency-Selective Zero-Permeability Metamaterial Shield for Reduction of Near-Field Electromagnetic Energy. *IEEE Antennas Wirel. Propag. Lett.* **2016**, *15*, 654–657. [CrossRef]

35. Lu, C.; Rong, C.; Huang, X.; Hu, Z.; Tao, X.; Wang, S.; Chen, J.; Liu, M. Investigation of Negative and Near-Zero Permeability Metamaterials for Increased Efficiency and Reduced Electromagnetic Field Leakage in a Wireless Power Transfer System. *IEEE Trans. Electromagn. Compat.* **2018**, *99*, 1–9. [CrossRef]

36. Das, R.K.; Basir, A.; Yoo, H. A Metamaterial-Coupled Wireless Power Transfer System Based on Cubic High Dielectric Resonators. *IEEE Trans. Ind. Electron.* **2018**, *66*, 7397–7406. [CrossRef]

37. Kim, M.; Kim, H.; Kim, D.; Jeong, Y.; Park, H.H.; Ahn, S. A Three-Phase Wireless-Power-Transfer System for Online Electric Vehicles with Reduction of Leakage Magnetic Fields. *IEEE Trans Micro. Theory. Tech.* **2015**, *63*, 3806–3813. [CrossRef]
38. Choi, S.Y.; Gu, B.W.; Lee, S.W.; Lee, W.Y.; Huh, J.; Rim, C.T. Generalized active EMF cancel methods for wireless electric vehicles. *IEEE Trans. Power Electron.* **2014**, *29*, 5770–5783. [CrossRef]

Article

Unilateral Route Method to Estimate Practical Mutual Inductance for Multi-Coil WPT System

Seon-Jae Jeon, Sang-Hoon Lee and Dong-Wook Seo *

Department of Radio Communication Engineering, Korea Maritime and Ocean University (KMOU), Busan·49112, Korea; seonjae@kmou.ac.kr (S.-J.J.); shlee@kmou.ac.kr (S.-H.L.)
* Correspondence: dwseo@kmou.ac.kr; Tel.: +82-51-410-4427

Received: 20 January 2020; Accepted: 21 February 2020; Published: 24 February 2020

Abstract: Multi-coil WPT systems require mutual inductance information between coils to increase the power transmission efficiency. However, in the high frequency (HF) bands such as 6.78 MHz and 13.56 MHz, the presence of surrounding coils changes the value of the mutual inductance between the two coils due to the parasitic element effect of the coils. These parasitic effects make it harder to estimate the mutual inductance among three or more coils. In contrast to ideal mutual inductance, which has a constant value regardless of frequency and surrounding coils, we define the practical mutual inductance as the mutual inductance varied by parasitic elements. In this paper, a new method is presented to estimate the practical mutual inductance between multiple coils in the HF band. The proposed method simply configures the expression of practical mutual inductance formula because only one of two bilateral dependent voltage sources generated by mutual inductance is considered. For several coils placed along the same axis, the practical mutual inductances between coils were measured with respect to the distance between them to validate the proposed method. The practical mutual inductance obtained from the proposed method was consistent with the simulated and measured values in HF band.

Keywords: multiple coils; mutual inductance; parasitic effect; practical mutual inductance; transfer impedance; wireless power transfer

---

1. Introduction

Recently, multi-coil wireless power transfer (WPT) systems using three or more coils have been proposed to achieve a high power transfer efficiency (PTE) at greater distance than in the case of a conventional two-coil WPT system [1–12]. The multi-coil WPT systems are categorized into the domino structure and multiple transmitting (or receiving) coils structure. The domino structure means that one or more coils are placed between transmitting and receiving coils [1–6]. To achieve a long transmission distance, the structure adjusts the capacitance of each coil or gaps among coils. On the other hand, multiple transmitting (or receiving) coils correspond to the multiple input single output (MISO) or single input multiple output (SIMO) system in the wireless communication field [7–12]. The WPT system with multiple transmitting coils usually maximizes the power in the receiving coil through entering power of different magnitude or phase into transmitting coils. Sometimes, this technique is called magnetic beamforming [7,8].

The trouble in implementing these multi-coil WPT systems is that the mutual inductances (or coupling coefficients) among coils should be basically known. For the domino structure, the coupling coefficient information is used to estimate the target free-resonant frequencies, that is the resonant capacitances, or target distances between adjacent coils. For the multiple transmitting coils, power can be transferred into a not-aligned receiving coil by controlling the phase and magnitude of signal components input to each transmitting coil, where the mutual inductance information is used to calculate the phase and magnitude of input power [7,8]. This technique is referred to as magnetic beamforming. Similarly, for the multiple receiving coils, the output power can be maximized by adjusting the phase of the received power based on the mutual inductance information. Therefore, the accurate value of mutual inductance between coils is important to obtain high efficiency of multi-coil WPT systems such as domino, MISO, and SIMO structures in common. On the other hand, inaccurate mutual inductance causes interruption of maximum performance for the multi-coil systems.

The ideal mutual inductance between coils is completely separated from the operating frequency or presence of surrounding coils. In practice, when the mutual inductance between coils is measured, the measurement frequency and the presence of surrounding coils affect the measurement results. These phenomena can be inferred by the effect of the parasitic elements of the coil. We refer to the mutual inductance affected by the parasitic elements as the practical mutual inductance.

As is well known, the parasitic effect of a component is commonly frequency dependent, which becomes large as the frequency increases [13]. That is, the parasitic effects are not significant and can be ignored in low frequency (LF) bands such as 100–205 kHz. Therefore, most previous researches on the mutual inductance for the WPT system have focused on the estimation of the ideal mutual inductance of two coils with respect to the alignment status [14–18]. For WPT systems used in the HF band such as 6.78 MHz and 13.56 MHz, the practical mutual inductances between the transmitting and receiving coils are different from the theoretical value due to the effect of the parasitic component. S. Hackl et al. proposed an equivalent circuit model for the radio frequency identification (RFID) system with two coils operating at a very close distance considering not only the parasitic capacitors but also the capacitive coupling between coils [19]. However, the capacitive coupling between coils is generally taken account for 1 cm or less gap between coils, while the WPT system in the HF band is mainly used for a distance of tens or several cm. In addition, the method is difficult to expand to three or more coils because of complicated formulation.

In HF band, the practical mutual inductance between two coils can be easily measured using a vector network analyzer (VNA). However, it cannot be used to measure the practical mutual inductance among three or more coils since public VNAs usually have only two ports. Moreover, it is obvious that the number of transmitting or receiving coils tends to increase steadily. As the number of coils increases, the parasitic effect becomes greater. Therefore, an effective method is highly required to estimate the practical mutual inductance among multiple coils with parasitic effects using a two-port VNA in HF band. In this paper, a new method is proposed to obtain the practical mutual inductance among multiple coils using a general two-port VNA under the assumption that the coils are unilateral instead of reciprocal characteristics. The proposed method is suitable for general HF band WPT systems used at several cm distance and the simple expression of the formula allows to determine the effect of the presence of surrounding coils through every possible route among coils. We also verify the proposed method by comparison with measured and simulated mutual inductance among multiple coils.

2. Theoretical Analysis

This section presents theoretical analysis for the proposed unilateral route method.

2.1. Practical and Ideal Mutual Inductance between Two Coils

Figure 1a,b shows the equivalent circuit model of two ideal coils and two coils with parasitic element, respectively. If the two coils are considered as a two-port network, the two-port network can be represented by various two-port parameters such as impedance parameter and scattering parameter. For the two ideal coils of Figure 1a, the mutual inductance between the two coils is independent of the operating frequency. From the definition of the impedance parameter, the transfer impedance between two ideal coils is given by

$$Z_{21} = \left. \frac{V_2}{I_1} \right|_{I_2=0} = j\omega M_{21}, \tag{1}$$

where V_i and I_i are the voltage and current of the *i*th port, and ω is the angular frequency. M_{21} is the mutual inductance between the coils and the two-port network is reciprocal so that $Z_{ij} = Z_{ji}$ and $M_{ij} = M_{ji}$. Equation (1) means that Z_{ij} can be found by exciting port *j* with the current I_j, open-circuiting all other ports, and measuring the open-circuit voltage at port *i*. From Equation (1), the mutual inductance (M_{21}) between the coils can be easily obtained by measuring the transfer impedance (Z_{21}). This method is commonly used to measure the mutual inductance between the coils.

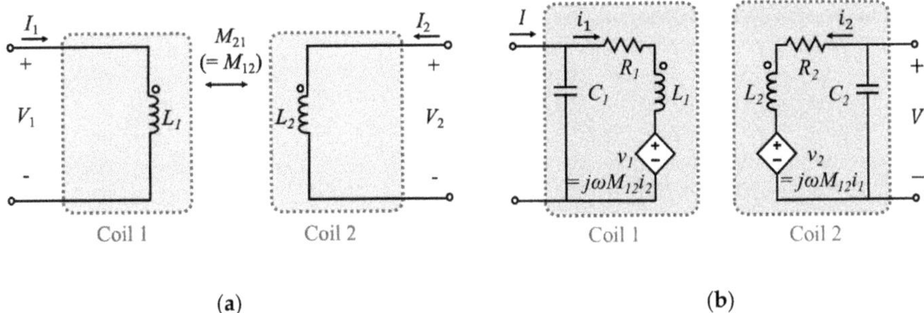

Figure 1. The equivalent circuit model: (**a**) Ideal two coils; and (**b**) real coils with parasitic capacitors.

On the other hand, as the operating frequency increases, the coil is not presented as a single inductor but combined with parasitic elements such as resistor and capacitor. However, the parasitic capacitor is only considered as shown in the equivalent model of Figure 1b. This is because the parasitic resistor of coils is usually much smaller than the impedance (ωL) of the inductor in HF band, $R << \omega L_1$, and then the reactance is dominant in determining the current flowing on the inductor and series parasitic resistor. Moreover, the parasitic capacitor of real coil connected in parallel makes a closed-loop, which causes the current path. Therefore, while the real coil is open-terminated in HF band, the practical mutual inductance between two coils is changed by the current path. From the circuit theory, the transfer impedance between the two coils with parasitic capacitors is given by

$$Z_{21} = \left.\frac{V_2}{I_1}\right|_{I_2=0} = \left.\frac{i_1}{I_1} \cdot \frac{v_2}{i_1} \cdot \frac{V_2}{v_2}\right|_{I_2=0} = \frac{\frac{1}{j\omega C_1}}{\frac{1}{j\omega C_1} + j\omega L_1 + \left(\frac{\omega^2 M_{12}^2}{\frac{1}{j\omega C_2}+j\omega L_2}\right)} \cdot j\omega M_{21} \cdot \frac{\frac{1}{j\omega C_2}}{\frac{1}{j\omega C_2} + j\omega L_2} \tag{2}$$

The transfer impedance Z_{21} is presented by multiplying the ideal transfer impedance $j\omega M$ and other terms. From Equation (2), causes of the parasitic effect are categorized into two. First, not all of the input current (I_1) of port 1 flows through L_1 due to the presence of C_1 and only part (i_1) of the current flows; then, the induced voltage (v_2) on L_2 decreases. In addition, the induced voltage (v_2) generates the current (i_2) flowing closed current path, which consists of L_2 and C_2, even when port 2 is opened. Then, the current (i_2) flowing through L_2 induces the voltage (v_1) on L_1 by the mutual inductance. Consequently, this lowers the amount of the current (i_1) flowing through L_1 and also lowers the induced voltage (v_2). This effect is expressed as the first term of the right side of Equation (2). Second, the voltage (v_2) induced in Coil 2 is divided into L_2 and C_2 so that only a part of the induced voltage (V_2) is delivered into port 2. This voltage distribution is expressed by the third term on the right side of Equation (2).

Similar to Equation (1), we define the transfer impedance using the practical mutual inductance ($M_{21(practical)}$) as

$$Z_{21} = j\omega M_{21(practical)}. \tag{3}$$

Comparing Equation (3) with Equation (2) yields that $M_{21(\text{practical})}$ is obviously different with M_{12} and strongly dependent on the coils' inductance, L_1 and L_2, and the parasitic capacitance, C_1 and C_2, as well as the angular frequency, ω.

The term in parentheses in the first term denominator of Equation (2) means that some of the voltage induced from Coil 1 to Coil 2 is induced again to Coil 1 through mutual inductance. Except for special cases where the coils are very close to each other or operate near the self-resonant frequency (SRF) of the coils, it is small enough to be negligible compared to other terms. Therefore, Equation (2) can be approximately expressed as follows in combination with Equation (3)

$$M_{21(\text{practical})} = \frac{Z_{21}}{j\omega} \simeq \frac{\frac{1}{j\omega C_1}}{j\omega L_1 + \frac{1}{j\omega C_1}} \cdot M_{21} \cdot \frac{\frac{1}{j\omega C_2}}{j\omega L_2 + \frac{1}{j\omega C_2}}. \tag{4}$$

Equation (4) is equivalent to the circuit diagram of Figure 2 that shows the dependent voltage source on L_1 removed in Figure 1b. That is, it is assumed that the coils are unilateral. If the coils' parameters, L_1, C_1, L_2 and C_2, are known, the ideal mutual inductance M_{21} can be obtained by measuring the transfer impedance Z_{21} and using Equation (4). Conversely, if the coil's parameters and ideal mutual inductance M_{21} are known, the practical mutual inductance $M_{21(\text{practical})}$ in the HF band can be estimated by using Equation (4).

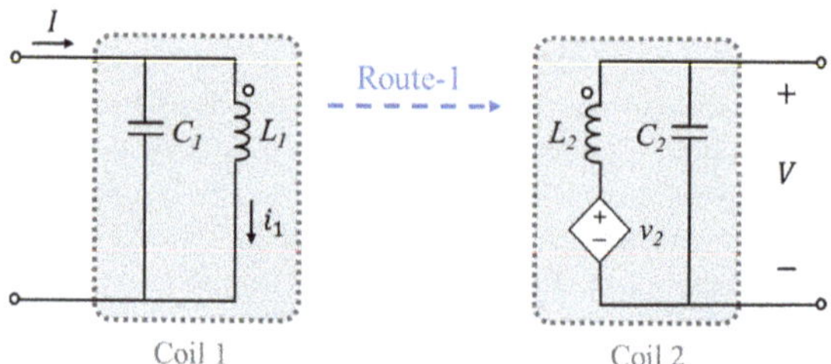

Figure 2. The equivalent circuit model of two coils with Route 1.

2.2. Measurement Method for the Coil's Parameters and Ideal Mutual Inductance between Two Coils

As mentioned above, the coil's parameters and ideal mutual inductance should be known in order to estimate the practical mutual inductance between two coils. Figure 3 shows the proposed procedure to obtain the coil's parameters and ideal mutual inductance. First, parameters of each single coil are measured without surrounding coils or the object. Since the parasitic effect rarely appears in the LF band, the input impedance Z_{11} of a single coil is simply expressed as $R_1 + j\omega L_1$. Therefore, the self-inductance and resistance of each coil are directly obtained by measuring the input impedance of the coil using a VNA at the lowest possible frequency in the LF band. Next, the parasitic capacitance C_1 of the coil can be obtained from the SRF of the coil as given by

$$C_1 = \frac{1}{\omega_0{}^2 L_1} = \frac{1}{(2\pi f_0)^2 L_1}, \tag{5}$$

where f_0 is the SRF at which the reactance of the coil is zero, and it can be easily measured from the Smith Chart of S_{11} or the input impedance plot on the VNA. Finally, the ideal mutual inductance M_{21} between two coils can also be obtained from measuring the transfer impedance Z_{21} between the coils using the VNA in the LF band. Therefore, the parameters of coils measured through the procedure of Figure 3 can be used to estimate the practical mutual inductance between two coils. Moreover, if the ideal mutual inductances among three or more coils are well defined, the measured parameters can also be used to estimate the practical mutual inductance.

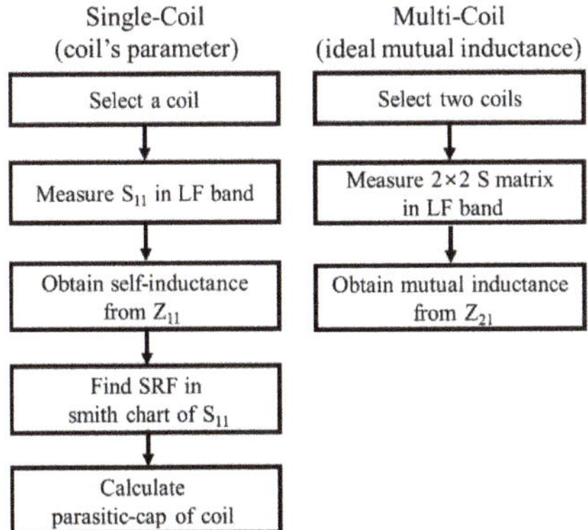

Figure 3. The procedures to obtain the coil's parameter and the ideal mutual inductance between coils.

2.3. Transfer Impedance of Multi-Coil Using the Proposed Circuit Model

For three coils shown in Figure 4a, we should estimate the transfer impedance Z_{31} between Coils 1 and 3 to obtain the practical mutual inductance between the coils. From the definition of impedance parameter, the current applied to the first port can directly induce the voltage on the opened third port and also indirectly induce the voltage via the second port. Strictly, the induced voltage makes a current in the closed loop formed by the parasitic capacitor, and the current in the third inductor can affect other inductors. However, since the high order mutual effects are weak, under the assumption that the coils are unilateral, the transfer impedance between Coils 1 and 3 is obtained as the sum of two routes in Figure 4b. Thus, the practical mutual inductance between Coils 1 and 3 is given by

$$M_{31(\text{practical})} = \frac{Z_{31}}{j\omega} \simeq \frac{Z_{R-1} + Z_{R-2}}{j\omega}, \tag{6}$$

where Z_{R-1} and Z_{R-2} are the transfer impedances of the direct and one-hop indirect routes between Coils 1 and 3, respectively, as shown in Figure 4b. Using Equation (4) the transfer impedance of the direct route is given by

$$Z_{R-1} = \frac{\frac{1}{j\omega C_1}}{j\omega L_1 + \frac{1}{j\omega C_1}} \cdot j\omega M_{31} \cdot \frac{\frac{1}{j\omega C_3}}{j\omega L_3 + \frac{1}{j\omega C_3}}. \tag{7}$$

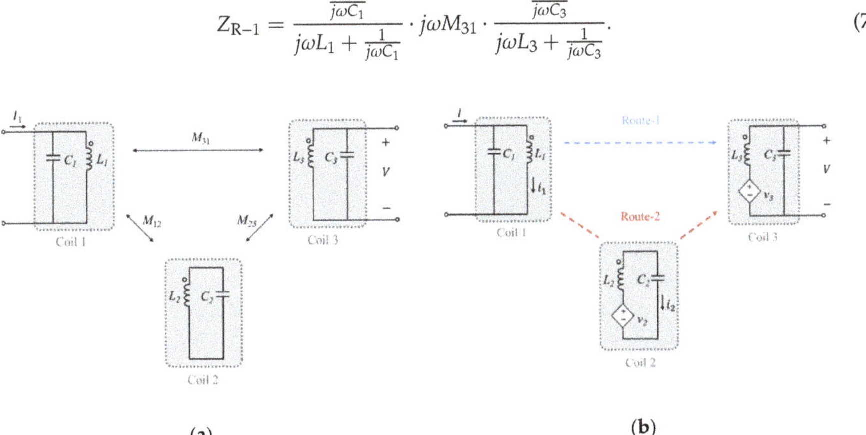

Figure 4. The equivalent circuit model: (**a**) Three coils with ideal mutual inductances; and (**b**) with two routes.

From unilateral characteristics, the transfer impedance of the indirect one-hop route is also easily obtained by

$$Z_{\text{R}-2} = \frac{V}{I} = \frac{i_1}{I} \cdot j\omega M_{21} \cdot \frac{-1}{Z_2} \cdot j\omega M_{32} \cdot \frac{V}{v_3} = \frac{\frac{1}{j\omega C_1}}{j\omega L_1 + \frac{1}{j\omega C_1}} \cdot j\omega M_{21} \cdot \frac{-1}{j\omega L_2 + \frac{1}{j\omega C_2}} \cdot j\omega M_{32} \cdot \frac{\frac{1}{j\omega C_3}}{j\omega L_3 + \frac{1}{j\omega C_3}}, \tag{8}$$

where M_{21}, M_{32} and M_{31} are ideal mutual inductances, and the minus sign is due to the dot notation between Coils 1 and 2.

For four coils shown in Figure 5, there are four possible routes from Coil 1 to Coil 4 under the unilateral assumption. The transfer impedance of the fourth route is given by

$$Z_{\text{R}-4} = \frac{V}{I} = \frac{i_1}{I} \cdot j\omega M_{21} \cdot \frac{-1}{Z_2} \cdot j\omega M_{32} \cdot \frac{-1}{Z_3} \cdot j\omega M_{43} \cdot \frac{V}{v_4}. \tag{9}$$

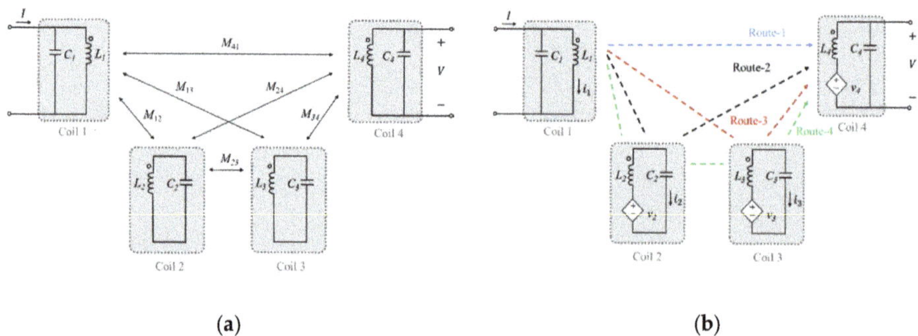

(a)   **(b)**

Figure 5. The equivalent circuit model: (**a**) Four coils with ideal mutual inductances; and (**b**) with four routes.

The practical mutual inductance is summarized as

$$M_{41(\text{practical})} = \frac{Z_{41}}{j\omega} \approx \frac{Z_{\text{R}-1} + Z_{\text{R}-2} + Z_{\text{R}-3} + Z_{\text{R}-4}}{j\omega}. \tag{10}$$

If the number of coils is n, the transfer impedance is determined as

$$Z_{n1} \approx \sum_{k=1}^{2^{n-2}} Z_{\text{R}-k}. \tag{11}$$

3. Experimental Results and Discussion

To verify the proposed method in the HF band, we analyzed the practical mutual inductance by changing the number of coils as shown in Figure 6. Two types of coils were used for the experiment, and parameters of the coils are summarized in Table 1. Coil$_a$ has three turns and a single layer printed on an FR-4 substrate and dimensions 95.7 mm × 105.7 mm. Coil$_b$ has six turns and two layers printed on an FR-4 substrate of the same dimensions as the Coil$_a$. All coils were arranged coaxially in order to maximize the parasitic effect and the effect of surrounding coils. The practical mutual inductance among multiple coils was measured with respect to the distance using the four-port VNA, Agilent's E5071, as shown in Figure 7. Therefore, the [S] matrix for three coils that are directly connected to ports of the multi-port VNA was measured and then converted into the [Z] matrix using the equation in [20] on the Keysight's ADS.

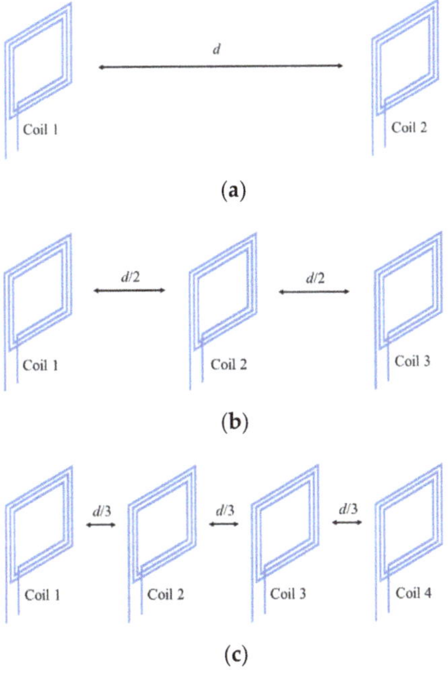

Figure 6. The experimental array of the coils: (**a**) Two coils; (**b**) three coils; and (**c**) four coils.

Table 1. The parameter for designed coils.

Parameter	Coil$_a$	Coil$_b$
Inductance (μH)	1.13	3.83
Parasitic cap. (pF)	8.22	15.17

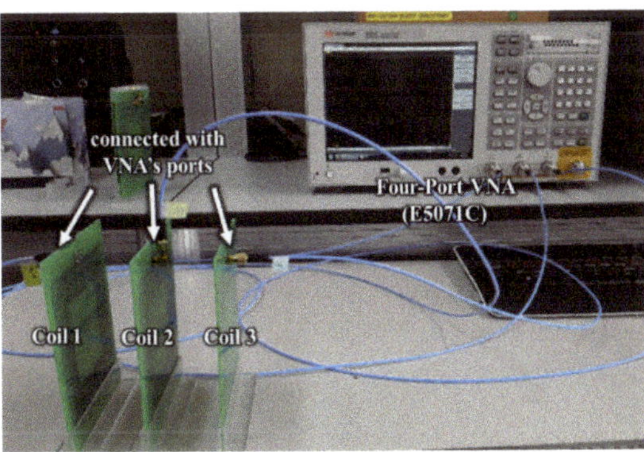

Figure 7. The experimental setup with a four-port vector network analyzer (VNA).

Additionally, we simulated the practical mutual inductance from the full circuit models with all parasitic components and dependent voltage sources using the ADS, and the measured practical mutual inductances are compared with the simulated results as well as those from the proposed method in several cases.

3.1. Two Coils Analysis

For two coils of Figure 6a, where the two coils used are identical to $Coil_a$ in Table 1, the S parameters were measured without surrounding coils and the practical mutual inductances were estimated by substituting the Z parameters converted from S parameters into Equation (4). Figure 8 shows the measured and simulated practical mutual inductances with those of the proposed estimation method. As the distance between the coils decrease, the ideal mutual inductance increases and then the practical mutual inductance also increases. On the other hand, at 13.56 MHz, the practical mutual inductance has a higher level than 6.78 MHz. This result comes from the first and third terms of the right-hand side of Equation (4), being a function of frequency. As the frequency increases, the magnitude of the denominators of the terms dramatically increases more than those of the numerators. The mutual inductances estimated by the proposed method are in good agreement with the simulated and measured values over the distance. The closer the distance between the coils and the higher the frequency, the larger the practical mutual inductance becomes due to the parasitic effect. The proposed method accurately simulates these effects.

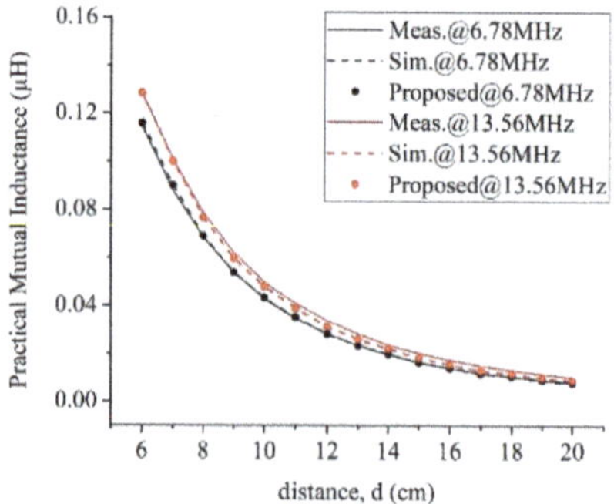

Figure 8. Practical mutual inductance between two coils with distance.

3.2. Three Coils Analysis

For three coils of Figure 6b, the experiments were conducted on the two cases of Table 2, where Coil 2 was located in the middle of Coils 1 and 3. The self-inductance of Coil 2 for Case#2 is larger than the value for Case#1. Figure 9 shows the simulated, measured, and estimated practical mutual inductances with respect to distance at 6.78 MHz and 13.56 MHz. It is noticed that the variation in the mutual inductance with Case#2 is greater than Case#1. This means that the larger the inductance of the surrounding coils, the greater the influence on the mutual inductance between other coils. In addition, the variation of the practical mutual inductance is greater at 13.56 MHz than at 6.78 MHz. That is, the parasitic effect is larger at a higher frequency. On the other hand, the proposed method exactly estimates the practical mutual inductance all over the distance, even if the parameters of the coil change.

Table 2. Configuration of three coils for two cases.

Two Cases	Coil 1	Coil 2	Coil 3
Case#1	$Coil_a$	$Coil_a$	$Coil_a$
Case#2	$Coil_a$	$Coil_b$	$Coil_a$

To show the accuracy of the proposed method, we calculate the mean and normalized root-mean-square error (NRMSE) of difference between the results of measurement and the proposed method shown in Figures 8 and 9, and summarize them in Table 3. While the mean and NRMSE increase as the number of coil and frequency increases, the mean and error are less than 20×10^{-10} (0.002 µH) and 3%, respectively, for all cases. This means that the proposed method accurately estimates the practical mutual inductance as measured by a multi-port VNA.

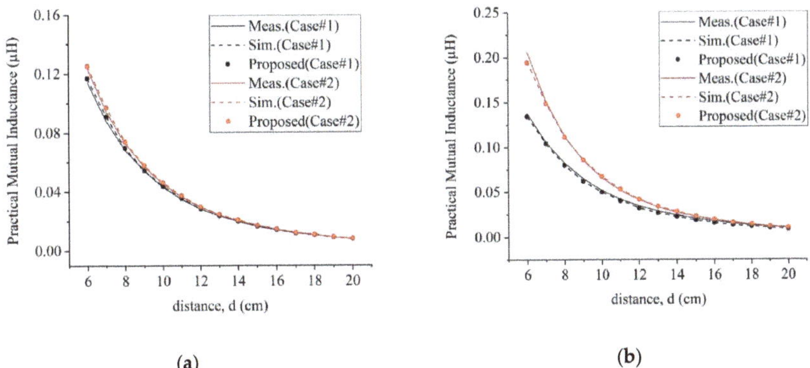

(a) **(b)**

Figure 9. Practical mutual inductance between Coils 1 and 3 with distance: (**a**) 6.78 MHz; and (**b**) 13.56 MHz.

Table 3. Mean and normalized root-mean-square error (NRMSE) of difference between the results of measurement and the proposed method for Figures 8 and 9.

	6.78 MHz		13.56 MHz	
	Mean (µH)	**NRMSE (%)**	**Mean (µH)**	**NRMSE (%)**
Two coils	-2.7×10^{-4}	0.68	16.9×10^{-4}	1.67
Case#1	-5.8×10^{-4}	1.03	20.0×10^{-4}	1.92
Case#2	-6.5×10^{-4}	0.84	7.3×10^{-4}	2.81

3.3. Analysis of Unilateral Method for the Number of Coils

Figure 10 shows the practical mutual inductance as a function of the frequency at several distances for two, three, and four coils as shown in Figure 6. The parameters of all arranged coils are identical with $Coil_a$ in Table 1. The ideal mutual inductance does not depend on the frequency, but the practical mutual inductances are proportional to frequency. In addition, as the number of coils increases, the mutual inductance increases further due to the parasitic effect. In these cases, the results of the proposed method are almost identical to the simulated value.

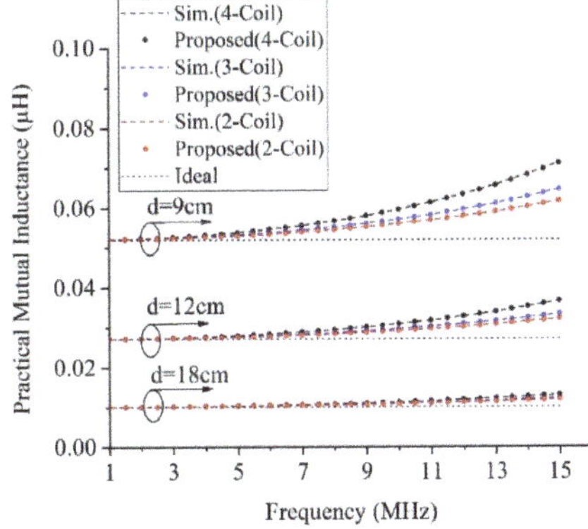

Figure 10. Practical mutual inductances between side coils.

Figure 11 shows the simulated transfer impedance of each route for four coils, which is normalized by the transfer impedance of the direct route, Z_{R-1}. The transfer impedances of one- and two-hop indirect routes are presented in Figure 11a,b, respectively. As the number of hops increase, the transfer impedance decreases dramatically. In other words, the normalized transfer impedance of one-hop routes have a value smaller than 0.1, while those of two-hop routes have a very small value of less than 0.005. Thus, the two-hop route is almost unaffected as an element to obtain the practical mutual inductance in a four-coil array. This means that among 2^{n-2} routes for n coils, only the direct and one-hop routes are practically dominant, so the transfer impedance between the primary and the nth coil can be approximately expressed as

$$Z_{n1} \simeq \sum_{k=1}^{2^{n-2}} Z_{R-k} \simeq \sum_{k=1}^{n-1} Z_{R-k}. \tag{12}$$

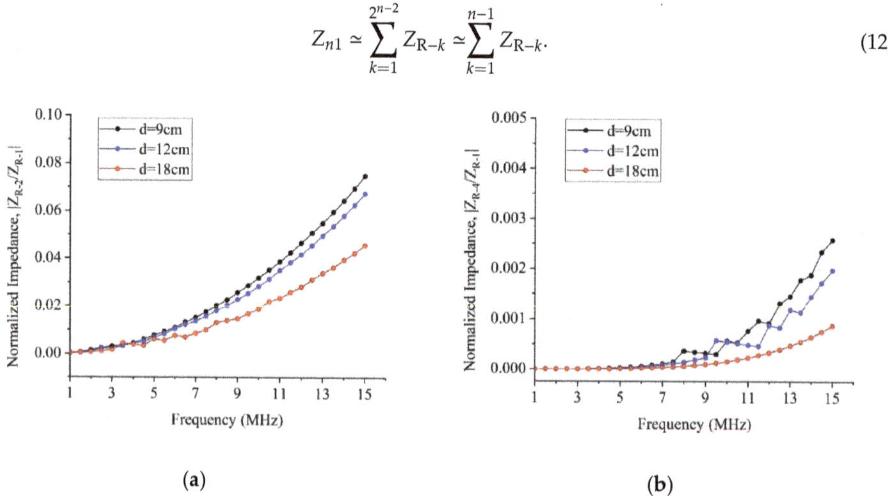

Figure 11. Normalized transfer impedance in the array of four coils: (**a**) One-hop; and (**b**) two-hop routes.

Finally, for four coils, the practical mutual inductances by all routes and the routes except two more hopping routes with respect to the frequency are as shown in Figure 12. Even if the two-hop routes are not taken into account, the calculated mutual inductances are in good agreement with the results of all the routes as well as the simulated results.

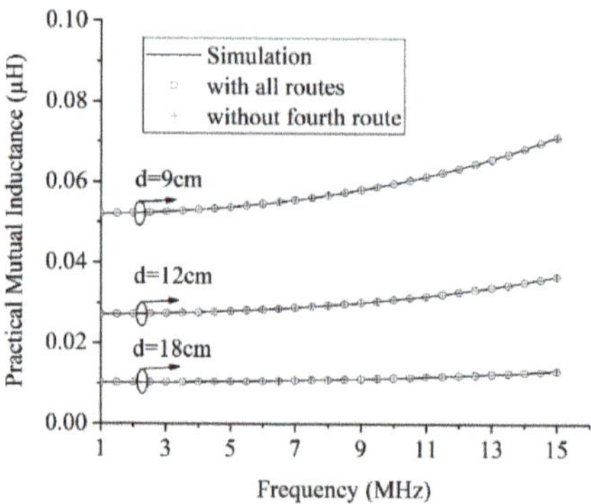

Figure 12. Simulated and calculated results for the array of four coils.

4. Conclusions

The multi-coil WPT systems commonly use three or more coils to improve efficiency in HF bands as well as LF bands. These systems are basically required to estimate or measure the practical mutual inductance between coils. In this paper, the unilateral approach method is proposed to estimate the mutual inductance among multiple coils. With the proposed method, it is not only simple to configure the expression of formulas but also possible to estimate the practical mutual inductance through several routes for multiple coils since the assumption that only one voltage generated with respect to the coil pair was considered. To verify the validity of the proposed method, we coaxially placed several coils along the distance and confirmed the consistent results with simulation and measurement. In reality, it is difficult to analyze the mutual inductance in multiple coils using a public VNA that has only two ports. However, the proposed method can provide the mutual inductance using a public two-port VNA. Furthermore, it will be quite useful for the multi-coil WPT systems with a regular array of identical coils.

Author Contributions: Conceptualization, S.-J.J. and S.-H.L.; methodology, S.-J.J.; software, S.-J.J.; validation, S.-J.J. and S.-H.L.; formal analysis, S.-J.J. and D.-W.S.; investigation, S.-J.J. and S.-H.L.; resources, S.-H.L.; data curation, S.-J.J.; writing—original draft preparation, S.-J.J.; writing—review and editing, D.-W.S.; visualization, S.-J.J.; supervision, D.-W.S.; project administration, D.-W.S.; funding acquisition, D.-W.S. All authors have read and agreed to the published version of the manuscript.

Funding: This work was supported by the National Research Foundation of Korea (NRF) grant funded by the Korea government (MSIT) (No. 2018R1C1B6003854).

Conflicts of Interest: The authors declare no conflicts of interest.

References

1. Zhong, W.X.; Lee, C.K.; Hui, S.Y.R. General analysis on the use of tesla's resonators in domino forms for wireless power transfer. *IEEE Trans. Ind. Electron.* **2013**, *60*, 261–270. [CrossRef]

2. Kim, J.H.; Park, B.C.; Lee, J.H. Optimum design of WPT relay system by controlling capacitance. *Microw. Opt. Technol. Lett.* **2014**, *56*, 1658–1661. [CrossRef]

3. Sun, Y.; Liao, Z.J.; Ye, Z.H.; Tang, C.S.; Wang, P.Y. Determining the maximum power transfer points for MC-WPT systems with arbitrary number of coils. *IEEE Trans Power Electron.* **2018**, *33*, 9734–9743. [CrossRef]

4. Seo, D.W. Design method of three-coil WPT system based on critical coupling condition. *IEEE J. Emerg. Sel. Top. Power Electron.* **2019**. [CrossRef]

5. Sun, T.; Xie, X.; Li, G.; Gu, Y.; Wang, Z. Indoor wireless power transfer using asymmetric directly-strong-coupling mechanism. *Microw. Opt. Technol. Lett.* **2013**, *55*, 250–253. [CrossRef]

6. Xu, B.; Li, Y. Investigation of surface wave propagation along a multi-coil wireless power transfer system. *Microw. Opt. Technol. Lett.* **2016**, *58*, 2261–2265. [CrossRef]

7. Jadidian, J.; Katabi, D. Magnetic MIMO: How to charge your phone in your pocket. In Proceedings of the 20th Annual International Conference on Mobile Computing and Networking (MOBICOM), Maui, HI, USA, 7–11 September 2014; pp. 495–506.

8. Shi, L.; Kabelac, Z.; Katabi, D.; Perreault, D. Wireless power hotspot that charges all of your devices. In Proceedings of the 21th Annual International Conference on Mobile Computing and Networking (MOBICOM), Paris, France, 7–11 September 2015; pp. 2–13.

9. Sritongon, C.; Wisestherrakul, P.; Hansupho, N.; Nutwong, S.; Sangswang, A.; Naetiladdanon, S.; Mujjalinvimut, E. Novel IPT multi-transmitter coils with increase misalignment tolerance and system Efficiency. In Proceedings of the 2018 IEEE International Symposium on Circuits and Systems (ISCAS), Florence, Italy, 27–30 May 2018; pp. 1–5.

10. Nguyen, M.Q.; Ta, K.; Dubey, S.; Chiao, J.C. Frequency modes in a MIMO wireless power transfer system. In Proceedings of the 2017 IEEE Asia Pacific Microwave Conference (APMC), Kuala Lumpar, Malaysia, 13–16 November 2017; pp. 146–149.

11. Kim, D.H.; Ahn, D. Maximum efficiency point tracking for multiple-transmitters wireless power transfer. *IEEE Trans. Power Electron.* **2019**. [CrossRef]

12. Fu, M.; Zhang, T.; Ma, C.; Zhu, X. Efficiency and optimal loads analysis for multiple-receiver wireless power transfer systems. *IEEE Trans. Microw. Theory Tech.* **2015**, *63*, 801–812. [CrossRef]

13. Alley, C.L.; Atwood, K.W. *Electronic Engineering*; Wiley: Hoboken, NJ, USA, 1973; p. 199.

14. Piri, M.; Jaros, V.; Fricaldsky, M. Verification of a mutual inductance calculation between two helical coils. In Proceedings of the 2015 16th International Scientific Conference on Electric Power Engineering (EPE), Kouty nad Desnou, Czech Republic, 20–22 May 2015; pp. 712–717.

15. Wu, L.; Lu, K.; Xia, Y. Mutual inductance calculation of two coaxial solenoid coils with iron core. In Proceedings of the 2018 21st International Conference on Electrical Machines and System (ICEMS), Jeju, Korea, 7–10 October 2018; pp. 1804–1808.

16. Ariffin, A.; Inamori, M. Influence of mutual inductance measurement for high efficiency wireless power transmission. In Proceedings of the 2018 7th International Conference on Renewable Energy Research and Applications (ICRERA), Paris, France, 14–17 October 2018; pp. 949–953.

17. Zhang, X.; Meng, H.; Wei, B.; Wang, S.; Yang, Q. Mutual inductance calculation for coils with misalignment in wireless power transfer. *IET J. Eng.* **2019**, *16*, 1041–1044. [CrossRef]

18. Rakhymbay, A.; Khamitov, A.; Bagheri, M.; Alimkhanuly, B.; Lu, M.; Phung, T. Precise analysis on mutual inductance variation in dynamic wireless charging of electric vehicle. *Energies* **2018**, *11*, 624. [CrossRef]

19. Hackl, S.; Lanschutzer, C.; Raggam, P.; Randeu, W.L. A Novel Method for Determining the Mutual Inductance for 13.56MHz RFID Systems. In Proceedings of the 2008 6th International Symposium on Communication Systems, Networks and Digital Signal Processing, Graz, Austria, 25–25 July 2008; pp. 297–300.

20. Pozar, D.M. *Microwave Engineering*, 4th ed.; Wiley: Hoboken, NJ, USA, 2011; p. 181.

 *electronics* 

Article

Energy Harvesting Maximizing for Millimeter-Wave Massive MIMO-NOMA

Shufeng Li *, Zelin Wan, Libiao Jin and Jianhe Du

School of Information and Communication Engineering, Communication University of China, Beijing 100024, China; 13787218609@163.com (Z.W.); libiao@cuc.edu.cn (L.J.); dujianhe1@163.com (J.D.)
* Correspondence: lishufeng@cuc.edu.cn; Tel.: +86-1355-2288-205

Received: 23 November 2019; Accepted: 24 December 2019; Published: 26 December 2019

Abstract: Multiple-Input Multiple-Output Non-Orthogonal Multiple Access (MIMO-NOMA) is considered a promising multiple access technology in fifth generation (5G) networks, which can improve system capacity and spectral efficiency. In this paper, we proposed two methods of user grouping and proposed a dynamic power allocation solution for MIMO-NOMA system. Then we proposed an algorithm to maximize energy harvest for MIMO-NOMA system by integrating Simultaneous Wireless Information and Power Transfer (SWIPT), known as maximizing energy harvesting. Specifically, we added a power splitter at the receiver and found the optimal power splitting factor for each user. The harvested power of the user is maximized under the premise of satisfying the minimum communication rate. The simulation results show that the proposed method is effective.

Keywords: MIMO; NOMA; precoding; power allocation; user-clustering; power splitter

1. Introduction

Non-Orthogonal Multiple Access (NOMA) is one of the key technologies of fifth generation (5G) networks, which can significantly improve the overall performance of the system. Through time and frequency resource reuse and user grouping for large-scale connection, NOMA can improve spectral efficiency [1]. Multiple-Input Multiple-Output (MIMO) is regarded as a promising technique for 5G wireless communication systems. The principle of MIMO is the use of multi-antenna technology to achieve spatial diversity. The multi-receiving and multi-transmitting mechanisms can effectively combat multipath interference and increase system capacity. The author proposed a new low complexity arrival direction estimation algorithm in MIMO system for meeting the needs of green communication in [2]; this algorithm is based on a new downlink transmission frame structure that can make full use of the prior information under the channel codebook feedback mechanism. NOMA superimposes multiple user signals in the power domain, superimposes coding on the transmitter, performs serial interference cancellation on the receiver, and eliminates inter-user interference for grouped users [3]. NOMA can be divided into two categories: NOMA in power domain and NOMA in code domain. The power domain NOMA scheme can provide service to multiple users with different channel conditions simultaneously in the same time, frequency, coding, and space [4–7]. Uplink and downlink NOMA transmission of single cellular network is studied in [8], the author also analyzed the effect of distance on performance of the system. The influencing factors of NOMA, such as user power allocation, new order cost, and Serial Interference Cancellation (SIC) error propagation were discussed in [9].

NOMA combined with MIMO technology has attracted considerable research interest. The basic principle of NOMA and MIMO combination in downlink transmission was studied in [10]. The MIMO-NOMA could improve spectrum reuse efficiency, transmission throughput, and energy efficiency. In MIMO-NOMA, it is essential to make user grouping efficient. If the user is grouped by appropriate methods, the error rate of the system can be reduced [11]. In the existing MIMO-NOMA system, users are divided into multiple groups. The group uses the NOMA principle to serve users. The precoding between groups is used to eliminate interference. The user grouping of the downlink NOMA system is studied in [12]; the user clustering problem is formalized into a semi-definite programming problem which can be solved using numerical toolbox [13]. The accuracy of power allocation affects the system performance. In [14], the author derived the closed expressions for the traversal, rate, and interrupt probability of the two-user NOMA system for static power allocation. A dynamic power allocation solution is provided in [15] with the goal of maximizing the total unit capacity.

Wireless communication devices still use electric cables or batteries to obtain electric energy. The battery storage capacity and usage period are often limited, which will cause the development and application of new technologies in specific scenarios to be deeply bounded. Harvesting energy from radio frequency (RF) signals has become an attractive strategy to address the critical challenges of limited battery life in wireless communication networks. An advanced technology called Simultaneous Wireless Information and Power Transfer (SWIPT) emerged in [16], by which the energy transmission and information transmission using RF signals can be achieved. Therefore, SWIPT is considered a potential energy-saving solution for 5G [17], which has attracted widespread attention in academia and industry. A capacity-energy function was defined and the receiver can perform both information decoding and energy harvesting (EH) without any restrictions [16]. The work [18] considered the sum rate and the per-user optimized data rate of the SWIPT-enabled NOMA system, in which two information decoding schemes are proposed, "fixed decoding order" and "time sharing", respectively, and proved that system performance could be significantly improved by integrating SWIPT on NOMA. The work of [19] jointly optimized the transmission power of the Base Station (BS), as well as the length of time for energy acquisition and data transmission. The application of SWIPT technology in NOMA is studied and a new cooperative SWIPT NOMA protocol is proposed in [20]. By jointly optimizing the power allocation coefficient of MIMO-NOMA and the power splitting factor of SWIPT, the achievable sum rate can be maximized [21].

The issue of energy conservation is also an issue that recently has attracted considerable attention. To meet the needs of green communication and realize the recycling of energy, we implement energy-saving wireless communication in MIMO-NOMA system integrated with SWIPT. Specifically, each user uses a power splitter to split the received signal into two parts. The receiver performs information retrieval and energy storage to implement SWIPT simultaneously. In this paper, we also studied the user clustering, precoding design, and power allocation to optimize the power-splitter factor of SWIPT. The harvested energy is maximized under the premise of satisfying the minimum communication rate of the user.

The main contributions of this paper are as follows:

1. In this paper, for reducing intra-cluster interference and complexity, we proposed two methods for user grouping. One is based on channel gain; the other is based on antenna grouping. The performance effects of the two grouping methods on the system are analyzed from the perspective of spectrum efficiency and energy efficiency.

2. We provided a dynamic power allocation solution for downlink multi-user MIMO-NOMA. Power allocation is divided into two steps: power allocation between clusters and power allocation within the cluster. In the power allocation between clusters, the power allocation of each beam is proportional to the number of users, and the power within the cluster is allocated according to the maximum communication rate of the cluster.

3. We proposed the energy harvesting maximizing method. We added a SWIPT split receiver in MIMO-NOMA system for each user. In this method, the optimal power split coefficient of each user is found with the optimization objective of harvesting the maximum energy and satisfying the minimum communication rate of users.

Notation: Within this paper, the upper-case boldface letters denote matrices; lower-case boldface letters are vectors. $(\cdot)^T, (\cdot)^{-1}, (\cdot)^H$ denotes the transpose, matrix inversion, and conjugate transpose. $CN(a, b)$ denote the complex Gaussian distribution with mean a and covariance b. $\|\cdot\|_n$ is l_n norm operation. $|\Gamma|$ denotes the number of elements in set Γ. $E\{\cdot\}$ denotes the expectation.

2. System Model

In this paper, we consider a single-cell downlink Millimeter-Wave massive MIMO-NOMA system, as shown in Figure 1. The base station is equipped with N antennas and N_{RF} RF chains, and K single antenna users are served by the base station. By using NOMA, each beam can support multiple users. S_g represents the set of users served by the gth beam. To fully realize the multiplexing gain, we assume that the beam number G is equal to the number of RF chains N_{RF}. The received signal at mth user in the nth beam is [21]:

$$
\begin{aligned}
y_{n,m} &= \mathbf{h}_{n,m}^H \sum_{j=1}^{G} \sum_{i=1}^{|S_n|} \mathbf{w}_j \sqrt{p_{i,j}} s_{i,j} + v_{n,m} \\
&= \mathbf{h}_{n,m}^H \mathbf{w}_n \sqrt{p_{n,m}} s_{n,m} + \mathbf{h}_{n,m}^H \mathbf{w}_n \sum_{j=1}^{m-1} \sqrt{p_{n,j}} s_{n,j} + \mathbf{h}_{n,m}^H \mathbf{w}_n \sum_{i=m+1}^{|S_i|} \sqrt{p_{n,i}} s_{n,i} \\
&\quad + \mathbf{h}_{n,m}^H \sum_{i \neq n}^{G} \sum_{j=1}^{|S_i|} \mathbf{w}_j \sqrt{p_{i,j}} s_{i,j} + v_{n,m}
\end{aligned}
\tag{1}
$$

where the interference within the cluster and the interference between the cluster are existing. $\mathbf{h}_{m,n}$ represents the channel of the mth user in the nth beam, $\mathbf{w}_n$ is precoding vector of the nth beam. $s_{n,m}$ is the transmitted signal and $p_{n,m}$ denotes transmitted power for the mth user in the nth beam, and $v_{n,m}$ is the noise following the distribution $CN(0, \sigma_u^2)$.

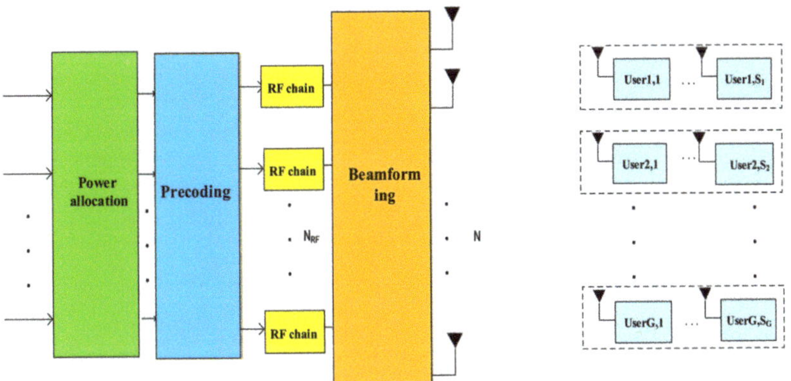

Figure 1. Millimeter-Wave massive Multiple-Input, Multiple-Output, Non-Orthogonal Multiple Access (MIMO-NOMA) system.

The mth user in the nth beam can eliminate the interference of the ith user (for all $i > m$) in the nth beam by performing Serial Interference Cancellation (SIC). The remaining signal received by the mth user in the nth beam can be rewritten as:

$$
\widetilde{y}_{n,m} = \left(\mathbf{h}_{n,m}^H \mathbf{w}_n \sqrt{p_{n,m}} s_{n,m} + \mathbf{h}_{n,m}^H \mathbf{w}_n \sum_{j=1}^{m-1} \sqrt{p_{n,j}} s_{n,j} + \mathbf{h}_{n,m}^H \sum_{i \neq n}^{G} \sum_{j=1}^{|S_i|} \mathbf{w}_j \sqrt{p_{i,j}} s_{i,j} + v_{n,m} \right)
\tag{2}
$$

The Signal to Interference plus Noise Ratio (SINR) at the mth user in the nth beam is:

$$\gamma_{n,m} = \frac{\left\|\mathbf{h}_{n,m}^{H}\mathbf{w}_n\right\|_2^2 p_{n,m}}{\xi_{n,m}} \tag{3}$$

where,

$$\xi_{n,m} = \left\|\mathbf{h}_{n,m}^{H},\mathbf{W}_n\right\|_2^2 \sum_{j=1}^{m-1} p_{n,j} + \sum_{i\neq n}^{G}\left\|\mathbf{h}_{n,m}^{H}\mathbf{W}_i\right\|_2^2 \sum_{j=1}^{|S_i|} p_{i,j} + \sigma_v^2 \tag{4}$$

The achievable rate of the mth user in the nth beam can be written as:

$$R_{n,m} = \log_2(1 + \gamma_{n,m}) \tag{5}$$

Finally, the achievable sum rate is:

$$R_{sum} = \sum_{n=1}^{G}\sum_{m=1}^{|S_i|} R_{m,n} \tag{6}$$

2.1. User-Clustering

We assume that all users in the downlink MIMO cellular system can utilize NOMA-based resource allocation. Users need to be grouped first and the grouped users share a set of codes in the same group by the precoding matrix. Low channel gain users of NOMA clusters are often subject to higher intra-cluster interference [22].

In this paper, we propose the following two methods for user clustering:

(1) The user clustering method based on channel gain. As mentioned in [23], the cluster head user with the highest channel gain can eliminate intra-cluster interference, thereby obtaining the maximum throughput gain. Therefore, the keys to maximize overall system capacity is to ensure that high channel gain users are selected as cluster heads for different MIMO-NOMA clusters in one unit. To improve system performance, we grouped users by assigning the user with largest channel gain as cluster head. As shown in Figure 2, the number of user groups M is equal to the number of beams G. In this way, users in the same group will suffer higher channel correlation, which is beneficial to eliminate interference between users. The lower equivalent channel correlation of users in different beams is beneficial to eliminate inter-beam interference, which improves the multiplexing gain. The proposed solution is described in Algorithm 1.

Algorithm 1: User Clustering based on Channel Gain (UC_CG)

Input:

The number of users K and the number of antennas N;

Channel vectors: $\mathbf{h}_k$ for $k = 1, 2, \cdots K, N_{RF}$

Output:

User-grounding T

1. Select number of cluster-heads;

$\mathbf{H} = [|\mathbf{h}_1|, |\mathbf{h}_2|, \cdots, |\mathbf{h}_K|], [\sim, order] = (sort(H), 'descend')$

$O = [order\,(1), \cdots, order\,(G)]$

2. Include other users into each cluster;

$O^C = K/O;$

$\max\left|\mathbf{h}_i^{H}\mathbf{h}_j\right|, \forall i \in O, j \in O^C$, Grouping the cluster channel with a large correlation.

Return T

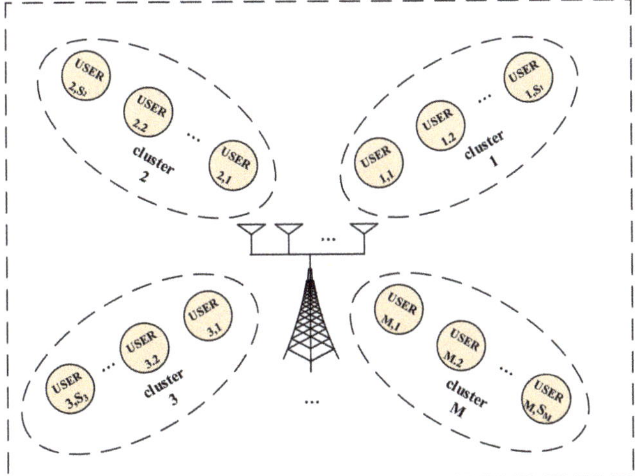

Figure 2. User clustering based on channel gain.

(2) The user clustering method based on antenna grouping. We consider the downlink MIMO-NOMA system, the number of users K is larger than the number of beams G. We provide a low complexity MIMO-NOMA user clustering algorithm, where the number of clusters G is equal to the number of RF chains N_{RF}. As shown in Figure 3, the antennas at the BS are sequentially grouped into G groups, there are Nt antennas in each group. We first select the user with the largest channel gain corresponding to each antenna group as the cluster head and find the correlation between the remaining users and each cluster head user. Then, we match the user with high channel correlation to the selected cluster head user. The proposed solution is described in Algorithm 2.

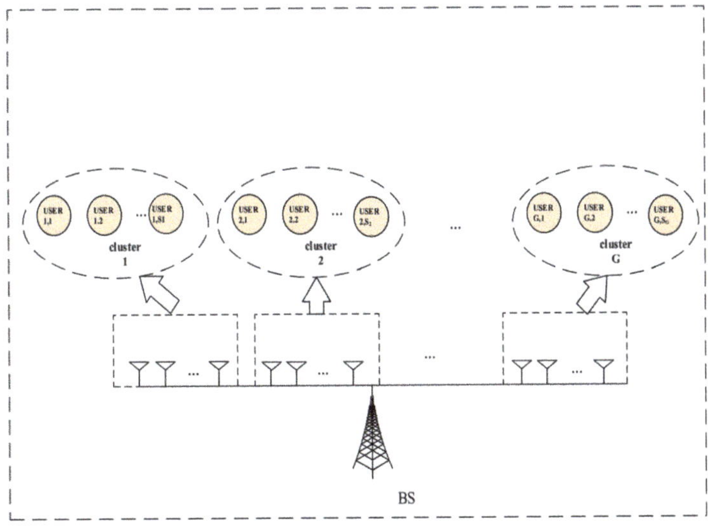

Figure 3. User clustering based on fixed antenna grouping.

Algorithm 2: User Clustering based on Fixed Antenna Grouping (UC_FAG)

Input:

The number of users K, and the number of antennas N;

Channel vectors: $\mathbf{h}_k$ for $k = 1, 2, \cdots K$ RF chains: N_{RF}

Output:

1. User-grounding T1

Select number of cluster-heads;

$N_t = \frac{N}{N_{RF}}$

For g=1:G

$\tilde{\mathbf{H}} = [\left|\tilde{\mathbf{h}}_1\right|, \left|\tilde{\mathbf{h}}_2\right|, \cdots, \left|\tilde{\mathbf{h}}_K\right|]$

$\tilde{\mathbf{h}}_n = \mathbf{h}_n((g-1) * N_t + 1 : g * N_t, :)$

[a,order] = (sort(H),'descend')

O(g)= [order (1)]

$\tilde{\mathbf{H}} = \tilde{\mathbf{H}} / \left|\tilde{\mathbf{h}}_{a(1)}\right|$

end

2. Include other users into each cluster;

$O^C = K/O$

$\max\left|\mathbf{h}_i^H \mathbf{h}_j\right|, \forall i \in O, j \in O^C$, Grouping the cluster channel with a high correlation

Return T1.

When operating the user grouping in Algorithm 1, the complexities of calculating the channel correlation and the norm channel vector are $O(KN)$ and $O(K^2N)$. The complexity of Algorithm 1 is $O(NK + K^2N)$. In Algorithm 2, the complexities of calculating channel correlation and norm effective channel vector are $O(N(K - \frac{1+N_{RF}}{2}))$ and $O\left(\frac{K^2N}{N_{RF}}\right)$. Then the complexity of Algorithm 2 is $O(N(K - \frac{1+N_{RF}}{2}) + \frac{K^2N}{N_{RF}})$. By comparing the complexity of the two algorithms, we can know that the complexity of algorithm 2 is lower than that of Algorithm 1.

2.2. Precoding

We consider the nth MIMO-NOMA cluster that contains m users, the channel matrix $\mathbf{H}_n \in \mathbb{C}^{m \times N}$ could be defined [24]:

$$\mathbf{H}_n = \left[\mathbf{h}_{n,1}, \mathbf{h}_{n,2}, \cdots, \mathbf{h}_{n,m}\right] \tag{7}$$

By taking the Singular Value Decomposition (SVD) of the channel matrix $\mathbf{H}_n$ we obtain:

$$\mathbf{H}_n^T = \mathbf{U}_n \sum_n \mathbf{V}_n^H \tag{8}$$

Each beam is utilized by a MIMO-NOMA cluster so that the channel corresponding to the nth beam is:

$$\tilde{\mathbf{h}}_n = \mathbf{H}_n \mathbf{u}_n^* \tag{9}$$

where $\mathbf{u}_n^*$ is the first column of $\mathbf{u}_n$. The equivalent channel matrix can be expressed as follows:

$$\tilde{\mathbf{H}} = [\tilde{\mathbf{h}}_1, \tilde{\mathbf{h}}_2, \cdots, \tilde{\mathbf{h}}_G] = [\mathbf{H}_1 \mathbf{u}_1^*, \mathbf{H}_2 \mathbf{u}_2^*, \cdots, \mathbf{H}_G \mathbf{u}_G^*] \tag{10}$$

Then, the precoding matrix can be written as:

$$\tilde{\mathbf{W}} = [\tilde{\mathbf{w}}_1, \tilde{\mathbf{w}}_2, \cdots, \tilde{\mathbf{w}}_G] = \tilde{\mathbf{H}}(\tilde{\mathbf{H}}^H \tilde{\mathbf{H}})^{-1} \tag{11}$$

After normalization of the precoding matrix, the precoding vector of the nth beam is:

$$\mathbf{w}_n = \frac{\tilde{\mathbf{w}}_n}{\left\|\tilde{\mathbf{w}}_n\right\|_2} \tag{12}$$

2.3. Power Allocation

In the NOMA system, the channel gain difference between users can be converted to a multiplexing gain by superposition coding. Therefore, power allocation has an important impact on system performance [25]. We proposed a dynamic power allocation method for the MIMO-NOMA system. Firstly, the transmission power is allocated according to the number of beams, which is proportional to the number of users served by the beam. Each beam is used by all users of the cluster and each MIMO-NOMA cluster contains users with near-similar channel differences. Therefore, the power allocation of users in the cluster is very important. We allocate power to users within the cluster for maximizing the cluster communication rate. The proposed power allocation method is described as:

$$P_g = P \times \frac{|S_g|}{|S_1| + |S_2| + \cdots + |S_G|} \tag{13}$$

The first step is to allocate the transmit power between beams. P_g is the transmitted power in the gth beam, $g = 1, 2 \cdots G$. P denotes the total transmitted power. After obtaining the transmit power of each beam, S_g is a set of the users served by the gth beam. The second step is to perform power allocation on the user cluster served by the beams. We assume that the interference between users is small within the same user cluster, and the problem can be defined as:

$$\max_{p_{g,1}, p_{g,2} \cdots p_{g,S_g}} C_g = \sum_{n=1}^{|S_g|} \log_2\left(1 + \frac{|h_{g,n}|^2 p_{g,n}}{\sigma}\right)$$

$$s.t.\ C_1 : \sum_{n=1}^{S_n} p_{g,n} = P_g \tag{14}$$

where $h_{g,n}$ is the channel of the nth user in the gth beam ($g = 1, 2, \cdots G, n = 1, 2, \cdots S_g$). $p_{g,n}$ denotes the transmitted power for the nth user in the gth beam. σ denotes noise power spectral density. To solve the convex optimization problem (14), we define the Lagrange function as:

$$L(\lambda, p_{g,1}, p_{g,2} \cdots p_{g,S_g}) = \sum_{n=1}^{|S_g|} \log_2\left(1 + \frac{|h_{g,n}|^2 p_{g,n}}{\sigma}\right) + \lambda\left(\sum_{n=1}^{|S_g|} p_{g,n} - P_g\right) \tag{15}$$

where $\lambda \geq 0$,

By calculating the derivative (15):

$$\frac{\partial L}{\partial p_{g,n}} = \frac{h_{g,n}}{(1 + p_{g,n} h_{g,n}) In2} - \lambda = 0 \tag{16}$$

we have:

$$p_{g,n} = \frac{1}{\tilde{\lambda}} - \frac{1}{|h_{g,n}|} \tag{17}$$

where $\tilde{\lambda} = \lambda In2$, $|h_{g,n}|$ is the channel gain of the nth user in the gth beam.

By substituting (17) into the constraint C1 in (14) we have:

$$\sum_{n=1}^{S_n} \frac{1}{\widetilde{\lambda}} - \frac{1}{|h_{g,n}|} = P_g \tag{18}$$

$\widetilde{\lambda}$ can be written as:

$$\widetilde{\lambda} = \frac{|S_g|}{P_g + \sum_{n=1}^{|S_g|} \frac{1}{|h_{g,n}|}} \tag{19}$$

Substituting (19) into (17), we have

$$p_{g,n} = \frac{P_g + \sum_{n=1}^{|S_g|} \frac{1}{|h_{g,n}|}}{|S_g|} - \frac{1}{|h_{g,n}|} = \frac{P_g}{|S_g|} + \frac{\sum_{n=1}^{|S_g|} \frac{1}{|h_{g,n}|}}{|S_g|} - \frac{1}{|h_{g,n}|} \tag{20}$$

where P_g is the transmitted power in the gth beam, $|S_g|$ represents the number of users served by the gth beam, and $|h_{g,n}|$ is the channel gain of the nth user in the gth beam. From (20) we obtain the transmitted power of the nth user in the gth beam and find that when the number of users in the group is larger, the power allocated to user would be reduced.

3. Energy Harvesting Maximizing

To maximize the harvested energy while meeting the minimum communication rate, we propose the addition of a power splitter for each user at the receiver to help implement SWIPT. This method is called SWIPT with power split [26], as shown in Figure 4. The signal received by each user is divided into two parts. One part is forwarded to the information decoder for information decoding, and the other part is subjected to Energy Harvesting (EH). The received signal to EH at the mth user in the nth beam can be can be formulated [27]:

$$y_{n,m}^{EH} = \sqrt{1 - \beta_{n,m}} y_{n,m} \tag{21}$$

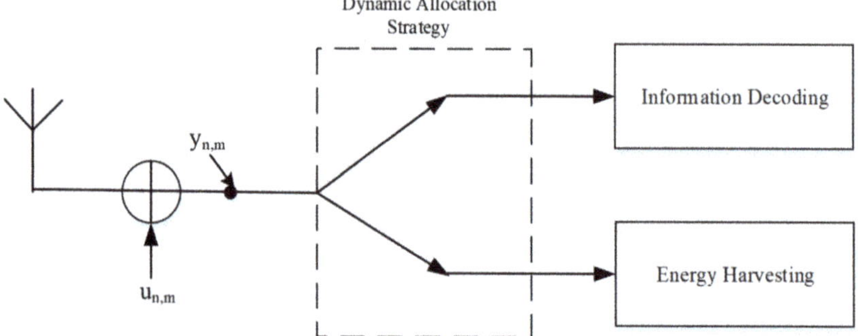

Figure 4. Simultaneous Wireless Information and Power Transfer (SWIPT) split receiver Power Split (PS) mode.

The harvested energy at the mth user in the nth beam is

$$P_{n,m}^{EH} = \eta(1 - \beta_{n,m})\left(\left\| h_{n,m}^H w_n \right\|_2^2 p_{n,m} + \sigma_v^2 \right) \tag{22}$$

where η is the energy conversion efficiency, $\beta_{n,m}$ is the power splitting factor at the mth user in the nth beam, $0 \leq \beta_{n,m} \leq 1$.

Meanwhile, the signal used to carry out the information decoding is expressed as

$$y_{n,m}^{ID} = \sqrt{\beta_{n,m}}y_{n,m} + u_{n,m} \tag{23}$$

Substituting (1) into (23), we have

$$y_{n,m}^{ID} = \sqrt{\beta_{n,m}}\left(\mathbf{h}_{n,m}^{H}\mathbf{w}_n \sqrt{p_{m,n}}s_{m,n} + \mathbf{h}_{n,m}^{H}\mathbf{w}_n \sum_{j=1}^{m-1} \sqrt{p_{i,n}}s_{i,n} \right.$$
$$\left. +\mathbf{h}_{n,m}^{H}\mathbf{w}_n \sum_{i=m+1}^{|S_i|} \sqrt{p_{i,n}}s_{i,n} + \mathbf{h}_{n,m}^{H} \sum_{i\neq n}^{G} \sum_{j=1}^{|S_i|} \mathbf{w}_j \sqrt{p_{i,j}}s_{i,j} + v_{n,m} \right) + u_{n,m} \tag{24}$$

where $u_{n,m}$ is the noise the distribution $CN(0,\sigma_u{}^2)$. By applying NOMA in each beam, intra-beam superposition coding of the transmitter and the receiver is realized. The mth user in the nth beam can eliminate the interference of the ith user (for all $i > m$) in the nth beam by performing SIC, and the remaining received signal of the mth user in the nth beam to information decoding can be rewritten as

$$\hat{y}_{n,m}^{ID} = \sqrt{\beta_{n,m}}\left(\mathbf{h}_{n,m}^{H}\mathbf{w}_n \sqrt{p_{n,m}}s_{n,m} + \mathbf{h}_{n,m}^{H}\mathbf{w}_n \sum_{j=1}^{m-1} \sqrt{p_{i,n}}s_{i,n} +\mathbf{h}_{n,m}^{H} \sum_{i\neq n}^{G} \sum_{j=1}^{|S_i|} \mathbf{w}_j \sqrt{p_{i,j}}s_{i,j} + v_{n,m} \right) + u_{n,m} \tag{25}$$

Then, according to (25), the SINR at the mth user in the nth beam can be written as

$$\gamma_{n,m} = \frac{\left\| \mathbf{h}_{n,m}^{H}\mathbf{w}_n \right\|_2^2 p_{n,m}}{\xi_{n,m}} \tag{26}$$

where,

$$\xi_{n,m} = \left\| \mathbf{h}_{n,m}^{H}\mathbf{W}_n \right\|_2^2 \sum_{j=1}^{m-1} p_{n,j} + \sum_{i\neq n}^{G} \left\| \mathbf{h}_{n,m}^{H}\mathbf{W}_i \right\|_2^2 \sum_{j=1}^{|S_i|} p_{i,j} + \sigma_v^2 + \frac{\sigma_u^2}{\beta_{n,m}} \tag{27}$$

The achievable rate of the mth user in the nth beam can be written as

$$R_{n,m} = \log_2(1 + \gamma_{n,m}) \tag{28}$$

We have grouped users, designed a precoding matrix, and allocated power to users in Sections 2 and 3. According to (22), we know we need to find the power splitting coefficient of each user for making the harvested energy at the receiver is maximized. We formulate the problem as

$$\begin{aligned} \max_{\{\beta_{n,m}\}} \quad & P^{EH} \\ s.t.\ C_1 : \ & R_{m,n} \geq R_{\min} \\ C_2 : \ & 0 \leq \beta_{n,m} \leq 1 \end{aligned} \tag{29}$$

Substituting (22), (28), into (29), we have

$$\begin{aligned} \max_{\{\beta_{n,m}\}} \quad & \sum_{i=1}^{G} \sum_{j=1}^{|S_j|} \eta(1 - \beta_{n,m})\left(\left\| \mathbf{h}_{n,m}^{H}\mathbf{w}_n \right\|_2^2 p_{n,m} + \sigma_v{}^2 \right) \\ s.t.\ C_1 : \ & \log_2(1 + \gamma_{n,m}) \geq R_{\min} \\ C_2 : \ & 0 \leq \beta_{n,m} \leq 1 \end{aligned} \tag{30}$$

where P^{EH} is the total harvested energy. $R_{\min}$ denotes the minimum achievable rate of the user.

To maximize the total harvested energy, the energy harvested by each user is maximal, the problem is converted to maximize the energy harvested by each user:

$$\max_{\{\beta_{n,m}\}} P_{n,m} = \eta(1 - \beta_{n,m})\left(\left\|\mathbf{h}_{n,m}^H \mathbf{w}_n\right\|_2^2 p_{n,m} + \sigma_v{}^2\right)$$
$$s.t.\ C_1 : \log_2(1 + \gamma_{n,m}) \geq R_{\min}$$
$$C_2 : 0 \leq \beta_{n,m} \leq 1$$

(31)

Substituting (26), (27) into (31), we have

$$\max_{\{\beta_{n,m}\}} P_{n,m} = \eta(1 - \beta_{n,m})\left(\left\|\mathbf{h}_{n,m}^H \mathbf{w}_n\right\|_2^2 p_{n,m} + \sigma_v{}^2\right)$$
$$s.t.\ C_1 : \log_2\left(1 + \frac{\left\|\mathbf{h}_{n,m}^H \mathbf{w}_n\right\|_2^2 p_{n,m}}{\left\|\mathbf{h}_{n,m}^H \mathbf{w}_n\right\|_2^2 \sum_{j=1}^{m-1} p_{n,j} + \sum_{i \neq n}^{G} \left\|\mathbf{h}_{n,m}^H \mathbf{w}_i\right\|_2^2 \sum_{j=1}^{|S_i|} p_{i,j} + \sigma_v^2 + \frac{\sigma_u^2}{\beta_{n,m}}}\right) \geq R_{\min}$$
$$C_2 : 0 \leq \beta_{n,m} \leq 1$$

(32)

By simplifying C1 in (32),

$$1 + \frac{\left\|\mathbf{h}_{n,m}^H \mathbf{w}_n\right\|_2^2 p_{n,m}}{\omega_{n,m} + \frac{\sigma_u^2}{\beta_{n,m}}} \geq 2^{R_{\min}}$$

(33)

where,

$$\omega_{n,m} = \left\|\mathbf{h}_{n,m}^H \mathbf{W}_n\right\|_2^2 \sum_{j=1}^{m-1} p_{n,j} + \sum_{i \neq n} \left\|\mathbf{h}_{n,m}^H \mathbf{W}_i\right\|_2^2 \sum_{j=1}^{|S_i|} p_{i,j} + \sigma_v^2$$

(34)

By simplifying (33),

$$\beta_{n,m} \geq \frac{\sigma_u^2(2^{R_{\min}} - 1)}{\left\|\mathbf{h}_{n,m}^H \mathbf{w}_n\right\|_2^2 p_{n,m} - \omega_{n,m}(2^{R_{\min}} - 1)}$$

(35)

Then, the constraint C_1 in (32) can be rewritten as

$$\max_{\{\beta_{n,m}\}} P_{n,m} = \eta(1 - \beta_{n,m})\left(\left\|\mathbf{h}_{n,m}^H \mathbf{w}_n\right\|_2^2 p_{n,m} + \sigma_v{}^2\right)$$
$$s.t.\ C_1 : \beta_{n,m} \geq \frac{\sigma_u^2(2^{R_{\min}} - 1)}{\left\|\mathbf{h}_{n,m}^H \mathbf{w}_n\right\|_2^2 p_{n,m} - \omega_{n,m}(2^{R_{\min}} - 1)}$$
$$C_2 : 0 \leq \beta_{n,m} \leq 1$$

(36)

According to (36), we know that when $\beta_{n,m}$ is the minimum that meets constraint C_1, $1 - \beta_{n,m}$ is the maximum. Then we obtain the maximum $P_{n,m}$. Accordingly, we get the optimal power splitting coefficient at the mth user in the nth beam:

$$\beta_{n,m} = \frac{\sigma_u^2(2^{R_{\min}} - 1)}{\left\|\mathbf{h}_{n,m}^H \mathbf{w}_n\right\|_2^2 p_{n,m} - \omega_{n,m}(2^{R_{\min}} - 1)}$$

(37)

Substituting (37) into (22), The maximal harvested energy at the mth user in the nth beam is

$$P_{n,m_{\max}} = \eta\left(1 - \frac{\sigma_u^2(2^{R_{\min}} - 1)}{\left\|\mathbf{h}_{n,m}^H \mathbf{w}_n\right\|_2^2 p_{n,m} - \omega_{n,m}(2^{R_{\min}} - 1)}\right)\left(\left\|\mathbf{h}_{n,m}^H \mathbf{w}_n\right\|_2^2 p_{n,m} + \sigma_v{}^2\right)$$

(38)

4. Simulation Results

We consider a typical downlink mmWave massive MIMO-NOMA system, the spectral efficiency is defined as the reachability rate in equation (6), and the energy efficiency is defined as the ratio of reachability to total power consumption [28],

$$EE = \frac{R_{sum}}{P + N_{RF}P_{RF} + P_{BB}} \tag{39}$$

where P is the total transmitted power, P_{RF} is the power consumed by each RF chain, P_{BB} is the baseband power consumption, N_{RF} is the number of the RF chain, R_{sum} is from (6). Simulation parameters are shown in Table 1.

Table 1. Simulation parameters.

Parameter	Value
Number of antennas at BS	256
Number of RF chain	8
Users	32
Antenna number of each user	1
Total transmitted power P (mW)	32
Minimum user communication rate (bps/Hz)	0.3
Each RF chain power consumption, P (mW)	300
Baseband power consumption, P (mW)	200

In the simulation, we first consider three kinds of mmWave massive MIMO systems and compare them by using two different user grouping methods proposed in Section 2.1:

(1) "Full-digital MIMO system" with one RF chain connected to each antenna ($N=N_{RF}$).
(2) "MIMO-NOMA under the UC_CG algorithm" grouping users according to the proposed UC_CG algorithm, and performing NOMA for the user in the beam.
(3) "MIMO-NOMA under the UC_FAG algorithm" grouping users according to the proposed UC_FAG algorithm and performs NOMA for the user in the beam.
(4) MIMO-OMA under the UC_CG algorithm": The user is grouped according to the UC_CG algorithm, and the OMA is executed for the user in the beam.
(5) "MIMO-OMA under the UC_FAG algorithm": The users are grouped according to the proposed UC_FAG algorithm, and the OMA is executed for the user in the beam.

The power allocation method proposed in this paper is applied to the MIMO-NOMA system and we compare the performance with the system using traditional average power allocation method that it allocated equal power to all users. Finally, the SWIPT technology is integrated into the system to compare the power harvested by the MIMO-NOMA and MIMO-OMA.

Figure 5 shows the spectral efficiency against Signal to Noise Ratio(SNR)of the considered five schemes mentioned above, where the number of users K is set to 32 and the number of antennas is set to 256. From the figure, we can see that the proposed MIMO-NOMA scheme has a higher spectral efficiency than the MIMO-OMA scheme. It is intuitive that the fully digital MIMO can achieve the best spectrum efficiency, as shown in Figure 5. However, the number of RF chains required in the full digital MIMO scheme is equal to the number of antennas ($N_{RF} = N$), and the number of RF chains required in MIMO-NOMA is 8. The full digital MIMO scheme needs higher hardware costs and overhead. Through the simulation diagram, we can obtain that the UC_CG algorithm gives higher spectral efficiency than the UC_FAG algorithm. Given the users that are matched according to the correlation between all channels of the user and the cluster-head user in UC_CG algorithm, in the UC_FAG algorithm, the users are matched according to the correlation between the part of the channel of the user and the cluster-head user. In comparison with the UC_FAG algorithm, the interference between users in the group is smaller in the UC_CG algorithm.

correlation between all channels of the user and the cluster-head user in UC_CG algorithm, in the UC_FAG algorithm, the users are matched according to the correlation between the part of the channel of the user and the cluster-head user. In comparison with the UC_FAG algorithm, the interference between users in the group is smaller in the UC_CG algorithm.

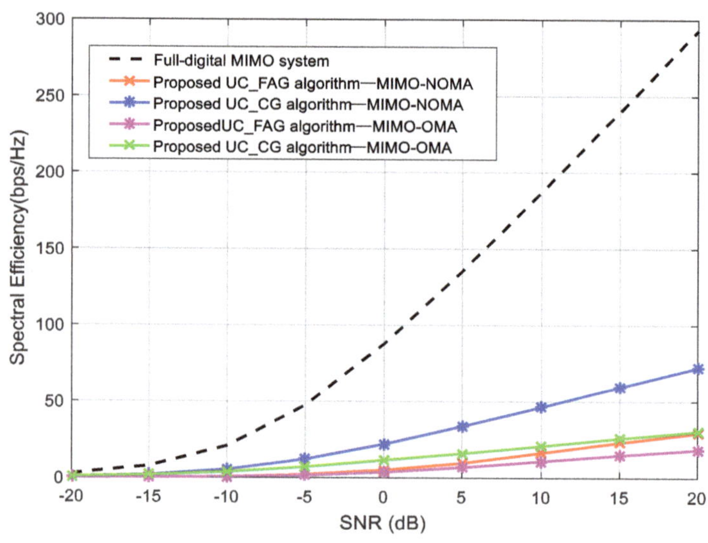

Figure 5. Spectral efficiency of the considered five schemes system against SNR

Figure 6 shows the energy efficiency of the five schemes considered under different SNR, where the number of users is set to 32, the number of antennas is set to 256. From Figure 6, we know that the MIMO-NOMA scheme has a higher energy efficiency than MIMO-OMA and fully digital MIMO, where the number of RF chains of the fully digital MIMO is equal to the number of base station antennas, which results in very high energy consumption. In contrast, in the MIMO-NOMA scheme, the number of RF chains is much smaller than the number of antennas. Therefore, the energy consumption of the RF chain can be significantly reduced when compared with the fully digital MIMO scheme.

According to Figure ... is higher than UC_FAG with MIMO ... same, UC_CG with MIMO-NOMA ... according to (38). Therefore, whe ... UC_CG with MIMO-NOMA ... the ...

A comparison ... rs is shown in Figure 7 in which the ... ses, the energy efficiency is gradually ... MIMO-NOMA scheme is more energ ...

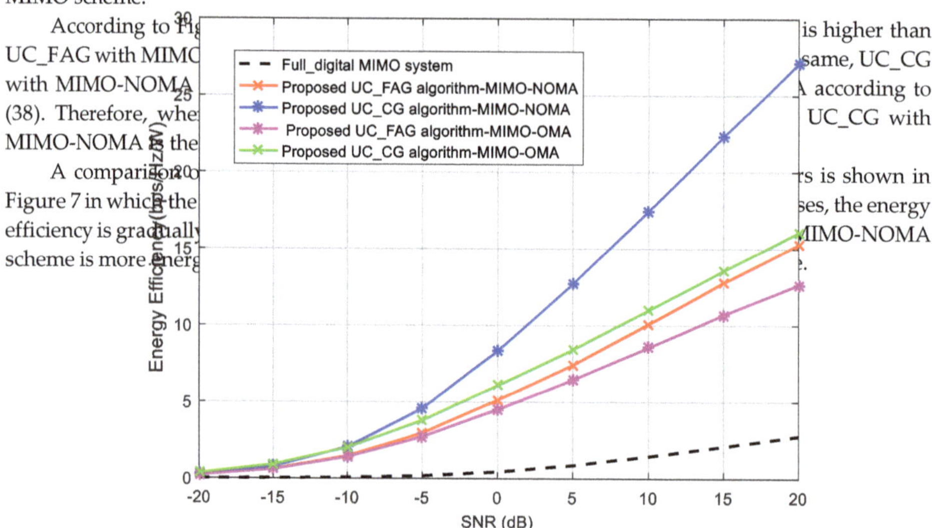

Figure 6. Energy efficiency against SNR

According to Figure 5, the communication rates of UC_CG with MIMO-NOMA is higher than UC_FAG with MIMO-NOMA. When the total power consumption of the system is the same, UC_CG

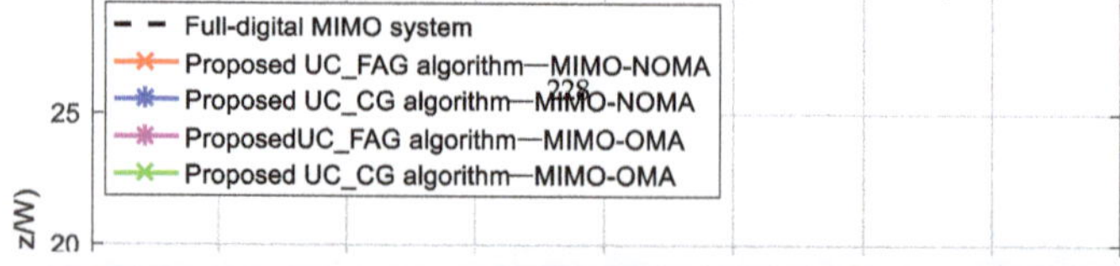

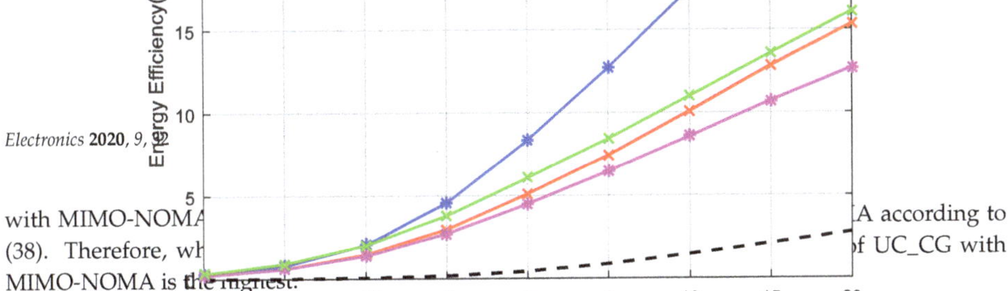

Figure 6. Energy efficiency against SNR.

Electronics **2020**, *9,*

with MIMO-NOMA [...] A according to
(38). Therefore, wh [...] of UC_CG with
MIMO-NOMA is the highest.

A comparison of the performance of energy efficiency with the number of users is shown in Figure 7 in which the SNR is set to 10dB. We can see that, as the number of users increases, the energy efficiency is gradually reduced. Even with a very large number of users, the proposed MIMO-NOMA scheme is more energy efficient than MIMO-OMA and the fully digital MIMO scheme.

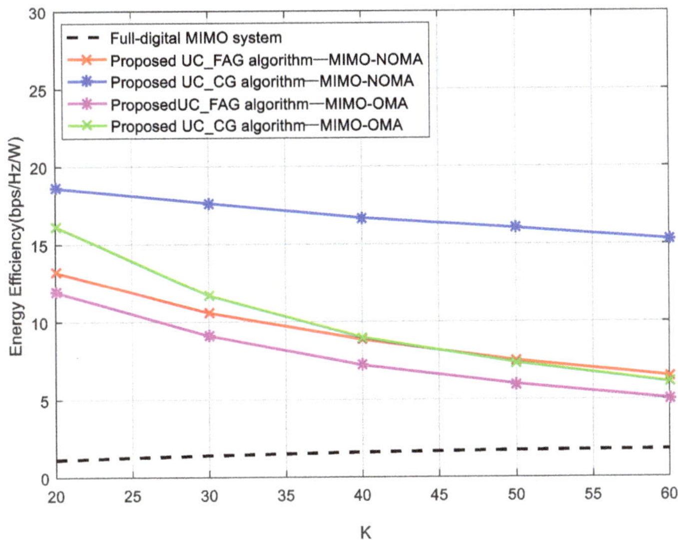

Figure 7. Energy efficiency against the number of users K.

The next experiment considers the spectral efficiency of the SNR under two different power allocation algorithms. From Figure 8, we obtain that the power allocation algorithm proposed in this paper has higher spectral efficiency than the traditional average power allocation algorithm. We understand that the power allocation algorithm proposed is better **than the** traditional average allocation algorithm.

Figure 9 shows the energy harvesting performance against SNR. To enable the user to maximize harvested power and meet the communication requirement, in Section 3, we proposed a method that finds the power splitting optimization. From Figure 9, we can see that, when signal power is low, the received signal performs information decoding. The receiver can start harvesting energy when the signal becomes larger. In comparison with the MIMO-OMA scheme, MIMO-NOMA can harvest more energy. Therefore, the proposed MIMO-NOMA scheme with SWPIT is superior to MIMO-OMA scheme, which can realize the recycling of energy.

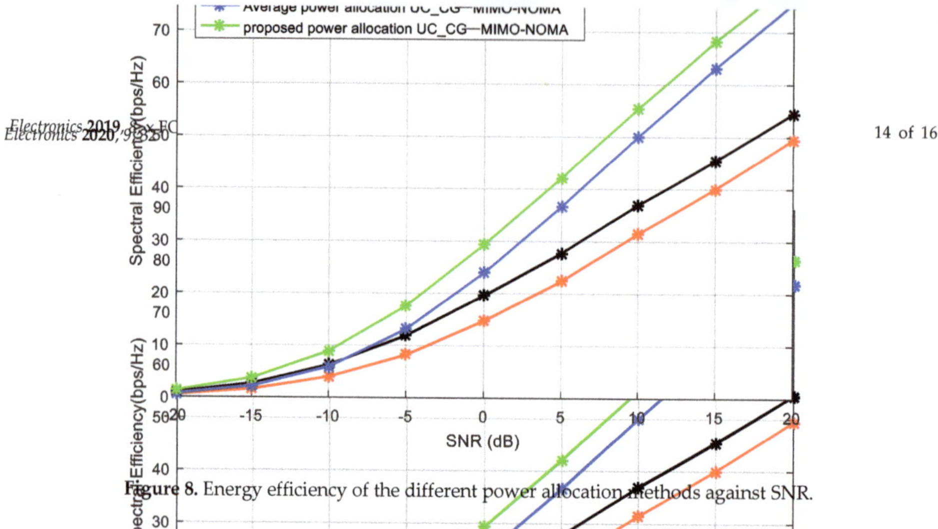

Figure 8. Energy efficiency of the different power allocation methods against SNR.

Figure 9 shows the energy harvesting performance against SNR. To enable the user to maximize harvested power and meet the communication requirement, in Section 3, we proposed a method that finds the power splitting optimization. From Figure 9, we can see that, when signal power is low, the received signal performs information decoding. The receiver can start harvesting energy when the signal becomes larger. In comparison with the MIMO-OMA scheme, MIMO-NOMA can harvest more energy. Therefore, the proposed MIMO-NOMA scheme with SWPIT is superior to MIMO-OMA scheme, which can realize the recycling of energy.

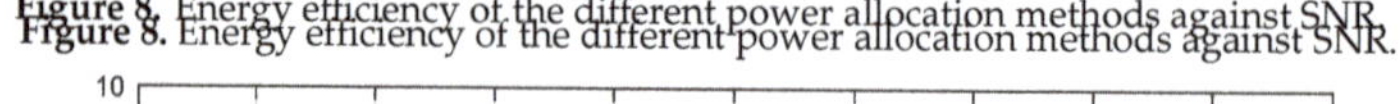

Figure 8. Energy efficiency of the different power allocation methods against SNR.

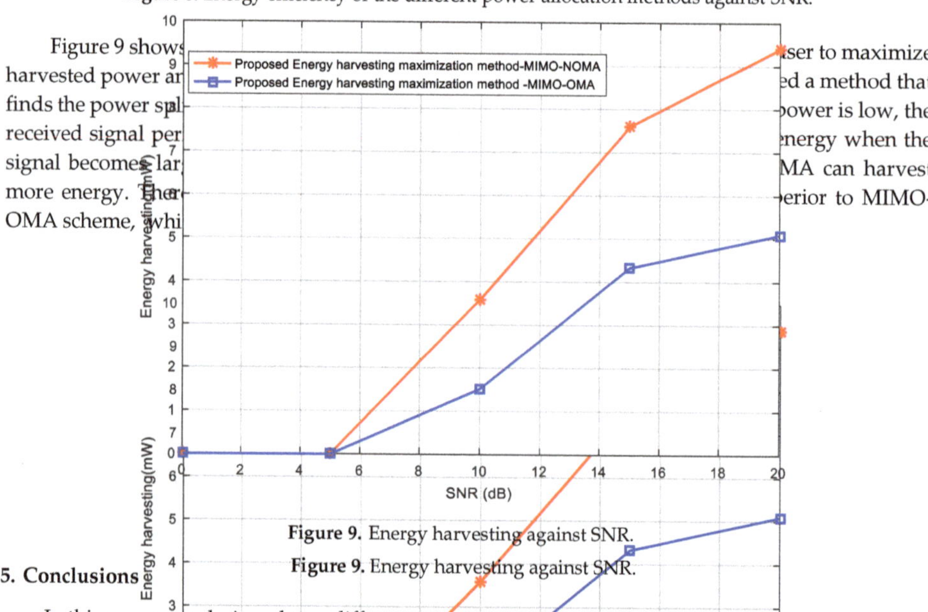

Figure 9. Energy harvesting against SNR.

5. Conclusions

In this paper, we designed two different user grouping methods for MIMO-NOMA system: the UC_CG algorithm and the UC_FAG algorithm. From the simulation, we can see that the UC_CG algorithm is better than the UC_FAG algorithm, which improves spectral efficiency. We proposed a new power allocation method. The simulation results show that the algorithm is superior to the traditional average power allocation algorithm. Finally, we apply the SWIPT for MIMO-NOMA system. Under the premise of satisfying the minimum communication rate of each user, we proposed the method based on maximizing the harvested energy to find the optimal power splitting factor for each user. This method allows the system to harvest more energy and meets the user's minimum communication rate, thereby achieving the recycling of energy and green communication.

Author Contributions: Conceptualization, S.L.; Methodology, Z.W.; Supervision, L.J. and J.D.; Writing—review & editing, S.L. All authors have read and agreed to the published version of the manuscript.

Funding: This work was supported by National Nature Science Funding of China (NSFC): 61401407, 61601414 and the Fundamental Research Funds for the Central Universities.

Conflicts of Interest: The authors declare no conflict of interest.

Abbreviations

The following abbreviations are used in this manuscript:

MIMO	Multiple Input Multiple Output
NOMA	Non-Orthogonal Multiple Access
SIC	Serial Interference Cancellation
SWIPT	Simultaneous Wireless Information and Power Transfer
OMA	Orthogonal Multiple Access
RF	Radio Frequency
EH	Energy Harvesting
UC_CG	User Clustering based on Channel Gain UC_CG
UC_FAG	User Clustering based on Fixed Antenna Grouping
SINR	Signal to Interference plus Noise Ratio
SNR	Signal to Noise Ratio
PS	Power Split
BS	Base Station

References

1. Dai, L.; Wang, B.; Ding, Z.; Wang, Z.; Chen, S.; Hanzo, L. A Survey of Non-Orthogonal Multiple Access for 5G. *IEEE Commun. Surv. Tutor.* **2018**, *20*, 2294–2323. [CrossRef]
2. Li, S.; Wu, H.; Jin, L. Codebook-Aided DOA Estimation Algorithms for Massive MIMO System. *Electronics* **2018**, *8*, 26. [CrossRef]
3. Sari, H.; Vanhaverbeke, F.; Moeneclaey, M. Multiple access using two sets of orthogonal signal waveforms. *IEEE Commun. Lett. Jpn.* **2000**, *4*, 4–6. [CrossRef]
4. Yang, Z.; Ding, Z.; Fan, P.; Al-Dhahir, N. A general power allocation scheme to guarantee quality of service in downlink and uplink NOMA systems. *IEEE Trans. Wirel. Commun.* **2016**, *15*, 7244–7257. [CrossRef]
5. Zhang, D.; Zhou, Z.; Xu, C.; Zhang, Y.; Rodriguez, J.; Sato, T. Capacity Analysis of NOMA With mmWave Massive MIMO Systems. *IEEE J. Sel. Areas Commun.* **2017**, *35*, 1606–1618. [CrossRef]
6. Zhang, N.; Wang, J.; Kang, G.; Liu, Y. Uplink nonorthogonal multiple access in 5G systems. *IEEE Commun. Lett. Mar.* **2016**, *20*, 458–461. [CrossRef]
7. Chen, X.; Zhang, Z.; Zhong, C.; Jia, R.; Ng, D.W.K. Fully non-orthogonal communication for massive access. *IEEE Trans. Commun.* **2018**, *66*, 1717–1731. [CrossRef]
8. Benjebbour, A.; Saito, Y.; Kishiyama, Y.; Li, A.; Harada, A.; Nakamura, T. Concept and practical considerations of non-orthogonal multiple access (NOMA) for future radio access. In Proceedings of the 2013 International Symposium on Intelligent Signal Processing and Communication Systems, Naha, Japan, 12–15 November 2013; pp. 770–774.
9. Zhang, Z.; Sun, H.; Hu, R.Q. Downlink and Uplink Non-Orthogonal Multiple Access in a Dense Wireless Network. *IEEE J. Sel. Areas Commun.* **2017**, *35*, 2771–2784. [CrossRef]
10. Saito, Y.; Kishiyama, Y.; Benjebbour, A.; Nakamura, T.; Li, A.; Higuchi, K. Non-orthogonal multiple access (NOMA)for cellular future radio access. In Proceedings of the 2013 IEEE 77th Vehicular Technology Conference (VTC Spring), Dresden, Germany, 2–5 June 2013; pp. 1–5.
11. Xu, Y.; Yue, G.; Mao, S. User grouping for massive MIMO in FDD systems: New design methods and analysis. *IEEE Access J.* **2014**, *2*, 947–959. [CrossRef]
12. Kang, J.-M.; Kim, I.-M. Optimal user grouping for downlink NOMA. *IEEE Wirel. Commun. Lett.* **2018**, *7*, 724–727. [CrossRef]

13. Zaw, C.W.; Tun, Y.K.; Hong, C.S. User clustering based on correlation in 5G using semidefinite programming. In Proceedings of the 2017 19th Asia-Pacific Network Operations and Management Symposium (APNOMS), Seoul, Korea, 27–29 September 2017; pp. 342–345.

14. Ding, Z.; Yang, Z.; Fan, P.; Poor, H.V. On the performance of nonorthogonal multiple access in 5G systems with randomly deployed users. *IEEE Signal Process Lett.* **2014**, *21*, 1501–1505. [CrossRef]

15. Al-Hussaibi, W.A.; Ali, F.H. Efficient User Clustering, Receive Antenna Selection, and Power Allocation Algorithms for Massive MIMO-NOMA Systems. *IEEE Access* **2019**, *7*, 31865–31882. [CrossRef]

16. Varshney, L. Transporting information and energy simultaneously. In Proceedings of the IEEE International Symposium on Information Theory (IEEE ISIT'08), Toronto, ON, Canada, 6–11 July 2008; pp. 1612–1616.

17. Zhang, J.; Yuen, C.; Wen, C.; Jin, S.; Wong, K.; Zhu, H. Large System Secrecy Rate Analysis for SWIPT MIMO Wiretap Channels. *IEEE Trans. Inf. Forensics Secur.* **2016**, *11*, 74–85. [CrossRef]

18. Diamantoulakis, P.D.; Pappi, K.N.; Ding, Z.; Karagiannidis, G.K. Wireless-powered communications with non-orthogonal multiple access. *IEEE Trans. Wirel. Commun.* **2016**, *15*, 8422–8436. [CrossRef]

19. Chingoska, H.; Hadzi-Velkov, Z.; Nikoloska, I.; Zlatanov, N. Resource allocation in wireless powered communication networks with non-orthogonal multiple access. *IEEE Wirel. Commun. Lett.* **2016**, *5*, 684–687. [CrossRef]

20. Xu, Y.; Shen, C.; Ding, Z.; Sun, X.; Yan, S.; Zhu, G.; Zhong, Z. Joint Beamforming and Power-Splitting Control in Downlink Cooperative SWIPT NOMA Systems. *IEEE Trans. Signal Process.* **2017**, *65*, 4874–4886. [CrossRef]

21. Dai, L.; Wang, B.; Peng, M.; Chen, S. Hybrid Precoding-Based Millimeter-Wave Massive MIMO-NOMA With Simultaneous Wireless Information and Power Transfer. *IEEE J. Sel. Areas Commun.* **2019**, *37*, 131–141. [CrossRef]

22. Yang, Z.; Xu, W.; Pan, C.; Pan, Y.; Chen, M. On the Optimality of Power Allocation for NOMA Downlinks With Individual QoS Constraints. *IEEE Commun. Lett.* **2017**, *21*, 1649–1652. [CrossRef]

23. Ali, M.S.; Tabassum, H.; Hossain, E. Dynamic user clustering and power allocation for uplink and downlink non-orthogonal multiple access (NOMA) systems. *IEEE Access* **2016**, *4*, 6325–6343. [CrossRef]

24. Ali, S.; Hossain, E.; Kim, D.I. Non-Orthogonal Multiple Access (NOMA) for Downlink Multiuser MIMO Systems: User Clustering, Beamforming, and Power Allocation. *IEEE Access* **2017**, *5*, 565–577. [CrossRef]

25. Wang, H.; Zhang, R.; Song, R.; Leung, S. A Novel Power Minimization Precoding Scheme for MIMO-NOMA Uplink Systems. *IEEE Commun. Lett.* **2018**, *22*, 1106–1109. [CrossRef]

26. Zhou, X.; Zhang, R.; Ho, C.K. Wireless information and power transfer: Architecture design and rate-energy tradeoff. *IEEE Trans. Commun.* **2013**, *61*, 4754–4767. [CrossRef]

27. Zhang, H.; Dong, A.; Jin, S.; Yuan, D. Joint transceiver and power splitting optimization for multiuser MIMO SWIPT under MSE QoS constraints. *IEEE Trans. Veh. Technol.* **2017**, *66*, 7123–7135. [CrossRef]

28. Gao, X.; Dai, L.; Han, S.; Chih-Lin, I.; Heath, R.W. Energy-efficient hybrid analog and digital precoding for mmWave MIMO systems with large antenna arrays. *IEEE J. Sel. Areas Commun.* **2016**, *34*, 998–1009. [CrossRef]

Review

On-Site and External Energy Harvesting in Underground Wireless

Usman Raza * and Abdul Salam

Department of Computer and Information Technology, Purdue University, West Lafayette, IN 47907, USA; salama@purdue.edu
* Correspondence: uraza@purdue.edu

Received: 22 March 2020; Accepted: 16 April 2020; Published: 22 April 2020

Abstract: Energy efficiency is vital for uninterrupted long-term operation of wireless underground communication nodes in the field of decision agriculture. In this paper, energy harvesting and wireless power transfer techniques are discussed with applications in underground wireless communications (UWC). Various external wireless power transfer techniques are explored. Moreover, key energy harvesting technologies are presented that utilize available energy sources in the field such as vibration, solar, and wind. In this regard, the Electromagnetic (EM)- and Magnetic Induction (MI)-based approaches are explained. Furthermore, the vibration-based energy harvesting models are reviewed as well. These energy harvesting approaches lead to design of an efficient wireless underground communication system to power underground nodes for prolonged field operation in decision agriculture.

Keywords: soil sensing; decision agriculture; smart farming

1. Introduction

Wireless Underground Sensor Networks (WUSNs) is a subset of Wireless Sensor Network (WSN) paradigm. It is becoming a popular and developing area [1,2], and there is a large margin of improvements to resolve research challenges. WUSN have buried sensor nodes, which use communication technologies to communicate the geological data in real-time. It is considered as a promising paradigm for the monitoring of various underground applications [3]. Some of the important and valuable application areas of WUSNs include sports, agriculture, environment, border patrolling and health [1,4–6]. However, adoption of WUSNs faces many implementation challenges. WUSN accomplishes wireless communication via Electromagnetic (EM) waves propagation through soil [7]. EM waves, through soil, suffer more attenuation compared to when they are propagated through air [8–10]. Moreover, changes in soil parameters, e.g., soil moisture and temperature, further affect real-time communication [11–13]. For example, some studies [14,15] are researching and mitigating soil moisture effect using microwave heating. Similarly, in [16], authors have proposed a Microwave and Meteorological fusion (MMF) strategy to downscale soil moisture. Due to these challenges, many researchers are working to investigate WUSNs empirically and model the underground wireless communication channel [5,8,11,17–19].

One of the challengess in WUSN is the provision of sustainable renewable energy to the sensors deployed in the field [20]. In the last decade, there have been many technological advancements in natural and renewable energy sources with an aim to reduce climate change effects and extend the battery life of sensor nodes [21,22]. For any cost-efficient WUSN application, a WUSN device is expected to have a lifetime of several years [20]. There has been a lot of efforts in conserving energy by using energy-efficient communication protocols and hardware; however, communicating through soil is still a power hungry process. In some WUSN applications, sensors are either partly or completely buried in the soil. Researchers have investigated battery replacements as an alternative method to extend the network life. However, replacing the battery is a very time-consuming, laborious and expensive process. Moreover, it might not always be possible to easily access the sensor power source for replacement or maintenance. Therefore, replacing the battery of sensor nodes is not applicable to large-scale WUSN deployments [23].

From the above discussion, it is evident that a potential solution must be developed to overcome the energy challenges in WUSNs. To that end, two general approaches to power up the underground sensor networks are:

- Wireless Power Transfer (WPT): WPT converts wireless Radio Frequency (RF) signals into electrical energy to power up the buried nodes [24]. Numerous wireless methods, e.g., electromagnetic induction, electromagnetic resonance, or radiation, can be used to transfer energy [25–28]. WPT techniques have been reviewed in detail in Section 2.
- Energy Harvesting (EH): EH is a method to extract energy from natural energy resources, e.g., solar [29,30], Human Body Area Network (HBAN) [31,32], water and wind flow [33–35], radio frequencies [36,37], and vibrations generated from different objects [33,38,39]. The extracted energy is then converted into electrical energy to power up the sensor nodes. Underground sensor nodes can be integrated with an energy harvesting component. This harvesting component uses natural energy sources, e.g., solar energy [40] from the environment to preserve energy. Energy Harvesting techniques have been reviewed in detail in Section 3.

The purpose of this work is to educate about the above mentioned techniques and review the existing energy scavenging techniques in the literature. To that end, this paper is divided into two major sections (see Figure 1): WPT techniques have been reviewed in Section 2 along with recent advancements and future research possibilities. EH techniques have been reviewed in detail in Section 3 along-with techniques and future research considerations explained in subsequent subsections.

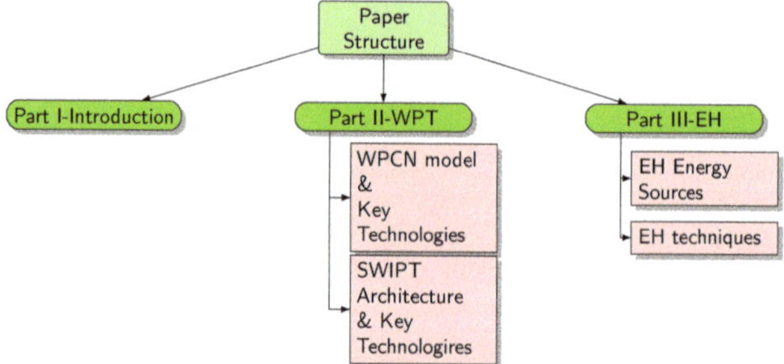

Figure 1. The organization of the article.

2. Wireless Power Transfer (WPT)

WPT is a method of transferring power using wireless technology. It was first introduced by Nikola Tesla in 1890s [41]. It is used to build a Wireless Powered Communication Network (WPCN). WPCN is a promising technology in which networks of Wireless Devices (WD) are wirelessly energized using RF signal with the help of dedicated power transmitters [42–45]. WPCN distinguishes itself from traditional battery-powered wireless networks in that it does not require manual replacement or recharging of battery. Instead, it completely controls its power transfer process by tuning various parameters such as waveform, transmit power, time and frequency domains. This reduces the operational cost while enhancing communication performance of the network.

Although the transfer range depends on the local emission regulations and differs from location to location, WPT can transfer tens of micro-watts of RF power to devices at a distance of more than 10 m (Please visit: htttp://www.powercastco.com for detailed product specifications.) [42]. It makes them suitable for implementation in low-powered applications such as WUSNs, Radio

Frequency Identification (RFID) and Internet of Underground Things (IOUT) [46,47]. WPCN, when deployed in WUSNs or IOUTs, has its own implementation challenges. As WDs are buried underground, the received energy can be very low because of reduced signal power (attenuation) in an underground environment [7]. This can cause severe performance degradation and unfairness between the devices [48,49]. This degradation of signals due to an unfavorable sub-surface environment has been studied in detail in the literature [6,50–53]. This literature investigates this problem by proposing different underground channel models based on Magnetic Induction (MI), EM, and acoustic technologies. However, it is out of the scope of this paper to discuss these techniques.

Another WUSN challenge is the joint transfer of E&I. It is important because information and energy are interrelated as underground WDs are required to harvest enough energy by WPT before transmitting the information. To that end, a WPT-related technology to wirelessly energize a sensor node known as Simultaneous Wireless Information and Power Transfer (SWIPT) [24,54] is also reviewed. Unlike traditional WPT, SWIPT simultaneously transfers information and power in the same signal instead of transferring power only. Moreover, energy transfer and communication channel may share the common channel which can cause co-channel interference. In coming sections, we have discussed both WPT and SWIPT techniques of wireless transfer of energy.

2.1. WPCN Model

Figure 2 shows the components of WPCN [55]. The basic building blocks of WPCN consist of: Energy Nodes (EN), which transmits energy; WDs, which harvest energy; Access Points (APs), which receives information and energy from WDs. ENs sends the energy to WDs in down-link, and WDs use this energy to send information by communicating with the APs in an uplink (Figure 2a).

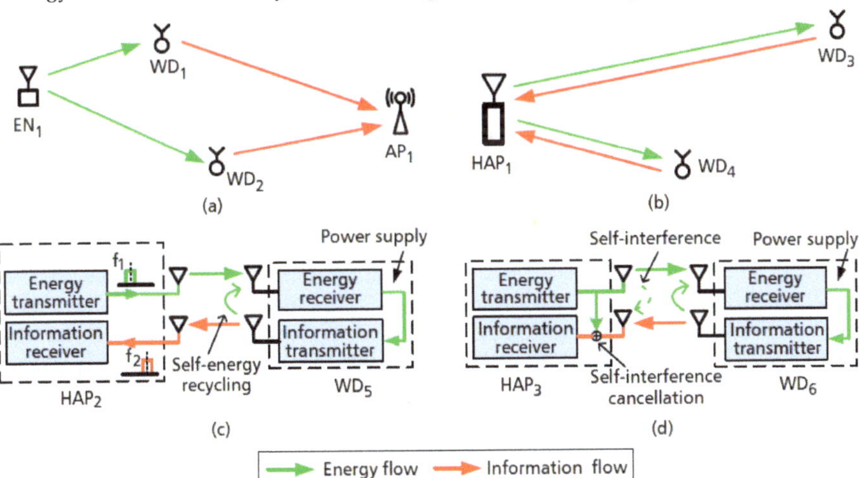

Figure 2. Wireless Powered Communication Network (WPCN) models with different transmitter and receiver schemes: (**a**) Physically different energy transmitter and information receiver, (**b**) Energy transmitter and information receiver co-located in same physical entity, (**c**) out-of-band transmission of information and energy and (**d**) full-duplex transmission of energy and information.

Normally, ENs and APs are separated [56]; however, they can be combined in a single device also known as Hybrid APs (HAP) as shown in (Figure 2b). HAP has an advantage of being cost-efficient in terms of production, operation, coordination and management of Energy and Information (E&I) transfer in a network. However, it can also introduce unfairness into the network. For example, in (Figure 2b), WD_3 will harvest less energy as compared to WD_4 because of the distance. On the contrary, in Figure 2a, WD_2 will harvest less energy from EN_1 than WD_1; however, it will also require less energy for communication because of less distance to AP. The circuitry for E&I transfer is different.

For example, sensitivity of received signal power, sent by HAP, for information receiver is $-60\,$dBm and that for energy receiver is $-10\,$dBm [42]. Therefore, a practical WPT-based WD has two antennas: one for harvesting energy and other for transmitting information. Similarly, HAP with combined energy transmitter and information receiver also have two sets of antennas.

The E&I transmission can either be done in-band or out-band. Out-band means using different frequencies for transmission to eliminate interference; however, it has to satisfy the additional constraints imposed by Federal Communication Commission (FCC) on its operational frequency. In-band means same frequency band for E&I transmission, which suffers from co-channel interference especially in the HAP case. This can be solved by using full-duplex WDs and full duplex HAPs. Full-duplex, in the current context, refers to the ability of HAP to simultaneously transmit energy and information. This duplex operation also enables WDs to harvest energy from its own transmitted information. This phenomenon is known as self-energy recycling and is shown in Figure 2c [57]. Similarly, full-duplex HAP has the ability to cancel the high self-interference (Figure 2d), which arises due to energy transmission and severely affects information decoding.

2.2. Key WPCN Technologies

The performance of WPCN is limited by the low efficiency and range of WPT. There are some techniques that can be used to enhance the performance of WPCNs. This section discusses those techniques. These technologies can be used to extend the operation of WPCN to make it a viable solution for a broader set of applications.

2.2.1. Energy Beamforming

WPT in LOS links uses conventional large antennas with large aperture (dish or horn antennas). However, in an IOUT/WUSN environment with mobile application and dynamically changing environment, a more suitable solution is to use an electronically steered array antenna with energy beamforming enabled [42,58]. Energy beamforming superimposes the weighted signals from different antennas at Energy Receivers (ERs). One requirement, to maximize the level of received signal, is to have complete Channel State Information (CSI). The CSI includes magnitude and phase shift from transmit to receive antenna of ERs. Figure 3c shows one of the methods for CSI at Energy Transmitters (ETs) is via froward-link (ET to ER) training and reverse-link (ER to ET). This training design of WPT channel is limited by energy available to ER for channel estimation and sending feedback rather than bandwidth/time. Hence, accuracy of CSI knowledge at ET is highly dependent upon the energy used by ERs for the channel estimation and feedback, i.e., more available energy will result in more accurate CSI. However, energy cost at ER can increase the energy gain at ET, especially for the ET with a large number of antennas because CSI overhead increases with the increase in antennas. Therefore, reverse-link training seems a more feasible option where ER sends a training signal in reverse, and ET determines the CSI. This procedure saves ER from channel estimation and feedback, and training overhead is not dependent on the number of antennas. In addition to energy constraints of ER, WPT training design may also face challenges of hardware processing abilities, e.g., having sensors with not enough base-band processing units for estimating CSI [59,60].

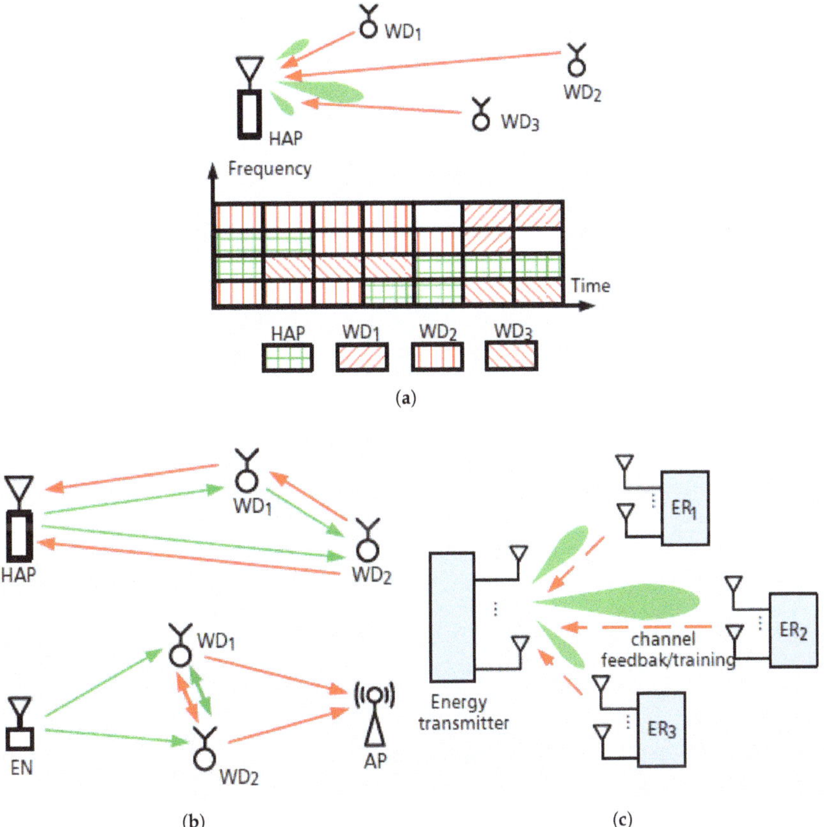

Figure 3. Major performance improving mechanisms for WPCNs; energy transfer is denoted by green lines and information by red lines: (**a**) wireless powered cooperative communication, (**b**) wireless powered cooperative communication joint scheduling for communication and energy transfer and (**c**) energy beamforming.

2.2.2. Joint Communication and Energy Scheduling

In WPCN, communication and energy transmission are dependent on each other. EN makes energy requirements from WDs to transmit downlink energy as per demand of WDs. Similarly, the uplink information transmitted by WD is highly constrained by available energy from WPT energy harvesting. Given this interdependence of E&I transmission, there is a need of combined scheduling to prevent co-channel interference for an efficient system. As shown in Figure 3a, there are multiple frequency-time resource blocks available to WPCN, which can be scheduled dynamically to HAP (for energy transmission) or WDs (for information transmission). Multiple factors are considered for the scheduling such as communication requests, channel conditions, battery conditions and fairness among WDs. For example, because WD_2 separated from HAP at large distance, more resources can be allocated to WD_2 than other WDs to ensure fairness (Figure 3a). This resource allocation method can also be used for WPCN with separate EN and AP. Although it seems effective, dynamic scheduling is very challenging to implement because of temporal dependence of wireless channels and cause–effect relation between WPT and future wireless communication techniques [61].

2.2.3. Wireless Powered Cooperative Communication

Wireless powered cooperative communication is a collaborative mechanism where users, i.e., WDs, share their resources with other WDs. The resources could be energy, time and collaboration with neighboring APs. In Figure 3b, user close to HAP, WD_1, shares its energy and time to transmit by relaying data transmission for distant user WD_2. The relay protocol can be designed with three time slots: one for downlink transmission from HAP to WD_1, a second one for WD_2 sending data to WD_1, and a third one for W_1 sending a message with its own and WD_2 data to HAP. In this way, a long-distant node can overcome the disadvantage of having short-range. To ensure fairness, WD_1 can be compensated by allowing more time to transmit because cooperation enables HAP to devote more time to communication than WPT [62]. This cooperation can be extended from communication to energy, i.e., WD_1 transmitting available excess energy to WD_2. This cooperation mechanism makes the WPCN a low-cost and efficient system for communication and energy harvesting [63].

2.2.4. Future Research Considerations

In addition to the above methods, there are other important areas in WPCN, which can be further studied for improvement of the overall system. Some of those are discussed below:

Green WPCN [55,64]: In WPCN, EN draws their energy from fixed aboveground energy sources. There is a potential to improve WPCN significantly by combining WPT-based energy harvesting methods with green and sustainable energy sources. Figure 4a shows the architecture of green WPCN. It can be seen that energy harvesting methods can be implemented at EN and WD. WDs can harvest energy from renewable energy sources and store them in rechargeable batteries. When there is enough renewable energy at WDs, ENs can stop transmitting energy because of battery constraint. On the contrary, WPT methods can be used when energy is not enough at WDs. With hybrid energy sources, i.e., fixed and renewable energy sources, the challenge is to change operation modes and using less amount of fixed energy source without degradation in communication performance. An optimal green WPCN depends on many factors: intensity of renewable energy, wireless channel and battery state.

Cognitive WPCN [55]: In practical implementations, WPCNs are expected to co-exist with various communication networks, which can lead to co-interference when they are simultaneously operating in same frequency bands. For example, in Figure 4b, there are two networks: a traditional communication network and a WPCN. WPCN can cause interference at WD_3, and similarly AP, can cause interference at HAP in WPCN, while decoding information. Note that in case of interference from AP to HAP, HAP harvests energy from AP for the use of WD_1 and WD_2. Therefore, with limited available spectrum of frequencies, WPCN must be able to cognitively share the spectrum with other networks. A cognitive WPCN can be cooperative or non-cooperative. A cooperative cognitive WPCN and existing network can work similarly as a traditional primary/secondary setup of cognitive radio network where cognitive WPCN (secondary network) operates in the most optimal way possible while not affecting the operation of the existing network (primary network) [59,65]. A non-cooperative cognitive WPCN works in a similar way, except that it prioritizes its own well-being and gives secondary consideration to minimizing its effect on the primary network [66].

Transmission Range: Transmission range of devices is one of the important issues in WPT. It is evident from the fact that a maximum of 50% energy efficiency is achieved in a far field region [67–69]. Some systems can achieve up to 90% of efficiency using the high antennas and microwave signals; however, they are constrained by transmitter-receiver alignment, which requires both of them to be in Line-of-Sight (LoS). Therefore, there is a need to design omni-directional antenna for WPT systems. Moreover, WPT range in far field regions also needs to be investigated to enhance directivity and efficiency of energy transmission.

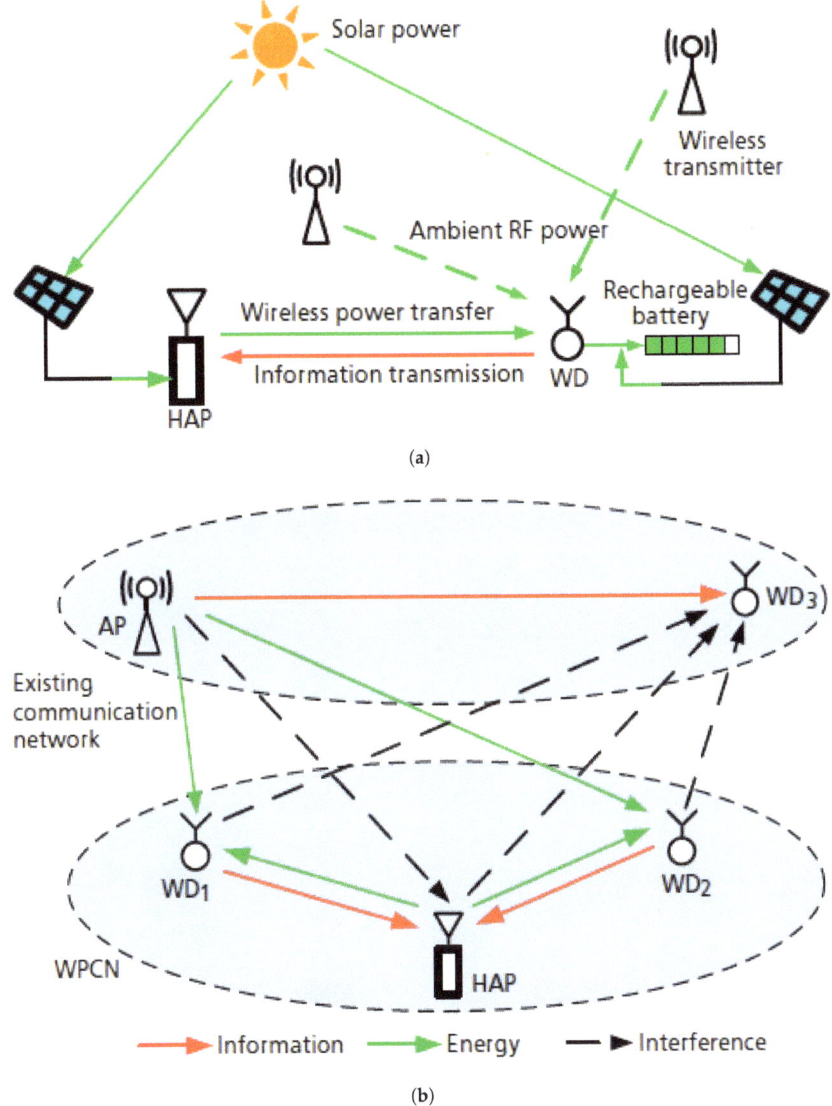

Figure 4. (**a**) WPCN implementation using hybrid energy sources, (**b**) WPCN using cognitive radio to efficiently use frequency spectrum.

2.3. SWIPT

SWIPT was first studied by [70] as a technology that stems out of various WPT technologies [71]. It allows simultaneous transmission of information and power (Figure 5) by exploiting the EM waves emitted in WPT. There are three basic advantages of SWIPT [72]: (1) WDs in SWIPT are able to harvest energy while receiving data, thus extending their lifetime; (2) it improves the transmission efficiency from traditional Time Division Multiplexing Access (TDMA) methodology where information and power are transmitted separately and finally, (3) SWIPT gives a controlled communication interference, which is beneficial for EH.

Reference [70] uses an ideal receiver to for EH and Information Retrieval (IR). Reference [73] extended [70] by implementing SWIPT in for frequency selective channels with Additive White Gaussian Noise (AWGN). In [74] author uses two separate antenna circuits, i.e., one each for EH and IR, and [75] uses asmart antenna and the concept of relaying in SWIPT Multiple Inputs Multiple Outputs (MIMO). In [76], authors evaluate SWIPT EH capabilities in presence of multiple users. EH with cognitive radios is studied in [77] where secondary radios harvest energy for primary users. Reference [78] uses energy constrained relay nodes to perform communication between two devices. However, practical SWIPT systems do not perform EH and IR using the same signal because of the possibility of information loss from the same RF signal performing EH. Therefore, SWIPT either divides the received signal into two separate parts (i.e., energy and information) or uses two different antennas for both tasks [74].

An efficient SWIPT system warrants some changes in a wireless communication system. Along with the traditional performance metrics (i.e., reception reliability and data rate) of a wireless communication network, trade-off between harvested power energy and received information data is an important factor to consider while evaluating SWIPT performance [70]. There are two major types of SWIPT architectures (Figure 6): (1) Time Switching [75] and (2) Power Splitting [79]. In time switching (Figure 6a), antenna at receiving node switches periodically between energy and information receiving modes. In power splitting (Figure 6b), signal at receiver is separated into two separate streams of information and energy.

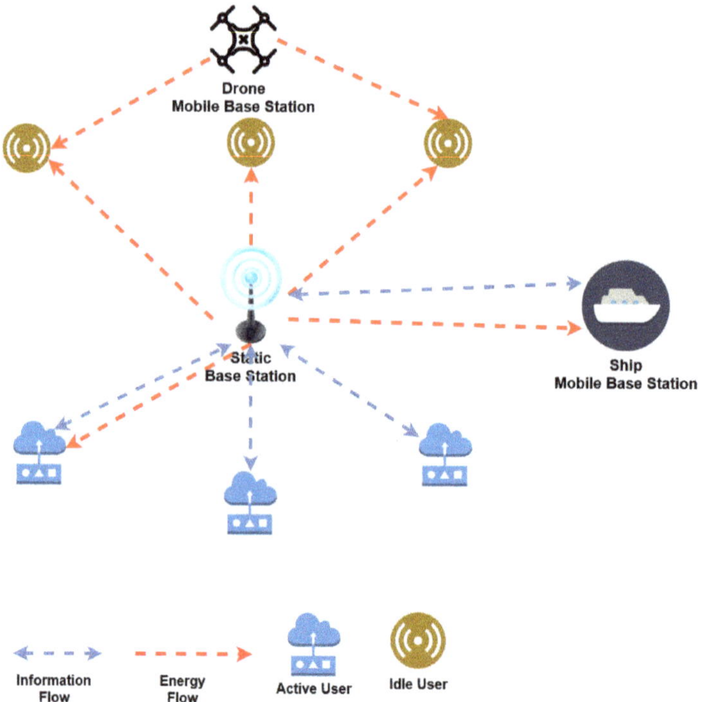

Figure 5. Simultaneous Wireless Information and Power Transfer (SWIPT) using static and mobile base stations where arrow heads represent the direction of information and power flow. Idle users only harvest energy from base stations. Active users transmit and receive energy and information.

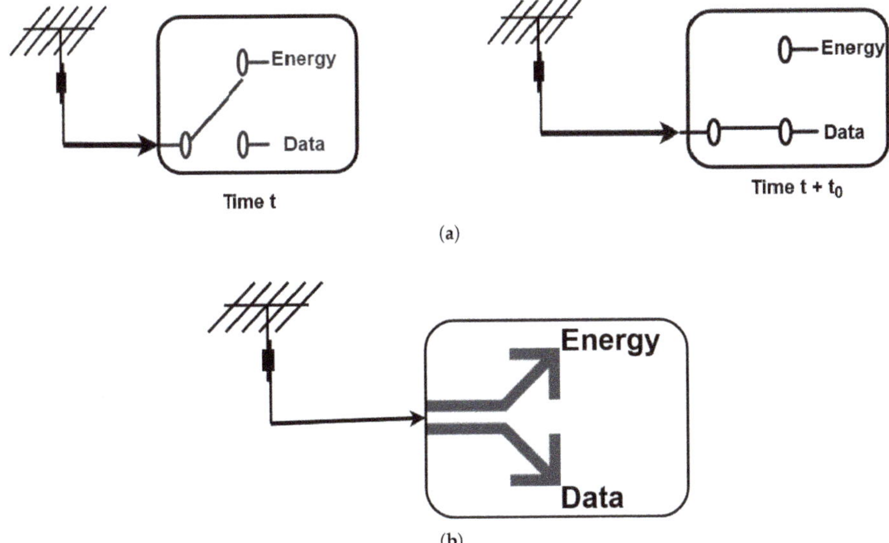

(a)

(b)

Figure 6. SWIPT Architecture: (**a**) Time Switching, (**b**) Power Splitting.

2.3.1. SWIPT-Enabled Wireless Systems

This section classifies SWIPT enabled wireless networks into the following types:

A   WSN: In WUSNs or IOUTs, the underlying buried sensors are connected via some WSN. The sensors have limited battery life. In some cases this network is huge, and it is almost impossible or very difficult to replace the batteries [80]. SWIPT is an enabling technology that can improve the WUSN/IOUT paradigm by prolonging the life of underlying WSN. The simultaneous exchange of energy and information can increase the performance of systems where sensors are frequently communicating with each other.

B   Relayed Networks: Relay networks use intermediate nodes to transmit signal or data in cooperative way. This improves performance by reducing fading and signal attenuation. SWIPT can be applied to a relayed network to power up the relay nodes in an effort to compensate them for helping in data transmission [81]. There are two types of scenarios in relayed networks for energy harvesting: SWIPT-based and Multihop-based. In the former, both relay nodes and source nodes harvest energy from each other whereas in the latter relay nodes are used to transfer energy to remote nodes [72]. The SWIPT relays are also studied in the context of physical, data and network layer where issues like relay operation, relay selection and power allocation, are addressed.

C   Cognitive Radio Networks: Cognitive network is a spectrum sharing network where high priority users (primary users-PUs) share their underutilized spectrum with secondary users (SUs) such that the SUs do not cause interference to the transmission of PUs. It aims to solve spectrum scarcity [82]. SWIPT-based cognitive network [83–85] can increase the spectrum sharing and EH efficiency. Extra energy from SUs can be utilized to transfer energy between PUs.

D   Collaborative Mobile Clouds (CMC): CMC is a cooperative way of sharing multimedia content in mobile computing in peer-to-peer manner [86]. In contrast to traditional cloud computing, CMC consist of mobile terminals that collaborate and cooperate to complete a task in a distributed manner. SWIPT can introduce energy efficiency to current CMC paradigm by allowing mobile terminals to receive information and harvest energy simultaneously. Moreover, as transmitting data consumes a large amount of energy, users may become selfish and do not join the network.

SWIPT can be used as an incentive for the users to motivate them to join the CMC network, hence, improving the overall performance of the network.

2.3.2. SWIPT Technologies

Implementation of SWIPT in wireless systems requires integration of multiple technologies. This section discusses some of these state of the art technologies below:

A   Multi-antenna Transmission: Limited communication is one of the major challenges in SWIP-based wireless systems. To that end, multiple antennas can be used to increase the antenna aperture and gain [72] higher communication frequency with multiple antenna arrangement in small devices. One of the challenges in multiple antenna design is the co-channel interference due to the presence of multiple users. Reference [74] attempts to solve this problem by block diagonalization precoding. This technique selectively transmits data to receivers with no interference only and energy to all other users.

B   Resource Allocation: SWIPT resource allocation is the optimal allocation of the resources available to the system. The resources for wireless systems include energy, time, bandwidth and space. The dual function of a transmitted signal needs an optimum method of scheduling and power allocation mechanisms. To that end, opportunistic power control uses the channel fading feature to improve energy and information transmission. Moreover, higher gain users, which are not transmitting the data, can be used to transfer power. Moreover, SWIPT systems can use the interference signals to their favor by directing it towards power hungry nodes. In [87], authors proposed resource allocation for SWIPT-based multi-user Orthogonal Frequency Division Multiplexing (OFDM) systems, which maximizes the total information rate under the constraint of minimum harvested energy. Reference [88] extends [87] by implementing a sub-optimal resource allocation technique for OFDM to balance the downlink and uplink communication rate. A SWIPT protocol is given in [89] for a massive MIMO antenna array system, which performs the scheduling of sensor nodes on the basis of beam-domain channel distributions to increase the transmission rate and decrease interference between the sensors.

C   Signal Processing: Another concern for the SWIPT is the signal attenuation due to path loss when distance is increased. Beamforming signal processing solutions [90–92] are presented as one of the viable methods to solve this problem. In [92], a SWIPT-MIMO system uses multi-antenna APs to collaboratively transmit a beam to multi-antenna active users. Reference [91] gives a hybrid approach of SWIPT-beamforming combining both analogue and digital beamforming for efficient energy harvesting. Moreover, received power over a SWIPT wireless system varies over time; however, the goal is to keep the received signal power below some threshold. To that end, energy modulation scheme can be used. In energy modulation, information can be encoded in the energy signal to ensure continuous information transmission. Reference [93] presents a modulation scheme, which uses multiple-antenna architecture to transfer an information encoded energy signal.

2.3.3. Future Research Considerations

In an effort to improve a complex SWIPT technology, this section presents some of the areas that are worth exploring in the context of SWIPT.

Mobility: Mobility is one of the desired features of SWIPT systems. The information transfer, energy harvesting and status of the network is time-dependent. Therefore, mobility affects the wireless channel, and it is not easy to obtain the channel information in a SWIPT system with mobility. To that end, it is important to device efficient beam formers to solve the mobility issue.

Security: Increase in energy of transmitted signal is desirable because it enhances the energy transfer from one source to destination. However, it also exposes various eavesdroppers stealing the information and causing a serious security breach. Therefore, it is important to devise SWIPT systems that adapt their performance to the channel state by decreasing the signal power on the legitimate channel while decreasing on a wiretap channel.

Multi-hop Networks: There is a trade-off between information transmission and energy harvesting in relay networks [72], i.e., a relay node which is suitable for information transfer may not be good for energy transfer. Therefore, optimal selection of relay node is an important issue in SWIPT. Network coding allows increased data transmission for multiple receivers. It would be interesting to explore the combination of SWIPT, network coding and multi-hop networks.

3. Energy Harvesting

Wireless methods have been studied for mobile and aboveground applications. However, one of the challenges in this method is that an aboveground energy source must be available to provide adequate energy when needed. For this purpose, either a separate facility can be build aboveground or a flying object, e.g., Unmanned Aerial Vehicle (UAV), can be used as an energy carrier at site on the regular basis. Moreover, efficiency of transferring power wirelessly is yet to be understood completely. On the contrary, energy harvesting is a promising solution as an energy harvester can be deployed underground. This energy harvester can use existing vibration sources, e.g., agricultural machinery and vehicles on road); hence, it requires no separate and dedicated aboveground facility [60,94].

As per the International Energy Agency (IEA), the energy demands of the world are mainly fulfilled by non-renewable fossil fuels. With the world's energy demand incessantly increasing, these non-renewable resources are also being depleted and expected to exhaust in few years [95,96]. Besides the environmental drawbacks of using fossil fuels, such as pollution, being harmful to human health, greenhouse gas emissions are also an important issue to consider while using these resources [97]. Therefore, it is important to shift the energy supply from fossil fuels to renewable and sustainable energy resources [98]. To that end, this section reviews the energy harvesting methods from renewable and sustainable energy resources.

WUSN's research mainly focuses on achieving advance functionalities within constrained energy and battery capacities. It is important to distinguish between energy sources and harvesting techniques in order to understand the energy harvesting paradigm completely. Energy sources are the natural and environmental phenomenon that one can use to harvest energy, e.g., radiant energy sources, kinetic energy sources. Next, we discuss some of the important energy sources that can be used to extract energy in WUSN [99,100].

3.1. Energy Harvesting Sources

3.1.1. Kinetic Energy Sources

Kinetic energy is the energy generated from motion. It is the work required for moving an object of a certain mass from rest to a certain speed. Kinetic energy may take number of forms such as vibrations and air or water flows. Some of these are discussed below:

Vibrations: Different manufacturing machines, mechanical stress and sound wave machines produce vibrations when used in various applications. The vibrations produce high-density energy, and devices used for producing energy are readily available off-the-shelf. The methods using these sources are majorly based on piezoelectric, electrostatic and electromagnetic effect [101,102]. These solutions use vibrations to generate electrical energy through two sub-systems: (1) a mass-spring system that transforms vibration into motion between two elements, and (2) a mechanical-to-electrical converter that changes the motion to electrical energy using any of the above mentioned effects. Vehicle vibrations are used as source of kinetic energy [103] and may also power on-board sensors [46,104]. For example, in [105], the authors propose a prototype design for energy harvesting in roadway pavements using the vibration produced by vehicles passing from speed bumps. Kinetic energy from human, e.g., energy from heart-beats [106], is also used to generate energy.

Air or Water Flows: One of the oldest methods to extract energy from natural resources such as air and water is by using windmills and hydroelectric turbines. Turbines transform a regular flow to rotational movement that powers the electromagnetic generator. Physical properties of turbine (e.g., number of blades, type of blades and axis of rotation) affect the amount of energy harvested. For example, for high-speed flows, few short blades work well, and for low-speed flows, a large number of long and wide blades serve the purpose. Due to low robustness and high maintenance costs, they are limited to specific applications. A system for self sustained fire-monitoring in the forest was designed by [107]. It uses a microwind turbine generator for energy harvesting. The output of wind turbine used by [107] depends upon the area, cube of wind speed and air density. The contact area of the wind turbine was 28.3 cm and the radius of its blade was 3 cm. They were able to harvest a power of 7.7 mW for the lowest wind speed of $3 \, \mathrm{m \, s^{-1}}$. A vertical-axis water turbine is used by [9,108] for energizing IOUT nodes in the application of water pipe monitoring.

3.1.2. Radiant Energy Sources

Radiant energy is the energy of electromagnetic waves. The most common radiant energy example is solar energy. Solar energy is an abundant, clean and readily available energy source. Photovoltaic cells, also known as solar cells, are used to extract energy from the EM radiations below the infrared spectrum. These cells work on the principle of photoelectric effect. Photoelectric effect is the phenomenon in which EM radiations, after hitting the material, emit electrons. The cell's material (e.g., Monocrystalline, Polycrystalline and Amorphous) plays an important role in efficiency, form actor and cost. Efficiency also depends on the load and operating temperature of the cell. For example, a cell generates maximum power with a load of 10,000 Ω. under office lightning and 1000 Ω. under PowerLED torch. A solar cell incorporates Maximum Power Point Tracking (MPPT) module to take advantage of this phenomenon. MPPT keeps a record of output power and applies the required load to optimize the performance. Radiant energy harvesting sources are popular because of their off-shelf availability and low cost. However, their performance relies heavily on duration and intensity of light. Some of the examples of energy harvesting using radiant sources for IOUT are given in [109], which designs an automated irrigation system using solar cells.

Solar energy harvesting is most suitable for outdoor implementations; however, some indoor implementations for hospitals and stadiums, etc., are also presented in [110]. For outdoor settings, Reference [111] extends the life of agricultural sensor nodes using the combination of wireless technology and solar energy. The achieved significant performance increases by combining solar energy and wireless communication technology. Similarly, Reference [112] also uses solar energy harvesting to propose an efficient energy management technique. They achieved minimum network delay and optimal throughput through this management technique.

Many solar EH implementations differ on the basis of solar cells and battery type. Some of the platforms include Batteryless Solar Harvester (BLSH) [113], Long-Term Solar Powered Node (LTSN) [114] and Micro-Scale Indoor Light (MSIL) energy harvesting system [115]. Reference [116] reviews solar energy based harvesting solutions. The wide availability of solar energy makes it an ideal option to be used in WSNs [117,118]. Many of the approaches are adaptive in nature and change their transmission range [119], routing [120], MAC [121] and scheduling [122] in order to adapt to solar energy harvesting. There are many other approaches [123,124], which ensure energy optimization and prolonged life of the network.

3.1.3. Energy from RF Transmission

RF energy harvesters are reliable, controllable, predictable and are available in both indoor and outdoor environments, which makes them more suitable than a solar energy harvester for some applications [125]. However, they have relatively low power density, i.e., $0.2 \, \mathrm{nW \, cm^{-1}}$ – $1 \, \mathrm{\mu W \, cm^{-1}}$ [126] and harvest a very low amount of energy, which makes them unsuitable for applications with a large number of nodes expecting low energy consumption [127].

There has been widespread use of cellular stations, WiFi networks and FM radios in recent years. This motivated the energy extraction from available RF transmissions [128]. The major component of RF-energy harvesting is the use of *rectenna*. Rectenna is a special type of antenna that converts the energy from electromagnetic waves into electrical energy [129]. Historically, the first microwave rectenna was introduced by Raytheon in 1964 [130], and it was first used to convert solar power to electrical power in 1972 [131]. Rectennas have been used for transmission of microwave power [132] and satellite power to analyze the performance of a rectenna array [133]. It consists of an antenna for capturing waves in the from of AC current, and a rectifying circuit to perform AC-DC conversion [134,135]. Multiple antenna types and rectifying circuits have been used in order to design an efficient antenna. Multiband and broadband rectennas have been discussed in [136]. Lack of mechanical process in conversion of RF transmissions into electrical current make it a highly robust technique.

RF energy harvesting is gaining momentum [137]. Some of the studies are very encouraging [125,138,139] while some are pessimistic [140–142]. A single frequency GSM base station provides the power density of $0.1\,\mathrm{mW\,m^{-1}}$ to $1\,\mathrm{mW\,m^{-1}}$ to users in its proximity [142]. For a total GSM downlink frequency band, it increases the power density up to the factor of 1 to 3 [143]. However, for WLAN, lower power densities were observed (i.e., one order of magnitude lower) [140]. It shows that both, WLAN and GSM, fail to provide enough energy unless a large antenna is being used. This can be complicated in routing and scheduling decisions [137]. However, improvements are possible and more energy can be harvested using beam steering approaches at transmitter and efficient antenna design at the receiver [140]. Moreover, use of rectenna can provide stable DC electricity for the electric field ranging from $0.5\,\mathrm{V\,m^{-1}}$ to $1\,\mathrm{V\,m^{-1}}$.

3.1.4. Thermal Energy Sources

Internal energy of an object under thermodynamic equilibrium is known as thermal energy. Thermal energy can be harvested using Thermo Electric Generators (TEG). TEG work on the principle of Seebeck's effect given by Thomas Seebeck. Seebeck effect is the phenomenon that generates a voltage difference from the temperature difference between conductors/semi-conductors. Basically, a connection point of two metal elements (connected in series) is brought into contact with a hot spot to generate electrical energy as voltage. An important issue with the Seebeck effect is that it produces very low output voltage. Most of the electrical circuits are unable to operate at that voltage. To that end, more metal elements are connected in series to increase the output power [144].

Although the Seebeck effect is a very old technique (more than a decade), researchers have started investigating solutions based on this technique because devices, circuits, processors and system-on-chips (SoC) have a low power consumption [145,146]. Therefore, thermoelectric harvesters found their application as voltage sources [60,147]. The temperature of pyroelectric harvesters continuously changes; therefore, they use the material with ability to generate temporary voltage. Absence of mechanical motion makes these thermal harvesters robust and prolongs their lifetime without any maintenance. References [108,148] developed self-powered WUSN underground nodes, which harvest energy from the environment. Similarly, an underground health monitoring system for oil, gas and water pipes generates energy by leveraging thermal sources such as hot water and steam [149,150].

There have been efforts made to harvest energy from thermal sources. For example, the human body can generate power of around $30\,\mathrm{\mu W\,cm^{-1}}$ [151]. This energy can be used as an energy source for wearable devices in the healthcare sector. Thermal energy harvesting is being used in many fall detection systems for elderly people [144]. Some hybrid energy harvesting approaches combining thermal and energy sources are reported in [152–154]. Similarly, in [155], thermal and optical energy harvesting is combined by fabricating a micro-TEG (μ-TEG) and a solar cell on a single chip. Reference [156] propose a special antenna design, nanoantenna, to harvest thermal energy from an automobile exhaust system.

4. Energy Harvesting Techniques

4.1. EM-Based Approach

According to Faraday's Law of Electromagnetic (EM) Induction, an electric voltage is generated by any conductor that moves towards a magnetic field. EM-based approaches use Faraday's Law of EM induction [157]. Wireless power transfer (WPT) is being accomplished using RF over long-ranges. RF energy transfer enables wireless transfer of energy form power source to remote devices using electromagnetic waves as a medium [43,59]. Due to characteristics of EM waves, it has added advantage of energy transfer over long distance over other competing technologies [43,158]. EM harvesters are reliable, require no external voltage source and suffer from lower mechanical damping. However, they also suffer from rapid power dissipation as they travel in space resulting in very low end-to-end energy transfer, e.g., few watt of transmitted power will be received as few microwatt or milliwatt.

EM-based energy transfer method has been regarded as one of the possible ways of transferring energy for a very long time. However, it has recently been proved to be feasible because of reduction in power requirements of modern electronic devices, which according to Koomey's law [159], will further decrease in the future (by the factor of 10,000 over the next 20 years) [160,161]. This explosive decrease in power calls for rethinking and redesigning wireless networks.

In [158,162], authors have proposed an EM-based wireless powered sensor network (WPSN) that uses a power beacon to transfer energy to a sensor node. An efficient WPSN uses adaptive energy beamforming to change the direction of a microwave beam towards a sensor node in real-time. They proposed two algorithms: (1) a beamforming algorithm, which adapts to received power, and (2) an adaptive algorithm for controlling duty cycle. The purpose of the duty cycle algorithm is to prevent sensor nodes from depleting their energy. A testbed is created, which consists of one power beacon and one sensor node. WPSN protocol, beamforming and duty-cycle control algorithms are implemented in the testbed. The power beacon comprised six dipole antennas, six universal software radio peripherals (USRPs), an OctoClock, an Ethernet switch and a laptop. They conducted detailed experiments on the testbed and empirically evaluated the feasibility of multi-antenna WPSNs. The beamforming algorithm was able to receive 6 times more power than any random beamforming technique. The results also showed that the efficiency of the EM-based energy transfer (using RF), along with the applying beamforming, is directly proportional to the number of antennas used. This is because beamforming in large number of antennas makes the microwave beam sharper, resulting in better end-to-end efficiency. However, they considered only one sensor node for their experiments.

In [163], authors have developed EM harvesters, which have an energy conversion efficiency of 65 % and generate power of 22.5 µW at 10 Hz. Another study [157] developed a $4.5 \times 4.5 \times 1 \, \text{mm}^3$ EM harvester, which delivers 20.9 µV. Byung et al. [164] developed a self-powered system consisting of a permanent magnet, a planar spring and a copper coil. It was able to generate a power of 1.52 mW. Similarly, another self-powered EM harvesting system is developed in [165], which has the capability of generating 140 mW. This system was used in portable electronic devices. A bi-stable EM vibration energy harvester is developed by Podder et al., which generates a power of 22 µW at the frequency of 35 Hz [166].

In [167], a WPSN has been extended using multi-nodes multi-antenna. The system consists of multiple sensor nodes and a power beacon. The power beacon has multiple transmitting antennas, and each sensor node is equipped with a receiving antenna. The sensor uses this antenna to harvest energy from the power beacon. There is no other power source, except power beacon, for the sensor nodes. The beamforming algorithm enables the power beacon (connected to the power grid) to divide the microwave and direct them towards multiple sensor nodes for simultaneous charging. Two types of beamforming algorithms are used: (1) Time-sharing (TS) and (2) Beam-splitting (BS). For TS, the energy beam from the beacon is concentrated towards a single sensor for charging, and this beam is shared between all sensors in different time slots, hence maximizing the power of one sensor node at a time. In BS, a single beam is shared between multiple sensor nodes to charge them simultaneously, achieving a Pareto optimal point in the region of received power. BS performs better than the TS. The technique aims to solve the problem of sending power to multiple nodes and prolonging their operation.

4.2. MI-Based Approach

MI-based WUSNs were introduced in [1,6], which uses magnetic antennas implemented as coils. Magnetic Induction (MI)-based communication systems have gained popularity in recent years. Battery life is an important issue in IOUT, and many nodes require to be charged wirelessly. Wireless Power Transfer (WPT) of traditional WSNs can be directly implemented in IOUT; however, few changes will be required because of difference in medium and transceiver design (as coils are used instead of antenna). Alignment of coils also plays an important role in transfer efficiency [61,168].

MI-based schemes are famous for Near-Field Communication (NFC) [169], WPT, and IOUT [7,46,47,63]. These works study the design of point-to-point MI-based information transmission. For example, [170] performs the channel characterization of point-to-point transmission. There have been many efforts to extend a point-to-point MI-based transmission system into multiple transmitters [63,171], receivers [172] and even relays [7,173]. MI-based networks involving multiple transceivers and relays have also been in analyzed in underground WSNs [174]. Various multiple-input multiple-output (MIMO) methods are used in different configurations of MI-based WPT and communication systems [58,175].

In [176], a MI-based SWIPT system with a three-coil transmitter and multiple single-coil receiver has been proposed. It divides the number of users into two groups: data receivers (DRs) and power receivers (PRs). From a pool of receiver devices, it randomly selects a device as a DR to receive information and uses the transmitter signal for WPT to PRs (remaining users). It further investigates two beamforming problems: (1) maximization of total power received for PRs (a max-sum problem) and (2) maximization of minimum power received among all PRs (max-min problem). High gains were observed in the received power as compared to the baseline technique, which validates the accuracy of proposed accuracy mechanism. However, there is a need of studying MI transmitter with more than three coils.

4.3. Vibration-Based Approach

Vibrations can be defined as the mechanical oscillation of an object whose equilibrium has been disturbed. These oscillations can be shown in any of the two functions of an object: (1) displacement and (2) frequency. Another method of energy harvesting uses the vibration sources from the environment. Vibration EHs are able to convert mechanical vibration into electricity to power up low-powered electrical equipment [177]. Vibration energy harvesters use energy sources such as sun [178], biomass [179] and wind [180] to harvest energy. Vibration energy harvesting has many advantages in WSNs. Some of these include: (1) no requirement of voltage source, (2) highly efficient power generation, (3) producing high power voltage, (4) suitable for using with resonant devices and (5) quick response time. Some of the studies [19,38,39] have considered vibration energy harvesting in the application of WUSNs. These solutions have been sought as promising alternatives to battery replacement [181,182].

This method is based on piezoelectric effect where vibrations due to stress are converted to electricity. The main challenge in this technology is to extract enough energy needed to fulfill the energy requirement of a certain application. Furthermore, piezoelectric is frequency dependent, therefore, to generate the desired power it should be set on a right frequency. Another challenge using vibration harvesting is having a wide range of frequency in the environment. It is difficult to get the right frequency, from this wide range, which can generate highest power. The amount of power generated by vibration energy harvesting method has been investigated by many studies [183–186].

In WUSNs/IOUT, vibration energy can be generated from several sources. It depends upon the WUSN application. For example, for an agricultural IOUT, these sources can be seeders, harvesters, irrigation systems and other machinery. the vibration generated by this equipment, above the ground, should propagate underground reaching buriedpiezoelectric energy harvesters. The amount of power generated is then dependent upon the intensity of vibration reached at deployment depth. Therefore, it is very important to understand and study mechanisms of wave propagation from the soil and other underground mediums.

In [187,188], authors perform a three-step theoretical analysis and propose a mathematical model for estimating the amount of power generated by an underground vibration energy harvester, buried at some depth (d_h), in response to the amount of vibration generated by the above-ground vibrating object. First, they formulate the intensity of vibration generated. Second, they model how most of that vibration propagates to the soil. Finally, the amount of energy generated from this UG vibration and energy harvester is captured. Figure 7a illustrates the procedure. They tested their research on an agricultural setup by performing various experiments in South Central Agricultural Laboratory, one of the agricultural research divisions (ARD) near Clay Center, Nebraska. Figure 7b shows the schematic of the devices for the experiments. The intensity and frequency of vibration is measured for agricultural machinery. The agricultural machinery are the frequently used center pivot irrigation system and the four-wheeler used in farms. These vibrations were measured at varying depths to measure the feasibility of UG energy harvesting. Three DLP-TILT-G accelerometer sensors were used for the experiments (see Figure 7b). As a result of the experiments, a maximum output power of 17 mW was calculated. However, this power might not be achieved practically. For example, in one of the studies [183], the energy harvester achieved 3.5 mW. For the practical purpose of generating energy form a vibration harvester, high acceleration is required from the vibration sources (seeders, harvesters and sprayers). Furthermore, there is need of further advancement and investigation in vibration energy harvesting for studying environmental conditions, e.g., rain and temperature changes, and their impacts.

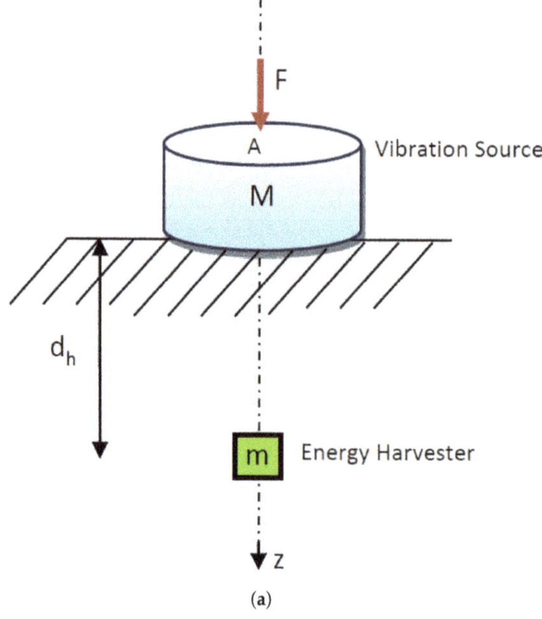

(a)

Figure 7. *Cont.*

(**b**)

Figure 7. Vibration energy harvester: (**a**) source for the vibration is on the surface with energy harvester deployed in the soil, (**b**) deployment in agriculture field.

Future Research Considerations

This section presents some of the areas that are worth exploring in the context of Energy Harvesting approaches.

Generic Harvester: Harvesting energy from multiple sources is a challenging task that needs more advanced power management techniques. To this end, it is important to develop a plug-n-play energy harvesting method, which uses multiple energy sources to harvest energy. Such generic harvesters may even eliminate the energy storage systems.

Efficient Networking Protocol: Research in WUSN has mostly been focused on efficient networking protocol [189]. A good EH approach can shift this focus from energy-efficient to information centric protocols.

Simulation Environment: To the best of our knowledge, there is no simulation environment for evaluating energy harvesting in WUSN. Such simulation environment would help in prediction performance of EH approaches in large-scale deployment of WUSNs.

Auxiliary Energy Storage: Large-scale WUSNs affect different parameters of batteries, e.g., charge cycles, self-discharge and environmental conditions. Therefore, it is important for researchers to analyze and improve the performance of rechargeable batteries. Moreover, battery-capacitor trade-off as a storage device is also an important area to improve.

Miniaturization: Large-scale WUSNs along-with heavy and bulky energy harvesting systems increase the deployment costs several times, which is not feasible when the budget is low. Therefore, it is important to develop mini energy harvesting systems. Such miniaturized harvesting systems can also be used to empower implants and for the monitoring of the human body.

Author Contributions: Conceptualization, U.R. and A.S.; methodology, U.R.; software, A.S.; validation, A.S. and U.R.; formal analysis, A.S.; investigation, U.R. and A.S.; resources, A.S.; data curation, A.S.; writing—original draft preparation, A.S.; writing—review and editing, U.R.; visualization, A.S.; supervision, A.S.; project administration, A.S.; funding acquisition, A.S. All authors have read and agreed to the published version of the manuscript.

Funding: This research received no external funding.

Acknowledgments: In this section you can acknowledge any support given which is not covered by the author contribution or funding sections. This may include administrative and technical support, or donations in kind (e.g., materials used for experiments).

Conflicts of Interest: The authors declare no conflict of interest.

Abbreviations

APs	Access Points
AWGN	Additive White Gaussian Noise
CSI	Channel State Information
EH	Energy Harvesting
EM	Electromagnetic
EN	Energy Nodes
ERs	Energy Receivers
ETs	Energy Transmitters
FCC	Federal Communication Commission
HAP	Hybrid APs
HBAN	Human Body Area Network
IOUT	Internet of Underground Things
LoS	Line-of-Sight
MI	Magnetic Induction
MIMO	Multiple Inputs Multiple Outputs
NFC	Near-Field Communication
OFDM	Orthogonal Frequency Division Multiplexing
RF	Radio Frequency
RFID	Radio Frequency Identification
SWIPT	Simultaneous Wireless Information and Power Transfer
TDMA	Time Division Multiplexing Access
UAV	Unmanned Aerial Vehicle
WD	Wireless Devices
WPCN	Wireless Powered Communication Network
WPT	Wireless Power Transfer
WSN	Wireless Sensor Network
WUSNs	Wireless Underground Sensor Networks

References

1. Akyildiz, I.F.; Stuntebeck, E.P. Wireless underground sensor networks: Research challenges. *Ad Hoc Netw.* **2006**, *4*, 669–686. [CrossRef]
2. Salam, A.; Vuran, M.C. Smart Underground Antenna Arrays: A Soil Moisture Adaptive Beamforming Approach. In Proceedings of the IEEE INFOCOM 2017, Atlanta, GA, USA, 1–4 May 2017.
3. Liu, G.; Wang, Z.; Jiang, T. QoS-aware throughput maximization in wireless powered underground sensor networks. *IEEE Trans. Commun.* **2016**, *64*, 4776–4789. [CrossRef]
4. Bogena, H.; Herbst, M.; Huisman, J.; Rosenbaum, U.; Weuthen, A.; Vereecken, H. Potential of wireless sensor networks for measuring soil water content variability. *Vadose Zone J.* **2010**, *9*, 1002–1013. [CrossRef]
5. Dong, X.; Vuran, M.C.; Irmak, S. Autonomous precision agriculture through integration of wireless underground sensor networks with center pivot irrigation systems. *Ad Hoc Netw.* **2013**, *11*, 1975–1987. [CrossRef]
6. Salam, A. An Underground Radio Wave Propagation Prediction Model for Digital Agriculture. *Information* **2019**, *10*, 147. [CrossRef]
7. Salam, A. Wireless Underground Communications in Sewer and Stormwater Overflow Monitoring: Radio Waves through Soil and Asphalt Medium. *Information* **2020**, *11*, 98.

8. Silva, A.R.; Vuran, M.C. Communication with aboveground devices in wireless underground sensor networks: An empirical study. In Proceedings of the 2010 IEEE international conference on communications, Cape Town, South Africa, 23–27 May 2010; IEEE: Washington, DC, USA, 2010; pp. 1–6.

9. Salam, A.; Vuran, M.C. EM-Based Wireless Underground Sensor Networks. In *Underground Sensing*; Academic Press: Cambridge, MA, USA, 2018; pp. 247–285. [CrossRef]

10. Silva, A.R.; Vuran, M.C. Empirical evaluation of wireless underground-to-underground communication in wireless underground sensor networks. In *International Conference on Distributed Computing in Sensor Systems*; Springer: Berlin/Heidelberg, Germany, 2009; pp. 231–244.

11. Vuran, M.C.; Akyildiz, I.F. Channel model and analysis for wireless underground sensor networks in soil medium. *Phys. Commun.* **2010**, *3*, 245–254. [CrossRef]

12. Salam, A. Internet of Things for Sustainable Forestry. In *Internet of Things for Sustainable Community Development: Wireless Communications, Sensing, and Systems*; Springer International Publishing: Cham, Switzerland, 2020; pp. 147–181.

13. Salam, A. Internet of Things for Environmental Sustainability and Climate Change. In *Internet of Things for Sustainable Community Development: Wireless Communications, Sensing, and Systems*; Springer International Publishing: Cham, Switzerland, 2020; pp. 33–69. [CrossRef]

14. Fanaria, F.; Dachenab, C.; Cartaa, R.; Desogusa, F. Heat Transfer Modeling in Soil Microwave Heating. *Chem. Eng.* **2019**, *76*. [CrossRef]

15. Macana, R.J. Disinfestation of Rusty Grain Beetle (Cryptolestes Ferrungineus) in Stored Wheat Grain Using 50-Ohm Radio Frequency (RF) Heating System. Ph.D. Thesis, University of Saskatchewan, Saskatoon, SK, Canada, 2019.

16. Sun, H.; Cai, C.; Liu, H.; Yang, B. Microwave and Meteorological Fusion: A method of Spatial Downscaling of Remotely Sensed Soil Moisture. *IEEE J. Sel. Top. Appl. Earth Obs. Remote Sens.* **2019**, *12*, 1107–1119. [CrossRef]

17. Dong, X.; Vuran, M.C. A channel model for wireless underground sensor networks using lateral waves. In Proceedings of the 2011 IEEE Global Telecommunications Conference-GLOBECOM 2011, Houston, TX, USA, 5–9 December 2011; IEEE: Washington, DC, USA, 2011; pp. 1–6.

18. Li, L.; Vuran, M.C.; Akyildiz, I.F. Characteristics of underground channel for wireless underground sensor networks. *Proc. Med-Hoc-Net* **2007**, *7*, 13–15.

19. Salam, A. Internet of Things in Sustainable Energy Systems. In *Internet of Things for Sustainable Community Development: Wireless Communications, Sensing, and Systems*; Springer International Publishing: Cham, Switzerland, 2020; pp. 183–216. [CrossRef]

20. Tooker, J.; Vuran, M.C. Mobile data harvesting in wireless underground sensor networks. In Proceedings of the 2012 9th Annual IEEE Communications Society Conference on Sensor, Mesh and Ad Hoc Communications and Networks (SECON), Seoul, Korea, 18–21 June 2012; IEEE: Washington, DC, USA, 2012; pp. 560–568.

21. Matos, C.R.; Carneiro, J.F.; Silva, P.P. Overview of large-scale underground energy storage technologies for integration of renewable energies and criteria for reservoir identification. *J. Energy Storage* **2019**, *21*, 241–258. [CrossRef]

22. Flores-Quintero, R.R.; Flores-Verdad, G.E.; Gonzalez-Diaz, V.R.; Carrillo-Martinez, L.A. A parallel auto-adaptive topology for integrated energy harvesting system. *Microelectron. J.* **2020**, *98*, 104736. [CrossRef]

23. Zhang, S.; Qian, Z.; Wu, J.; Kong, F.; Lu, S. Wireless charger placement and power allocation for maximizing charging quality. *IEEE Trans. Mob. Comput.* **2017**, *17*, 1483–1496. [CrossRef]

24. Hossain, M.A.; Noor, R.M.; Yau, K.L.A.; Ahmedy, I.; Anjum, S.S. A survey on simultaneous wireless information and power transfer with cooperative relay and future challenges. *IEEE Access* **2019**, *7*, 19166–19198. [CrossRef]

25. Castelvecchi, D. Wireless energy may power electronics: Dead cell phone inspired research innovation. *MIT TechTalk* **2006**, *51*, 1.

26. Karalis, A.; Joannopoulos, J.D.; Soljačić, M. Efficient wireless non-radiative mid-range energy transfer. *Ann. Phys.* **2008**, *323*, 34–48. [CrossRef]

27. Kurs, A.; Moffatt, R.; Soljačić, M. Simultaneous mid-range power transfer to multiple devices. *Appl. Phys. Lett.* **2010**, *96*, 044102. [CrossRef]

28. Salam, A. Internet of Things for Sustainable Human Health. In *Internet of Things for Sustainable Community Development: Wireless Communications, Sensing, and Systems*; Springer International Publishing: Cham, Switzerland, 2020; pp. 217–242. [CrossRef]

29. Sharma, H.; Haque, A.; Jaffery, Z.A. Maximization of wireless sensor network lifetime using solar energy harvesting for smart agriculture monitoring. *Ad Hoc Netw.* **2019**, *94*, 101966. [CrossRef]

30. Mohammadnia, A.; Rezania, A.; Ziapour, B.M.; Sedaghati, F.; Rosendahl, L. Hybrid energy harvesting system to maximize power generation from solar energy. *Energy Convers. Manag.* **2020**, *205*, 112352. [CrossRef]

31. Ghomian, T.; Mehraeen, S. Survey of energy scavenging for wearable and implantable devices. *Energy* **2019**, *178*, 33–49. [CrossRef]

32. Boumaiz, M.; El Ghazi, M.; Mazer, S.; Fattah, M.; Bouayad, A.; El Bekkali, M.; Balboul, Y. Energy harvesting based WBANs: EH optimization methods. *Procedia Comput. Sci.* **2019**, *151*, 1040–1045. [CrossRef]

33. Wang, J.; Geng, L.; Ding, L.; Zhu, H.; Yurchenko, D. The state-of-the-art review on energy harvesting from flow-induced vibrations. *Appl. Energy* **2020**, *267*, 114902. [CrossRef]

34. Chen, X.; Foley, A.; Zhang, Z.; Wang, K.; O'Driscoll, K. An assessment of wind energy potential in the Beibu Gulf considering the energy demands of the Beibu Gulf Economic Rim. *Renew. Sustain. Energy Rev.* **2020**, *119*, 109605. [CrossRef]

35. Dandy, G.C.; Marchi, A.; Maier, H.R.; Kandulu, J.; MacDonald, D.H.; Ganji, A. An integrated framework for selecting and evaluating the performance of stormwater harvesting options to supplement existing water supply systems. *Environ. Model. Softw.* **2019**, *122*, 104554. [CrossRef]

36. Nintanavongsa, P. A survey on RF energy harvesting: Circuits and protocols. *Energy Procedia* **2014**, *56*, 414–422. [CrossRef]

37. Cansiz, M.; Altinel, D.; Kurt, G.K. Efficiency in RF energy harvesting systems: A comprehensive review. *Energy* **2019**, *174*, 292–309 . [CrossRef]

38. Pan, J.; Xue, B.; Inoue, Y. A self-powered sensor module using vibration-based energy generation for ubiquitous systems. In Proceedings of the 2005 6th International Conference on ASIC, Shanghai, China, 24–27 October 2005; IEEE: Washington, DC, USA, 2005; Volume 1, pp. 403–406.

39. Roundy, S.; Wright, P.K.; Rabaey, J. A study of low level vibrations as a power source for wireless sensor nodes. *Comput. Commun.* **2003**, *26*, 1131–1144. [CrossRef]

40. Raghunathan, V.; Kansal, A.; Hsu, J.; Friedman, J.; Srivastava, M. Design considerations for solar energy harvesting wireless embedded systems. In Proceedings of the IPSN 2005. Fourth International Symposium on Information Processing in Sensor Networks, Boise, ID, USA, 15 April 2005; IEEE: Washington, DC, USA, 2005; pp. 457–462.

41. Perera, T.D.P.; Jayakody, D.N.K.; Sharma, S.K.; Chatzinotas, S.; Li, J. Simultaneous wireless information and power transfer (SWIPT): Recent advances and future challenges. *IEEE Commun. Surv. Tutor.* **2017**, *20*, 264–302. [CrossRef]

42. Bi, S.; Ho, C.K.; Zhang, R. Wireless powered communication: Opportunities and challenges. *IEEE Commun. Mag.* **2015**, *53*, 117–125. [CrossRef]

43. Lu, X.; Wang, P.; Niyato, D.; Kim, D.I.; Han, Z. Wireless networks with RF energy harvesting: A contemporary survey. *IEEE Commun. Surv. Tutor.* **2014**, *17*, 757–789. [CrossRef]

44. Krikidis, I.; Timotheou, S.; Nikolaou, S.; Zheng, G.; Ng, D.W.K.; Schober, R. Simultaneous wireless information and power transfer in modern communication systems. *IEEE Commun. Mag.* **2014**, *52*, 104–110. [CrossRef]

45. Butt, M.M.; Nauryzbayev, G.; Marchetti, N. On maximizing information reliability in wireless powered cooperative networks. In *Physical Communication*; Elsevier: Amsterdam, The Netherlands, 2020; in press.

46. Vuran, M.C.; Salam, A.; Wong, R.; Irmak, S. Internet of underground things in precision agriculture: Architecture and technology aspects. *Ad Hoc Netw.* **2018**, *81*, 160–173. [CrossRef]

47. Saeed, N.; Al-Naffouri, T.Y.; Alouini, M.S. Towards the Internet of Underground Things: A Systematic Survey. *arXiv* **2019**, arXiv:1902.03844

48. Ju, H.; Zhang, R. Throughput maximization in wireless powered communication networks. *IEEE Trans. Wirel. Commun.* **2013**, *13*, 418–428. [CrossRef]

49. Salam, A.; Vuran, M.C.; Irmak, S. Di-Sense: In situ real-time permittivity estimation and soil moisture sensing using wireless underground communications. *Comput. Netw.* **2019**, *151*, 31–41. [CrossRef]

50. Saeed, N.; Alouini, M.S.; Al-Naffouri, T.Y. Toward the internet of underground things: A systematic survey. *IEEE Commun. Surv. Tutor.* **2019**, *21*, 3443–3466. [CrossRef]
51. Sun, Z.; Akyildiz, I.F. Channel modeling and analysis for wireless networks in underground mines and road tunnels. *IEEE Trans. Commun.* **2010**, *58*, 1758–1768. [CrossRef]
52. Salam, A.; Vuran, M.C.; Dong, X.; Argyropoulos, C.; Irmak, S. A theoretical model of underground dipole antennas for communications in internet of underground things. *IEEE Trans. Antennas Propag.* **2019**, *67*, 3996–4009. [CrossRef]
53. Kisseleff, S.; Akyildiz, I.F.; Gerstacker, W.H. Survey on advances in magnetic induction-based wireless underground sensor networks. *IEEE Internet Things J.* **2018**, *5*, 4843–4856. [CrossRef]
54. Liang, Y.; He, Y.; Qiao, J.; Hu, A.P. Simultaneous Wireless Information and Power Transfer in 5G Mobile Networks: A Survey. In Proceedings of the 2019 Computing, Communications and IoT Applications (ComComAp), Shenzhen, China, 26–28 October 2019; IEEE: Washington, DC, USA, 2019; pp. 460–465.
55. Bi, S.; Zeng, Y.; Zhang, R. Wireless powered communication networks: An overview. *IEEE Wirel. Commun.* **2016**, *23*, 10–18. [CrossRef]
56. Huang, K.; Lau, V.K. Enabling wireless power transfer in cellular networks: Architecture, modeling and deployment. *IEEE Trans. Wirel. Commun.* **2014**, *13*, 902–912. [CrossRef]
57. Zeng, Y.; Zhang, R. Full-duplex wireless-powered relay with self-energy recycling. *IEEE Wirel. Commun. Lett.* **2015**, *4*, 201–204. [CrossRef]
58. Salam, A. Sensor-Free Underground Soil Sensing. In Proceedings of the 2019 ASA-CSSA-SSSA International Annual Meeting, San Antonio, TX, USA, 10–13 November 2019.
59. Salam, A.; Karabiyik, U. A Cooperative Overlay Approach at the Physical Layer of Cognitive Radio for Digital Agriculture. In Proceedings of the Third International Balkan Conference on Communications and Networking 2019 (BalkanCom'19), Skopje, North Macedonia, 10–12 June 2019.
60. Salam, A. Subsurface MIMO: A Beamforming Design in Internet of Underground Things for Digital Agriculture Applications. *J. Sens. Actuator Netw.* **2019**, *8*, 41. [CrossRef]
61. Salam, A. Design of Subsurface Phased Array Antennas for Digital Agriculture Applications. In Proceedings of the 2019 IEEE International Symposium on Phased Array Systems and Technology (IEEE Array 2019), Waltham, MA, USA, 15–18 October 2019.
62. Ju, H.; Zhang, R. User cooperation in wireless powered communication networks. In Proceedings of the 2014 IEEE Global Communications Conference, Austin, TX, USA, 8–12 December 2014; IEEE: Washington, DC, USA, 2014; pp. 1430–1435.
63. Salam, A.; Hoang, A.D.; Meghna, A.; Martin, D.R.; Guzman, G.; Yoon, Y.H.; Carlson, J.; Kramer, J.; Yansi, K.; Kelly, M.; et al. The Future of Emerging IoT Paradigms: Architectures and Technologies. *Preprints* **2019**. [CrossRef]
64. Salam, A.; Vuran, M.C.; Irmak, S. Pulses in the Sand: Impulse Response Analysis of Wireless Underground Channel. In Proceedings of the The 35th Annual IEEE International Conference on Computer Communications (INFOCOM 2016), San Francisco, CA, USA, 10–14 April 2016.
65. Lee, S.; Zhang, R. Cognitive wireless powered network: Spectrum sharing models and throughput maximization. *IEEE Trans. Cogn. Commun. Netw.* **2015**, *1*, 335–346. [CrossRef]
66. Xu, J.; Bi, S.; Zhang, R. Multiuser MIMO wireless energy transfer with coexisting opportunistic communication. *IEEE Wirel. Commun. Lett.* **2015**, *4*, 273–276. [CrossRef]
67. Hui, S.Y.R.; Zhong, W.; Lee, C.K. A critical review of recent progress in mid-range wireless power transfer. *IEEE Trans. Power Electron.* **2013**, *29*, 4500–4511. [CrossRef]
68. Brown, W.C. The history of power transmission by radio waves. *IEEE Trans. Microw. Theory Tech.* **1984**, *32*, 1230–1242. [CrossRef]
69. McSpadden, J.O.; Mankins, J.C. Space solar power programs and microwave wireless power transmission technology. *IEEE Microw. Mag.* **2002**, *3*, 46–57. [CrossRef]
70. Varshney, L.R. Transporting information and energy simultaneously. In Proceedings of the 2008 IEEE International Symposium on Information Theory, Toronto, ON, Canada, 6–11 July 2008; IEEE: Washington, DC, USA, 2008; pp. 1612–1616.
71. Jameel, F.; Faisal.; Haider, M.A.A.; Butt, A.A. A technical review of simultaneous wireless information and power transfer (SWIPT). In Proceedings of the 2017 International Symposium on Recent Advances in Electrical Engineering (RAEE), Islamabad, Pakistan, 24-26 October 2017; pp. 1–6.

72. Huang, J.; Xing, C.C.; Wang, C. Simultaneous wireless information and power transfer: Technologies, applications, and research challenges. *IEEE Commun. Mag.* **2017**, *55*, 26–32. [CrossRef]

73. Grover, P.; Sahai, A. Shannon meets Tesla: Wireless information and power transfer. In Proceedings of the 2010 IEEE international symposium on information theory, Austin, TX, USA, 13–18 June 2010; IEEE: Washington, DC, USA, 2010; pp. 2363–2367.

74. Ding, Z.; Zhong, C.; Ng, D.W.K.; Peng, M.; Suraweera, H.A.; Schober, R.; Poor, H.V. Application of smart antenna technologies in simultaneous wireless information and power transfer. *IEEE Commun. Mag.* **2015**, *53*, 86–93. [CrossRef]

75. Zhang, R.; Ho, C.K. MIMO broadcasting for simultaneous wireless information and power transfer. *IEEE Trans. Wirel. Commun.* **2013**, *12*, 1989–2001. [CrossRef]

76. Fouladgar, A.M.; Simeone, O. On the transfer of information and energy in multi-user systems. *IEEE Commun. Lett.* **2012**, *16*, 1733–1736. [CrossRef]

77. Lee, S.; Zhang, R.; Huang, K. Opportunistic wireless energy harvesting in cognitive radio networks. *IEEE Trans. Wirel. Commun.* **2013**, *12*, 4788–4799. [CrossRef]

78. Shah, S.; Choi, K.; Hasan, S.; Chung, M. Energy harvesting and information processing in two-way multiplicative relay networks. *Electron. Lett.* **2016**, *52*, 751–753. [CrossRef]

79. Zhou, X.; Zhang, R.; Ho, C.K. Wireless information and power transfer: Architecture design and rate-energy tradeoff. *IEEE Trans. Commun.* **2013**, *61*, 4754–4767. [CrossRef]

80. Huang, K.; Zhong, C.; Zhu, G. Some new research trends in wirelessly powered communications. *IEEE Wirel. Commun.* **2016**, *23*, 19–27. [CrossRef]

81. Guo, S.; Wang, F.; Yang, Y.; Xiao, B. Energy-efficient cooperative tfor simultaneous wireless information and power transfer in clustered wireless sensor networks. *IEEE Trans. Commun.* **2015**, *63*, 4405–4417. [CrossRef]

82. Zhao, N.; Yu, F.R.; Sun, H.; Li, M. Adaptive power allocation schemes for spectrum sharing in interference-alignment-based cognitive radio networks. *IEEE Trans. Veh. Technol.* **2015**, *65*, 3700–3714. [CrossRef]

83. Liu, Y.; Mousavifar, S.A.; Deng, Y.; Leung, C.; Elkashlan, M. Wireless energy harvesting in a cognitive relay network. *IEEE Trans. Wirel. Commun.* **2015**, *15*, 2498–2508. [CrossRef]

84. Singh, S.; Modem, S.; Prakriya, S. Optimization of cognitive two-way networks with energy harvesting relays. *IEEE Commun. Lett.* **2017**, *21*, 1381–1384.

85. Ashraf, M.; Shahid, A.; Jang, J.W.; Lee, K.G. Optimization of the overall success probability of the energy harvesting cognitive wireless sensor networks. *IEEE Access* **2016**, *5*, 283–294.

86. Chang, Z.; Gong, J.; Ristaniemi, T.; Niu, Z. Energy-efficient resource allocation and user scheduling for collaborative mobile clouds with hybrid receivers. *IEEE Trans. Veh. Technol.* **2016**, *65*, 9834–9846.

87. Xu, D.; Li, Q. Optimization of wireless information and power transfer in multiuser ofdm systems. *AEU-Int. J. Electron. Commun.* **2018**, *90*, 171–174.

88. Xu, D.; Li, Q. Resource allocation in OFDM-based wireless powered communication networks with SWIPT. *AEU-Int. J. Electron. Commun.* **2019**, *101*, 69–75.

89. Xu, K.; Shen, Z.; Wang, Y.; Xia, X. Resource allocation for hybrid TS and PS SWIPT in massive MIMO system. *Phys. Commun.* **2018**, *28*, 201–213.

90. Zhang, J.; Zheng, G.; Krikidis, I.; Zhang, R. Specific Absorption Rate-Aware Beamforming in MISO Downlink SWIPT Systems. *IEEE Trans. Commun.* **2019**. [CrossRef]

91. Li, A.; Masouros, C. Energy-efficient SWIPT: From fully digital to hybrid analog–digital beamforming. *IEEE Trans. Veh. Technol.* **2017**, *67*, 3390–3405.

92. Qin, C.; Ni, W.; Tian, H.; Liu, R.P.; Guo, Y.J. Joint beamforming and user selection in multiuser collaborative MIMO SWIPT systems with nonnegligible circuit energy consumption. *IEEE Trans. Veh. Technol.* **2017**, *67*, 3909–3923.

93. Zhang, R.; Yang, L.L.; Hanzo, L. Energy pattern aided simultaneous wireless information and power transfer. *IEEE J. Sel. Areas Commun.* **2015**, *33*, 1492–1504.

94. Salam, A. Internet of Things for Sustainability: Perspectives in Privacy, Cybersecurity, and Future Trends. In *Internet of Things for Sustainable Community Development: Wireless Communications, Sensing, and Systems*; Springer International Publishing: Cham, Switzerland, 2020; pp. 299–327. [CrossRef]

95. World Energy Outlook. 2015. Available online: https://www.iea.org/reports/world-energy-outlook-2015 (accessed on 22 March 2020).

96. Chen, F.; Taylor, N.; Kringos, N. Electrification of roads: Opportunities and challenges. *Appl. Energy* **2015**, *150*, 109–119. [CrossRef]

97. Ahmad, S.; Abdul Mujeebu, M.; Farooqi, M.A. Energy harvesting from pavements and roadways: A comprehensive review of technologies, materials, and challenges. *Int. J. Energy Res.* **2019**, *43*, 1974–2015.

98. Buhaug, H.; Urdal, H. An urbanization bomb? Population growth and social disorder in cities. *Glob. Environ. Chang.* **2013**, *23*, 1–10. [CrossRef]

99. Bhatti, N.A.; Alizai, M.H.; Syed, A.A.; Mottola, L. Energy harvesting and wireless transfer in sensor network applications: Concepts and experiences. *ACM Trans. Sens. Netw.* **2016**, *12*, 1–40.

100. Salam, A.; Vuran, M.C. Wireless Underground Channel Diversity Reception with Multiple Antennas for Internet of Underground Things. In Proceedings of the IEEE ICC 2017, Paris, France, 21–25 May 2017.

101. De Queiroz, A.C.M. Electrostatic generators for vibrational energy harvesting. In Proceedings of the 2013 IEEE 4th Latin American Symposium on Circuits and Systems (LASCAS), Cusco, Peru, 27 Febryary–1 March 2013; IEEE: Washington, DC, USA, 2013; pp. 1–4.

102. Chye, W.C.; Dahari, Z.; Sidek, O.; Miskam, M.A. Electromagnetic micro power generator—A comprehensive survey. In Proceedings of the 2010 IEEE Symposium on Industrial Electronics and Applications (ISIEA), Penang, Malaysia, 3–5 October 2010; IEEE: Washington, DC, USA, 2010; pp. 376–382.

103. Sazonov, E.; Li, H.; Curry, D.; Pillay, P. Self-powered sensors for monitoring of highway bridges. *IEEE Sens. J.* **2009**, *9*, 1422–1429. [CrossRef]

104. Tolentino, I.; Talampas, M. Design, development, and evaluation of a self-powered GPS tracking system for vehicle security. In Proceedings of the SENSORS, 2012 IEEE, Taipei, Taiwan, 28–31 October 2012; IEEE: Washington, DC, USA, 2012; pp. 1–4.

105. Gholikhani, M.; Nasouri, R.; Tahami, S.A.; Legette, S.; Dessouky, S.; Montoya, A. Harvesting kinetic energy from roadway pavement through an electromagnetic speed bump. *Appl. Energy* **2019**, *250*, 503–511. [CrossRef]

106. Kumar, A.; Kiran, R.; Kumar, S.; Chauhan, V.S.; Kumar, R.; Vaish, R. A comparative numerical study on piezoelectric energy harvester for self-powered pacemaker application. *Glob. Chall.* **2018**, *2*, 1700084. [CrossRef]

107. Tan, Y.K.; Panda, S.K. Self-autonomous wireless sensor nodes with wind energy harvesting for remote sensing of wind-driven wildfire spread. *IEEE Trans. Instrum. Meas.* **2011**, *60*, 1367–1377. [CrossRef]

108. Zhao, C.; Yisrael, S.; Smith, J.R.; Patel, S.N. Powering wireless sensor nodes with ambient temperature changes. In Proceedings of the 2014 ACM International Joint Conference on Pervasive and Ubiquitous Computing, Seattle, WA, USA, 13–17 September 2014; pp. 383–387.

109. Gutiérrez, J.; Villa-Medina, J.F.; Nieto-Garibay, A.; Porta-Gándara, M.Á. Automated irrigation system using a wireless sensor network and GPRS module. *IEEE Trans. Instrum. Meas.* **2013**, *63*, 166–176. [CrossRef]

110. Hande, A.; Polk, T.; Walker, W.; Bhatia, D. Indoor solar energy harvesting for sensor network router nodes. *Microprocess. Microsyst.* **2007**, *31*, 420–432. [CrossRef]

111. Wang, C.; Li, J.; Yang, Y.; Ye, F. Combining solar energy harvesting with wireless charging for hybrid wireless sensor networks. *IEEE Trans. Mob. Comput.* **2017**, *17*, 560–576. [CrossRef]

112. Sharma, V.; Mukherji, U.; Joseph, V.; Gupta, S. Optimal energy management policies for energy harvesting sensor nodes. *IEEE Trans. Wirel. Commun.* **2010**, *9*, 1326–1336. [CrossRef]

113. Brunelli, D.; Moser, C.; Thiele, L.; Benini, L. Design of a solar-harvesting circuit for batteryless embedded systems. *IEEE Trans. Circuits Syst. I Regul. Pap.* **2009**, *56*, 2519–2528. [CrossRef]

114. Corke, P.; Valencia, P.; Sikka, P.; Wark, T.; Overs, L. Long-duration solar-powered wireless sensor networks. In Proceedings of the 4th Workshop on Embedded Networked Sensors, Cork, Ireland, 25–26 June 2007; pp. 33–37.

115. Yu, H.; Yue, Q. Indoor light energy harvesting system for energy-aware wireless sensor node. *Energy Procedia* **2012**, *16*, 1027–1032. [CrossRef]

116. Sudevalayam, S.; Kulkarni, P. Energy harvesting sensor nodes: Survey and implications. *IEEE Commun. Surv. Tutor.* **2010**, *13*, 443–461. [CrossRef]

117. Magno, M.; Porcarelli, D.; Benini, L.; Brunelli, D. A power-aware multi harvester power unit with hydrogen fuel cell for embedded systems in outdoor applications. In Proceedings of the 2013 International Green Computing Conference Proceedings, Arlington, VA, USA, 27–29 June 2013; IEEE: Washington, DC, USA, 2013; pp. 1–6.

118. Buchli, B.; Sutton, F.; Beutel, J.; Thiele, L. Towards enabling uninterrupted long-term operation of solar energy harvesting embedded systems. In *European Conference on Wireless Sensor Networks*; Springer: Berlin/Heidelberg, Germany, 2014; pp. 66–83.

119. Noh, D.K. Transmission range determination with a timeslot-based energy distribution scheme for solar-energy harvesting sensor systems. In *Future Information Communication Technology and Applications*; Springer: Berlin/Heidelberg, Germany, 2013; pp. 661–669.

120. Kawashima, K.; Sato, F. A routing protocol based on power generation pattern of sensor node in energy harvesting wireless sensor networks. In Proceedings of the 2013 16th International Conference on Network-Based Information Systems, Gwangju, Korea, 4–6 September 2013; IEEE: Washington, DC, USA, 2013; pp. 470–475.

121. Tadayon, N.; Khoshroo, S.; Askari, E.; Wang, H.; Michel, H. Power management in SMAC-based energy-harvesting wireless sensor networks using queuing analysis. *J. Netw. Comput. Appl.* **2013**, *36*, 1008–1017. [CrossRef]

122. Imoto, N.; Yamashita, S.; Yamamoto, K.; Morikura, M. Experiment of power and data transmission scheduling for single wireless LAN sensor. In Proceedings of the 2013 International Symposium on Electromagnetic Theory, Hiroshima, Japan, 20–24 May 2013; IEEE: Washington, DC, USA, 2013; pp. 834–837.

123. El Korbi, I.; Zeadally, S. Energy-aware sensor node relocation in mobile sensor networks. *Ad Hoc Netw.* **2014**, *16*, 247–265. [CrossRef]

124. Chilamkurti, N.; Zeadally, S.; Vasilakos, A.; Sharma, V. Cross-layer support for energy efficient routing in wireless sensor networks. *J. Sens.* **2009**, *2009*, 1–9. [CrossRef]

125. Ugwuogo, J. On-Demand Energy Harvesting Techniques-a System Level Perspective. Master's Thesis, University of Waterloo, Waterloo, ON, Canada, 2012.

126. Sidhu, R.K.; Ubhi, J.S.; Aggarwal, A. A Survey Study of Different RF Energy Sources for RF Energy Harvesting. In Proceedings of the 2019 International Conference on Automation, Computational and Technology Management (ICACTM), London, UK, 24–26 April 2019; IEEE: Washington, DC, USA, 2019; pp. 530–533.

127. Baroudi, U. Robot-assisted maintenance of wireless sensor networks using wireless energy transfer. *IEEE Sens. J.* **2017**, *17*, 4661–4671. [CrossRef]

128. Salam, A. A Comparison of Path Loss Variations in Soil using Planar and Dipole Antennas. In Proceedings of the 2019 IEEE International Symposium on Antennas and Propagation, Atlanta, GA, USA, 20 July 2019; IEEE: Washington, DC, USA, 2019.

129. Brown, W.C. The History of the Development of the Rectenna. [Solar Power Satellites]. Available online: https://ntrs.nasa.gov/archive/nasa/casi.ntrs.nasa.gov/19810008041.pdf (accessed on 22 March 2020).

130. Brown, W.C. The Receiving Antenna and Microwave Power Rectification. *J. Microw. Power* **1970**, *5*, 279–292. [CrossRef]

131. Bailey, R.L. A Proposed New Concept for a Solar-Energy Converter. *J. Eng. Gas Turbines Power* **1972**, *94*, 73–77. [CrossRef]

132. Shimanuki, Y.; Adachi, S. Theoretical and experimental study on rectenna array for microwave power transmission. *Electron. Commun. Jpn. Part I Commun.* **1985**, *68*, 110–118. [CrossRef]

133. Dickinson, R.M. Rectenna Array Measurement Results. [Satellite Power Transmission And Reception]. Available online: https://ntrs.nasa.gov/archive/nasa/casi.ntrs.nasa.gov/19810008045.pdf (accessed on 22 March 2020).

134. Nimo, A.; Beckedahl, T.; Ostertag, T.; Reindl, L. Analysis of passive RF-DC power rectification and harvesting wireless RF energy for micro-watt sensors. *AIMS Energy* **2015**, *3*, 184–200. [CrossRef]

135. Salam, A. Underground Soil Sensing Using Subsurface Radio Wave Propagation. In Proceedings of the 5th Global Workshop on Proximal Soil Sensing, Columbia, MO, USA, 28–31 May 2019.

136. Song, C.; Huang, Y.; Zhou, J.; Carter, P. Recent advances in broadband rectennas for wireless power transfer and ambient RF energy harvesting. In Proceedings of the 2017 11th European Conference on Antennas and Propagation (EUCAP), Paris, France, 19–24 March 2017; IEEE: Washington, DC, USA, 2017; pp. 341–345.

137. Baroudi, U. Management of RF Energy Harvesting: A Survey. In Proceedings of the 2019 16th International Multi-Conference on Systems, Signals & Devices (SSD), Istanbul, Turkey, 21–24 March 2019; IEEE: Washington, DC, USA, 2019; pp. 44–49.

138. Kawahara, Y.; Lakafosis, V.; Sawakami, Y.; Nishimoto, H.; Asami, T. Design issues for energy harvesting enabled wireless sensing systems. In Proceedings of the 2009 Asia Pacific Microwave Conference, Singapore, 7–10 December 2009; IEEE: Washington, DC, USA, 2009; pp. 2248–2251.

139. Kawahara, Y.; Tsukada, K.; Asami, T. Feasibility and potential application of power scavenging from environmental RF signals. In Proceedings of the 2009 IEEE Antennas and Propagation Society International Symposium, Charleston, SC, USA, 1–5 June 2009; IEEE: Washington, DC, USA, 2009; pp. 1–4.

140. Vullers, R.; van Schaijk, R.; Doms, I.; Van Hoof, C.; Mertens, R. Micropower energy harvesting. *Solid-State Electron.* **2009**, *53*, 684–693. [CrossRef]

141. Mekid, S.; Qureshi, A.; Baroudi, U. Energy harvesting from ambient radio frequency: Is it worth it? *Arab. J. Sci. Eng.* **2017**, *42*, 2673–2683. [CrossRef]

142. Ghor, H.E.; Chetto, M.; Chehade, R.H. A real-time scheduling framework for embedded systems with environmental energy harvesting. *Comput. Electr. Eng.* **2011**, *37*, 498–510. [CrossRef]

143. Chan, K.Y.; Phoon, H.J.; Ooi, C.P.; Pang, W.L.; Wong, S.K. Power management of a wireless sensor node with solar energy harvesting technology. *Microelectron. Int.* **2012**, *29*, 76–82. [CrossRef]

144. Lynggaard, P. A fall-detection system that uses body area network and thermal energy harvesting technologies. In Proceedings of the 2018 11th CMI International Conference: Prospects and Challenges Towards Developing a Digital Economy within the EU, Copenhagen, Denmark, 29–30 November 2018; IEEE: Washington, DC, USA, 2018; pp. 67–73.

145. Chandrakasan, A.P.; Verma, N.; Daly, D.C. Ultralow-power electronics for biomedical applications. *Annu. Rev. Biomed. Eng.* **2008**, *10*, 247–274. [CrossRef]

146. Alhawari, M.; Mohammad, B.; Saleh, H.; Ismail, M. A survey of thermal energy harvesting techniques and interface circuitry. In Proceedings of the 2013 IEEE 20th International Conference on Electronics, Circuits, and Systems (ICECS), Abu Dhabi, UAE, 8–11 December 2013; IEEE: Washington, DC, USA 2013; pp. 381–384.

147. Kang, K. Multi-Source Energy Harvesting for Wireless Sensor Nodes. Available online: http://www.diva-portal.org/smash/get/diva2:556732/FULLTEXT01.pdf (accessed on 22 March 2020).

148. Salam, A. A Path Loss Model for Through the Soil Wireless Communications in Digital Agriculture. In Proceedings of the 2019 IEEE International Symposium on Antennas and Propagation, Atlanta, GA, USA, 20 July 2019; IEEE: Washington, DC, USA, 2019; pp. 1–2.

149. Martin, P.; Charbiwala, Z.; Srivastava, M. DoubleDip: Leveraging thermoelectric harvesting for low power monitoring of sporadic water use. In Proceedings of the 10th ACM Conference on Embedded Network Sensor Systems, Toronto, ON, Canada, 6–9 November 2012; pp. 225–238.

150. Zhang, C.; Syed, A.; Cho, Y.; Heidemann, J. Steam-powered sensing. In Proceedings of the 9th ACM Conference on Embedded Networked Sensor Systems, Seattle, WA, USA, 1–4 November 2011; pp. 204–217.

151. Vullers, R.J.; Van Schaijk, R.; Visser, H.J.; Penders, J.; Van Hoof, C. Energy harvesting for autonomous wireless sensor networks. *IEEE Solid-State Circuits Mag.* **2010**, *2*, 29–38. [CrossRef]

152. Gu, X.; Guo, L.; Harouna, M.; Hemour, S.; Wu, K. Accurate Analytical Model for Hybrid Ambient Thermal and RF Energy Harvester. In Proceedings of the 2018 IEEE/MTT-S International Microwave Symposium-IMS, Philadelphia, PA, USA, 10–15 June 2018; IEEE: Washington, DC, USA, 2018; pp. 1122–1125.

153. Weddell, A.S.; Magno, M.; Merrett, G.V.; Brunelli, D.; Al-Hashimi, B.M.; Benini, L. A survey of multi-source energy harvesting systems. In Proceedings of the 2013 Design, Automation & Test in Europe Conference & Exhibition (DATE), Grenoble, France, 18–22 March 2013; IEEE: Washington, DC, USA, 2013; pp. 905–908.

154. Virili, M.; Georgiadis, A.; Collado, A.; Niotaki, K.; Mezzanotte, P.; Roselli, L.; Alimenti, F.; Carvalho, N.B. Performance improvement of rectifiers for WPT exploiting thermal energy harvesting. *Wirel. Power Transf.* **2015**, *2*, 22–31. [CrossRef]

155. Yan, J.; Liao, X.; Ji, S.; Zhang, S. A novel multi-source micro power generator for harvesting thermal and optical energy. *IEEE Electron. Dev. Lett.* **2018**, *40*, 349–352. [CrossRef]

156. Sandeep, R.; Sai, R.; Baskaradas, J.A.; Chandramouli, R. Design of Nanoantennas for harvesting waste thermal energy from hot automobile exhaust system. In Proceedings of the 2018 IEEE Indian Conference on Antennas and Propogation (InCAP), Hyderabad, India, 16–19 December 2018; IEEE: Washington, DC, USA, 2018; pp. 1–4.

157. Dixit, N.K.; Rangra, K.J. A Survey of Energy Harvesting Technologies. In Proceedings of the 2017 International Conference on Innovations in Control, Communication and Information Systems (ICICCI), Greater Noida, India, 12–13 August 2017; IEEE: Washington, DC, USA, 2017; pp. 1–7.

158. Salam, A. Internet of Things in Agricultural Innovation and Security. In *Internet of Things for Sustainable Community Development: Wireless Communications, Sensing, and Systems*; Springer International Publishing: Cham, Switzerland, 2020; pp. 71–112. [CrossRef]

159. Koomey, J.; Berard, S.; Sanchez, M.; Wong, H. Implications of historical trends in the electrical efficiency of computing. *IEEE Ann. Hist. Comput.* **2010**, *33*, 46–54. [CrossRef]

160. Smith, J.R. *Wirelessly Powered Sensor Networks and Computational RFID*; Springer Science & Business Media: Berlin/Heidelberg, Germany, 2013.

161. Hemour, S.; Wu, K. Radio-frequency rectifier for electromagnetic energy harvesting: Development path and future outlook. *Proc. IEEE* **2014**, *102*, 1667–1691. [CrossRef]

162. Choi, K.W.; Ginting, L.; Rosyady, P.A.; Aziz, A.A.; Kim, D.I. Wireless-powered sensor networks: How to realize. *IEEE Trans. Wirel. Commun.* **2016**, *16*, 221–234. [CrossRef]

163. Rahimi, A.; Zorlu, Ö.; Muhtaroglu, A.; Külah, H. A vibration-based electromagnetic energy harvester system with highly efficient interface electronics. In Proceedings of the 2011 16th International Solid-State Sensors, Actuators and Microsystems Conference, Beijing, China, 5–9 June 2011; IEEE: Washington, DC, USA, 2011; pp. 2650–2653.

164. Lee, B.C.; Rahman, M.A.; Hyun, S.H.; Chung, G.S. Low frequency driven electromagnetic energy harvester for self-powered system. *Smart Mater. Struct.* **2012**, *21*, 125024. [CrossRef]

165. Yang, X.; Zhu, G.; Wang, S.; Zhang, R.; Lin, L.; Wu, W.; Wang, Z.L. A self-powered electrochromic device driven by a nanogenerator. *Energy Environ. Sci.* **2012**, *5*, 9462–9466. [CrossRef]

166. Podder, P.; Amann, A.; Roy, S. FR4 based bistable electromagnetic vibration energy harvester. *Procedia Eng.* **2014**, *87*, 767–770. [CrossRef]

167. Choi, K.W.; Rosyady, P.A.; Ginting, L.; Aziz, A.A.; Setiawan, D.; Kim, D.I. Theory and experiment for wireless-powered sensor networks: How to keep sensors alive. *IEEE Trans. Wirel. Commun.* **2017**, *17*, 430–444. [CrossRef]

168. Flynn, B.W.; Fotopoulou, K. Rectifying loose coils: Wireless power transfer in loosely coupled inductive links with lateral and angular misalignment. *IEEE Microw. Mag.* **2013**, *14*, 48–54. [CrossRef]

169. Bansal, R. Near-field magnetic communication. *IEEE Antennas Propag. Mag.* **2004**, *46*, 114–115. [CrossRef]

170. Kisseleff, S.; Gerstacker, W.; Schober, R.; Sun, Z.; Akyildiz, I.F. Channel capacity of magnetic induction based wireless underground sensor networks under practical constraints. In Proceedings of the 2013 IEEE Wireless Communications and Networking Conference (WCNC), Shanghai, China, 7–10 April 2013; IEEE: Washington, DC, USA, 2013; pp. 2603–2608.

171. Casanova, J.J.; Low, Z.N.; Lin, J. A loosely coupled planar wireless power system for multiple receivers. *IEEE Trans. Ind. Electron.* **2009**, *56*, 3060–3068. [CrossRef]

172. Yoon, I.J.; Ling, H. Investigation of near-field wireless power transfer under multiple transmitters. *IEEE Antennas Wirel. Propag. Lett.* **2011**, *10*, 662–665. [CrossRef]

173. Shamonina, E.; Kalinin, V.; Ringhofer, K.; Solymar, L. Magneto-inductive waveguide. *Electron. Lett.* **2002**, *38*, 371–373. [CrossRef]

174. Kisseleff, S.; Akyildiz, I.F.; Gerstacker, W.H. Throughput of the magnetic induction based wireless underground sensor networks: Key optimization techniques. *IEEE Trans. Commun.* **2014**, *62*, 4426–4439. [CrossRef]

175. Nguyen, H.; Agbinya, J.I.; Devlin, J. FPGA-based implementation of multiple modes in near field inductive communication using frequency splitting and MIMO configuration. *IEEE Trans. Circuits Syst. I Regul. Pap.* **2014**, *62*, 302–310. [CrossRef]

176. Kisseleff, S.; Akyildiz, I.F.; Gerstacker, W.H. Magnetic induction-based simultaneous wireless information and power transfer for single information and multiple power receivers. *IEEE Trans. Commun.* **2016**, *65*, 1396–1410. [CrossRef]

177. Firoozy, P.; Khadem, S.E.; Pourkiaee, S.M. Broadband energy harvesting using nonlinear vibrations of a magnetopiezoelastic cantilever beam. *Int. J. Eng. Sci.* **2017**, *111*, 113–133. [CrossRef]

178. Broderick, L.Z.; Albert, B.R.; Pearson, B.S.; Kimerling, L.C.; Michel, J. Design for energy: Modeling of spectrum, temperature and device structure dependences of solar cell energy production. *Sol. Energy Mater. Sol. Cells* **2015**, *136*, 48–63. [CrossRef]

179. Yılmaz, S.; Selim, H. A review on the methods for biomass to energy conversion systems design. *Renew. Sustain. Energy Rev.* **2013**, *25*, 420–430. [CrossRef]

180. Cheng, M.; Zhu, Y. The state of the art of wind energy conversion systems and technologies: A review. *Energy Convers. Manag.* **2014**, *88*, 332–347. [CrossRef]
181. Yildirim, T.; Ghayesh, M.H.; Li, W.; Alici, G. A review on performance enhancement techniques for ambient vibration energy harvesters. *Renew. Sustain. Energy Rev.* **2017**, *71*, 435–449. [CrossRef]
182. Daqaq, M.F.; Masana, R.; Erturk, A.; Dane Quinn, D. On the role of nonlinearities in vibratory energy harvesting: A critical review and discussion. *Appl. Mech. Rev.* **2014**, *66*, 040801. [CrossRef]
183. Guyomar, D.; Sebald, G.; Kuwano, H. Energy harvester of 1.5 cm^3 giving output power of 2.6 mW with only 1 G acceleration. *J. Intell. Mater. Syst. Struct.* **2011**, *22*, 415–420. [CrossRef]
184. Ottman, G.K.; Hofmann, H.F.; Bhatt, A.C.; Lesieutre, G.A. Adaptive piezoelectric energy harvesting circuit for wireless remote power supply. *IEEE Trans. Power Electron.* **2002**, *17*, 669–676. [CrossRef]
185. Shu, Y.; Lien, I. Analysis of power output for piezoelectric energy harvesting systems. *Smart Mater. Struct.* **2006**, *15*, 1499. [CrossRef]
186. Sodano, H.A.; Inman, D.J.; Park, G. Generation and storage of electricity from power harvesting devices. *J. Intell. Mater. Syst. Struct.* **2005**, *16*, 67–75. [CrossRef]
187. Kahrobaee, S.; Vuran, M.C. Vibration energy harvesting for wireless underground sensor networks. In Proceedings of the 2013 IEEE International Conference on Communications (ICC), Budapest, Hungary, 9–13 June 2013; IEEE: Washington, DC, USA, 2013; pp. 1543–1548.
188. Salam, A. Internet of Things for Sustainable Mining. In *Internet of Things for Sustainable Community Development: Wireless Communications, Sensing, and Systems*; Springer International Publishing: Cham, Switzerland, 2020; pp. 243–271. [CrossRef]
189. Shaikh, F.K.; Zeadally, S. Energy harvesting in wireless sensor networks: A comprehensive review. *Renew. Sustain. Energy Rev.* **2016**, *55*, 1041–1054. [CrossRef]

 *electronics*

Article

Analysis of an Operating State of the Innovative Capacitive Power Transmission System with Sliding Receiver Supplied by the Class-E Inverter

Wojciech Ludowicz *, Wojciech Pietrowski and Rafal M. Wojciechowski

Division of Mechatronics and Electrical Machines, Institute of Electrical Engineering and Electronics, Faculty of Control, Robotics and Electrical Engineering, Poznan University of Technology, 60-965 Poznan, Poland; wojciech.pietrowski@put.poznan.pl (W.P.); rafal.wojciechowski@put.poznan.pl (R.M.W.)
* Correspondence: wojciech.r.ludowicz@doctorate.put.poznan.pl

Received: 18 March 2020; Accepted: 13 May 2020; Published: 19 May 2020

Abstract: In recent years, there has been a significant increase in interest in applications of wireless power transmission (WPT). The development of electromobility and consumer electronics creates a need for newer and more effective systems for wireless charging. In spite of the unceasing progress in the research area of wireless energy transmission systems, the significant problem of the motion of the receiver has still not been solved. In this work, an innovative capacitive power transmission (CPT) system with a sliding receiver has been proposed, which enables efficient energy transmission regardless of the location of the receiver system relative to the transmitter system. The analysis of the operating states of the considered system has been conducted with the use of an in-house algorithm, in which the parameters of the equivalent circuit of the CPT system have been implemented. The parameters have been determined by the finite element method (FEM) using the electric scalar potential V. The system of capacitive power transmission was supplied from an E-class inverter. Selected results of calculations of the considered system have been presented and discussed.

Keywords: capacitive power transfer (CPT) systems; wireless power transfer (WPT) systems; e-class inverter; wireless resonance energy link system

---

1. Introduction

With the progress of the power electronics system, more and more attention is paid to the development of wireless power transmission (WPT) [1,2]. Nowadays, many types of new and effective WPT systems can be found in the open literature [3–5]. Particularly popular today are inductive power transfer (IPT) systems, the operation principle of which is based on the transmission of electricity through an alternating magnetic field and the use of the phenomenon of electromagnetic induction. Unfortunately, due to the relatively high value of the leakage flux [6], deteriorating efficiency of the system and problems related to electromagnetic compatibility (EMC), IPT systems are mostly used in low power applications like mobile phones wireless charging [7]. However, the research is underway on the applications of IPT systems for charging the batteries of high-power devices like, for example, electric vehicles [8,9]. Currently, in the research concerning WPT, the capacitive power transmission (CPT) systems are often considered in the open literature as an effective alternative solution to IPT [10,11]. The most common structure of the CPT system consists of four separated galvanic conductive plates and a circuit with appropriate parameters [12], the energy transfer is achieved through the coupling capacitances of the plates working at a resonance condition forced by the supply and receiver circuits. The indisputable advantage of CPT systems is the ability to transfer a relatively large amount of energy over considerable distances while maintaining high transmission efficiency [13]. This is possible due to the usage of dedicated compensation systems, which are made up of a set of inductors and capacitors in appropriate configuration and allow to increase the voltage between capacitor plates [14,15]. The third group of WPT systems are newly developed systems combining the features of both IPT and CPT approaches. An example proposition of such a system has been discussed by Y. Achour and J. Starzyński in [16]. This kind of system can make a significant contribution to the development of high-frequency WPT systems.

Analyzing the features of the systems discussed above, it could be noticed that all of them reach the maximum efficiency at fixed relative positions between the receiver and transmitter, while maintaining the same distance between them. Usually, the change of this relative position leads to deterioration of the transmission efficiency. Unfortunately, it is difficult to find in the open literature systems in which a change of the position of the receiver relative to the transmitter does not result in a significant decrease in transmitting power and transmission efficiency. In the group of IPT systems in which energy transmission proceeds due to capacitive coupling, i.e., CPT systems, the proposition presented in [17] is particularly interesting. Authors of the work propose an application of two types of electrodes that differ in length. This approach enables the possibility of movement of the receiving electrodes relative to the transmitting ones without a coupling capacitance change. However, it should be noted, that in this type of solution the receiving electrodes can be moved only along the length of the transmitting electrodes. In the case of a change of the arrangement of electrodes along the width, the decrease in capacitance can be observed, which directly results in a significant decrease of the efficiency of the system.

In the present paper, a novel concept of a wireless capacitive power transfer system with sliding receiver has been proposed. The system is characterized by a high value of transmission efficiency for any (even accidental) arrangement of transmitting electrodes. Therefore, the proposed system can work properly in any position of the receiving electrode relative to the transmitting electrode, i.e., for any position both along the x- and y-axis. It also allows the receiving plate (electrode) to be rotated relative to the transmitting one by any angle (Section 2). Moreover, it should also be noted, that contrary to many solutions presented in open literature, the proposed CPT system contains of only two plates (electrodes). Undoubtedly, this solution will be especially valued wherever a highly efficient power transmission is required and the proper arrangement of the transmitting electrodes of the CPT system is difficult to implement. The system discussed here has been designed primarily for charging mobile devices such as smartphones and battery powered cordless screwdrivers. The basic configuration of the elaborated CPT system with sliding receiver has been presented in Figure 1. The parameters of equivalent CPT circuits were calculated on the basis of the field model of the capacitive power transfer. The model as well as calculations were realized in ANSYS Maxwell. Results of field model calculations have been then implemented in the authored software to analyze operating states of the CPT system co-working with an E-class inverter [18]. In the paper, authors have limited themselves to discuss and present the analysis of the impact of the plate arrangement on the value of transmitted power and transmission efficiency. The results of simulation calculations for the selected operating state of the proposed CPT system with sliding receiver are given. An analysis of the content of higher harmonics was also carried out for the obtained current waveforms, i.e., supply current and receiver current. The results of simulation calculations were compared with the results of the measurements obtained for the prototype CPT system. In the paper, the influence of the rotation of electrodes relative to each other is not taken into account.

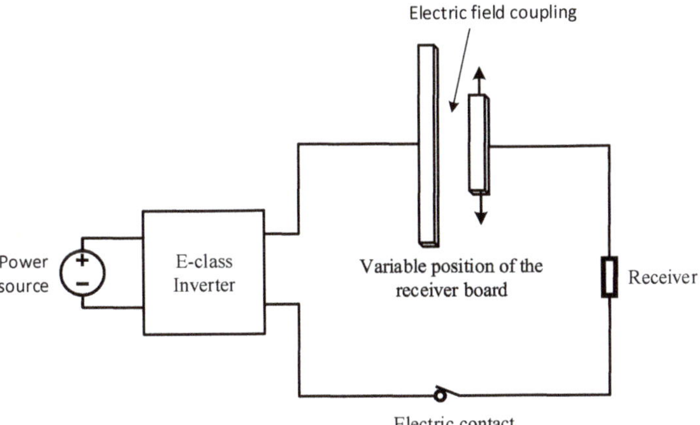

Figure 1. The configuration of elaborated capacitive power transmission (CPT) system.

2. The Field Model of the Elaborated CPT System

In order to study the proposed capacitive power transmission system, the performance of the field model has been developed in the professional finite element method (FEM) package ANSYS Maxwell. The structure of the considered system has been shown in Figure 2.

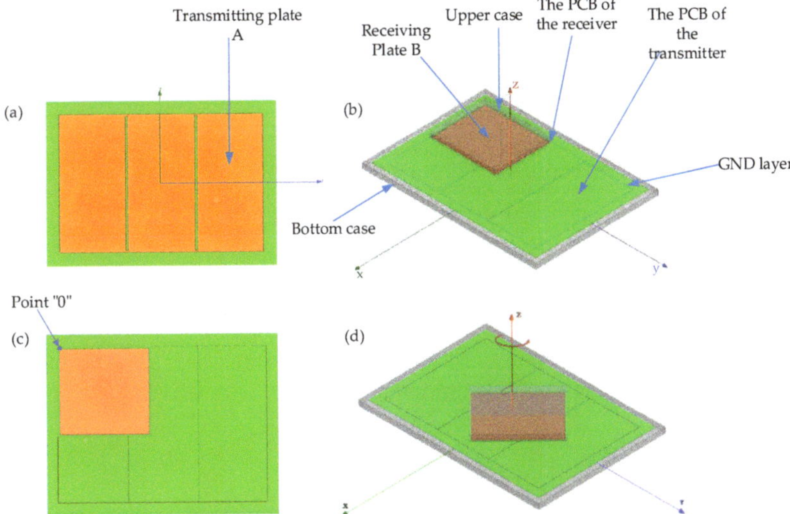

Figure 2. The structural views of the proposed CPT system, (a) the bottom view of the transmitting board; (b) the isometric view of the CPT system; (c) the top view of the CPT system; (d) with rotating receiving plate.

The CPT system consists of two parallel transmission plates (A) and (B) i.e., a transmitting (transmitter) and receiving plate (receiver), respectively, see Figure 2a,b. Both boards have been made as double-layer printed circuit board (PCB) technology. Referring to the transmission plate (A), its bottom layer made of copper, see Figure 2, is the cover of the transmission capacitor. Whereas the upper layer covered by the path made of copper (Figure 2b) is the ground of the system (GND). The receiving plate is also made as a two-layer plate, in which its bottom side with the conductive

conductive path is the GND layer, while the upper side of the plate (B) is the second of the transmission capacitor linings. The dimensions of the transmitting plate (A) are 330 mm × 230 mm while the receiving plate (B) is 130 mm× 110 mm. The bottom view of the elaborated model is shown in Figure 3.

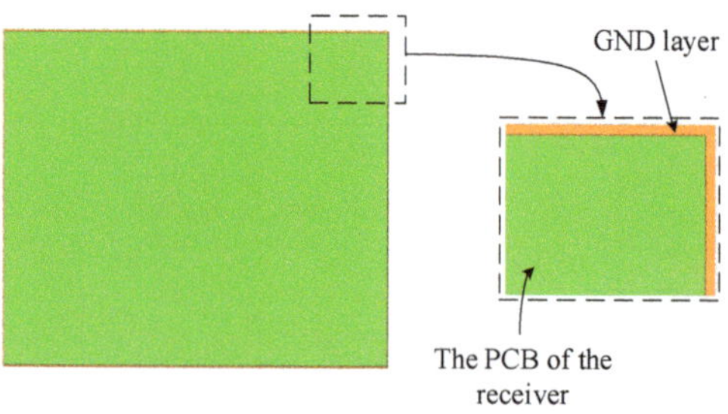

Figure 3. View of the bottom side of the receiver board.

The developed system was designed to supply receivers of arbitrary position of the receiver relative to the transmitter. In order to maximize the transmission efficiency and the resultant capacity value between the plates, it was decided to use a contact connection of layers constituting the ground of the system (GND).

In the work, as mentioned earlier, to determine the value of the resultant capacitance C_s between the plates of the considered system and the capacitances constituting the parasitic capacitances C_p, i.e., the capacitances between the linings of the transmission capacitor and the layers (linings) constituting the GND layer of the system, professional software was used in which to analyze the electrostatic field the popular FE method was implemented using the formulation of the electric potential V. The capacitance values obtained in the software as a function of the position of the receiver relative to the transmitter are given in Section 4.

3. The Circuit Model of the Elaborated CPT System

For the purposes of analysis of the operating states of the capacitive power transmission system, a circuit model was developed. Due to the resonance operation characteristics of the CPT system, an E-type inverter was chosen to supply the system. The power supply system together with the power transmission circuit and the receiver is shown in Figure 4. The power supply system consists of the U_d voltage source, C_d reactor and switching transistor T_1, D_1 return diode and output capacitor C_T. The transmission circuit the field model, which has been presented in Section 2, includes the maintained capacity C_g two parasitic capacities C_{p1} and C_{p2}, as well as the contact surface S_1. The L_r inductor together with the transmission system and the receiver form a resonant circuit

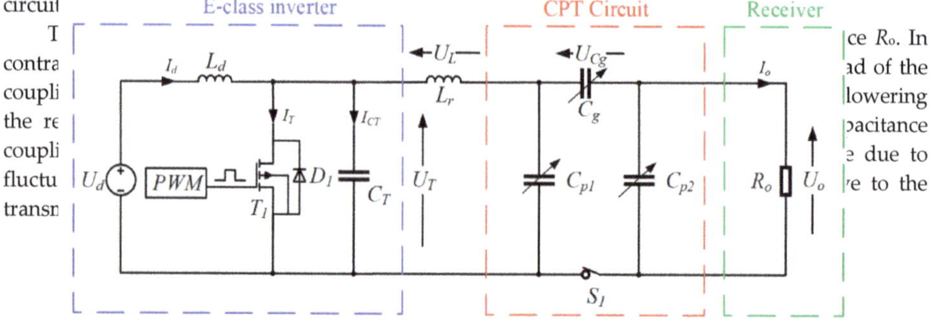

Figure 4. The structural schema of the considered CPT system.

The presence of C_{p1} and C_{p2} parasitic capacitances negatively affects the efficiency of energy transmission. Total leveling of these capacitances is unfortunately impossible. By using solutions such as limiting parallel surfaces or reducing the width of layers constituting the ground of the CPT system, it was only possible to limit the values of the considered capacitances to a minimum. The simulation that has been carried out as a part of the study takes into account the presence of parasitic

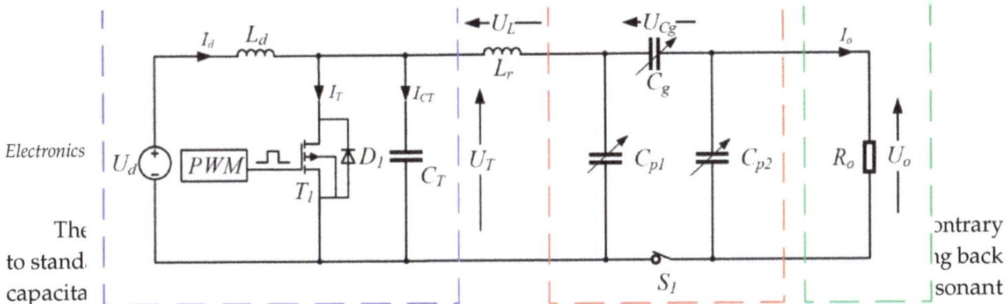

Figure 4. The structural schema of the considered CPT system.

The [...] ontrary to stand[...]ng back capacita[...]sonant frequency of the power supply system by increasing the equivalent capacitance coupling the system. All capacitances included in the transmission circuit are variable due to fluctuations caused by the change of the position of the receiver system board relative to the transmitter board.

The presence of C_{p1} and C_{p2} parasitic capacitances negatively affects the efficiency of energy transmission. Total leveling of these capacitances is unfortunately impossible. By using solutions such as limiting parallel surfaces or reducing the width of layers constituting the ground of the CPT system, it was only possible to limit the values of the considered capacitances to a minimum. The simulation that has been carried out as a part of the study takes into account the presence of parasitic capacities of the transmission system, as well as their variability depending on the position of the receiver relative to the transmitter. The operation of the system is based on the resonance phenomenon, which enables compensation of the coupling capacitance of the system. Formulas that allow to determine the value of resonance parameters are presented in Section 4.

4. Determining the Resonance Circuit's Parameters

For the purpose of calculation of individual parameters of the supply system, the circuit from Figure 4 has been brought to the simpler form and shown in Figure 5.

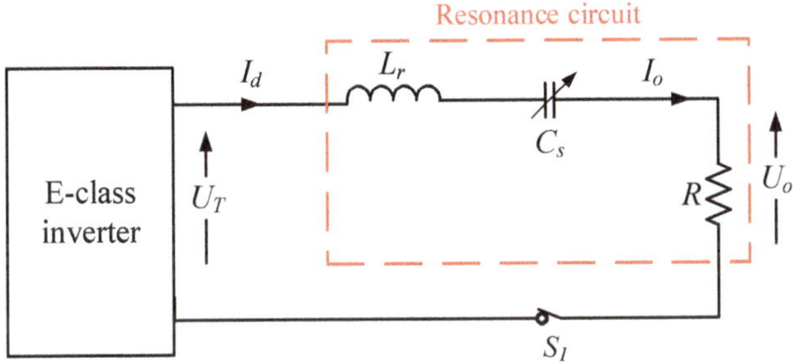

Figure 5. Simplified structural circuit of the mobile CPT system.

In order to obtain resonance state between L_r and C_s it is necessary to perform the following calculations. According to the circuit theory, the resultant coupling capacitance C_s value has been obtained by using Equation (1):

$$\underline{Z} = R + jX_{Lr} - jX_{Cs} = jX_{Lr} + \cfrac{(-jX_{Cp1}) \cdot \left(-jX_{Cg} + \cfrac{-jX_{Cp2}R_o}{R_o - jX_{Cp2}}\right)}{-jX_{Cg} + \cfrac{-jX_{Cp2}R_o}{R_o - jX_{Cp2}} - jX_{Cp1}} \tag{1}$$

in which R can be calculated as follows:

$$R = \frac{X_{Cp1}^2 X_{Cp2}^2 R_o}{R_o^2\left(X_{Cp1} + X_{Cp2} + X_{Cg}\right)^2 + \left(X_{Cp1} X_{Cp2} + X_{Cg} X_{Cp2}\right)^2} \tag{2a}$$

whereas X_{Cs} can be calculated as follows:

$$X_{Cs} = \frac{1}{\omega C_s} = \frac{X_{Cp1}\left(R_o^2\left(X_{Cp1}\left(X_{Cg} + X_{Cp2}\right) + \left(X_{Cg}^2 + X_{Cp2}^2\right)\right) + X_{Cp2}X_{Cg}\left(X_{Cp1}X_{Cp2} + X_{Cp2}X_{Cg} + 2R_o^2\right)\right)}{R_o^2\left(X_{Cp1} + X_{Cp2} + X_{Cg}\right)^2 + \left(X_{Cp1}X_{Cp2} + X_{Cg}X_{Cp2}\right)^2} \tag{2b}$$

where X_{Lr} is the reactance of the resonance inductor L_r, X_{Cs} is the reactance of the resultant capacitor C_s, X_{Cp1} and X_{Cp2} are the reactances of the parasitic capacitors C_{p1} and C_{p2}, respectively, X_{Cg} is the reactance of the coupling capacitor C_g, R is the resistance of the equivalent circuit (i.e., the resistance seen by the inverter resulting from bringing the circuit in Figure 4 into equivalent circuit as shown in Figure 5) and ω is the pulsation of the resonance circuit.

The inappropriate value of load resistance results in non-optimal conditions of the transistor switching circuit. We can distinguish three fundamental working states of the circuit: optimal, sub-optimal and non-optimal [18]. Due to fluctuations of the resultant capacitance C_s in case of movement of the receiver board, the considered circuit has been designed to operate in sub-optimal and optimal states. The value of the optimal resistance value has been calculated as follow:

$$R_o = \frac{1}{\eta \cdot \omega \cdot C_s \cdot \left(Q - \frac{\pi \cdot (\pi^2 - 4)}{16}\right)} \tag{3}$$

In Equation (3), η is the estimated value of system transmission efficiency adopted for the purpose of the design (in the design process the value η was assumed to be 0.95) and Q is the loaded quality factor at the operating frequency of circuit. In the work, it has been assumed that the value of the Q_L coefficient is equal to 10 [19–21]. According to [21] the choice of value of $Q_L \approx 10$ enables to construct the inverter, which is characterized by low value of power losses. Moreover, due to application of the condition of $Q_L > 7$, for switch-on duty ratio value $D = 0.5$; the waveform of the load current I_o is similar in shape to the sine wave that results in a low value of higher harmonic distortion [20,21].

The value of transmitted power to the receiver can be calculated by usage of previously calculated optimal values of the load resistance and following relation [22]

$$P_o = \frac{8 \cdot U_d^2}{(\pi^2 + 4) \cdot R_o} \tag{4}$$

The value of the inductance for the resonance circuit can be calculated by following formula:

$$L_r = \frac{Q \cdot R}{\omega} \tag{5}$$

Besides the parameters of the resonance and transmitting circuit, it is necessary to specify the parameters of the supplying inverter. The Equations (6)–(8) enable to determine the values of inductance for the choke L_d and the output capacitance C_T of the transistor T_1, i.e.,

$$T = \frac{1}{f} \tag{6}$$

$$L_d = \frac{U_d \cdot T}{2 \cdot \Delta I_d} \tag{7}$$

$$C_T = \frac{P_o}{\pi \cdot U_d^2 \cdot \omega} \cdot \eta \tag{8}$$

where ΔI_d is the maximum pulsation of the input current I_d, T is the switching period of the inverter and f is the switching frequency of the inverter.

In order to simplify the calculation and shorten the convergence time, additional values of the output capacitance and the series resistance of transistors, inductors and also capacitors can be neglected. In this work, due to the high complexity of the issue, the equivalent series resistance (ESR) of designed transmitting capacitor has not been considered.

In order to simplify the calculation and shorten the convergence time, additional values of the output capacitance would be too high stamped. Additional values of the neglected capacitors would due to the high complexity of the issue, the equivalent series resistance (ESR) of designed transmitting capacitor has not been considered.

5. Selected Results

In this work, to determine values of parameters as well as to conduct the analysis of discussed CPT system, the field model presented in Section 2 was used. At first, the calculations of values of coupling capacitances and parasitic capacitances were performed for different locations of receiving board relative to the transmitting board, starting from "0" (see Figure 2c). According to Equation (9) resultant parasitic capacitance C_p is equal to the sum of parasitic capacitances C_{p1} and C_{p2}.

$$C_p = C_{p1} + C_{p2} \tag{9}$$

Obtained results in the form of 3D plots of coupling capacitance C_g and resultant parasitic capacitance C_p are presented in Figures 6 and 7, respectively.

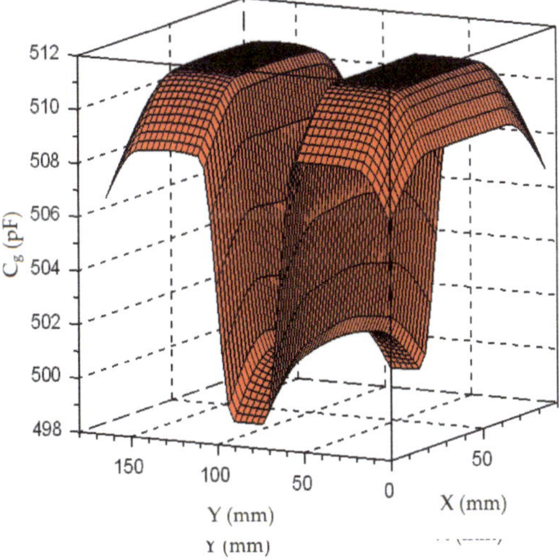

Figure 6. The 3D plot of the coupling capacitance C_g as a function of the relative position.

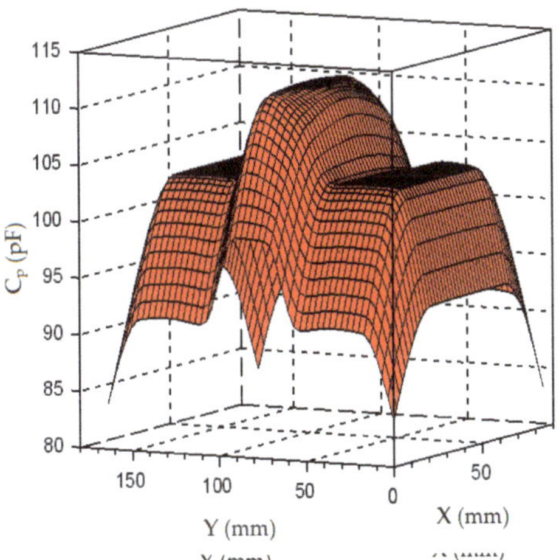

Figure 7. The 3D plot of the resultant parasitic capacitance C_p as a function of the relative position.

In Figure 6 the relation between the coupling capacitance C_g and the relative position of receiving board in relation to transmitter has been given. By analyzing Figures 6 and 7 it can be concluded that the extremum for coupling capacitance C_g is located in the middle of the transmitting board, whereas

Figure 7. The 3D plot of the resultant parasitic capacitance C_p as a function of the relative position.

In Figure 6 the relation between the coupling capacitance C_g and the relative position of receiving board in relation to transmitter has been given. By analyzing Figures 6 and 7 it can be concluded that the extremum for coupling capacitance C_g is located in the middle of the transmitting board, whereas for the resultant parasitic capacitance C_p it is on the borders of the transmitting board. The range of capacitance changes is equal to 12.4 pF (2.4%) and 27.7 pF (24.2%) for coupling capacitance C_g and resultant parasitic capacitance C_p, respectively. The equivalent capacitance C_s describing the resultant capacitance of the transmitting system has been shown in Figure 8.

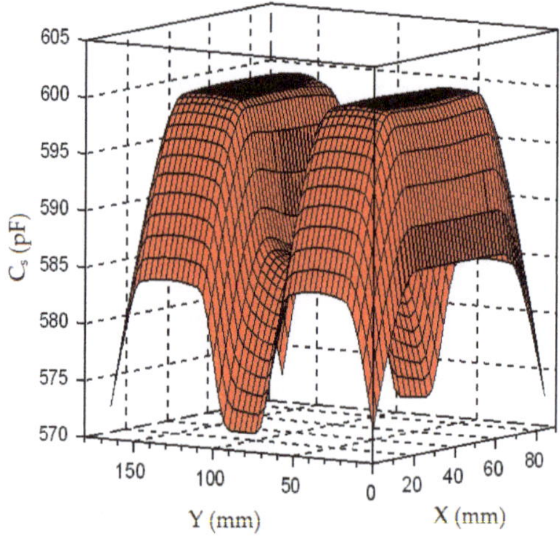

Figure 8. The 3D plot of the resultant coupling capacitance C_s as a function of the relative position.

The relationship presented in Figure 8 shows that the highest values of the resultant capacity C_s are obtained for positions where the center of the receiving system board is located at a distance corresponding to approximately 1/4 of the length of the transmitter from the outer borders in the y direction for the adopted reference system and half of the length of the plate in the x direction of the reference system.

In the next step of studies, the analysis of operating states of the CPT system co-working with an E-class inverter has been performed. The results in form of 3D plots of transmitted power and transmission efficiency have been shown in Figures 9 and 10. The calculation of the transmitted power P_T, the input electrical power P_d supplied to the CPT system and total efficiency η_t as a function of receiver position have been calculated using the formulas:

$$P_T = \frac{1}{T}\int_0^T R_0 I_0^2(t)\,dt \tag{10a}$$

$$P_d = \frac{1}{T}\int_0^T U_d I_d(t)\,dt \tag{10b}$$

$$\eta_t = \frac{P_t}{P_d} \tag{10c}$$

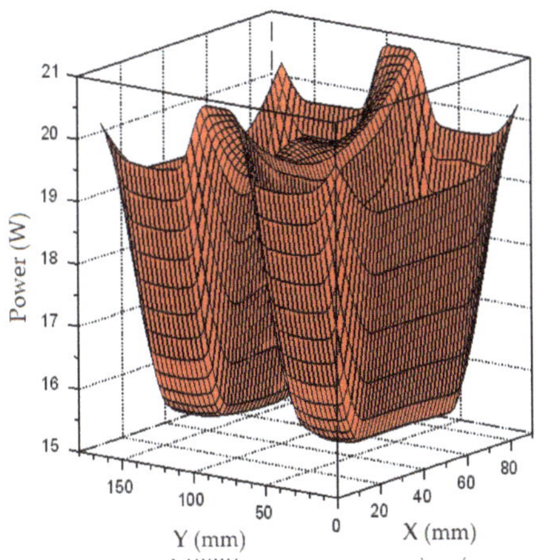

Figure 9. The 3D plot of the transmitted power as a function of the relative position of the receiver.

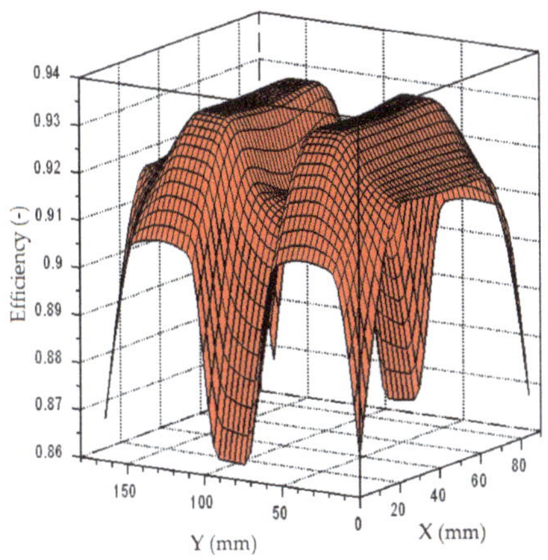

Figure 10. The 3D plot of the total efficiency as a function of the relative position of the receiver.

Changes in the values of transmitted power and transmission efficiency result directly from fluctuation in the values of individual system capacitances. It can be seen that decreases in efficiency of the CPT system (about 2%) are observed for the central location of the receiving board. However, in this position it can be observed that the system obtains the maximum power values that can be transferred between the transmitter and the receiver, obtaining values of 20.5 W. The highest values of the system efficiency, as could be expected, are obtained for the positions of the receiver located at one quarter of the distance from the outer borders of the transmitter for the adopted direction of the reference system.

The last step of the research was to conduct a simulation in order to verify the current and voltage waveforms on individual elements of the power supply circuit from Figure 4. Simulation has been performed for the optimal working point, located approximately a quarter of the length of the transmitter from the outer borders in *y* direction for the adopted reference system and half of the length of the plate in the *x* direction of the reference system. Parameters of the system for that point are as follows: $R_0 = 31.56\ \Omega$, $L_r = 50\ \mu H$, $C_T = 1\ nF$, $L_d = 1.15\ mH$, $C_g = 511.25\ pF$, $C_{p1} = 89.07\ pF$ and $C_{p2} = 14.77\ pF$. The simulation has been carried out under following conditions: $\Delta I_d = 20\ mA$, $Q = 10$, $U_d = 40\ V$ and $f = 1\ MHz$. Figures 11–13 show the current and voltage waveforms corresponded to the elements marked in Figure 4.

$U_{k} = 40$ V and of $f = 1$ MHz. Figures 11–13 show the current and voltage waveforms corresponded to the elements marked in Figure 4.

The waveforms in Figure 11 confirm the validity of performed calculations presented in Section 4. By analyzing the waveforms of current I_{d} and voltage U_{t} on the switching transistor T_{r}, it can be clearly stated that the system worked in a sub-optimal/near-optimal stage. Due to the fluctuation of circuit parameters related to the change of location of the receiving board, it is impossible to obtain optimal operating conditions for each position of the receiver. However, by maintaining the sub-optimal working conditions in the whole working area, it is possible to obtain high parameters of energy transmission regardless of the fluctuation of the value of individual capacities.

Figures 12 and 13 present the waveforms of input current I_{d} and output current (receiver current) I_{o} as well as voltage drops on the resonant elements of the system U_{l} and U_{k} respectively. First of all, attention should be paid to the noticeable presence of zero order harmonic in the output current I_{o} and in the voltage waveforms U_{l} and U_{k}. This phenomenon is characteristic of systems supplied by E class inverters, due to the necessity of separating two resonance systems of different parameters depending on the state of the transistor's operation.

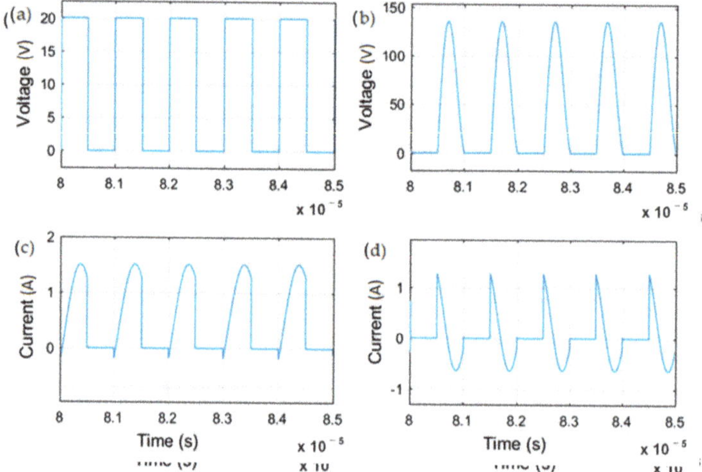

Figure 11. Waveforms of: (**a**) the pulse width modulation (PWM) signal controlling the gate of the transistor T_{r}; (**b**) the voltage drop U_{t} on transistor T_{r}; (**c**) the current I_{t} flowing through the transistor T_{r}; (**d**) the current I_{CT} flowing through the output capacitor C_{T}.

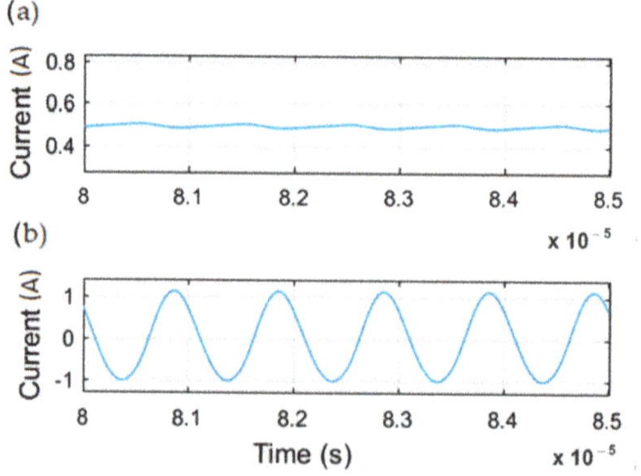

Figure 12. Waveforms of (**a**) input current I_{d} and (**b**) output current I_{o}.

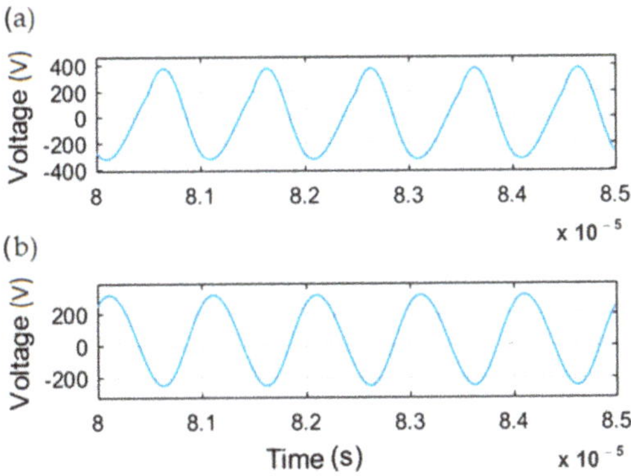

Figure 13. Waveforms of (a) voltage drop on inductor U_L and (b) voltage drop on capacitor U_C.

In order to verify simulation results, the experimental stand of elaborated CPT system was set up and appropriate measurements were conducted. The laboratory stand is shown in Figure 14. The values of parameters of individual elements that have been used in the laboratory system as well as the position of the receiving plate relative to the transmitting one are corresponded to ones used in the simulation. During tests, values of voltage and current on individual elements of the system have been measured and summarized in the form of waveforms shown in Figures 15 and 16.

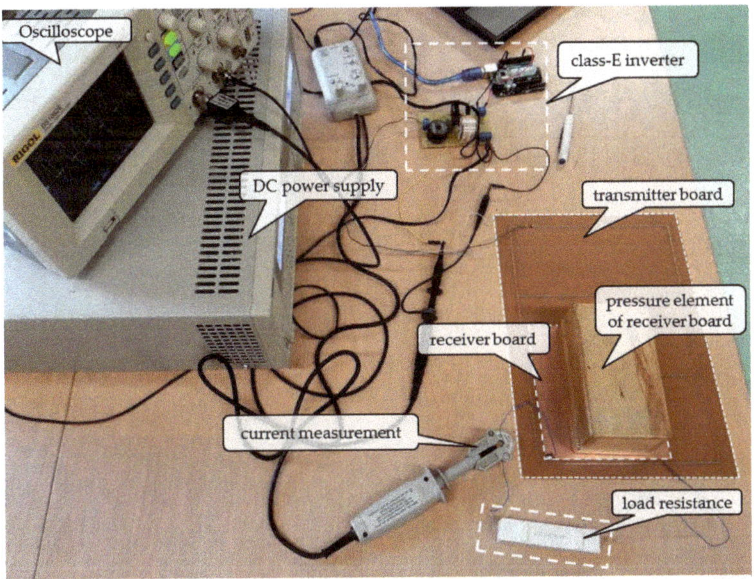

Figure 14. The experimental stand of the elaborated CPT system.

Electronics **2020**, *9*, x FOR PEER REVIEW 12 of 15

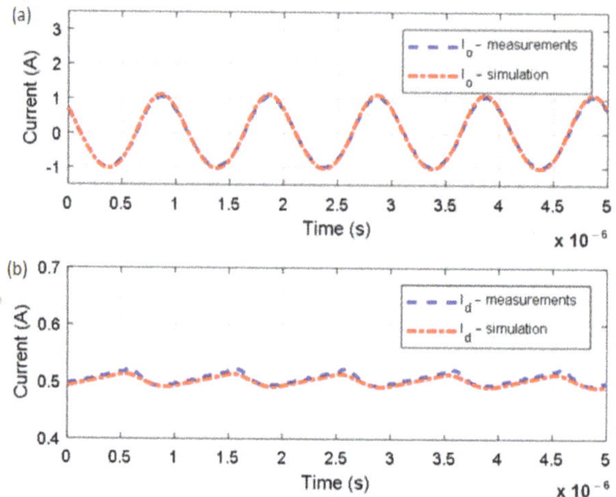

Figure 15. The comparative waveforms of simulation and experimental results: (**a**) output current I_o and (**b**) input current I_d.

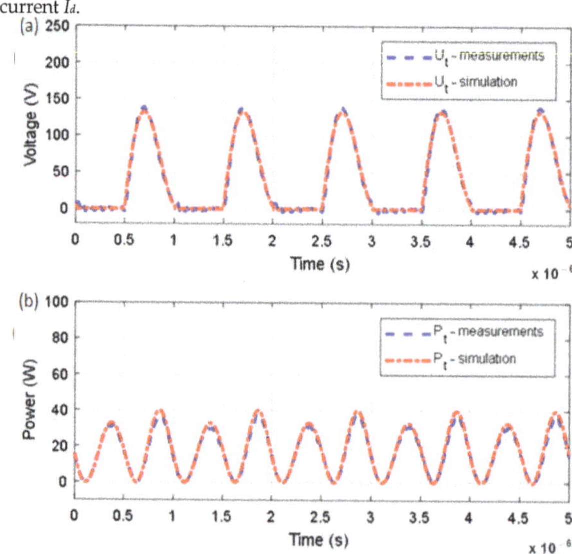

Figure 16. The comparative waveforms of simulation and experimental results: (**a**) transistor voltage drop U_T and (**b**) transmitted power P_T.

The quality of the obtained waveforms was verified by means of Fast Fourier Transform (FFT) analysis. The obtained results are presented in Figure 17. By analyzing the results presented in Figure 15, it can be noticed that in the receiver current, apart from the first harmonic, the influence of the zero and second harmonic can be also observed. The presence of these harmonics is directly related to the change in configuration of the resonant circuit resulting from the change of operation states of the transistor [22].

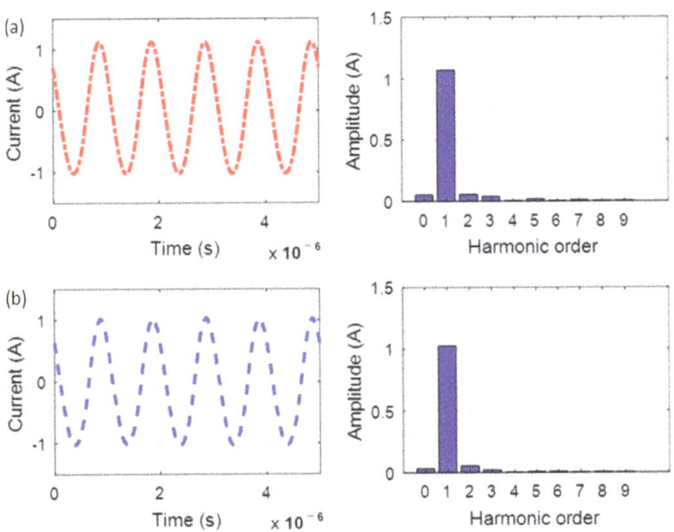

Figure 17. The FFT analyses of: (a) receiver current waveform I_o obtained from the simulation and (b) receiver current waveform I_o obtained from the measurements.

Analyzing the waveforms presented in Figures 15–17, the convergence between the results obtained on the basis of the computed simulation and the ones obtained from measurements of the system can be clearly noticed. Undoubtedly, it proves the validity of the assumptions made and the reliability of the elaborated model. However, it should be noted, that there are differences between the efficiency value obtained by simulation (ca. 92%) and measurements (ca. 84%) that result from the negligence of values of coupled capacitor series resistances ESR in simulation process. Therefore, it can be stated that the ESR resistance of the coupled capacitor contributes to a significant decrease in the transmission efficiency. During the tests some observations were made. Both pressure and equilibrium of the receiving plate (i.e., its distribution) is particularly important and have significant impact on operation of the system. Uneven distribution of the receiving plate contributes to the fluctuation of the coupled capacitance value due to formation of the irregular air gap between the transmitting and receiving plates.

6. Conclusions

The purpose of the work was to analyze working parameters of the developed simulation model of a capacitive power transmission system. The main advantage of the developed system is the ability to work efficiently with the receiver's changing position relative to the transmitter. Based on analysis of the obtained results, it can be clearly stated that the goal has been met. Based on individual capacitances as a function of a relative position of the receiver, it can be concluded that the fluctuation of the transmission parameters of the system is acceptable and that it enables stable operation of the system. In addition, referring to the transmitted power and system efficiency as a function of the position of the receiving board in relation to the transmitting board, it can be shown that the elaborated system maintains a stable efficiency close to the assumed one in the entire working area. Moreover, the paper presents and discusses the waveforms of current and voltage of individual system components. The system has been designed so that it can work in both optimal and sub-optimal working conditions of class-E inverter for the considered working area. This assumption allows for a stable operation and relatively high transmission parameters regardless of the relative position of the receiving board. The test that has been conducted on the laboratory stand of the elaborated CPT system have proven the validity of assumptions made and the reliability of the

The authors of the work intend to continue research on the presented system. Presently, work related to the construction of the discussed system is in progress. Moreover, authors consider that the possibility of the application of feedback from coupling equivalent capacitance C_s and active resonance frequency regulation would allow multiple receivers to be supplied from one transmitting board. Undoubtedly, the proposed solution would be a significant contribution to the development of consumer electronics.

Author Contributions: Conceptualization, W.L. and W.P.; methodology, W.L.; software, W.L.; formal analysis, R.M.W.; writing—original draft preparation, W.L.; writing—review and editing, R.M.W. and W.P.; supervision, R.M.W. All authors have read and agreed to the published version of the manuscript.

Funding: This research received no external funding.

Conflicts of Interest: The authors declare no conflict of interest.

References

1. Liu, Y.; Li, B.; Huang, M.; Chen, Z.; Zhang, X. An Overview of Regulation Topologies in Resonant Wireless Power Transfer Systems for Consumer Electronics or Bio-Implants. *Energies* **2018**, *11*, 1737. [CrossRef]
2. Zhang, X.; Ni, X.; Wei, B.; Wang, S.; Yang, Q. Characteristic Analysis of Electromagnetic Force in a High-Power Wireless Power Transfer System. *Energies* **2018**, *11*, 3088. [CrossRef]
3. Karagozler, M.E.; Goldstein, S.C.; Ricketts, D.S. Analysis and Modeling of Capacitive Power Transfer in Microsystems. *IEEE Trans. Circuits Syst. I* **2012**, *59*, 1557–1566. [CrossRef]
4. Fu, M.; Yin, H.; Liu, M.; Ma, C. Loading and Power Control for a High-Efficiency Class E PA-Driven Megahertz WPT System. *IEEE Trans. Ind. Electron.* **2016**, *63*, 6867–6876. [CrossRef]
5. Lu, F.; Zhang, H.; Mi, C. A Review on the Recent Development of Capacitive Wireless Power Transfer Technology. *Energies* **2017**, *10*, 1752. [CrossRef]
6. Hong, J.; Guan, M.; Lin, Z.; Fang, Q.; Wu, W.; Chen, W. Series-Series/Series Compensated Inductive Power Transmission System with Symmetrical Half-Bridge Resonant Converter: Design, Analysis, and Experimental Assessment. *Energies* **2019**, *12*, 2268. [CrossRef]
7. Li, Q.; Liang, Y.C. An Inductive Power Transfer System with a High-Q Resonant Tank for Mobile Device Charging. *IEEE Trans. Power Electron.* **2015**, *30*, 6203–6212. [CrossRef]
8. Beh, H.Z.Z.; Covic, G.A.; Boys, J.T. Investigation of Magnetic Couplers in Bicycle Kickstands for Wireless Charging of Electric Bicycles. *IEEE J. Emerg. Sel. Topics Power Electron.* **2015**, *3*, 87–100. [CrossRef]
9. Luo, B.; Shou, Y.; Lu, J.; Li, M.; Deng, X.; Zhu, G. A Three-Bridge IPT System for Different Power Levels Conversion under CC/CV Transmission Mode. *Electronics* **2019**, *8*, 884. [CrossRef]
10. Jegadeesan, R.; Guo, Y.X.; Je, M. Electric near-field coupling for wireless power transfer in biomedical applications. In Proceedings of the 2013 IEEE MTT-S International Microwave Workshop Series on RF and Wireless Technologies for Biomedical and Healthcare Applications (IMWS-BIO), Singapore, 9–11 December 2013; pp. 1–3.
11. Lu, F.; Zhang, H.; Mi, C. A Two-Plate Capacitive Wireless Power Transfer System for Electric Vehicle Charging Applications. *IEEE Trans. Power Electron.* **2018**, *33*, 964–969. [CrossRef]
12. Lee, I.-O.; Kim, J.; Lee, W. A High-Efficient Low-Cost Converter for Capacitive Wireless Power Transfer Systems. *Energies* **2017**, *10*, 1437. [CrossRef]
13. Li, S.; Liu, Z.; Zhao, H.; Zhu, L.; Shuai, C.; Chen, Z. Wireless Power Transfer by Electric Field Resonance and Its Application in Dynamic Charging. *IEEE Trans. Ind. Electron.* **2016**, *63*, 6602–6612. [CrossRef]
14. Lu, F.; Zhang, H.; Hofmann, H.; Mi, C.C. A Double-Sided LC-Compensation Circuit for Loosely Coupled Capacitive Power Transfer. *IEEE Trans. Power Electron.* **2018**, *33*, 1633–1643. [CrossRef]
15. Lu, F.; Zhang, H.; Hofmann, H.; Mi, C. A Double-Sided LCLC-Compensated Capacitive Power Transfer System for Electric Vehicle Charging. *IEEE Trans. Power Electron.* **2015**, *30*, 6011–6014. [CrossRef]
16. Achour, Y.; Starzyński, J. High-frequency displacement current transformer with just one winding. *COMPEL* **2019**. [CrossRef]
17. Lu, F.; Zhang, H.; Hofmann, H.; Mei, Y.; Mi, C. A dynamic capacitive power transfer system with reduced power pulsation. In Proceedings of the 2016 IEEE PELS Workshop on Emerging Technologies: Wireless Power Transfer (WoW), Knoxville, TN, USA, 4–6 October 2016; pp. 60–64.

18. Rybicki, K.; Wojciechowski, R.M. Inverter E-class in system of wireless power transfer design software, simulationcalculation. *Pozn. Univ. Technol. Acad. J.* **2017**, *91*, 263–275. (In Polish)

19. Kaczmarczyk, K. Improvement of power capabilities of class E inverters by maximizing transistor utilization. *Sci. J. Sil. Univ. Technol.* **2007**, *200*, 1–177. (In Polish)

20. Kazimierczuk, M.; Puczko, K. Exact analysis of class E tuned power amplifier at any Q and switch duty cycle. *IEEE Trans. Circuits Syst.* **1987**, *34*, 149–159. [CrossRef]

21. Rybicki, K. Design of the E-Class Inverter for Supplying of the Wireless Power Transmission System. Bachelor's Thesis, Poznan University of Technology, Poznan, Poland, 2017. (In Polish).

22. Warren, J.R. Cell Modulated DC/DC Converter. Master's Thesis, Massachusetts Institute of Technology, Cambridge, MA, USA, 2005.

Article

Reduction of Human Interaction with Wireless Power Transfer System Using Shielded Loop Coil

Akihiko Kumazawa [1], Yinliang Diao [2], Akimasa Hirata [2] and Hiroshi Hirayama [2,*]

[1] Tokai Rika, 3-260 Toyota, Oguchi-cho, Niwa-gun, Aichi 480-0195, Japan; hirayama_hiroshi@m.ieice.org

[2] Graduate School of Engineering, Nagoya Institute of Technology, Gokiso-cho, Showa-ku, Nagoya,
    Aichi 466-8555, Japan; diao.yinliang@nitech.ac.jp (Y.D.); ahirata@nitech.ac.jp (A.H.)

* Correspondence: hirayama@nitech.ac.jp; Tel.: +81-52-735-5448

Received: 13 May 2020; Accepted: 1 June 2020; Published: 8 June 2020

Abstract: The impedance variation of wireless power transfer (WPT) coils owing to the presence of the human body may result in mismatches, resulting in a decrease of the transmission efficiency. In addition, one of the decisive factors of the permissible transfer power in WPT systems is a compliance assessment with the guidelines/standards for human protection from electromagnetic fields. In our previous study, we reported that a shielded loop coil can potentially reduce human interaction with WPT coils. In this study, first, the rationale for this reduction is investigated with equivalent circuit models for a WPT system using a shielded loop coil operated in close proximity to the human body. We then conducted an equivalent circuit analysis considering the capacitance between the inner and outer conductors of the shielded loop coil, suggesting the stability of the impedance matching. From computational results, the mitigation capability of the shielded loop coil on impedance matching and transmission efficiency owing to the presence of the human body was verified for 6.78 MHz wireless power transfer. Additionally, the reduction of the specific absorption rate (SAR) with coils comprised of the shielded loop structure was confirmed in the presence of anatomically realistic human body models. The maximum transferable power was increased from 1.5 kW to 2.1 kW for the restrictions of the local SAR limit prescribed in the international safety guidelines/standard.

Keywords: shielded loop coil; wireless power transfer; SAR; coupled resonance

1. Introduction

Wireless power transfer (WPT) technology is being put to practical use in various places [1–3]. For example, it will be deployed for the wireless charging of electric vehicles in parking areas [4–7]. When new emerging technology may appear, there has been concern about electromagnetic compatibility issues, which are mainly classified into two categories: potential health effects of electromagnetic fields and unwanted field emissions. Unlike other devices used in wireless communications, the leaked field strength from WPT systems is non-negligible even for high transmission efficiency because of potentially high power transfer [8,9]. The above mentioned two issues should thus be considered carefully.

Unwanted field emissions are related to the assessment of the far-field strength, whereas the compliance assessment of human safety standards is more related to near-field issues. Our interest here focuses on the latter. Human safety assessment is also related to the performance of power transfer in close proximity to the human body. Specifically, the presence of the human body may result in a mismatch in the input impedance of the transmitting/receiving coils, resulting in degraded transmission efficiency [10,11]. The method of reducing this interaction may be related to human safety because compliance assessments are, in general, conducted in a worst-case exposure, and thus in a certain scenario where the impedance mismatch occurs.

In the International Commission on Non-Ionizing Radiation Protection (ICNIRP) guidelines and IEEE standard [12–14] for human protection from electromagnetic fields, two metrics are mentioned: basic restrictions and reference levels. The former involves the internal physical quantities which are related to adverse health effects with a certain reduction factor. The latter is the permissible external field strength, which is derived from basic restrictions in a conservative manner. Practically, the reference level for a near-field regime is conservative by one to two orders of magnitude. At frequencies higher than 100 kHz, the physical quantity representing the basic restriction is the specific absorption rate (SAR) averaged over 10 g of tissue. Note that the internal electric field strength should also be evaluated up to 10 MHz, although the former would be dominant for continuous (non-pulse) exposure in the MHz frequency range. Thus, the evaluation of the local spatial-averaged SAR and internal electric field is crucial in a WPT system design.

Many studies have investigated electromagnetic safety issues owing to the exposure to magnetic fields generated by WPT systems [15–22]. Chen et al. [15] evaluated the induced electric field in a human body model exposed to a prototype WPT system of 5 W which received power from 0.1 MHz to 10 MHz. Shimamoto et al. [19] investigated the discrepancies in the electric field in a human body with different postures. The differences in the SAR between adult and child models were studied for a WPT system at 6.78 MHz [20]. In [21,22], the coupling factor, which relates the peak in situ electric field with an applied nonuniform magnetic field, was calculated. Zhan et al. [23], Campi et al. [24], Park et al. [25] and Kim et al. [26] proposed methods such as compensation topology and a passive shield to reduce the leakage magnetic field of WPT systems, while Nadakuduti et al. [27] and Ishihara et al. [28] investigated the relevant measurement methodologies. The above mentioned studies mainly focused on the safety aspects of WPT systems but did not consider the effects of the presence of the human body on the WPT performance based on a worse-case scenario.

A shielded loop antenna has two significant functions. One is a self-balancing effect to decrease sensibility to an electric field of its feeding line [29]. To take advantage of this, the shielded loop antenna has been widely used for broadcasting reception to eliminate electric field noise [30,31]. The shielded loop antenna is also used as a sensor of magnetic fields [32–34] to ensure measurement accuracy [35]. The other is a mitigation effect from surrounding metallic objects [36]. The shielded loop antenna was first invented for direction finding systems to overcome degradation of directivity pattern due to the proximity of an external object (such as a metallic wall) [37]. To take advantage of this, the shielded loop antenna is also widely used for magnetic resonance imaging (MRI) to prevent the detune of resonant frequency due to a presence of human body [37–42]. Considering this evidence, it is expected that the use of a shielded loop coil in a WPT system has a capability of mitigating the detune effect due to presence of a human body.

Previously, we empirically presented a near-field WPT using a shielded loop coil [42]. This WPT had potential features of high transmission efficiency and reduction of external electric fields, which may be expected with the presence of the human body. However, the rationale for these features is unclear, especially for a realistic exposure scenario where the human body exists. In this study, we first develop an equivalent circuit model of a 6.78 MHz WPT system composed of shielded loop coils to demonstrate its capability. We then demonstrate that the impedance characteristics and the SAR reduction of a shielded loop coil are improved as compared to a conventional loop coil without a shield. The permissible field strength is evaluated based on computations using an anatomically based human model.

2. Simulation Model

To evaluate the WPT coil performance in the presence of a human body model, the commercial software FEKO (Altair, hyperworks, Michigan, MI, USA) was used. The structure of a shielded loop coil is shown in Figure 1a,b. The shielded loop coil is well-known and commonly used as a magnetic field sensor because of its ability to suppress the sensibility of electric fields [29]. A gap was made to suppress only the electric field without suppressing the magnetic field, as is common in a shielded loop coil [43]. Because of the gap, there was no current flow in the outer conductor, and thus current flows only in the inner conductor. The outer conductor acted as a shield for the electric field.

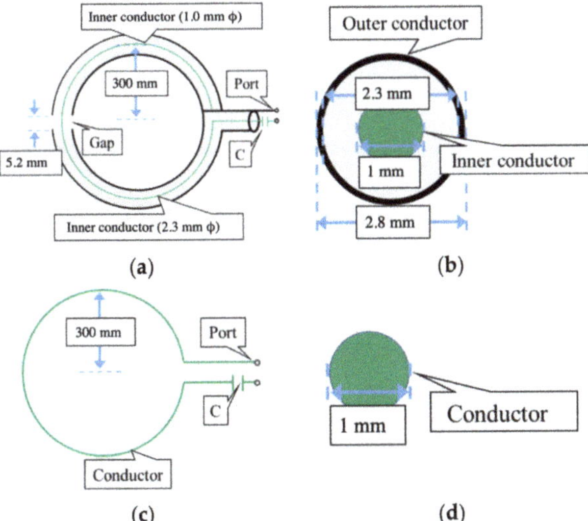

Figure 1. Schematic explanation of a shielded loop coil. (**a**) Overall view of a shielded loop coil, (**b**) sectional view of a shielded loop coil, (**c**) overall view of a one-turn loop coil, and (**d**) sectional view of a one-turn loop coil.

To verify the effectiveness of the WPT system composed of a shielded loop coil, a one-turn loop coil was also considered. The size of the one-turn loop coil is identical to that of the shielded loop coil, as shown in Figure 1c. The outer diameter of the conductor in the one-turn loop coil is identical to that of the shielded loop coil, as shown in Figure 1d.

The simple one-turn shielded loop structure was employed in this paper in order to evaluate basic characteristics of using a shielded loop structure for WPT. In a practical use, a multi-turn shielded loop [37] is effective to improve inductance and the coupling coefficient.

In this paper, the shielded loop structure is assumed to be made of shielded cable. In actual use, a shielded loop structure can be realized by using a printed circuit board (PCB) [44].

To discuss the electric field suppressing effect of the shielded loop coil, a one-turn loop coil with a parallel capacitance was also considered. This parallel capacitance simulates the capacitance between the inner and the outer conductors of the shielded loop coil.

For each model, capacitors were connected in series to each coil to resonate at 6.78 MHz. The same structure was used on the transmitting (Tx) and receiving (Rx) sides. By using a method-of-moment (MoM) simulation, the input impedances of these models were calculated. In a numerical simulation, the copper was assumed as a conductor, and then its conductivity of 5.81×10^6 S/m was assigned. An AC power supply with an output impedance of 50 Ω was connected to the port of the Tx side. A resistance of 50 Ω was loaded to the port of the Rx side. Delta-gap feeding was applied at the port in the MoM simulation.

An equivalent circuit was employed to understand the impedance matching mechanism of the shielded loop coil. Equivalent circuits of the coils are shown in Figure 2. For the one-turn loop coil shown in Figure 2a, the coil was modeled by an inductor whose inductance was derived by the MoM simulation at a low frequency (100 kHz). A series capacitance $C_{S1} = 223.3$ pF was determined so that the imaginary part of the input impedance becomes zero at 6.78 MHz. For the shielded loop coil shown in Figure 2b, the inductance of the equivalent circuit was calculated from the MoM simulation at a low frequency (100 kHz); this was identical to the value of the one-turn loop coil because the current flows only in the inner conductor of the shielded loop, which is the same structure as that of the one-turn loop coil. A series capacitance C_{s2} = 157.2 pF represents the stray capacitance between the inner and outer conductors, and was calculated by the self-resonant frequency through the MoM simulation.

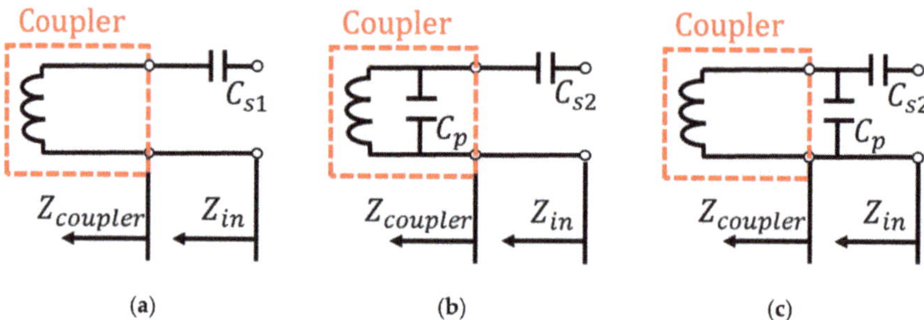

Figure 2. Equivalent circuits of different coils. (**a**) One-turn loop coil, (**b**) shielded loop coil, and (**c**) one-turn loop with capacitor.

For the one-turn loop coil with a shunt capacitor shown in Figure 2c, the equivalent circuit is identical to the shielded loop. It is noteworthy that the shunt capacitor of the one-turn loop was realized by a discrete capacitor, while the shunt capacitor of the shielded loop was realized by a stray capacitance. The aim of the equivalent circuit analysis is to understand the impedance matching mechanism of the shielded loop coil. Thus, only the imaginary part of the impedance was considered.

C_{S1} and C_{S2} are optimized without the human body. These values are maintained when the human body is present. Since C_p represents the stray capacitance of the shield structure itself, this value is not varied by the human body. Input impedance Z_{in} for these models, calculated by MoM and the equivalent circuit, is shown in Figure 3. The estimation error of the equivalent circuit was calculated by using the root mean square percentage error (RMSPE) [45]:

$$\text{RMSPE} = \sqrt{\frac{1}{N} \sum \left(\frac{Im\{Z_{in}^{MoM}\} - Im\{Z_{in}^{Eq}\}}{Im\{Z_{in}^{MoM}\}} \right)^2} \times 100 \tag{1}$$

where Z_{in}^{MoM} and Z_{in}^{Eq} are the input impedances calculated by MoM and the equivalent circuit, respectively. N is the number of calculation frequencies. Consequently, the RMSPE of the one-turn loop, one-turn loop with C_p, and shielded loop was 1.61%, 3.97%, and 4.75%, respectively. Considering that the estimation error becomes large when the denominator of Equation (1) (i.e., $Im\{Z_{in}^{MoM}\}$) is zero, the input impedance calculated by the equivalent circuit is in good accordance with that calculated by MoM.

A schematic explanation of the power transfer scenario is shown in Figure 4. To calculate the input impedance and transmission efficiency, a homogeneous cylinder model composed of 2/3 muscle tissue was used as a canonical human body [46], as shown in Figure 4a.

To calculate the internal electric field and SAR, Japanese male model "TARO" developed by National Institute of Information and Communications Technology (Japan), shown in Figure 4b, was also considered [47]. The height and weight were 1.73 m and 65 kg, respectively; and the number of tissues is 51. The spatial resolution of the body model was 2 mm. The separation between the coils and human body was set to be 2 cm. This separation distance assuming the worst-case where the receiving coil was set on a laptop computer and the transmitting coil was set on a desk. The tissues conductivities were adopted from [48], and those of selected tissues at 6.78 MHz are listed in Table 1.

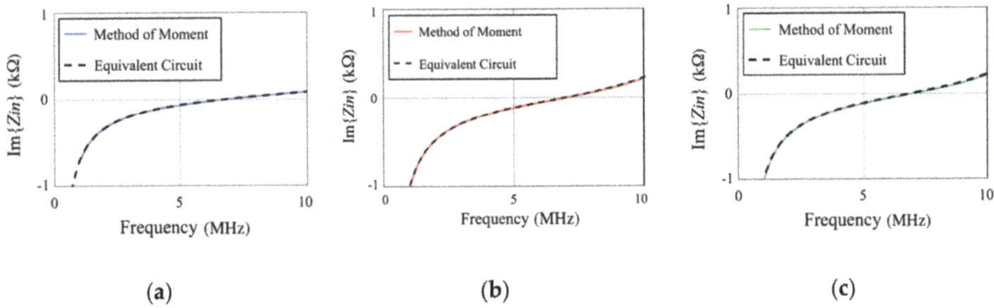

Figure 3. Input impedances calculated by Method-of-Moment (MoM) and equivalent circuit. (a) One-turn loop coil, (b) shielded loop coil, and (c) one-turn loop with capacitor.

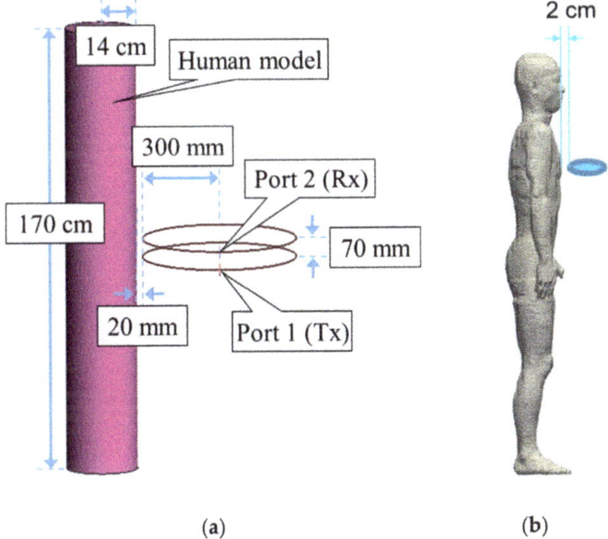

Figure 4. Simulation model of a shielded loop coil in the presence of a cylindrical model imitating a human body. (a) Simple cylindrical model and (b) realistic human body model.

Table 1. Electrical Conductivities of selected tissues of TARO model at 6.78 MHz.

Tissue	Conductivity (S/m)	Tissue	Conductivity (S/m)
Skin	0.1471	Heart	0.4713
Muscle	0.6021	Liver	0.2936
Fat	0.0278	Lung	0.3157
Bone(cortical)	0.0393	Kidney	0.4663
Bone(cancellous)	0.1159	Stomach	0.7575
Cartilage	0.3501	Cerebellum	0.3147
Nerve	0.2064	Tendon	0.4021
Grey matter	0.2520	Gall bladder	0.9013
White matter	0.1415	CSF	0.1415

At frequencies up to 10 MHz, Maxwell's equations can be simplified with a quasistatic approximation [48]. As a calculation method, an in-house multigrid-based scalar potential finite difference method was used [49,50] to solve the following partial differential equation:

$$\nabla \cdot [\sigma(-\nabla\varphi - \mathrm{j}\omega A_0)] = 0 \tag{2}$$

where boundary condition $\mathbf{n}\cdot(\nabla\varphi + j\omega A_0) = 0$. A_0 and σ denote the magnetic vector potential of the applied magnetic field and the tissue conductivity, respectively. The vector potential A_0 was obtained directly from the magnetic field, which was calculated by the method of moment. In this study, the scalar potential was computed iteratively via the successive-over-relaxation and multigrid methods [50]. The iteration stopped when the relative residual was less than 10^{-6}. When Equation (1) was solved, the in situ electric field (E) was calculated as $E = -\nabla\varphi - j\omega A_0$. The in situ electric field and SAR averaged over 10 g of tissue, which are metrics for the compliance assessment prescribed in the exposure guidelines/standards, and were analyzed.

3. Simulation Result

3.1. Mitigation Effect of a Shielded Loop Coil on Impedance Matching Owing to the Existence of a Human Body

By using the scenario shown in Figure 4a, the input impedances of the coils were calculated. The calculated results are shown in Figure 5. From this figure, better impedance matching was achieved in the shield structure than that of the conventional one-turn loop coil. In addition, the impedance of the one-turn loop coil with C_p presents almost identical behavior to that of the shielded loop coil, suggesting that impedance matching of the shielded loop coil was achieved by the stray capacitance between the inner and outer conductors.

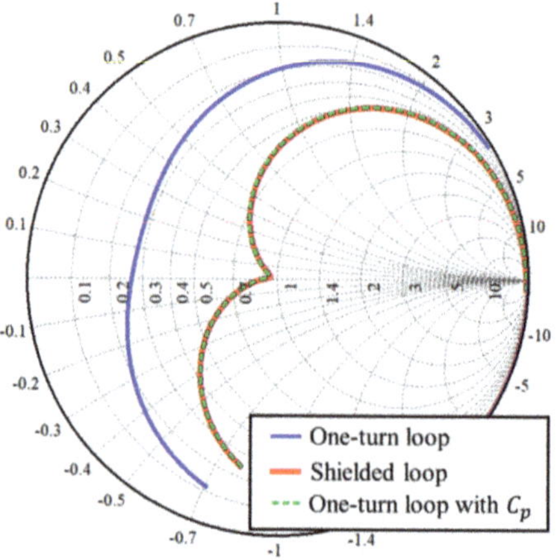

Figure 5. Smith chart of input impedance of loop coils.

To precisely discuss the effect of the human body, the reflection coefficients of the loop coil with C_p and the shielded loop coil with and without the simple cylindrical model are listed in Table 2. The variation of the reflection coefficient owing to the existence of the human body for the shielded loop coil was 0.14 dB, whereas that of a loop with C_p was 0.16 dB. From this result, the shielded loop coil can mitigate the impedance mismatch effect because the shielded loop coil can suppress the electric field.

Table 2. Reflection coefficients with and without simple cylindrical model.

Model	Shielded	Loop with C_p
Without human body model	−30.69 dB	−29.34 dB
Human body model	−30.83 dB	−29.50 dB
Variation owing to the human body	0.14 dB	0.16 dB

3.2. Transmission Efficiency

To evaluate the transmission efficiency, S parameters were calculated by using method-of-moment simulation. The scenario shown in Figure 4a was assumed, in which the transmitting and receiving coils had port 1 and 2, respectively. The frequency characteristics of the S parameters of these coils are shown in Figure 6 when the cylindrical model was present. Assuming a source and load impedance of 50 Ω, S_{21} represents the transmission efficiency. The transmission efficiency of the shielded loop coil was improved compared to the one-turn loop coil.

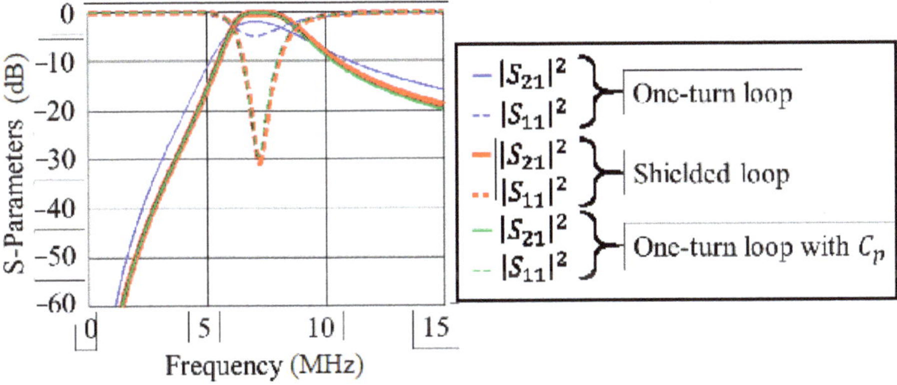

Figure 6. S parameters for frequencies of different coils.

The transmission efficiencies of the shielded loop coil and the one-turn loop with C_p were almost identical. Therefore, the transmission efficiency can be confirmed to be improved by the capacitance between the inner and outer conductors of the shield structure.

The transmission efficiencies of these coils calculated from the S parameters are listed in Table 3. The transmission efficiency of the shielded loop coil was improved by 49.5% compared to that of the one-turn loop coil. This is because of the impedance matching effect of the shielded loop coil. The transmission efficiency of the shielded loop coil was improved by 0.004 points compared to that of the one-turn loop coil with C_p. This is because of the mitigation effect on the impedance matching of the shielded loop coil.

Table 3. Transmission efficiency and transmitting power (P_T) for 1 W reception at a resonant frequency of 6.78 MHz.

Type of Coil	$|S_{21}|^2$	P_T (W)
One-turn loop coil	0.648	4.80
Shielded loop coil	0.969	2.12
One-turn loop with C_p	0.965	2.14

Transmitting power (P_T) was supplied from the power source to the Tx coil so that power of 1 W was received at a 50 Ω load at the terminal of the Rx coil. P_T is listed in Table 3. P_T of the shielded loop coil (2.12 W) was reduced by 55.8% compared to the one-turn loop coil (4.80 W). P_T of the shielded loop coil was reduced by 0.93% compared to the one-turn loop with C_p (2.14 W).

3.3. Reduction Effect of SAR

By using a realistic human body model, the induced electric field distribution and local SAR distribution were calculated at a 1 W reception power for all three types of coils at a resonant frequency of 6.78 MHz. The efficiencies and transmitting powers are listed in Table 3.

Figure 7 shows the induced electric field distributions. In all cases, hotspots appear around the outmost layers of the chest, which was the body part closest to the transmission coil. High electric fields could also be observed in the armpit region; this is attributable to the skin-to-skin contact [51]. When the electric fields passed through the interface of tissue layers with high-low-high conductivity contrast, a high electric field could be observed in the less conductive tissue layer. At 6.78 MHz, the conductivity of dry skin was about five times that of the hypodermis, resulting in a high in situ electric field strength in the hypodermis of the armpit region.

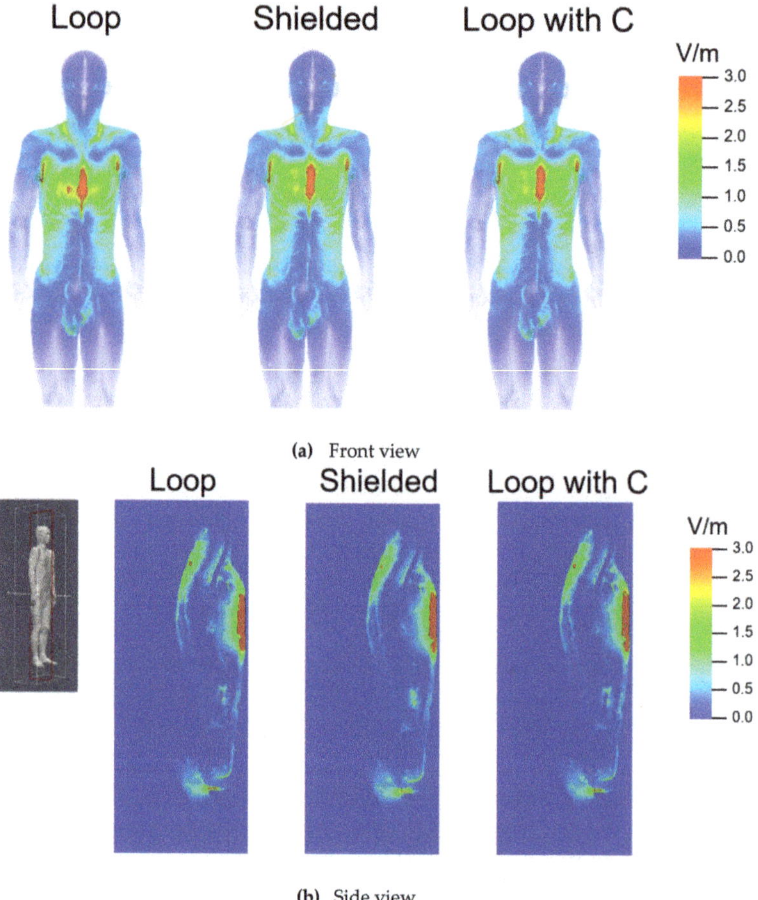

Figure 7. Internal electric field distribution in human body.

The maximum electric field strengths are 13.78 V/m, 11.88 V/m, and 11.85 V/m for the one-turn loop, shielded loop, and loop with capacitor, respectively. The maximum electric field strengths were reduced by 14.0% and 13.8% for the shielded loop coil and the loop coil with a capacitor compared to the loop coil. Figure 8 shows the SAR distributions in the human body model. The peak 10 g averaged SARs were 1.3 mW/kg, 0.94 mW/kg, and 0.93 mW/kg for the one-turn loop, shielded loop, and loop with capacitor, respectively. The SARs for the shielded loop coil and loop with capacitor were reduced by 27.9% and 27.8%, respectively, compared to that of the loop coil.

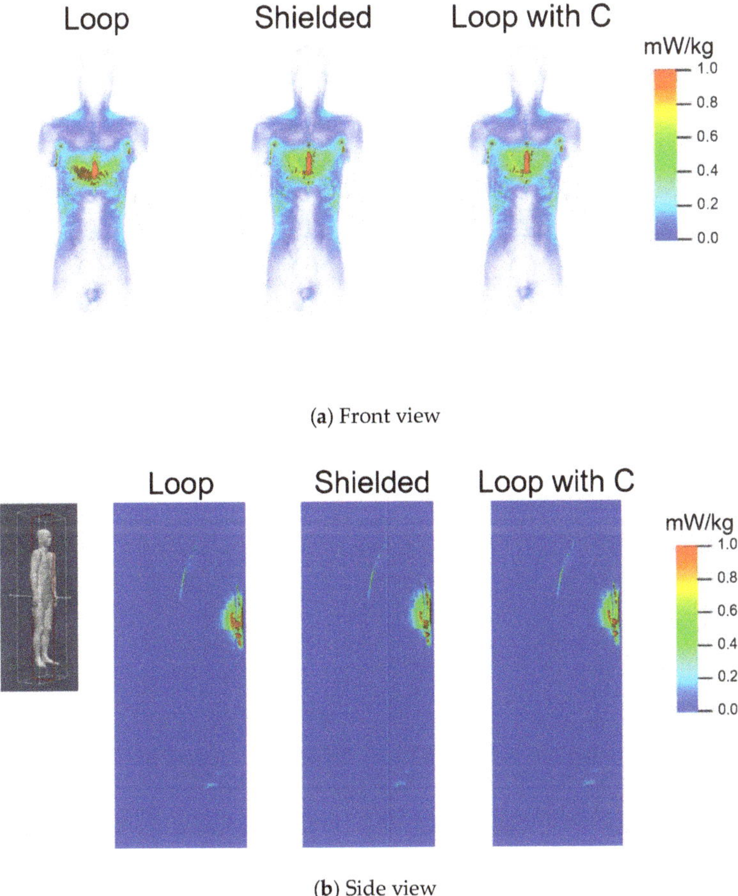

(**a**) Front view

(**b**) Side view

Figure 8. Local SAR (Specific Absorption Rate) distribution in the human body.

4. Discussion

In this study, we proposed using a shielded loop structure as a coil in a WPT system. Our motivation was based on the fact that this structure is known to suppress the electric field distribution, resulting in interaction with the human body model. We then clarify this with an equivalent circuit model explaining why this structure may suppress the human–coil interaction. For comparison, we presented a one-turn loop coil model with a capacitance, demonstrating that the structure is comparable to that of the shielded loop coil. We also computationally demonstrated that the proposed structure can mitigate the interaction of the human body.

Although the equivalent circuits of the shielded loop coil and the one-turn loop coil with capacitance are identical, their electric field distributions are not identical. The shielded loop structure can mitigate input impedance fluctuations. Thus, the difference between the shielded loop structure and one-turn loop structure with the capacitance becomes much larger. In addition, a one-turn loop structure was considered in this study for practical application as well as for simplicity. If a helix structure is considered, the contribution of the electric field becomes large [52], resulting in greater interaction with the human body model. Again, the shielded loop structure is more stable than a conventional loop coil.

For human safety compliance, there are two metrics to be evaluated at 6.78 MHz: the internal electric field for preventing electrostimulation and the SAR for heating. For compliance analysis in terms of the in situ electric field, the metric is the 99th-percentile value of a 2 mm cubically averaged rms electric field strength in a specific tissue. Recent studies [53–56] revealed that for nonuniform exposure, the 99th-percentile value may underestimate the in situ field, and a value >99.9 was suggested based on statistical analysis. In this study, the 99.9th-percentile value of the electric field in skin is considered as the quantity to be compared with the basic restriction for a slightly more conservative estimation. The result for the shielded loop coil is found to be 2.97 V/m.

According to ICNIRP [56], the in situ electric field at 6.78 MHz should not exceed 915.3 V/m, and the 10 g averaged SAR should be below 2 W/kg for the general public. Based on the computed values for the shielded loop coil, we found that the maximum transferable powers are 94.98 kW and 2.1 kW. These calculations applied basic restrictions of the in situ electric field and SAR, respectively, suggesting that the heating effect is dominant for the scenario considered here. This is consistent with the findings of previous studies [15,20,57]. For the one-turn loop coil, the maximum transferable power is 1.5 kW when applying the SAR limit.

In our simulations, the separation between the torso and the border of the coil is set to be 2 cm, which is much smaller compared to the generally used measuring distance (mostly 20 or 30 cm) specified in IEC (International Electrotechnical Commission) 62,233 [58] for determining the electromagnetic field around household appliances. For a WPT, such standardization is currently undergoing, and it has not been well defined. Nonetheless, the calculated values can provide a rough (more conservative) estimate of the exposure doses for a WPT device with similar configurations.

5. Conclusions

In this study, the mitigation capability of a shielded loop coil on impedance matching and transmission efficiency owing to the presence of the human body for a 6.78 MHz wireless power transfer was verified based on the suppression of the electric field. Additionally, an induced electric field and SAR reduction effect of the shielded loop coil were confirmed.

Author Contributions: Funding acquisition, H.H.; Investigation, A.K. and Y.D.; Methodology, H.H.; Project administration, H.H.; Supervision, A.H. and H.H.; Visualization, A.K. and Y.D.; Writing—original draft, A.K. and Y.D.; Writing—review and editing, A.H. and H.H. All authors have read and agreed to the published version of the manuscript.

Funding: This research was supported by JSPS KAKENHI Grant Number 18K04137.

Conflicts of Interest: The authors declare no conflict of interest.

References

1. Shinohara, N. Power without wires. *IEEE Microw. Mag.* **2011**, *12*, 64. [CrossRef]
2. Kim, H.-J.; Hirayama, H.; Kim, S.; Han, K.J.; Zhang, R.; Choi, J.-W. Review of Near-Field Wireless Power and Communication for Biomedical Applications. *IEEE Access* **2017**, *5*, 21264–21285. [CrossRef]
3. Arai, T.; Hirayama, H. Folded Spiral Resonator with Double-Layered Structure for Near-Field Wireless Power Transfer. *Energies* **2020**, *13*, 1581. [CrossRef]
4. Laakso, I.; Hirata, A. Evaluation of the induced electric field and compliance procedure for a wireless power transfer system in an electrical vehicle. *Phys. Med. Boil.* **2013**, *58*, 7583–7593. [CrossRef] [PubMed]
5. Kim, S.; Park, H.H.; Kim, J.; Kim, J.; Ahn, S. Design and Analysis of a Resonant Reactive Shield for a Wireless Power Electric Vehicle. *IEEE Trans. Microw. Theory Tech.* **2014**, *62*, 1057–1066. [CrossRef]
6. Buja, G.; Bertoluzzo, M.; Mude, K.N. Design and Experimentation of WPT Charger for Electric City Car. *IEEE Trans. Ind. Electron.* **2015**, *62*, 7436–7447. [CrossRef]
7. Campi, T.; Cruciani, S.; De Santis, V.; Maradei, F.; Feliziani, M. EMC and EMF safety issues in wireless charging system for an electric vehicle (EV). In Proceedings of the 2017 International Conference of Electrical and Electronic Technologies for Automotive, Torino, Italy, 15–16 June 2017; pp. 1–4.

8. Kong, S.; Bae, B.; Jung, D.H.; Kim, J.J.; Kim, S.; Song, C.; Kim, J.J.; Kim, J. An Investigation of Electromagnetic Radiated Emission and Interference from Multi-Coil Wireless Power Transfer Systems Using Resonant Magnetic Field Coupling. *IEEE Trans. Microw. Theory Tech.* **2015**, *63*, 833–846. [CrossRef]

9. Suzuki, M.; Ogawa, K.; Moritsuka, F.; Shijo, T.; Ishihara, H.; Kanekiyo, Y.; Ogura, K.; Obayashi, S.; Ishida, M. Design method for low radiated emission of 85 kHz band 44 kW rapid charger for electric bus. In Proceedings of the 2017 IEEE Applied Power Electronics Conference and Exposition (APEC), Tampa, FL, USA, 26–30 March 2017; pp. 3695–3701.

10. Hirayama, H.; Ozawa, T.; Hiraiwa, Y.; Kikuma, N.; Sakakibara, K. A consideration of electro-magnetic-resonant coupling mode in wireless power transmission. *IEICE Electron. Express* **2009**, *6*, 1421–1425. [CrossRef]

11. Laakso, I.; Tsuchida, S.; Hirata, A.; Kamimura, Y. Evaluation of SAR in a human body model due to wireless power transmission in the 10 MHz band. *Phys. Med. Boil.* **2012**, *57*, 4991–5002. [CrossRef]

12. International Commission on Non-Ionizing Radiation Protection. (ICNIRP)1,2 Guidelines for Limiting Exposure to Electromagnetic Fields (100 kHz to 300 GHz). *Health Phys.* **2020**, *118*, 483–524. [CrossRef]

13. Bailey, W.H.; Harrington, T.; Hirata, A.; Kavet, R.R.; Keshvari, J.; Klauenberg, B.J.; Legros, A.; Maxson, D.P.; Osepchuk, J.M.; Reilly, J.P.; et al. Synopsis of IEEE Std C95.1™-2019 "IEEE Standard for Safety Levels with Respect to Human Exposure to Electric, Magnetic, and Electromagnetic Fields, 0 Hz to 300 GHz". *IEEE Access* **2019**, *7*, 171346–171356. [CrossRef]

14. IEEE Std. *C95.1-2019, IEEE Standard for Safety Levels with Respect to Human Exposure to Radio Frequency Electromagnetic Fields, 0 Hz to 300 GHz*; IEEE: New York, NY, USA, 2019; pp. 1–312.

15. Chen, X.L.; Umenei, A.E.; Baarman, D.W.; Chavannes, N.; De Santis, V.; Mosig, J.R.; Kuster, N. Human Exposure to Close-Range Resonant Wireless Power Transfer Systems as a Function of Design Parameters. *IEEE Trans. Electromagn. Compat.* **2014**, *56*, 1027–1034. [CrossRef]

16. Christ, A.; Douglas, M.; Nadakuduti, J.; Kuster, N. Assessing human exposure to electromagnetic fields from wireless power transmission systems. *Proc. IEEE* **2013**, *101*, 1482–1493. [CrossRef]

17. Hong, S.; Cho, I.; Choi, H.; Pack, J. Numerical anlaysis of human exposure to electromagnetic fields from wireless power transfer systems. In Proceedings of the 2014 IEEE Wireless Power Transfer Conference, Jeju, Korea, 8–9 May 2014; pp. 216–219.

18. Sunohara, T.; Hirata, A.; Laakso, I.; Onishi, T. Analysis ofin situelectric field and specific absorption rate in human models for wireless power transfer system with induction coupling. *Phys. Med. Boil.* **2014**, *59*, 3721–3735. [CrossRef] [PubMed]

19. Shimamoto, T.; Laakso, I.; Hirata, A. In-situelectric field in human body model in different postures for wireless power transfer system in an electrical vehicle. *Phys. Med. Boil.* **2014**, *60*, 163–173. [CrossRef] [PubMed]

20. Shimamoto, T.; Iwahashi, M.; Sugiyama, Y.; Laakso, I.; Hirata, A.; Onishi, T. SAR evaluation in models of an adult and a child for magnetic field from wireless power transfer systems at 6.78 MHz. *Biomed. Phys. Eng. Express* **2016**, *2*, 027001. [CrossRef]

21. Sunohara, T.; Hirata, A.; Laakso, I.; De Santis, V.; Onishi, T. Evaluation of nonuniform field exposures with coupling factors. *Phys. Med. Boil.* **2015**, *60*, 8129–8140. [CrossRef]

22. Wake, K.; Laakso, I.; Hirata, A.; Chakarothai, J.; Onishi, T.; Watanabe, S.; De Santis, V.; Feliziani, M.; Taki, M. Derivation of Coupling Factors for Different Wireless Power Transfer Systems: Inter- and Intralaboratory Comparison. *IEEE Trans. Electromagn. Compat.* **2017**, *59*, 677–685. [CrossRef]

23. Zhang, W.; White, J.C.; Malhan, R.K.; Mi, C.C.; Mi, C.C. Loosely Coupled Transformer Coil Design to Minimize EMF Radiation in Concerned Areas. *IEEE Trans. Veh. Technol.* **2016**, *65*, 4779–4789. [CrossRef]

24. Campi, T.; Cruciani, S.; Maradei, F.; Feliziani, M. Near-Field Reduction in a Wireless Power Transfer System Using LCC Compensation. *IEEE Trans. Electromagn. Compat.* **2017**, *59*, 686–694. [CrossRef]

25. Park, J.; Kim, N.; Hwang, K.; Park, H.H.; Kwak, S.I.; Kwon, J.H.; Ahn, S. A Resonant Reactive Shielding for Planar Wireless Power Transfer System in Smartphone Application. *IEEE Trans. Electromagn. Compat.* **2017**, *59*, 695–703. [CrossRef]

26. Kim, M.; Kim, H.; Kim, N.; Jeong, Y.; Park, H.H.; Ahn, S. A Three-Phase Wireless-Power-Transfer System for Online Electric Vehicles with Reduction of Leakage Magnetic Fields. *IEEE Trans. Microw. Theory Tech.* **2015**, *63*, 3806–3813. [CrossRef]

27. Nadakuduti, J.; Douglas, M.; Lu, L.; Christ, A.; Guckian, P.; Kuster, N. Compliance Testing Methodology for Wireless Power Transfer Systems. *IEEE Trans. Power Electron.* **2015**, *30*, 1. [CrossRef]

28. Ishihara, S.; Onishi, T.; Hirata, A. Magnetic Field Measurement for Human Exposure Assessment near Wireless Power Transfer Systems in Kilohertz and Megahertz Bands. *IEICE Trans. Commun.* **2015**, *98*, 2470–2476. [CrossRef]

29. Libby, L.L. Special Aspects of Balanced Shielded Loops. *Proc. IRE* **1946**, *34*, 641–646. [CrossRef]

30. Swinyard, W.O. Measurement of Loop-Antenna Receivers. *Proc. IRE* **1941**, *29*, 382–387. [CrossRef]

31. IEEE. *186-1948-IEEE Standard Methods of Testing Amplitude-Modulation Broadcast Receivers*; IEEE: New York, NY, USA, 1949.

32. Hirayama, H.; Hayashi, H.; Kikuma, N.; Sakakibara, K. Estimation of poynting vector and wavenumber vector from near-magnetic-field measurement. In Proceedings of the 2006 12th International Symposium on Antenna Technology and Applied Electromagnetics and Canadian Radio Sciences Conference, Montreal, Canada, 16 July 2006; pp. 1–4.

33. Hirayama, H.; Kondo, H.; Kikuma, N.; Sakakibara, K. Visualization of emission from bend of a transmission line with Poynting vector and wave-number vector. *Proc. EMC Eur.* **2008**, *2008*, 1–4.

34. Hirayama, H.; Kikuma, N.; Sakakibara, K. An Estimation Method of Poynting Vector with Near-Magnetic-Field Measurement. *IEICE Trans. Electron.* **2010**, *E93c*, 66–73. [CrossRef]

35. Whiteside, H.; King, R. The loop antenna as a probe. *IRE Trans. Antennas Propag.* **1964**, *12*, 291–297. [CrossRef]

36. Harpen, M.D. The theory of shielded loop resonators. *Magn. Reson. Med.* **1994**, *32*, 785–788. [CrossRef]

37. Barfield, R. Some experiments on the screening of radio receiving apparatus. *J. Inst. Electr. Eng.* **1924**, *62*, 249–262. [CrossRef]

38. Gadian, D.; Robinson, F. Radiofrequency losses in NMR experiments on electrically conducting samples. *J. Magn. Reson.* **1979**, *34*, 449–455. [CrossRef]

39. Zabel, H.J.; Bader, R.; Gehrig, J.; Lorenz, W.J. High-quality MR imaging with flexible transmission line resonators. *Radiology* **1987**, *165*, 857–859. [CrossRef] [PubMed]

40. Vestergaard-Poulsen, P.; Thomsen, C.; Sinkjær, T.; Henriksen, O. Simultaneous31P NMR spectroscopy and EMG in exercising and recovering human skeletal muscle: Technical aspects. *Magn. Reson. Med.* **1994**, *31*, 93–102. [CrossRef]

41. Ruytenberg, T.; Webb, A.; Zivkovic, I. Shielded-coaxial-cable coils as receive and transceive array elements for 7T human MRI. *Magn. Reson. Med.* **2019**, *83*, 1135–1146. [CrossRef] [PubMed]

42. Kajiura, N.; Hirayama, H. Improvement of transmission efficiency using shielded-loop antenna for wireless power transfer. In Proceedings of the 2016 International Symposium on Antennas and Propagation (ISAP), Okinawa, Japan, 24–28 October 2016; pp. 52–53.

43. Thomas, E.M.; Heebl, J.D.; Pfeiffer, C.; Grbic, A. A Power Link Study of Wireless Non-Radiative Power Transfer Systems Using Resonant Shielded Loops. *IEEE Trans. Circuits Syst. Regul. Pap.* **2012**, *59*, 2125–2136. [CrossRef]

44. Tierney, B.B.; Grbic, A. Planar Shielded-Loop Resonators. *IEEE Trans. Antennas Propag.* **2014**, *62*, 3310–3320. [CrossRef]

45. Cagatay, B.; Serdar, D. Adaptive Weighted Performance Criterion for Artificial Neural Networks. In Proceedings of the International Conference on Artificial Intelligence and Data Processing (IDAP), Malatya, Turkey, 28–30 September 2018; pp. 1–4.

46. Gabriel, S.; Lau, R.W.; Gabriel, C. The dielectric properties of biological tissues: III. Parametric models for the dielectric spectrum of tissues. *Phys. Med. Boil.* **1996**, *41*, 2271–2293. [CrossRef]

47. Nagaoka, T.; Watanabe, S.; Sakurai, K.; Kunieda, E.; Watanabe, S.; Taki, M.; Yamanaka, Y. Development of realistic high-resolution whole-body voxel models of Japanese adult males and females of average height and weight, and application of models to radio-frequency electromagnetic-field dosimetry. *Phys. Med. Boil.* **2003**, *49*, 1–15. [CrossRef]

48. Hirata, A.; Ito, F.; Laakso, I. Confirmation of quasi-static approximation in SAR evaluation for a wireless power transfer system. *Phys. Med. Boil.* **2013**, *58*, N241–N249. [CrossRef]

49. Dawson, T.W.; Stuchly, M.A. Analytic validation of a three-dimensional scalar-potential finite-difference code for low-frequency magnetic induction. *Appl. Comput. Electromagnet. J.* **1996**, *11*, 72–81.

50. Laakso, I.; Hirata, A. Fast multigrid-based computation of the induced electric field for transcranial magnetic stimulation. *Phys. Med. Boil.* **2012**, *57*, 7753–7765. [CrossRef] [PubMed]

51. Reilly, J.P.; Hirata, A. Low-frequency electrical dosimetry: Research agenda of the IEEE International Committee on Electromagnetic Safety. *Phys. Med. Boil.* **2016**, *61*, R138–R149. [CrossRef] [PubMed]

52. Kraus, J.D.; Marhefka, R.J. *Antenna for All Applications*, 3rd ed.; McGraw Hill: New Jersey, NJ, USA, 2001.

53. Laakso, I.; Hirata, A. Reducing the staircasing error in computational dosimetry of low-frequency electromagnetic fields. *Phys. Med. Biol.* **2012**, *57*, N25–N34. [CrossRef] [PubMed]

54. Gomez-Tames, J.; Laakso, I.; Haba, Y.; Hirata, A.; Poljak, D.; Yamazaki, K. Computational Artifacts of the In Situ Electric Field in Anatomical Models Exposed to Low-Frequency Magnetic Field. *IEEE Trans. Electromagn. Compat.* **2017**, *60*, 589–597. [CrossRef]

55. Diao, Y.; Gomez-Tames, J.; Rashed, E.A.; Kavet, R.; Hirata, A. Spatial Averaging Schemes of In Situ Electric Field for Low-Frequency Magnetic Field Exposures. *IEEE Access* **2019**, *7*, 184320–184331. [CrossRef]

56. International Commission on Non-Ionizing Radiation Protection. Guidelines for limiting exposure to time-varying electric and magnetic fields for low frequencies (1 Hz–100 kHz). *Health Phys.* **2010**, *99*, 818–836.

57. Christ, A.; Douglas, M.G.; Roman, J.M.; Cooper, E.B.; Sample, A.; Waters, B.H.; Smith, J.R.; Kuster, N. Evaluation of Wireless Resonant Power Transfer Systems with Human Electromagnetic Exposure Limits. *IEEE Trans. Electromagn. Compat.* **2012**, *55*, 1–10. [CrossRef]

58. IEC. *Measurement Methods for Electromagnetic Fields of Household Appliances and Similar Apparatus with Regard to Human Exposure*; IEC 62233; International Electrotechnical Commission: Geneva, Switzerland, 2005.

Article

Energy-Efficient Wireless Hopping Sensor Relocation Based on Prediction of Terrain Conditions

Sooyeon Park [1], Moonseong Kim [2] and Woochan Lee [1,*]

[1]   Department of Electrical Engineering, Incheon National University, Incheon 22012, Korea; anisoo@inu.ac.kr
[2]   Department of Liberal Arts, Seoul Theological University, Bucheon 14754, Korea; moonseong@stu.ac.kr
*   Correspondence: wlee@inu.ac.kr; Tel.: +82-32-835-8436

Received: 9 December 2019; Accepted: 26 December 2019; Published: 28 December 2019

Abstract: It is inevitable for data collection that IoT sensors are distributed to interested areas. However, not only the proper placement of sensors, but also the replacement of sensors that have run out of energy is very difficult. As a remedy, wireless charging systems for IoT sensors have been researched recently, but it is apparent that the availability of charging system is limited especially for IoT sensors scattered in rugged terrain. Thus, it is important that the sensor relocation models to recover sensing holes employ energy-efficient scheme. While there are various methods in the mobile model of wireless sensors, well-known wheel-based movements in rough areas are hard to achieve. Thus, research is ongoing in various areas of the hopping mobile model in which wireless sensors jump. Many past studies about hopping sensor relocation assume that all sensor nodes are aware of entire network information throughout the network. These assumptions do not fit well to the actual environment, and they are nothing but classical theoretical research. In addition, the physical environment (sand, mud, etc.) of the area in which the sensor is deployed can change from time to time. In this paper, we overcome the theoretical-based problems of the past researches and propose a new realistic hopping sensor relocation protocol considering terrain conditions. Since the status of obstacles around the sensing hole is unknown, the success rate of the hopping sensor relocation is used to predict the condition of the surrounding environment. Also, we are confident that our team is uniquely implementing OMNeT++ (Objective Modular Network Testbed in C++) simulation in the hopping sensor relocation protocol to reflect the actual communication environment. Simulations have been performed on various obstacles for performance evaluation and analysis, and we are confident that better energy efficiency with later appearance of sensing holes can be achieved compared to well-known relocation protocols.

Keywords: mobile sensor; hopping sensor; relocation protocol; energy efficient protocol; internet of things (IoTs); wireless sensor networks (WSNs); simulation

---

1. Introduction

Recently, there have been active studies for big data analysis using networking to collect, process, and analyze a large amount of data [1]. In order to not only collect various data but also fast process information, the technologies of collecting and transmitting data have emerged as a very important issue. Here, there is no doubt that the transmission technology among wireless sensors is important for big data technology. Research on various wireless sensor networking technologies to transmit the collected data has been actively conducted for decades [2,3].

A lot of wireless sensors are properly distributed through the observation area to collect the interested data [4,5]. It is not difficult to distribute the sensors to accessible areas to human, however unmanned mobile devices (unmanned air vehicle, drone, etc.) could be used in large inaccessible areas. After deploying the sensors, let us consider that data collection occurs frequently in particular area.

There is an inherent problem that a small sensor has limited energy. To remedy this power limitation, wireless charging systems for IoT sensors have been researched actively recently [6]. However, charging system for the sensors is not always available, especially in inaccessible area like rugged terrain. Thus, it is necessary that the sensor relocation protocol to recover sensing hole employs energy efficient scheme as a preemptive measure. Some sensors may be quickly drained of energy while continuously collecting and transmitting data in the particular area. This case is called a sensor node failure, the communication of the entire network may be disconnected, and the desired data could not be collected in the worst case. The particular area, where a certain number of sensors become faulted and cannot collect the interested data anymore, is called as a sensing hole [7]. In order to prevent a sensing hole occurred, various studies such as schemes of minimizing energy consumption by adjusting the active/idle states and energy efficient routing protocols have been conducted [8–10]. Although the energy limit of the sensor node can be reduced by various applications, it is impossible to solve the sensor node failure problem completely. Therefore, it is necessary to recover it by moving other sensors to the sensing hole occurred, as the most realistic solution. For this reason, researches on mobile sensors have received a lot of attention. The failure sensors depleted of energy could be replaced by mobile sensors moved from other area. The authors of [11] considered a case that a wheel-based mobile node could temporarily replace the role of a fixed node by moving if a fixed sensor node is failed in the interested area. However, there are limitations in the migration of wheel-based sensors. In other words, energy consumption has to be further considered in order to move, however wheel-based movement is essentially inadequate in very rough areas.

In order to overcome the limitation of wheel-based mobility, a hopping-based moving model is introduced. A hopping-based sensor node is bionically designed to jump itself, like a frog. For example, a mobile unit implemented by DARPA (Defense Advanced Research Projects Agency, USA) is able to do up to 100 jumps with one full fuel injection and jump up to 10 m high [12,13]. The paper [14] by MSU (Michigan State University, USA) describes the performance comparisons for the maximum jump height, jumping distance, and so on for various jumping robots, in detail. Researches on not only the implementation of jump-based mobility but hopping sensor relocation algorithm to recover the sensing holes that occurred, have been actively conducted [15–18]. However, most of studies so far require that every cluster header sensor node has to know all information for the current whole network area and set up routes to supply/request hopping sensors to recover sensing holes occurred. In fact, no matter how small the observation field to obtain interested data is, the exchange of information and establishment of paths among the cluster header sensors are difficult in real world, and involve a storm of numerous control messages. Recently, our research team has solved these problems drastically [19]. First of all, every cluster header does not need to know all information of other cluster headers or all networks. It is a distributed networking-based relocation protocol that recovers a sensing hole by simply requesting necessary sensor nodes from neighbor cluster headers. A cluster header node selects appropriate hopping sensors based on information communicated among all sensor nodes in its cluster, and they are moved to the requesting neighbor cluster. Up to now, however, most relocation protocols have assumed that the observation fields, in which mobile sensors are scattered, are ideal environments. In other words, the areas where sensors move are likely to be irregular, under the existence of obstacles like stones or mud. Reference [18] is the first study of hopping sensor relocation based on probabilities of existence of obstacles, however there is a limitation that the assumption that every cluster header sensor grasps all information of rugged terrains in all regions is not very realistic.

In this paper, a novel relocation protocol that considers the probabilities of the level of obstacles using the information of success rate for migration is proposed. Here, it is not necessary for each cluster header sensor node to know all information of entire network; the cluster header only knows the information such as the probabilities between neighbor clusters. In addition, as far as we know, it can be very meaningful to have the first realistic simulation with OMNeT++ [20], similar to the actual environment.

The rest of this paper is organized as follows. Section 2 summarizes the mobility model of hopping sensor and the related relocation protocol. In Section 3, the proposed relocation protocol is explained along with the scenario in detail. The simulation results are analyzed in Section 4, and Section 5 concludes this paper.

2. Previous Work

2.1. Characteristics and Movement Models of Hopping Sensors

Typical mobile sensors are based on wheels. This structure has the disadvantage that it is very difficult to move in rough terrain such as gravel, sand, and so on. In order to overcome the mobility problem, hopping sensors that mimic movements such as a grasshopper or a frog have been devised. In addition, the hopping sensor enables communication among sensor nodes in environments where it is difficult for a sensor to communicate each other because the distance between sensor nodes is farther than a normal communication radius of a sensor, or where there are various obstacles which block communication propagation. Up to now, various researchers have shown that the connectivity between hopping sensors could be improved by the jumping of each hopping sensor. The papers [21,22] have shown that the communication radius is increased about six times compared to the one on the ground when a hopping sensor jumps 1 m from the ground. The paper [23] actually implemented a hopping sensor using a separate launcher for jumping and measured the transmission radius varied according to the height of the jump. The changed transmission radii were compared with the results of [21], and it was confirmed that they had similar results.

Compared to a wheel-based sensor node, a hopping sensor, which is moving on jump, might be inherently less accurate to migrate to the desired area. In fact, it would be more desirable to have a positive explanation for enabling movement in an environment that cannot be moved with wheels, rather than a negative idea that movement accuracy is poor. The paper [15] first attempted a performance analysis of how a hopping sensor node was affected by wind when jumping and moving. It is obvious that the movement of hopping model is more vulnerable to air resistance than that of wheel-based model. Mathematical modeling of this hopping movement can be explained as follows.

As shown in Figure 1, the hopping sensor does not land exactly the targeted location (T) after jumping, and it is more likely to land near it (i.e., actual landing location; L). Let the variables T and L be vectors of the target and actual landing locations. The error vector D with respect to the difference between the locations can be expressed as $L - T$. Here, we assume that D follows the two-dimensional standard normal distribution with means $(0, 0)$, standard deviations (σ_x, σ_y), and correlation ρ. The probability density function can be defined as

$$f_{XY} = \frac{1}{2\pi\sigma_x\sigma_y\sqrt{1-\rho^2}}e^{-\frac{1}{2(1-\rho^2)}\left(\frac{x^2}{\sigma_x^2}+\frac{y^2}{\sigma_y^2}-\frac{2\rho xy}{\sigma_x\sigma_y}\right)} \tag{1}$$

Let us consider that the hopping range is 2, standard deviations are 0.3, and the correlation is zero. The movement of the hopping sensor is as shown in Figure 1. The movement of the red circle is an ideal movement, but the movement of the blue circle is a movement that reflects the mentioned constraints. It can be easily seen that the movement of the blue circle more appropriately reflects the movement in rough environment suitable for the hopping sensor than that of the red circle. Therefore, the above error vector of the movement would be preferably applied to the simulation in Section 4.

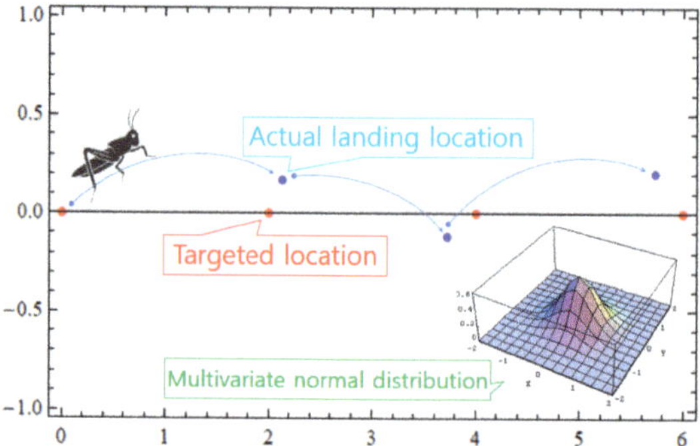

Figure 1. An example of the movement of a hopping sensor considering movement-error.

2.2. Basic Assumptions about Hopping Sensors and Relocation Protocol

In the traditional research fields of wireless sensor networks (WSNs), there are various methods for clustering and cluster header selection of sensor nodes randomly distributed in the interested areas [24]. In fact, since the purpose of this paper is to study the relocation of hopping sensor nodes, it is assumed that the clustering and selection of the cluster headers have already been completed by various existing methods. For example, hopping sensor nodes are randomly scattered firstly on the interested area in which we want to collect data. After that, the area is divided by an appropriate clustering algorithm, and a cluster header node is selected for each cluster zone. The cluster header periodically communicates with member sensor nodes in its cluster zone, and it manages representative information of each member. Considering the network connectivity problem, communication between each cluster header and all member sensor nodes could jump to send and receive messages with its maximum transmission radius. Moreover, every hopping sensor node contains a GPS unit capable of knowing its current location [25].

The terms used in this paper are explained in Figure 2. Every hopping sensor node can be either a cluster header node or a member sensor node. Since the maximum transmission radius is defined as a cluster zone when the cluster header maximally jumps, there is a high possibility that direct communication between the cluster headers could be impossible. However, some member sensor nodes in the area intersecting with the cluster zones (that is, near the maximum transmission radius of each cluster header) may communicate with two or more cluster headers. This hopping sensor node is called as a relay node, and the role of the node could help to facilitate communication between cluster headers [18].

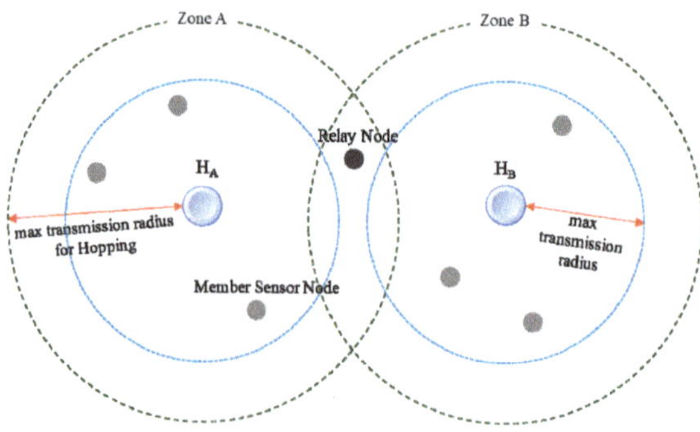

Figure 2. The defined terms for hopping sensor network.

Unlike hopping sensor relocation protocols, which were only theoretically studied, the paper [19] proposed a relocation protocol suitable for a distributed environment so that it could be applied to a real environment. The protocol proposed in the mentioned paper is briefly summarized using an example in Figure 3. If the cluster zone C is a sensing hole, the protocol strategy to recover the problem is as follows.

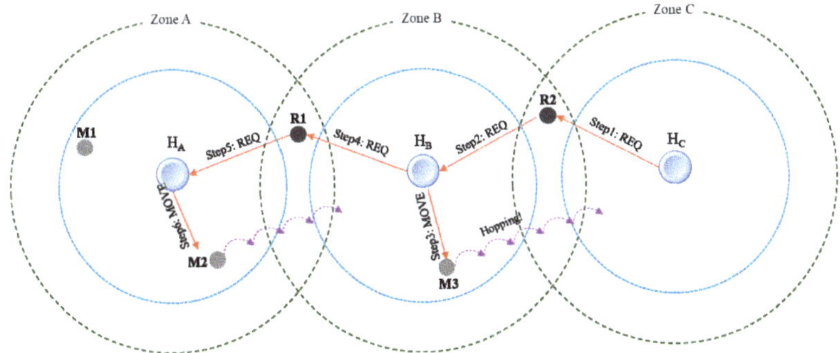

Figure 3. An example of recovering a sensing hole with cross-clusters.

Step 1. The cluster header H$_C$ sends a REQ message requesting one hopping sensor to the relay node R2.

Step 2. The relay node R2 forwards the REQ message received from H$_C$ to the cluster header H$_B$ of the cluster zone B.

Step 3. The cluster header H$_B$ sends a MOVE message for moving to neighbor cluster zone C to the hopping sensor M3 selected as a moveable member in its zone B.

Step 4. At the same time, the cluster header H$_B$ predicts that its zone could also be a sensing hole, and it sends a REQ message for requesting one hopping sensor to relay node R1.

Step 5. The relay node R1 forwards the message REQ to its another cluster header H$_A$.

Step 6. The cluster header H$_A$ chooses M2 among hopping member sensor nodes of its zone A in its zone and commands the movement. As a result, one sensor is properly allocated to each cluster zone, and the sensing zone (cluster zone C) could be recovered.

3. The Proposed Relocation Protocol for Rugged Terrains

Although the distributed-based relocation protocol [19] has made up for the drawbacks of the previous central-based relocation protocols, questions still remain whether it is suitable for the real world. In fact, it should be noted that the environment in which hopping sensors are deployed would be different from the general terrain. In the relocation protocol [19], topographical information on obstacles around the cluster zones were not taken into account. The movement of hopping sensors fails due to obstacles and the relocation cannot satisfy the movement requested from the cluster header of sensing hole. Thus, the number of messages continuously requested may increase to overcome the persisting sensing hole. This could put a heavy load on the entire network, and it would be very negative in terms of energy. Therefore, we propose a relocation protocol for more realistic hopping sensor networks by indirectly predicting the topographic information around sensing holes.

3.1. Basic Operation of the Proposed Relocation Protocol

In this section, we propose a novel relocation protocol that takes into account the environment surrounding the sensing hole occurred. Figure 4 shows the message flows between hopping sensors that the types are cluster header, relay node, and member sensor node. The detailed formats of the messages for the proposed relocation protocol and the message flows for each sensor node are described in detail as follows.

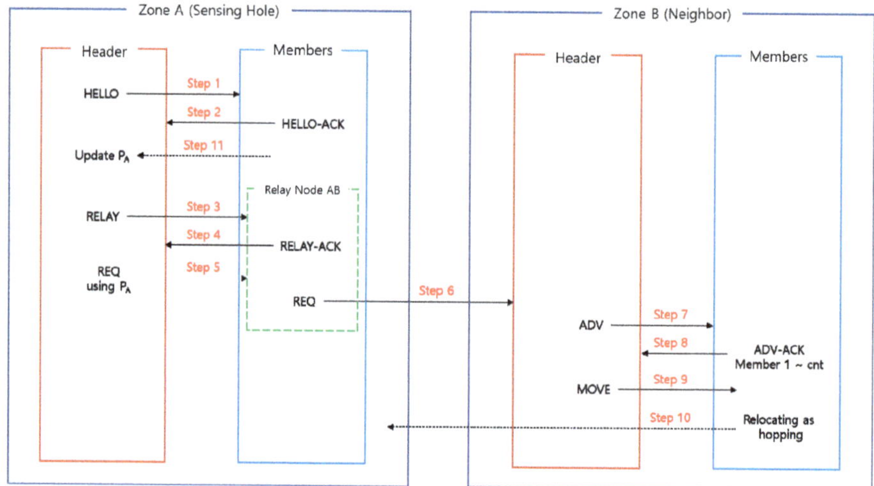

Figure 4. Scenario and message flow diagram for describing the proposed protocol.

Step 1: The cluster header of the cluster zone A periodically broadcasts a HELLO message inside its zone to identify member sensor nodes. If a member node receives HELLO for the first time, the source address of the received HELLO message is memorized as its cluster header.

HELLO: = {message type, source address, destination address (broadcasting)}

Step 2: The member that received the HELLO message responds to the cluster header with a HELLO-ACK message to notify its health. If different HELLO messages are received from several cluster headers in Step 1, the member node is aware of that it has become a relay node for each cluster header. Thus, it also indicates whether a relay node or not, when the HELLO-ACK message is replied.

HELLO-ACK: = {message type, source address, destination address, relay node? (T/F)}

Step 3: The cluster header determines whether or not, currently, its zone is a sensing hole using the number of addresses of HELLO messages received. In the case of a sensing hole, the cluster header sends a relay message to all its relay nodes.

RELAY: = {message type, source address, destination address (multicasting)}

Step 4: As soon as the relay nodes receive the RELAY message, they send a RELAY-ACK message back to the cluster header. The cluster header may sequentially receive several RELAY-ACK messages. Among them, the relay node of the RELAY-ACK message, which is first received at the cluster header, is selected and other RELAY-ACK messages are ignored. Also, if a relay node overhears a RELAY-ACK message from another relay node while preparing to send a RELAY-ACK message, it immediately stops sending its message.

RELAY-ACK: = {message type, source address, destination address}

Step 5: So far, the cluster header is able to detect that a sensing hole has occurred (Step 3), and it could choose a relay node to forward a message to the neighbor cluster header for requesting member hopping sensor nodes needed (Step 4). Here, the required number of members is the difference $(T - C)$ between the threshold value (T) to determine the sensing hole and the current number of members

(*C*). However, the environment around the detected sensing hole may be rough terrain in reality, the sensing hole would continue because it might be high probability that the number of relocated hopping sensors is smaller than the requested number (*T* − *C*). Therefore, the cluster header checks the numbers of previously requested members and successfully moved members and calculates the movement success rate (*p* = number of members successfully moved/number of members requested) of the surrounding environment for the current cluster zone. Here, the number of requesting member sensors (*cnt*) that can overcome the sensing hole can be calculated as follows.

$$cnt: = \mathrm{Ceiling}[(T - C) * (1 + (1 - p))] \tag{2}$$

The cluster header sends a REQ message to the selected relay node.

REQ: = {message type, source address, destination address, *cnt*, sensing hole address, sensing hole GPS info.}

Step 6: When the relay node receives the REQ message from the cluster header of the sensing hole occurred, it forwards the received REQ message to another cluster header. Here, the source and destination addresses of the currently received REQ message are modified to its own address and the address of the another cluster header to be delivered, respectively.

Step 7: When the neighbor cluster header of the occurred sensing hole receives the REQ message forwarded, it broadcasts an ADV message to determine movable member hopping sensor nodes among the current members in its zone.

ADV: = {message type, source address, destination address (broadcasting) }

Step 8: Each member node that receives the ADV message sends an ADV-ACK message containing its current physical information to the cluster header.

ADV-ACK:= {message type, source address, destination address, some info.}

Here, above 'some info.' indicates several type of information for relocation strategy [19], including the current energy state, capability of hopping movement, GPS coordinate information, etc.

Step 9: The cluster header receives ACK messages and selects the appropriately movable hopping sensor nodes. Then, it sends MOVE message, which includes the location information about the sensing hole, to the hopping sensors selected.

ADV: = {message type, source address, destination address (multicasting),
sensing hole address, sensing hole GPS info.}

Step 10: The member nodes receiving the MOVE message have to hop to the neighbor zone using the GPS information of the cluster header of the sensing hole. Here, as every hopping member is taking into account the usual transmission radius of the neighbor cluster header, it could calculate the coordinates of where to move properly. Thus, each member hopping sensor node no longer moves when the member node has moved as desired location.

Step 11: As mentioned in Step 1, the cluster header of the sensing hole broadcasts a HELLO message at periodic times. After a predetermined time, when the relocated member sensor node receives the new HELLO message, the member updates its previous cluster header information as the current one. If the relocated member sensor node receives multiple HELLO messages at this time, it would be changed to a relay node. In addition, each cluster header is able to update the current movement success rate p through the number of HELLO-ACK messages to properly reflect the level of obstacles. All hopping sensors have modules that can perform Steps 1 to 11 independently. Each module of a hopping sensor could be activated depending on whether it is currently playing a role of {cluster header, relay node, member sensor node}. This is summarized as follows.

$$\text{Cluster header: } = \{\text{Steps } 1, 3, 5, 7, 9, 11\}$$

$$\text{Relay node: } = \{\text{Steps } 3, 4, 5, 6\}$$

$$\text{Member sensor node: } = \{\text{Steps } 2, 8, 10\}$$

3.2. Descriptions of the Proposed Protocol

So far, most previous studies on hopping sensor relocation are central-based manner which each cluster header know all network information, but this is not a very practical study. The research of [19] only conducted a distributed-based approach to consider the realistic environment, but it still failed to take into account the surrounding obstacle information. That is, it is important to consider the environment using hopping sensors, not using the wheel-based ones. The following examples of simple scenarios in Figure 5 are described to understand the difference between cases reflected the surrounding obstacle information or not.

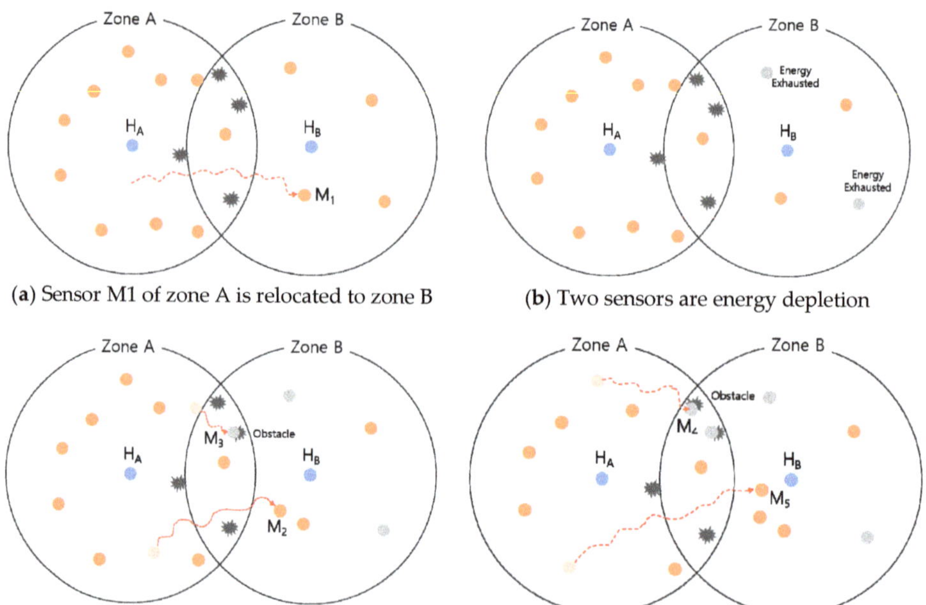

(**a**) Sensor M1 of zone A is relocated to zone B   (**b**) Two sensors are energy depletion

(**c**) Sensor M3 is failed to move due to obstacle   (**d**) Sensor M4 is also failed to move due to obstacle

Figure 5. Examples of comparison between the previous and proposed relocations.

In Figure 5a, we assume that the cluster zone B has to maintain five member sensor nodes, except for the cluster header H_B, for data collection. The cluster header determines that the sensing hole occurs when the number of members is smaller than 5. Here, the cluster zone B was a sensing hole due to a lack of one member (i.e., before moving M1). In order to recover the sensing hole, after the cluster header H_B of zone B requests one member from zone A (the first REQ), the member sensor node M1 of zone A moves to zone B.

In the first scenario, let us consider the relocation of hopping sensors without taking into account obstacles as mentioned previous literatures. In Figure 5b, two member sensor nodes fail due to energy exhausted in the cluster zone B. The cluster header H_B is aware of the state of sensing hole, and it requests two members from the neighbor cluster zone A (the second REQ). Two members, which were ordered to move to the cluster zone B, are moving as shown in Figure 5c. One member M2 is successfully moved to the cluster zone B, while the other M3 is failed to move due to obstacle. The cluster header of the cluster zone B perceives the state of sensing hole again, and it requests one member from the cluster zone A (the third REQ). In Figure 5d, since the selected member M4 fails to move because of unexpected obstacle, the cluster zone B is still a sensing hole. In order to overcome the sensing hole, the cluster header H_B requests one member from the cluster zone A repeatedly (the fourth REQ). The member M5 is successfully moved, and the sensing hole is finally recovered.

In the second scenario, let us take into account the relocation based on probabilities of existence of obstacles. In Figure 5b, the cluster header H_B should recognize the state of the sensing hole and request some members from the neighbor cluster zone A. First of all, let the initial value of the movement success rate p be 1. The success rate p is used to calculate the requested number of member sensors. According to Equation (2), the number of the requested members is 2, Ceiling[$(5 - 3) * (1 + (1 - 1))$], the cluster header H_B requests two members from the cluster zone A (the second REQ). In Figure 5c, the cluster header H_A commands two members to move to the cluster zone B. As described in the above first scenario, one member M2 moves successfully, but the other member M3 fails to move due to the obstacle. Thus, the cluster header H_B updates the value of p as 1/2, and it is also aware of the sensing hole. The cluster H_B calculates the number of requesting members using Equation (2), Ceiling[$(5 - 4)$ $* (1 + (1 - 1/2))$], so two members are requested to the cluster H_A (the third REQ). One of the two members of A, the member M3 fails to move to the zone B, as in the first previous scenario. However, the other member M2 can successfully move to the zone B to overcome the state of sensing hole.

As we are looking at the mentioned two scenarios, it might be thought that the probability of successful movement represents the state of obstacles between cluster zones. Therefore, it is a good example to see that the use of the above probability could reduce the number of request messages, when the cluster header of sensing hole is requesting members to the neighbor zones. Actually, the number of REQ messages of the first scenario is greater than that of the second scenario.

Message flow charts for the above described two scenarios could be explained easily, as shown in Figure 6. The cluster header H_B of the zone B first detects its sensing hole occurred, and it transmits the first REQ message to the cluster header H_A of the neighboring zone A, here the REQ includes that the number of member sensor node needed is one (i.e., $cnt = 1$). In fact, a direct communication between the cluster headers is impossible, but for convenience, transmission and reception of their messages with a RELAY node are omitted in the middle. The header H_A selects M1 appropriately among its member nodes, and it sends a MOVE message to relocate to the sensing hole, zone B. Member node M1, which received MOVE, was able to recover the sensing hole without any problems for moving.

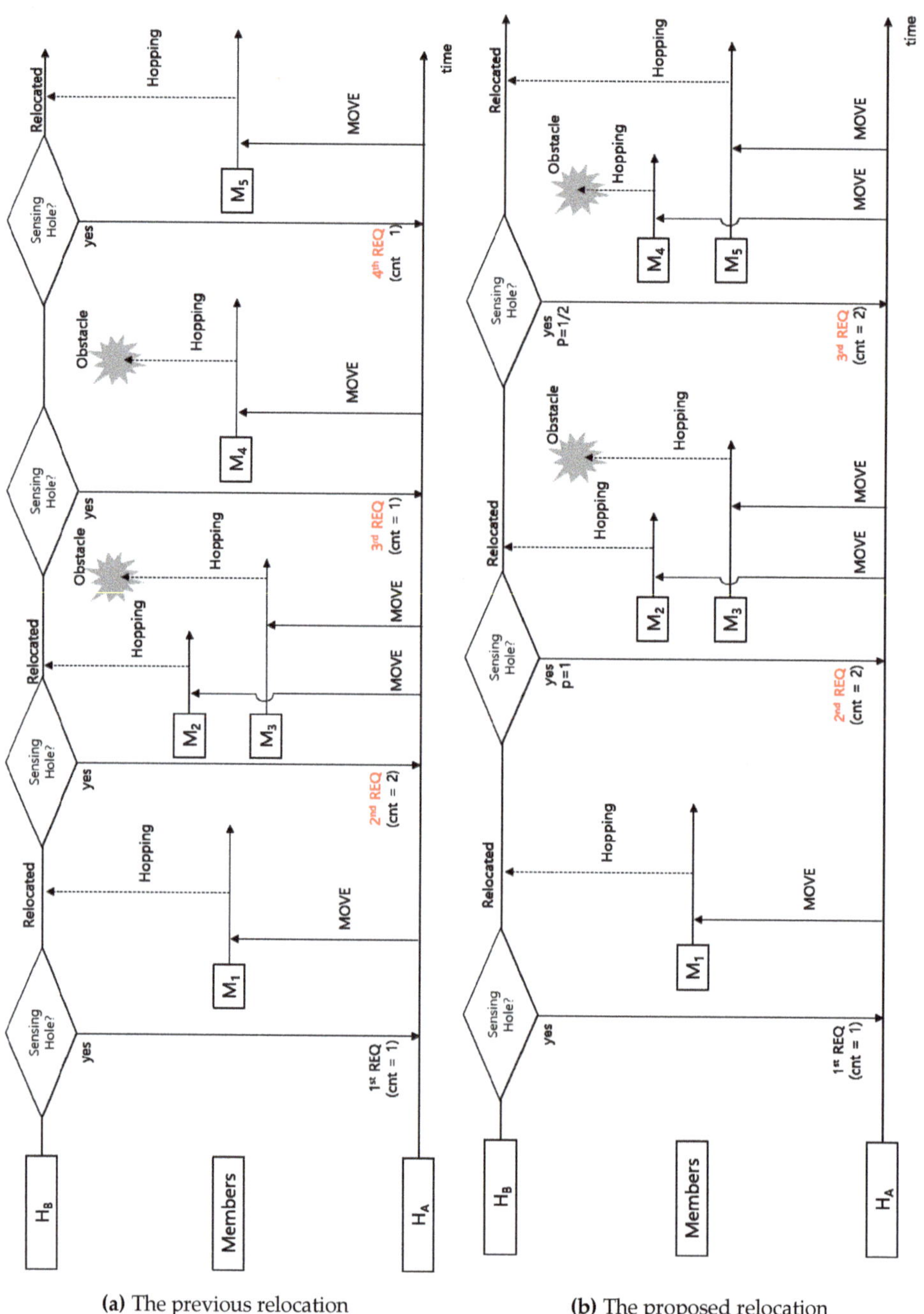

(a) The previous relocation   (b) The proposed relocation

Figure 6. Message flows for two scenarios.

First, the message flows for the first scenario are depicted in Figure 6a. The cluster header of B, H_B, detects a sensing hole occurred again, and it sets two members ($cnt = 2$) for recovery in the REQ message, and sends the message (the second REQ) to the neighbor header H_A. H_A appropriately selects two members (M2, M3) and transmits MOVE message to them to relocate, and the member node M2 of them is moved to the zone B without any problems. However, the other member M3 is not able to move to the zone B due to an unexpected obstacle. After a certain time, the cluster header H_B determines that the sensing hole is not recovered. H_B transmits a message (the third REQ) to request one member needed to the zone A. H_A directly sends MOVE message to the member M4, but M4 unfortunately crashes into an obstacle and fails to recover the sensing hole. After a certain amount of time again, the header H_B checks the current status of zone B, and it inserts that the value of *cnt* is 1 (i.e., requires one member) in the requesting message and sends the message (the fourth REQ). The member M5 receiving MOVE from its cluster header H_A successfully is moved to the sensing hole zone B without colliding with obstacles.

Next, the message flows for the second scenario are depicted in Figure 6b. The header H_B detects a sensing hole occurred again and sends a message to request some hopping sensors. Here, H_B appropriately calculates the number of members required using that the movement success rate p is one and sets the value of *cnt* to two. H_A chooses two members, M2 and M3, to transmit MOVE for relocation, and then M2 is successfully moved to the neighbor zone B. However, another hopping sensor M3 is unable to move because of unexpected obstacles. The cluster header H_B determines that the sensing hole is not recovered after broadcasting HELLO message repeatedly. The header of zone B transmits a message (the third REQ) to request some members needed from the header of zone A.

At this time, since the movement success rate for the previous REQ message is $p = 1/2$ (50%), the header of B could request two members calculated by Ceiling[(5 − 4) ∗ (1 + (1 − 1/2))], even though the zone B currently needs one member to recover the sensing hole. That is, H_B recognizes that the level of obstacles between cluster zones is high and requests a large number of sensors from H_A for rapid recovery. As expected, member M4 collided again with an obstacle, but member M5 successfully moved to neighbor zone B. Since the proposed relocation protocol reduces the number of REQ messages compared with the previous one, it could be verified that our protocol is energy efficient for wireless hopping sensor networks.

4. Simulation Results and Analysis

Unlike hopping sensor relocation studies so far, one of the most significant contributions this paper is the use of OMNeT++ [26], which can realistically reflect wireless communications situations in real-world environments. Table 1 describes the environment parameters used in the simulation for performance evaluation.

Table 1. Simulation environments.

network area	250 m × 150 m
number of all member hopping sensor nodes	285
number of cluster headers	15
minimum number of members for each cluster to properly gather data (i.e., a sensing hole occurs if number of current members lower than the value)	10
the probability that an obstacle exists	0%, 1%, 2%, 3%
maximum communication radius for each sensor node	20 m
maximum communication radius when highly jumping	29 m
maximum distance that a sensor node moves forward with one jump	2 m

As shown in Figure 7, 285 hopping sensors are randomly scattered to the whole area of 250 m × 150 m to collect data. As mentioned in the previous Sections, since the aim of our research is not clustering techniques, we assume that the 15 cluster zones and headers are properly pre-set up in well-known ways. If there are fewer than 10 member sensor nodes in each cluster zone, excluding the cluster headers, the cluster header determines that its zone is a sensing hole.

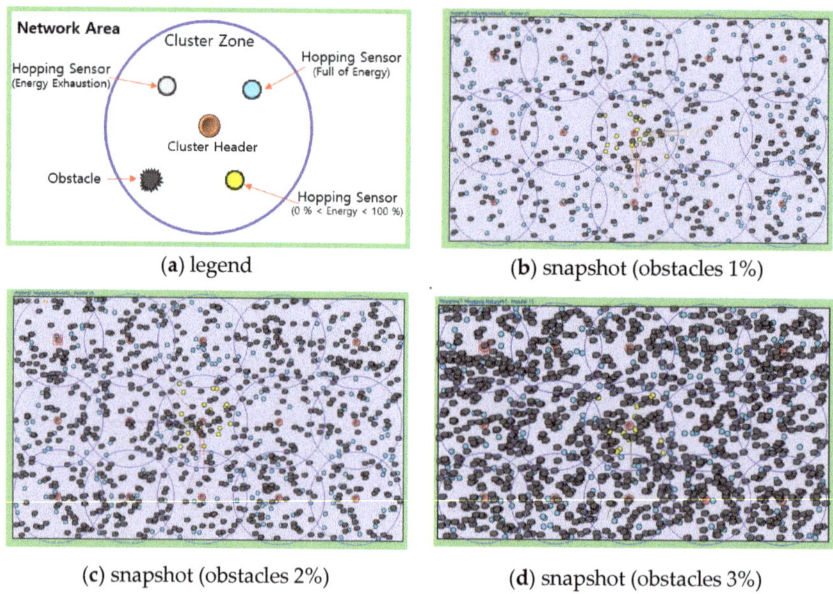

(**a**) legend

(**b**) snapshot (obstacles 1%)

(**c**) snapshot (obstacles 2%)

(**d**) snapshot (obstacles 3%)

Figure 7. Simulation snapshots for each environment.

As shown in Figure 7, obstacles are randomly generated with values of 1%, 2%, and 3% over the entire area. Although the size of the obstacle is very large in Figure 7d, the obstacle is actually 1 m × 1 m. Only for visual effects, we make the obstacle appear to be large. A communication model between sensors is used, IEEE 802.11, and the regular transmission radius is assumed to be 20 m. More specifically, *antennaType*, *transmitterType*, and *receiverType* parameters are set as IdealRadioMedium module environment in OMNeT++ [27]. While jumping as high as possible, each hopping sensor has a maximum radius of 29 m of transmission. It also can move forward 2 m with one jump.

In order to reflect the effect of the realistic environment (wind, etc.), the movement model of each hopping sensor node is assumed with two-dimensional standard normal distribution, where standard deviations are 0.3 and the correlation is zero, as shown in Figure 1. The movement of the hopping sensor is indicated by a solid line as shown in Figure 7. For the performance analysis of the proposed relocation protocol, the sensing hole occurred in the middle of area, as shown in Figure 7.

We assume the scenario that the sensor nodes continuously gather data in the middle cluster zone, and they consume energy rapidly. Each sensor in the middle cluster generates a data collection event with an exponential distribution (average of 5 min). For convenience, the initial energy value for sensing is set to 100, and the energy consumption is 1 for each event occurred. Any energy model can be adopted, but a specific energy model is not considered here for simplicity. The main flow of the paper is about the creation and restoration of the sensing hole. An occurrence of a sensing hole (less than 10 member sensor nodes) is determined by a cluster header after HELLO message for every 15-minute interval. In particular, to perform the simulation on relocation of member sensor nodes to recover the sensing hole occurred, sensor nodes in other zones are assumed to perform no data collection. It also indicates in yellow that a sensor has become faulted due to energy depletion after continuous data collection in Figure 7.

In the majority of the hopping sensor relocation protocols so far, each sensor node knows all network information; like a central-based relocation protocol, it only considers the shortest path between cluster zones. However, it is very difficult to implement in reality; therefore, we propose a distributed-based relocation protocol, in order to overcome the drawbacks of the previous central-based relocation protocols. The implementation of the proposed protocol and simulation for performance evaluation are based on OMNeT++, which takes all communication system layers into account and also reflects both distributed computing of all sensors and actual environments. As far as we know, our implementation of hopping movement model considering obstacles is the first attempt in this field of study.

As previously explained, a sensing event occurs only in the middle cluster zone. After a three-day-long simulation, the result is shown in Figure 8. The dead member sensor node is colored yellow and the path of each hopping sensor is plotted by a solid line. Also, the movement of the sensor is shown as a zigzag due to environmental factors such as wind. It can also be confirmed that sensors captured in obstacles can no longer move and become faulted.

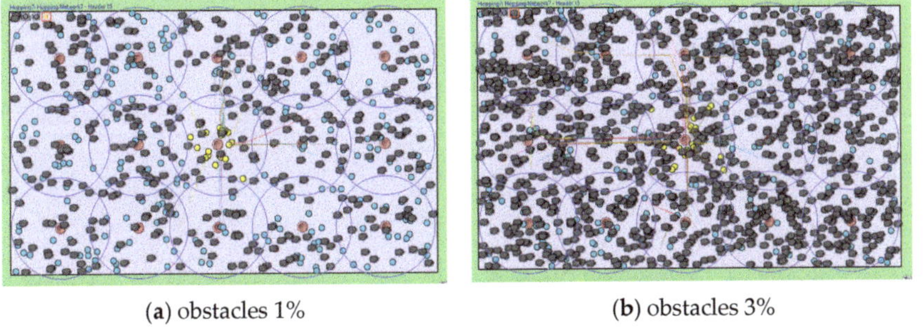

(**a**) obstacles 1%   (**b**) obstacles 3%

Figure 8. Simulation snapshots finished for each obstacle.

We have simulated the relocation protocols using a sensor network topology generated randomly, and the results are described in Figures 9 and 10. First, Figure 9 shows the sensor retention rate in the middle cluster. Since the minimum number of members for each cluster to properly gather data is 10, the sensor retention rate is 1 if there are more than 10 member sensor nodes. For example, if there are five sensor nodes, the sensor retention rate is 0.5 (= 5/10), and if there is only one sensor, the sensor retention rate is 0.1. Since the previous hopping sensor relocation protocol [19] ignores the existence of obstacles, the sensor retention rate of it tends to be much lower than that of the proposed protocol, as shown in Figure 9. Figure 9 shows the change of each sensor retention rate during about three days. Since the sensing hole of the middle cluster occurs continuously, the neighbors' cluster headers might provide redundant sensors to recover the middle cluster's sensing hole until around 1000 min. However, neighbors' nodes are also depleted at that moment, thus the sensor retention rates in both schemes start to drop as shown in the green circle of Figure 9a. The blue circle of Figure 9a indicates that the proposed method considering obstacles shows better retention rate than previous scheme.

Figure 9b illustrates the difference between the sensor retention rates of the proposed and the previous protocols. It is easy to see that the number of top marks is more than that of below marks in Figure 9b. Since the sensor retention rates are only considered in the middle cluster zone, the denominator is ten, which is the minimum number of sensors to recover a sensing hole. In other words, if the difference of the rates is the positive 0.5, the proposed method accommodates five more sensors than the previous method at the same instance. A higher retention rate would result in requesting a smaller number of sensors compared with another method and generating fewer messages for movements to neighboring cluster zones. Hence, the proposed scheme would consume smaller energy to generate messages compared to the previous scheme and can play a major role in extending the lifetime of the network.

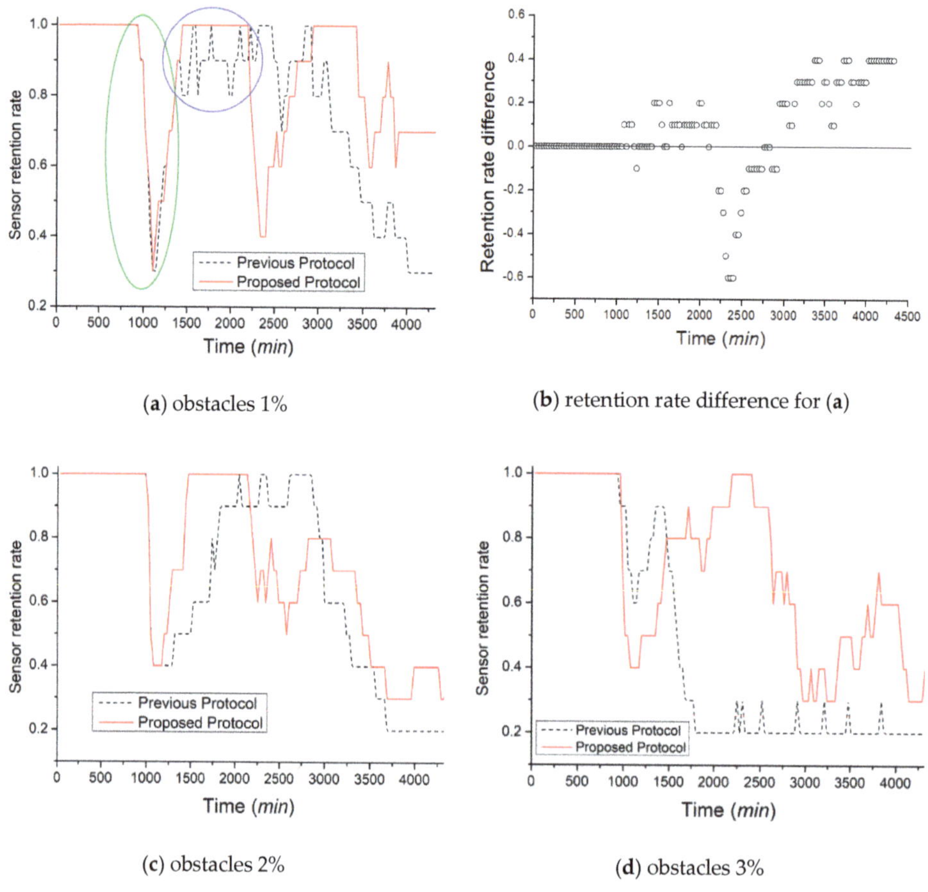

(**a**) obstacles 1%

(**b**) retention rate difference for (**a**)

(**c**) obstacles 2%

(**d**) obstacles 3%

Figure 9. Sensor retention rates at the middle cluster.

Specially, in Figure 9d of obstacles 3%, our protocol successfully recovers the sensing hole better than that of the previous one. In other words, it is possible to strongly predict that data collection could be better than the compared protocol, even if the cluster zone still cannot recover the sensing hole.

In the simulation, every HELLO message is periodically broadcasted by each cluster header. The interval is set to be 15 min, and the cluster header should determine whether or not its zone is a sensing hole for every interval time. Whenever each cluster header checks the state of the sensing hole and sends REQ message, the moments for sensing hole appearance are shown in Figure 10. The delayed occurrence of sensing holes of this protocol compared to previous one indicates that more successful data collection is achieved. In addition, as we mentioned earlier in the message flows, the short duration of each occurrence of the sensing hole means that numerous messages are generated for REQ throughout the entire network. This could lead to unnecessary energy consumption due to message storms. Furthermore, in order to successfully accomplish the continuous desired data correction at the end of the wireless sensor network (i.e., the sink nodes), a relocation protocol has to avoid unnecessary energy consumption. Therefore, our relocation protocol is appropriate to reduce unnecessary energy consumption.

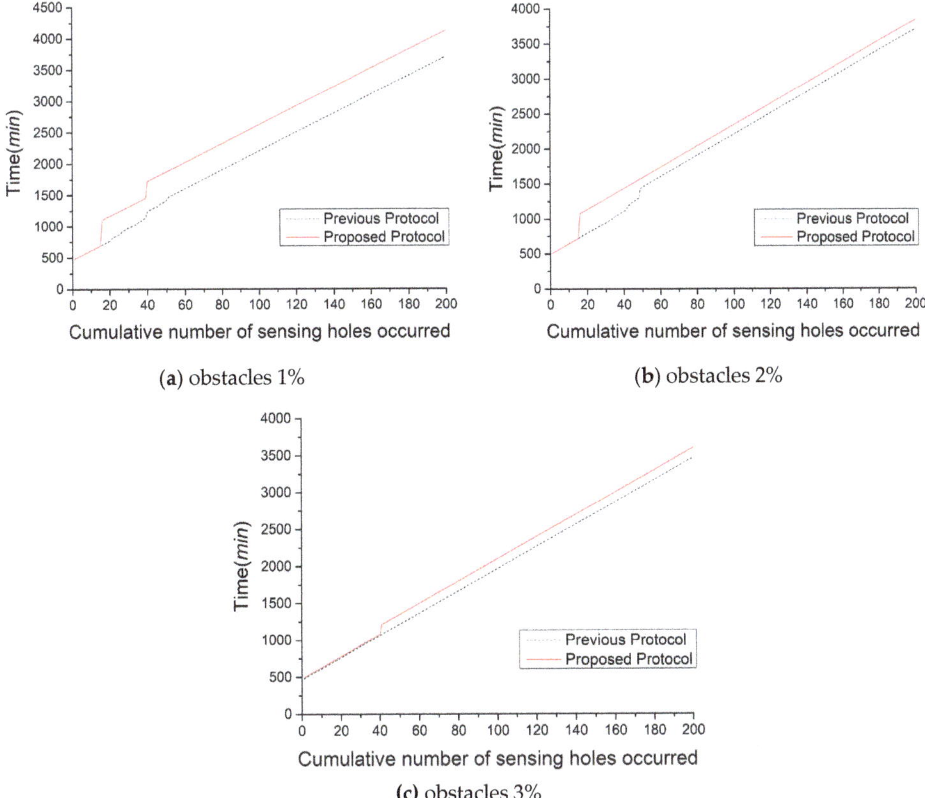

(**a**) obstacles 1%

(**b**) obstacles 2%

(**c**) obstacles 3%

Figure 10. The moments for sensing hole appearance at the middle cluster zone.

5. Conclusions

In the wireless sensor network, hundreds of sensors are dispatched in the desired area and sensors are continuously operating. Various previous studies were conducted, including the proper placement of sensors and the selection of cluster headers to transmit the collected data to the desired sink nodes. However, wireless sensors are inherently energy-limited, thus the appearance of the sensing hole cannot be avoidable and sensing hole recovery to maintain network operation is necessary. In particular, since the charging system of the sensor is not always available, it is essential to consider energy-efficiency from the time the sensor is relocated to recover the sensing hole.

While there has been vigorous study of hopping mobility models suitable for rough areas recently, the majority of past studies have been based on not practical classic theory (i.e., relocation schemes based on source-based routing). In this paper, overcoming these problems, a new hopping sensor relocation protocol is proposed to suit the real distributed environment. To reflect the real situations, obstacles of a given area (stones, mud, etc.) have been considered. Since the status of obstacles around the sensing hole is unknown, the condition of the surrounding environment is predicted through the success rate of the hopping sensor relocation.

We would strongly emphasize that the relocation protocol proposed by this research team is a protocol in a distributed environment and the successful implementation of OMNeT++ simulation in the hopping sensor relocation protocol is our unique skill in this field of research. Various obstacles were considered for performance evaluation and analysis. Finally, superior energy efficiency with delaying sensing hole appearance of the proposed scheme was demonstrated.

Author Contributions: S.P. and M.K. are the co-first authors and contributed equally. S.P., M.K., and W.L. designed the protocol and the simulation process. S.P., M.K., and W.L. coordinated the grant funding, conducted the study, analyzed the data, and drafted the first version of the manuscript. The simulation was conducted by S.P. and M.K., and W.L. supervised the software development. All authors have read and agreed to the published version of the manuscript.

Funding: This work was supported by Research Assistance Program (2019) in the Incheon National University and the National Research Foundation of Korea (NRF) grant funded by the Ministry of Science, ICT & Future Planning (No. NRF - 2019R1G1A1007832).

Conflicts of Interest: The authors declare no conflict of interest.

References

1. Yu, S.; Liu, M.; Dou, W.; Liu, X.; Zhou, S. Networking for Big Data: A Survey. *IEEE Commun. Surv. Tutor.* **2017**, *19*, 531–549. [CrossRef]
2. Chen, M.; Mao, S.; Liu, Y. Big Data: A Survey. *Mob. Netw. Appl.* **2014**, *19*, 171–209. [CrossRef]
3. Yaqoob, I.; Ahmed, E.; Hashem, I.A.T.; Ahmed, A.I.A.; Gani, A.; Imran, M.; Guizani, M. Internet of Things Architecture: Recent Advances, Taxonomy, Requirements, and Open Challenges. *IEEE Wirel. Commun.* **2017**, *24*, 10–16. [CrossRef]
4. Zhang, Y.; Xiong, Z.; Niyato, D.; Wang, P.; Kim, D.I. Toward a Perpetual IoT System: Wireless Power Management Policy With Threshold Structure. *IEEE Internet Things J.* **2018**, *5*, 5254–5270. [CrossRef]
5. Ray, P.P.; Mukherjee, M.; Shu, L. Internet of Things for Disaster Management: State-of-the-Art and Prospects. *IEEE Access* **2017**, *5*, 18818–18835. [CrossRef]
6. Chudzikiewicz, J.; Furtak, J.; Zielinski, Z. Fault-tolerant techniques for the Internet of Military Things. In Proceedings of the IEEE 2nd World Forum on Internet of Things (WF-IoT), Milan, Italy, 14–16 December 2015; pp. 496–501.
7. Kosar, R.; Onur, E.; Ersoy, C. Redeployment Based Sensing Hole Mitigation in Wireless Sensor Networks. In Proceedings of the IEEE 2009 Wireless Communications and Networking Conference (WCNC), Budapest, Hungary, 5–8 April 2009.
8. Kim, M.; Park, S.; Lee, W. A Robust Energy Saving Data Dissemination Protocol for IoT-WSNs. In *KSII Transactions on Internet and Information Systems (TIIS)*; 2018; pp. 5744–5764. [CrossRef]
9. Kim, M.; Jeong, E.; Bang, Y.-C.; Hwang, S.; Shin, C.; Jin, G.-J.; Kim, B. An Energy-aware Multipath Routing Algorithm in Wireless Sensor Networks. *IEICE Trans. Inf. Syst.* **2008**, *91*, 2419–2427. [CrossRef]
10. Elappila, M.; Chinara, S.; Parhi, D.R. Survivable Path Routing in WSN for IoT applications. *Pervasive Mob. Comput.* **2018**, *43*, 49–63. [CrossRef]
11. Luo, R.C.; Huang, J.-T.; Chen, O. A Triangular Selection Path Planning Method with Dead Reckoning System for Wireless Mobile Sensor Mote. In Proceedings of the 2006 IEEE International Conference on Systems, Man and Cybernetics, Taipei, Taiwan, 8–11 October 2006; pp. 162–168.
12. Chellappan, S.; Snyder, M.E.; Thakur, M. Distributed exploratory coverage with limited mobility. *Int. J. Space-Based Situated Comput.* **2014**, *4*, 114–124. [CrossRef]
13. Snyder, M.E. Foundations of Coverage Algorithms in Autonomic Mobile Sensor Networks. Ph.D. Thesis, Missouri University of Science and Technology, Rolla, MO, USA, 2014.
14. Zhao, J.; Xu, J.; Gao, B.; Xi, N.; Cintrón, F.J.; Mutka, M.W.; Xiao, L. MSU Jumper: A Single-Motor-Actuated Miniature Steerable Jumping Robot. *IEEE Trans. Robot.* **2013**, *29*, 602–614. [CrossRef]
15. Cen, Z.; Mutka, M.W. Relocation of Hopping Sensors. In Proceedings of the IEEE International Conference on Robotics and Automation (ICRA 08), Pasadena, CA, USA, 19–23 May 2008; pp. 569–574.
16. Kim, M.; Mutka, M.W. *On Relocation of Hopping Sensors for Balanced Migration Distribution of Sensors*; Springer: Berlin, Heidelberg, 2009; pp. 361–371.
17. Kim, M.; Mutka, M.W. Multipath-based Relocation Schemes Considering Balanced Assignment for Hopping Sensors. In Proceedings of the IEEE/RSJ International Conference on Intelligent Robots and Systems (IROS 09), St. Louis, MO, USA, 10–15 October 2009; pp. 5095–5100.
18. Kim, M.; Mutka, M.W.; Choo, H. On Relocation of Hopping Sensors for Rugged Terrains. In Proceedings of the IEEE International Conference on Computational Sciences and its Applications (ICCSA 10), Fukuoka, Japan, 23–26 March 2010; pp. 203–210.

19. Kim, M.; Park, S.; Lee, W. Energy and Distance-Aware Hopping Sensor Relocation for Wireless Sensor Networks. *Sensors* **2019**, *19*, 1567. [CrossRef] [PubMed]

20. OMNeT Web Site. Available online: https://www.omnetpp.org (accessed on 9 December 2019).

21. Cintr'on, F.; Pongaliur, K.; Mutka, M.W.; Xiao, L.; Zhao, J.; Xi, N. Leveraging height in a jumping sensor network to extend network coverage. *IEEE Trans. Wirel. Commun.* **2012**, *11*, 1840–1849. [CrossRef]

22. Cintr'on, F. Network Issues for 3D Wireless Sensors Networks. Ph.D. Thesis, Michigan State University, East Lansing, MI, USA, 2013.

23. Kim, M.; Kim, T.; Shon, M.; Kim, M.; Choo, H. Design of a Transmission Process for Hopping Sensors to Enhance Coverage. In Proceedings of the International Conference Wireless Networks (ICWN 10), Las Vegas, NV, USA, 12–15 July 2010; pp. 377–382.

24. Rostami, A.S.; Badkoobe, M.; Mohanna, F.; Keshavarz, H.; Hosseinabadi, A.A.R.; Sangaiah, A.K. Survey on clustering in heterogeneous and homogeneous wireless sensor networks. *J. Supercomput.* **2018**, *74*, 277–323. [CrossRef]

25. Sabor, N.; Sasaki, S.; Abo-Zahhad, M.; Ahmed, S.M. A Comprehensive Survey on Hierarchical-Based Routing Protocols for Mobile Wireless Sensor Networks: Review, Taxonomy, and Future Directions, Hindawi. *Wirel. Commun. Mob. Comput.* **2017**, *2017*, 23. [CrossRef]

26. Zarrad, A.; Alsmadi, I. Evaluating network test scenarios for network simulators systems. *International J. Distrib. Sens. Netw.* **2017**, *13*, 1550147717738216. [CrossRef]

27. Virdis, A.; Kirsche, M. *Recent Advances in Network Simulation: The OMNeT++ Environment and Its Ecosystem*; Springer: Cham, Switzerland, 2019. [CrossRef]

Article

An Ultra-Low-Power Area-Efficient Non-Coherent Binary Phase-Shift Keying Demodulator for Implantable Biomedical Microsystems

Milad Ghazi [1], Mohammad Hossein Maghami [1,*], Parviz Amiri [1] and Sotoudeh Hamedi-Hagh [2,*]

[1]   Faculty of Electrical Engineering, Shahid Rajaee Teacher Training University, Tehran 1678815811, Iran; milad.ghazi.sru@gmail.com (M.G.); pamiri@sru.ac.ir (P.A.)

[2]   Department of Electrical Engineering, San Jose State University, San Jose, CA 95192, USA

*   Correspondence: mhmaghami@sru.ac.ir (M.H.M.); sotoudeh.hamedi-hagh@sjsu.edu (S.H.-H.)

Received: 16 June 2020; Accepted: 7 July 2020; Published: 10 July 2020

Abstract: A novel non-coherent, low-power, area-efficient binary phase-shift keying demodulator for wireless implantable biomedical microsystems is proposed. The received data and synchronized clock signal are detected using a delayed digitized format of the input signal. The proposed technique does not require any kind of oscillator circuit, and due to the synchronization of all circuit signals, the proposed demodulator can work in a wide range of biomedical data telemetry common frequencies in different process/temperature corners. The presented circuit has been designed and post-layout-simulated in a standard 0.18 μm CMOS technology and occupies 17×27 μm^2 of active area. Post-layout simulation results indicate that with a 1.8 V power supply, power consumption of the designed circuit is 8.5 μW at a data rate of 20 Mbps. The presented demodulation scheme was also implemented on a proof-of-concept circuit board for verifying its functionality.

Keywords: implantable biomedical microsystems; data telemetry; low power; high data rate; binary phase-shift keying demodulation

---

1. Introduction

In Implantable Biomedical Microsystems (IBM), a wireless interface is used for transmission of power and data between the internal and external parts of the system (Figure 1). The most important issues in a wireless link for IBM are data transfer rate, power consumption, and chip area [1–3]. It is evident that in a forward data transfer, the maximum data rate is desired, especially in applications such as visual prostheses in which the implanted microsystem is in direct contact with the central neural system [4]. Furthermore, due to the power loss in the power-transmission circuitry and power dissipation in the tissue [5], the maximum frequency of the carrier signal for the IBMs is limited to a few tens of megahertz [1–5].

There have been proposed various modulation techniques for wireless data transfer [1,4,6]. Making comparisons between the various digital modulation schemes from different point of views, reveals that binary phase-shift keying (BPSK) structure is more appropriate in most IBMs [7–13]. Generally, this scheme provides proper insensitivity to amplitude noise, high data rate, and good power transfer efficiency. Moreover, if a capacitive link is utilized for data transmission, fast phase variations can be attained within every carrier cycle and data-rate-to-carrier-frequency (DRCF) ratios of as high as 100% are achievable in this scheme [1].

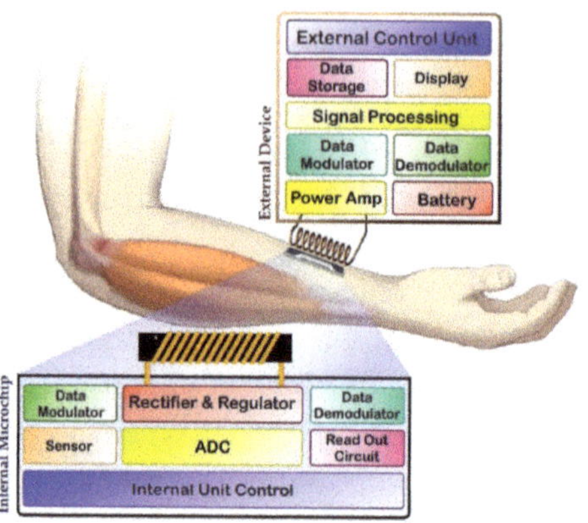

Figure 1. Simplified general block diagram of an Implantable Biomedical Microsystem (IBM).

BPSK demodulation can be realized in two ways: coherent and non-coherent structures. In coherent detection, phase synchronization between the received signal and the receiver is essential [8]. Most of the digital coherent demodulators consume lots of power and suffer from circuit complexity. Consequently, non-coherent BPSK demodulators are widely used in radio frequency applications and IBMs due to their lower circuit complexity, occupying less chip area, and lower power consumption, despite higher bit error rates [9,10].

Most of the non-coherent BPSK demodulation schemes are designed based on edge-detection technique [1,14], pulse width measurement [15], and filtering techniques [16]. Among all the benefits of these circuits, there are some disadvantages, like high power consumption and occupying a large area. This article introduces a new high-data-rate, low-power, non-coherent BPSK demodulator for fixed-in position implantable biomedical applications such as cochlear implants and visual prostheses, where the link properties such as distance between receiver and transmitter, orientation, and alignment are easily available [1]. The proposed demodulator is based on using a delayed digitized format of the input signal for the detection of the received data as well as for recovering the clock signal. The operation of the proposed circuit is closely comparable in terms of power consumption, occupied active area, and reliability to most of the existing similar works.

2. Proposed BPSK Demodulator and Clock Recovery Circuit

Figures 2 and 3 depict the block diagram of the designed demodulator as well as its essential theoretical waveforms, respectively. It should be noted that placing an automatic gain control as the first stage of the receiver is needed in applications where orientation and matching between the transmitting and receiving blocks are lost [17]. However, please note that as with fixed-in position implantable biomedical applications [1] where the link properties such as distance between receiver and transmitter, orientation, and alignment are easily available, it is not needed to use automatic gain control in the receiver. In these cases, the minimum required level of received signal for correct operation of the circuit is provided by the power amplifier in the external part according to the link properties. Moreover, in these cases, as the received BPSK signal level is high enough, power can also easily deliver wirelessly to the implant through the energy contained in the incoming BPSK data signal [18,19]. In this architecture, the rectified BPSK signal is digitized by logic inverters, to be used as an auxiliary clock with twice the BPSK frequency, and is called "2XClock". The digitized BPSK signal is then fed to the half-cycle delay block, containing a D flip-flop [20] which operates with 2XClock signal. For extracting data, a low-power data detector proposed in this work will retrieve original data from the digitized BPSK signal and a delayed digitized BPSK signal whose operation is described later in this section. In addition, through comparing the digitized BPSK and data recovered signals, only by a low-power XOR logic gate [21], clock recovery will take place.

In this section, in addition, through comparing the digitized BPSK and data recovered signals, only by allowing proper XOR operation, clock recovery will take place.

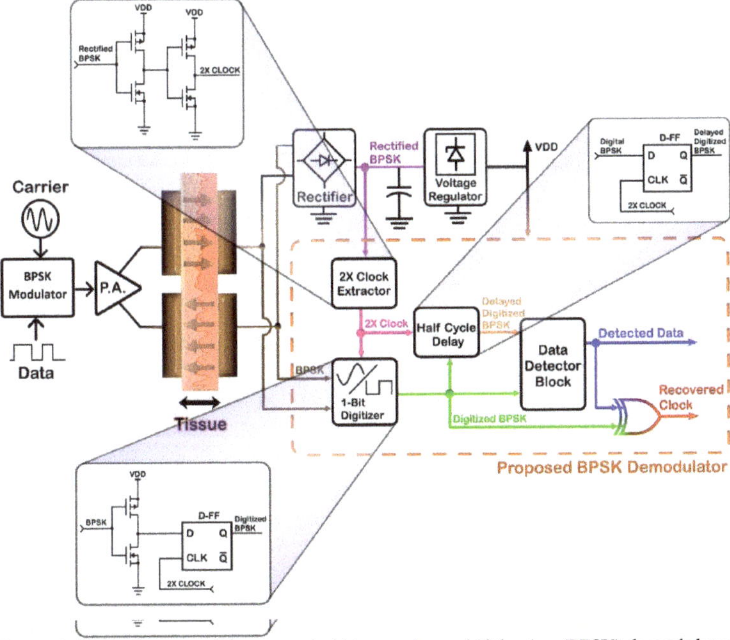

Figure 2. Block diagram of the presented binary phase-shift keying (BPSK) demodulator.

Figure 2. Block diagram of the presented binary phase-shift keying (BPSK) demodulator.

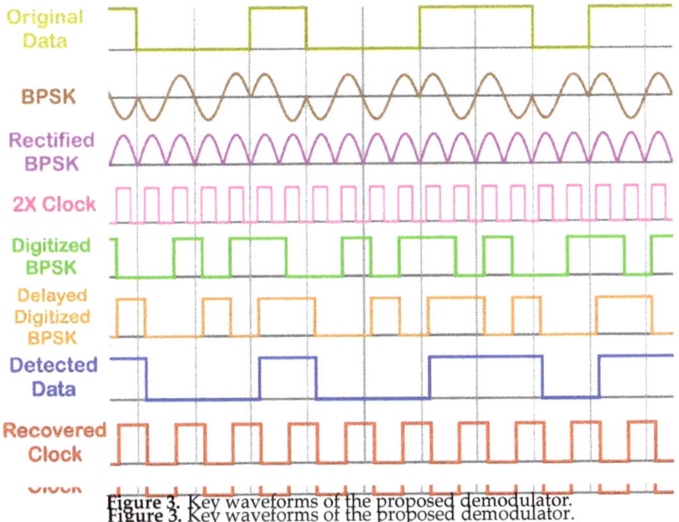

Figure 3. Key waveforms of the proposed demodulator.

Figure 3. Key waveforms of the proposed demodulator.

It can be understood from Figure 3 that to retrieve the original data from the digitized BPSK and delayed digitized BPSK signals, a circuit is needed so that when "0, 0" (digitized BPSK, delayed digitized BPSK, respectively) state is received, the output signal changes its value to "0" and remains unchanged while inputs are constant, and when "1, 1" state is received, the output signal changes its value to "1" and remains unchanged while inputs are not changed to "0, 0". It is important that in two another states ("0, 1" and "1, 0"), the output must not change and should retain its previous value. As this is shown in Figure 3, some logical states between the digitized BPSK signal and delayed digitized BPSK signal never happen, for example, "0, 0" after each other or "0, 1" after "0, 0" state. All situations that are possible to happen are illustrated in Figure 4.

value. As it is shown in Figure 3, some logical states between the digitized BPSK signal and delayed digitized BPSK signal never happen, for example, "0, 0" and "1, 1" after each other or "0, 1" after "0, 1" state. All situations that are possible to happen are illustrated in Figure 4.

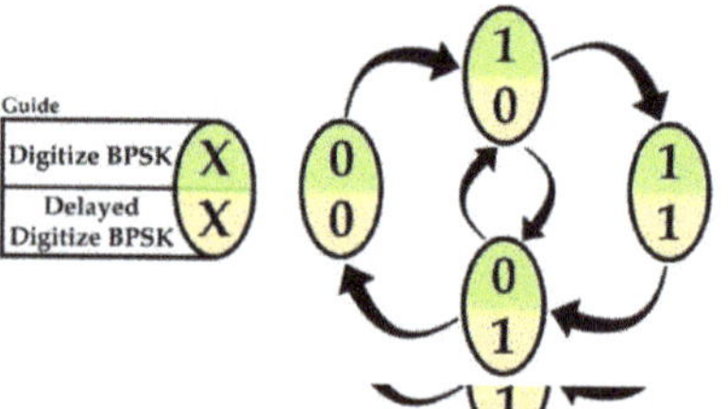

Figure 4. Possible states of data detector inputs.

According to above mentioned hints, a simple low-power circuit for retrieving data from existing signals is proposed (Figure 5). For understanding the operation of the proposed data detector, suppose that both digitized BPSK and delayed digitized BPSK signals are in low logic level. In this case, $M_{1,8,9}$ transistors are ON and $M_{3,4,6}$ devices are OFF. Therefore, "B" node will be shortened to ground and accordingly, "A" node is connected to V_{dd} via $M_{1,2}$ transistors. Likewise, M_{10} will be ON and M_5 will be OFF. In this state, the output level is low, and it will remain unchanged. According to Figure 4, after "0, 0" state, only "1, 0" state can happen. In this case, again, $M_{1,2,10}$ Mosfets are ON and the output voltage retains its previous value. Now, when "1, 1" signals are received at data detector inputs, opposite the "0, 0" state, $M_{3,4}$ transistors are ON and "A" node will be shortened to ground, and then M_5 will be ON and given that the M_6 was ON, "B" node will be connected to V_{dd}. Consequently, M_8 will be ON and makes "A" node unchanged at the low level. Similarly, when inputs change to "1, 0" state, the output level will not change. In this way, the original data can be detected by the circuit proposed in Figure 5. Transistor dimensions of the data detector circuit are reported in Table 1.

Table 1. Transistor dimensions of the data detector circuit.

Device Name	W/L
$M_{1,2,6,7}$	0.88 um/0.18 um
$M_{3,4,8,9}$	0.22 um/0.18 um
$M_{5,10}$	0.44 um/0.18 um

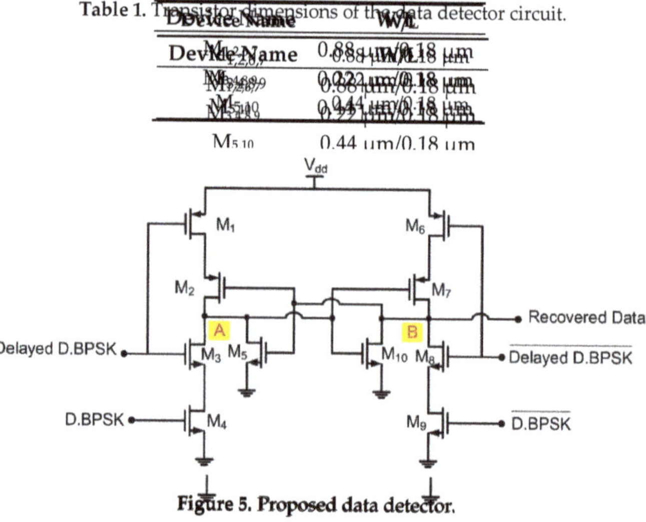

Figure 5. Proposed data detector.

3. Post-Layout Simulation and Experimental Results

The presented circuit has been designed and simulated in a 0.18 um CMOS process. The designed modulator occupies a total area of 17×27 um² (Figure 6). Important waveforms that resulted from post-layout simulations in the typical NMOS and PMOS corner case at 20 MHz carrier frequency are shown in Figure 7. All traces are named with the ones used in Figure 2. A power-on-reset signal makes the reset signal needed to initialize the flip-flops and all other circuits at the beginning and makes them ready for the correct operation. According to post-layout simulations, demodulated data has a latency of 5 ns from the input BPSK signal which can be ignored in noncoherent detection. data post-layout simulations in the typical NMOS and PMOS corner case at 20 MHz carrier frequency are shown in Figure 7. All traces are named with the ones used in Figure 2. A power-on-reset signal makes the reset signal needed to initialize the flip-flops and all other circuits at the beginning and makes them ready for the correct operation. According to post-layout simulations, demodulated data has a latency of 5 ns from the input BPSK signal which can be ignored in noncoherent detection.

Electronics **2020**, *9*, x FOR PEER REVIEW 5 of 11

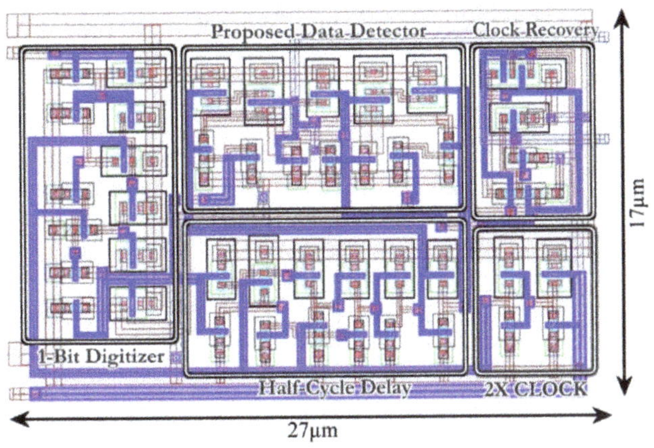

Figure 6. Layout view of the presented circuit.

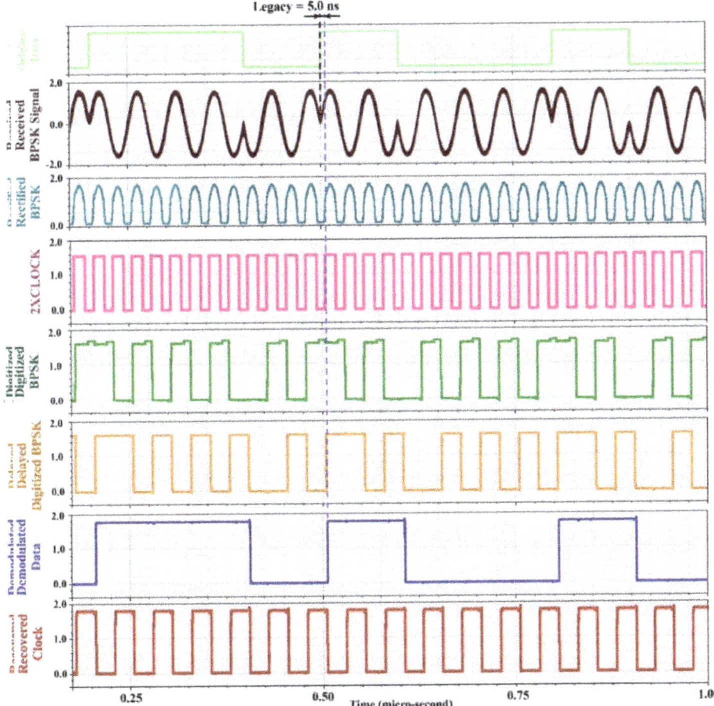

Figure 7. Transient response of the presented BPSK demodulator at 20 MHz carrier frequency.

The proposed circuit was post-layout simulated in various process/temperature corners, with ±10% changes in power supply, for all of which correct operation was seen. The demodulated data and the recovered clock, as well as some other waveforms, in some process/temperature corners are shown in Figures 8–10 at a 20 MHz carrier frequency that resulted from post-layout simulations. The results of post-layout simulations for different process corners and temperature variations in 0.18 µm CMOS process are summarized in Table 2, which indicates that the overall performance of the

The proposed circuit was post-layout-simulated in various process/temperature corners, with ±10% changes in power supply, for all of which correct operation was seen. The demodulated data and the recovered clock, as well as some other waveforms, in some process/temperature corners are shown in Figures 8–10 at a 20 MHz carrier frequency that resulted from post-layout simulations. The results of post-layout simulations for different process corners and temperature variations in 0.18 μm CMOS process are summarized in Table 2, which indicates that the overall performance of the designed circuit is robust against process/voltage/temperature (PVT) variations. Note that these results obtained from circuit simulations are in the typical NMOS typical PMOS (TT), slow NMOS slow PMOS (SS), and fast NMOS fast PMOS (FF) corner cases.

Electronics **2020**, *9*, x FOR PEER REVIEW 6 of 11

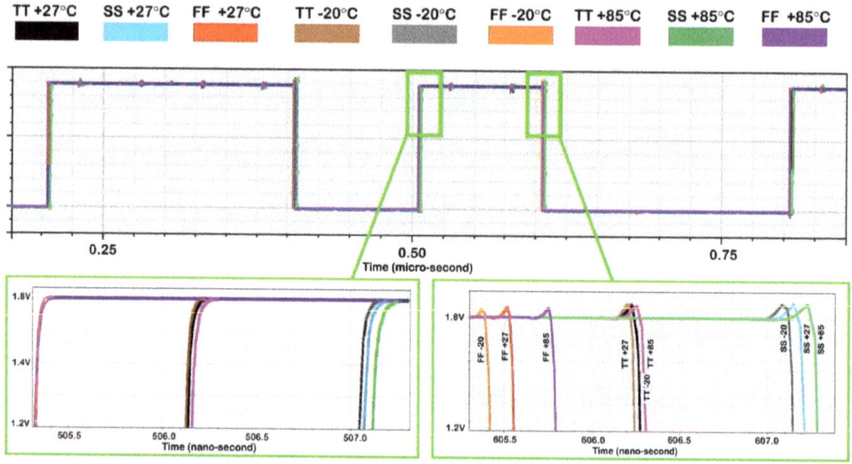

Figure 8. Process corners/temperatures simulation for demodulated data.

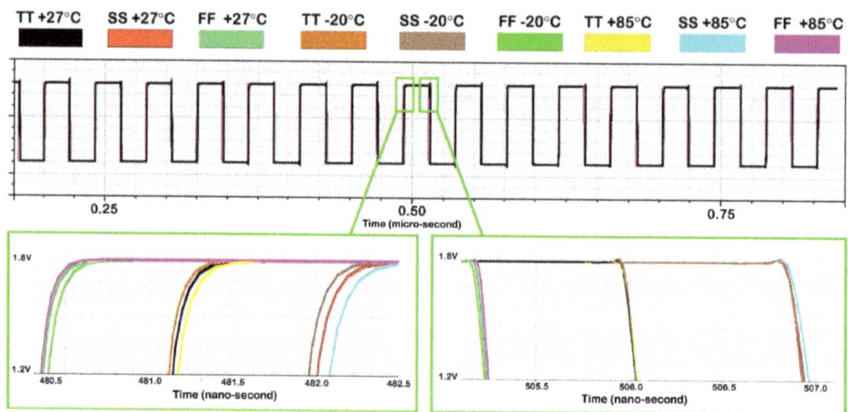

Figure 9. Process corners/temperatures simulation for the recovered clock.

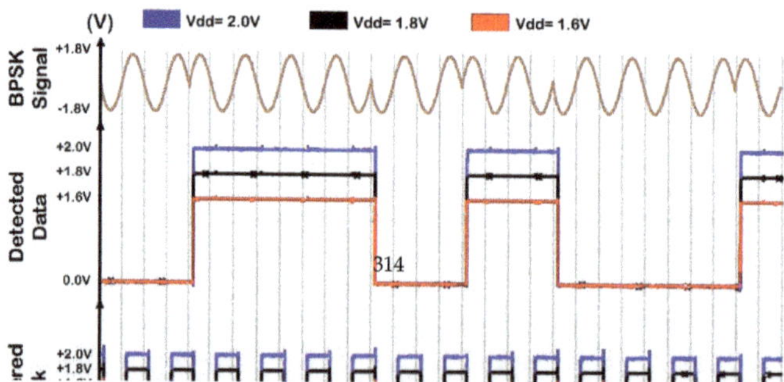

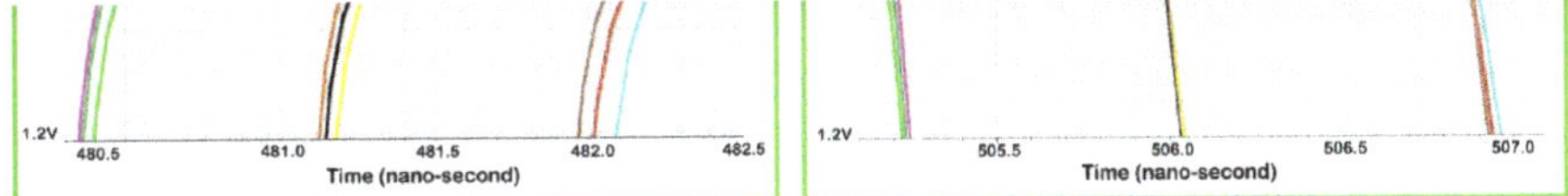

Figure 9. Process corners/temperatures simulation for the recovered clock.

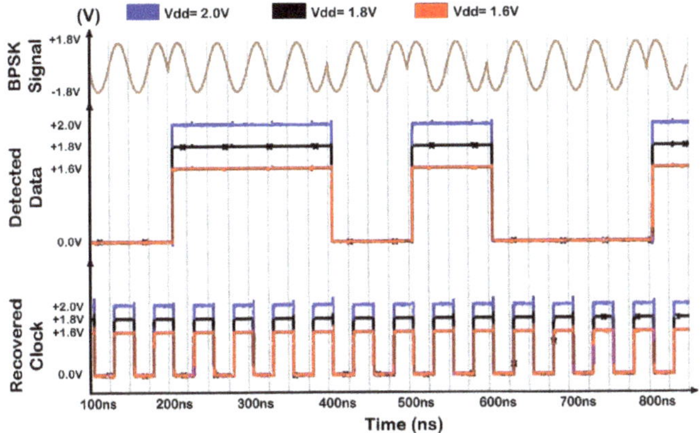

Figure 10. Demodulated data and the recovered clock with ±10% V_{dd} variation.

Table 2. Performance summary of the proposed demodulator.

Process Corner/Temperature	TT/27 °C	SS/85 °C	FF/−40 °C
Power Consumption * (1.8 V V_{dd})	8.7 µW	5.9 µW	13.2 µW

* @ 20 MHz carrier frequency and 20 Mbps data rate.

As stated in Section 2, there is no need to place automatic gain control in the receiver in fixed-in position implantable biomedical applications where the minimum required level of received signal for correct operation of the circuit is provided by the power amplifier in the external part according to the link properties. However, in order to show the correct operation of the proposed circuit against variations on received BPSK signal level, Figure 11 depicts the post-layout-simulated behavior of the proposed demodulator with ±30% changes in the received BPSK signal level, for all of which correct operation was seen.

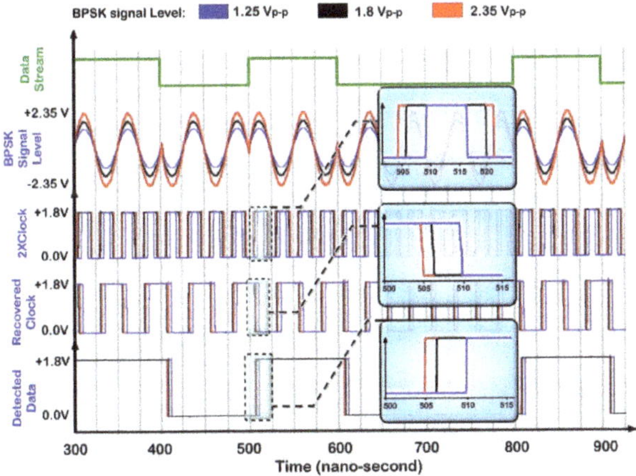

Figure 11. Demodulated data and the recovered clock with ±30% variations on received BPSK signal level.

In order to achieve accurate simulations considering fabrication process variations, the Monte Carlo analysis is performed on the designed BPSK demodulator by applying both process variations (such as variations on threshold voltage) and transistors' aspect ratio mismatches for all the devices employed. Figures 12 and 13 show the simulated transient response of the proposed circuit for demodulated data and the recovered clock, respectively, with 500 iterations. As it can be seen, the effect of non-idealities on the performance of the proposed circuit is negligible. Moreover, Figure 14 illustrates the Monte Carlo simulation of power consumption with 100 iterations. As it can be seen, the circuit consumes as low as 8.5 µW from the power supply of 1.8 V at the data rate of 20 Mbps, and the highest deviation of power consumption from its typical value is almost 30%.

employed. Figures 12 and 13 show the simulated transient response of the proposed circuit for demodulated data and the recovered clock, respectively, with 500 iterations. As it can be seen, the effect of non-idealities on the performance of the proposed circuit is negligible. Moreover, Figure 14 illustrates the Monte Carlo simulation of power consumption with 100 iterations. As it can be seen, the circuit consumes as low as 8.5 µW from the power supply of 1.8 V at the data rate of 20 Mbps, and the highest deviation of power consumption from its typical value is almost 30%.

Electronics **2020**, *9*, x FOR PEER REVIEW 8 of 11

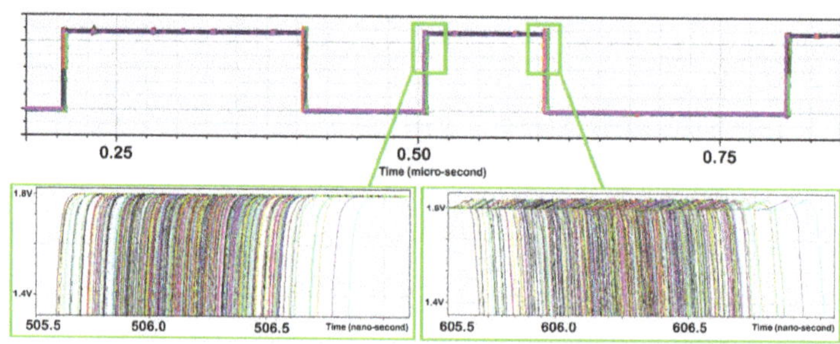

Figure 12. Monte Carlo simulations of demodulated data with 500 iterations.

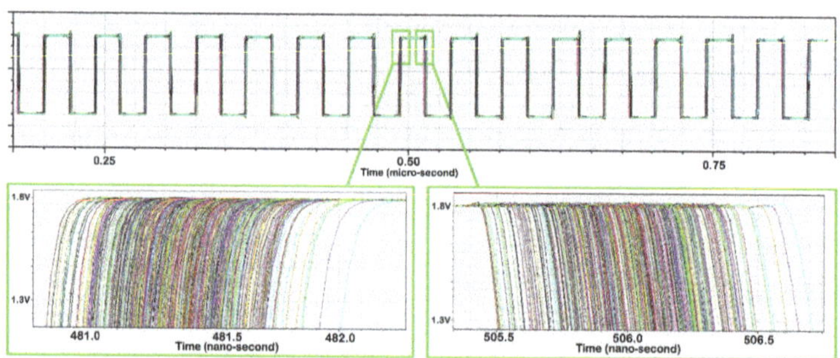

Figure 13. Monte Carlo simulation of the recovered clock with 500 iterations.

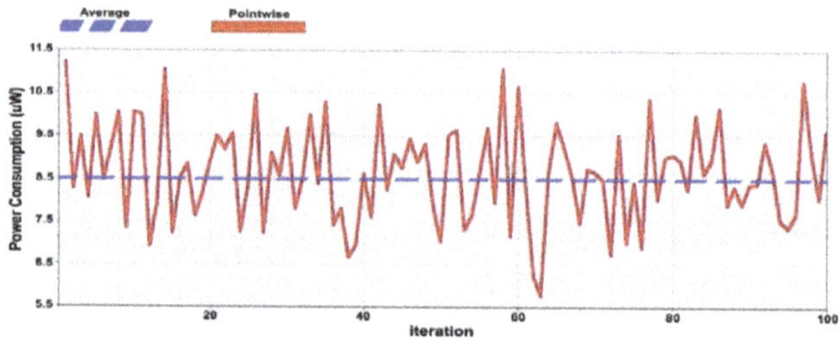

Figure 14. Monte Carlo simulation of power consumption with 100 iterations.

For verifying the functionality of the presented scheme in the circuit level, a proof-of-concept prototype was developed via TTL logic gates, a D flip-flop, an inductive coupling coil, and a diode rectification bridge, a photograph of which is shown in Figure 15. In this case, a 1 MHz, 1 Mbps BPSK signal with an amplitude of 5 V was employed as the input signal. Figure 16 shows an oscilloscope screen picture, presenting the important waveforms, which supports the proposed idea.

Performance comparisons between the proposed BPSK demodulator and other reported designs in the 0.18 µm CMOS process working at 1.8 V power supply are presented in Table 3. Note that the presented circuit is also simulated in different frequencies for making better comparisons. The simulation shows that static power dissipation is dominant at low operating frequencies. It can be seen that the designed circuit works with extremely low power consumption and within a small occupied active area, with a DRCF ratio of as high as 100%.

The simulation shows that static power dissipation is dominant at low operating frequencies. It can be seen that the designed circuit works with extremely low power consumption and within a small occupied active area, with a DRCF ratio of as high as 100%.

Electronics **2020**, *9*, x FOR PEER REVIEW　　9 of 11

Table 3. Performance comparisons with other similar works.

	[2]	[8]	[10]	[11]	[14]	[15]	[16] *	[22]	[23]	[24]	This Work		
Frequency (MHz)	10	13.56	10	16	4	10	1	8	10	10	2	10	20
Data rate (Mbps)	10	1.12	10	16	0.8	10	1	8	10	10	2	10	20
DRCF (%)	100	8.25	100	100	20	100	100	100	100	100	100	100	100
Chip Area (μm^2)	-	19×10^4	-	-	4300	3500	-	-	-	1120	459	459	459
Power Consumption (μW)	119	610	232	27	59	77.9	88.2	148	27.2	14	5.6	6.7	8.5
FOM $\left(\left(\dfrac{DRCF(\%)}{Power(\mu W)}\right) \times 10^{-3}\right)$	8.4	0.14	4.3	37	3.4	12.8	11.3	6.7	36.7	71.4	160	144	113

* Without clock recovery circuit.

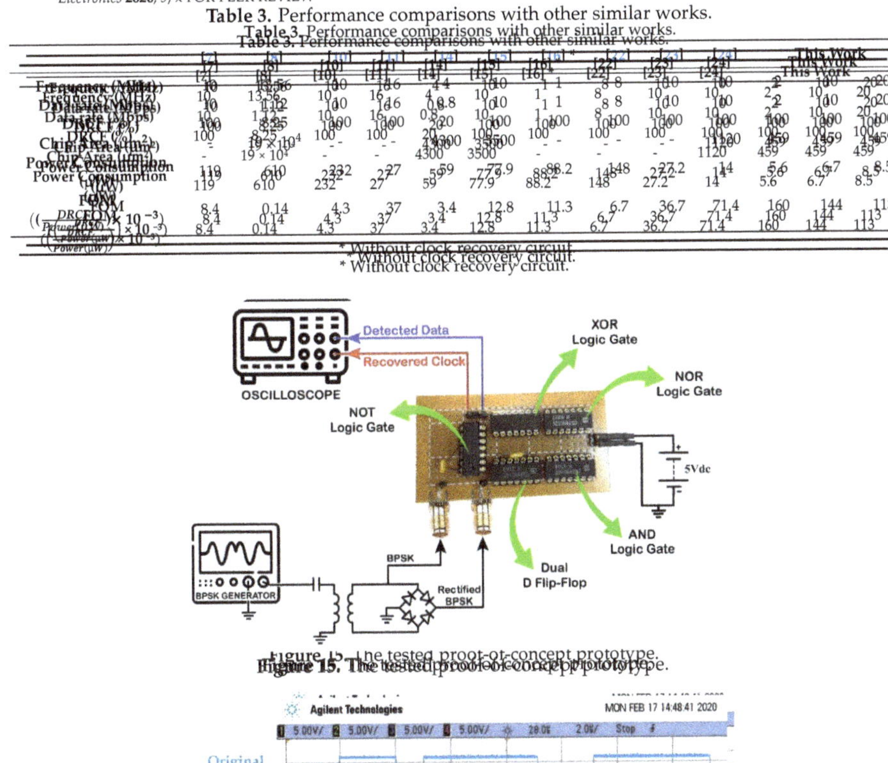

Figure 15. The tested proof-of-concept prototype.

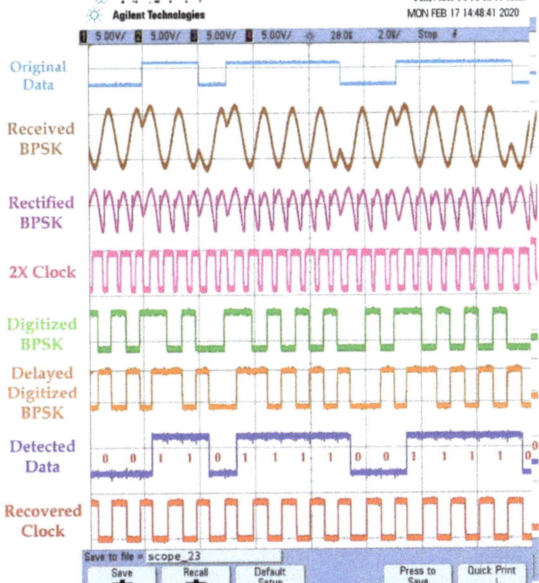

Figure 16. Experimental waveforms that resulted from the proof-of-concept prototype.

4. Conclusions

In this article, a low-power, high-data-rate, area-efficient BPSK demodulator utilizing a delayed digitized format of the input BPSK signal is presented. The presented circuit is designed and simulated in a standard 0.18 µm CMOS process using a 1.8 V power supply. Post-layout simulation and Monte Carlo analysis show that the presented circuit (including the clock recovery block) consumes 5.6 µW, 6.7 µW, and 8.5 µW at frequencies of 2 MHz, 10 MHz, and 20 MHz, respectively. The occupied active area of the whole circuit is 17×27 µm^2 and besides simplicity and low power consumption, the designed demodulator circuit benefits from a DRCF of 100%.

Author Contributions: Conceptualization and design, M.G.; formal analysis, M.G.; software, M.G.; investigation, M.H.M.; writing—original draft preparation, M.G.; writing—review and editing, M.H.M.; supervision, M.H.M., P.A., and S.H.-H.; funding acquisition, S.H.-H. All authors have read and agreed to the published version of the manuscript.

Funding: This research received no external funding.

Conflicts of Interest: The authors declare no conflict of interest.

References

1. Asgarian, F.; Sodagar, A.M. Wireless Telemetry for Implantable Biomedical Microsystems. In *Biomedical Engineering, Trends in Electronics, Communications and Software*; Laskovski, A.N., Ed.; InTech Press: Rijeka, Croatia, 2011; pp. 21–44.
2. Gati, E.; Kokosis, S.; Patsourakis, N.; Manias, S. Comparison of Series Compensation Topologies for Inductive Chargers of Biomedical Implantable Devices. *Electronics* **2020**, *9*, 8. [CrossRef]
3. Bazaka, K.; Jacob, M.V. Implantable Devices: Issues and Challenges. *Electronics* **2013**, *2*, 1–34. [CrossRef]
4. Zabihian, A.R.; Maghami, M.H.; Sodagar, A.M. Implantable Biomedical Devices. In *Biomedical Engineering, Technical Applications in Medicine*; Hudak, R., Penhaker, M., Majernik, J., Eds.; InTech Press: Rijeka, Croatia, 2012; pp. 157–190.
5. Ghovanloo, M.; Najafi, K. A wideband frequency-shift keying wireless link for inductively powered biomedical implants. *IEEE Trans. Circuit Syst. I* **2004**, *51*, 2374–2383. [CrossRef]
6. Lee, B.; Ghovanloo, M. An overview of data telemetry in inductively powered implantable biomedical devices. *IEEE Commun. Mag.* **2019**, *57*, 74–80. [CrossRef]
7. Asgarian, F.; Sodagar, A.M. A Carrier-Frequency-Independent BPSK Demodulator with 100% Data-Rate-To-Carrier-Frequency Ratio. In Proceedings of the 2010 IEEE Biomedical Circuits and Systems Conference, Paphos, Cyprus, 3–5 November 2010; pp. 29–32.
8. Hu, Y.; Sawan, M. A fully-integrated low power BPSK demodulator for implantable medical devices. *IEEE Trans. Circuit Syst. I* **2005**, *52*, 2552–2562.
9. Luo, Z.; Sonkusale, S. A novel BPSK demodulator for biological implants. *IEEE Trans. Circuit Syst. I* **2008**, *55*, 1478–1484.
10. Asgarian, F.; Sodagar, A.M. A low-power noncoherent BPSK demodulator and clock recovery circuit for high-data-rate biomedical applications. In Proceedings of the 31st International Conference of the IEEE/EMBS, Minneapolis, MN, USA, 3–6 September 2009; pp. 4840–4843.
11. Nabovati, G.; Ghafar-Zadeh, E.; Awwad, F.; Sawan, M. Fully digital low-power self-calibrating BPSK demodulator for implantable biosensors. In Proceedings of the IEEE 55th International Midwest Symposium on Circuits and Systems, Boise, ID, USA, 5–8 August 2012; pp. 354–357.
12. Wilkerson, B.P.; Jang, J.-K. A low power BPSK demodulator for wirelessly implantable biomedical devices. In Proceedings of the International Symposium on Circuits and Systems, Beijing, China, 19–23 May 2013; pp. 626–629.
13. Liu, H.-H.; Wu, Z.-H.; Li, B.; Zhao, M.-J. A fully digital low-power wide-speed-range BPSK demodulator. *Int. J. Electron. Lett.* **2014**, *2*, 158–165. [CrossRef]
14. Gong, C.-S.A.; Shiue, M.-T.; Yao, K.-W.; Chen, T.-Y. Low-power and area-efficient PSK demodulator for wirelessly powered implantable command receivers. *Electron. Lett.* **2008**, *44*, 841–842.

15. Karimi, M.; Maghami, M.H.; Faizollah, M.; Sodagar, A.M. A noncoherent low-power high-data-rate BPSK demodulator and clock recovery circuit for implantable biomedical devices. In Proceedings of the IEEE Biomedical Circuits and Systems Conference (BioCAS), Lausanne, Switzerland, 22–24 October 2014; pp. 372–375.

16. Beheshti, B.; Jalali, M. A low-power noncoherent BPSK demodulator for implantable medical devices. In Proceedings of the IEEE International Conference on Electronics Design, Systems and Applications (ICEDSA), Kuala Lumpur, Malaysia, 5–6 November 2012; pp. 203–206.

17. Eslampanah Sendi, M.S.; Nasiri, S.; Mousavi, N.; Sharifkhani, M.; Sodagar, A.M. A 3-d inductive powering approach dedicated to implantable/wearable biomedical microsystems. In Proceedings of the IEEE Biomedical Circuits and Systems Conference (BioCAS), Lausanne, Switzerland, 22–24 October 2014.

18. Sonkusale, S.; Luo, Z. A complete data and power telemetry system utilizing BPSK and LSK signaling for biomedical implants. In Proceedings of the International Conference of the IEEE Engineering in Medicine and Biology Society, Vancouver, BC, Canada, 20–25 August 2008; pp. 3216–3219.

19. Cheng, C.-H.; Tsai, P.-Y.; Yang, T.-Y.; Cheng, W.-H.; Yen, T.-Y.; Luo, Z.; Qian, X.-H.; Chen, Z.-X.; Lin, T.-H.; Chen, W.-H.; et al. A fully integrated 16-channel closed-loop neural-prosthetic CMOS SoC with wireless power and bidirectional data telemetry for real-time efficient human epileptic seizure control. *IEEE J. Solid-State Circuit* **2018**, *53*, 3314–3326. [CrossRef]

20. Prabhu Deva Kumar, S.V.S.; Nayini, L.; Ronald, E. Implementation of low power D-Flip Flop using 45nm Technology. *Int. J. Recent Sci. Res.* **2017**, *8*, 17729–17732.

21. Radhakrishnan, D. Low-voltage low-power CMOS full adder. *IEE Proc. Circuits Devices Syst.* **2001**, *148*, 19–24. [CrossRef]

22. Asgarian, F.; Sodagar, A.M. A high-data-rate low-power BPSK demodulator and clock recovery circuit for implantable biomedical devices. In Proceedings of the 4th International IEEE/EMBS Conference on Neural Engineering, Antalya, Turkey, 29 April–2 May 2009; pp. 407–410.

23. Javid, A.; Vahedian, H.; Sodagar, A.M.; Mofrad, M.E. Low-power, high data-rate, BPSK demodulator for implantable biomedical applications. In Proceedings of the IEEE ICECS, Marseille, France, 7–10 December 2014; pp. 415–418.

24. Hosseinnejad, M.; Erfanian, A. A VCO-free low-power fully digital BPSK demodulator for implantable biomedical microsystems. In Proceedings of the 24th IEEE International Conference on Electronics, Circuits and Systems (ICECS), Batumi, GA, USA, 5–8 December 2017; pp. 474–477.

 *electronics*

Article

In-Band Full-Duplex Relaying for SWIPT-Enabled Cognitive Radio Networks

Hieu V. Nguyen [1], Van-Dinh Nguyen [2] and Oh-Soon Shin [1,*]

[1] School of Electronic Engineering & Department of ICMC Convergence Technology, Soongsil University, Seoul 06978, Korea; hieuvnguyen@ssu.ac.kr
[2] Institute of Research and Development, Duy Tan University, Da Nang 550000, Vietnam; dinh.nguyen@uni.lu
* Correspondence: osshin@ssu.ac.kr

Received: 30 April 2020; Accepted: 15 May 2020; Published: 20 May 2020

Abstract: This paper studies sum rate maximization of a cognitive radio network, where a full-duplex relay (FDR) is considered to assist data transmission. An FDR equipped with multiple transmit/receive antennas is introduced to harvest energy from the radio frequency signal of the primary system to reuse the energy for its own data transmission. By exploiting the time-switching relaying protocol, we first formulate an optimization problem for the sum rate of primary and secondary receivers and then propose a low-complexity algorithm to find the optimal solution. Numerical results verify the effectiveness of the proposed technique for wireless information and power transfer in cognitive radio systems.

Keywords: cognitive radio; energy harvesting; full-duplex relay; simultaneous wireless information and power transfer (SWIPT); zero-forcing precoding

1. Introduction

In-band full-duplex (FD) radio has recently attracted much attention to improve the capacity of wireless communication systems. Theoretically, the use of FD radio is expected to double the spectral efficiency (SE) of a wireless channel [1]. However, the main challenge of an FD-based system is the self-interference (SI) due to the data transmission and reception operating at the same time and frequency. Accordingly, there have been many efforts to suppress SI, and thus, a small residual SI is usually taken into account [2–6]. Based on FD communications, relay-assisted networks have been studied in many works where the FD technique is used for relay nodes to forward the messages efficiently from the source to the destination nodes [7,8].

Another paradigm for radio resource sharing, known as cognitive radio (CR), allows an unlicensed secondary user (SU) to utilize the same spectrum that is allocated to a licensed primary user (PU), so that SE can be significantly enhanced [9–11]. To reap the benefits of both FD and CR, the authors in [12] proposed a new model, where an FD-enabled secondary transmitter (ST) is used to relay the messages from the primary transmitter (PT) to PU. The authors in [13] considered an FD relay (FDR) based on non-orthogonal multiple access (NOMA) to maximize the near user rate, while FDR selection applicable for multiple pairs of SUs and PUs was presented in [14].

Simultaneous wireless information and power transfer (SWIPT) has been accepted as a green solution for next-generation communication systems [15–21], in which the energy harvesting (EH) process enables the devices to recharge their own batteries through radio frequency (RF) signals. Consequently, SWIPT not only helps reduce the wasted energy, but also supports wireless networks with many devices, such as the Internet of Things (IoT) and sensor networks. There are four main types of SWIPT architectures: separate receiver, power splitting, time switching, and antenna switching. Among them, the time switching approach facilitates joint optimization in the signal processing under various conditions, i.e., interference management and quality-of-service (QoS) constraints.

In the context of SWIPT-integrated networks, there are many efforts to improve the system performance by applying SWIPT to both FD- and CR-based networks (CRNs). In [22], an FD-enabled system using SWIPT was proposed to improve spectral and energy efficiencies. In CRNs, the performance analysis for EH and system throughput was investigated in [23], indicating the effectiveness of SWIPT. To further enhance the performance for SWIPT-based CRN systems, the authors in [24] developed a model, where an FD-enabled access point simultaneously charges the battery of ST and receives the signals from PT in the first phase, and in the second phase, the ST transmits/forwards the messages to secondary/primary receivers (SR/PR). Although the CRN scheme in [23] could reuse the energy by sensing the appearance of the PT's signal, the conventional EH based on single-input single-output (SISO) cannot exploit the benefit of the multiple-antenna technique. In addition, the use of an energy beam in [22,24] might lead to inefficient power consumption at the energy beaming sources, i.e., base stations and APs, due to the strong attenuation caused by the path loss. The above discussions motivate us to develop an FD-based CRN that takes advantage of the beamforming technique for SWIPT.

This paper investigates a CRN where the data transmission between an ST and an SR is entirely performed via an FDR. Differently from previous works, the data transmission operates in two phases of a time block as in the time-switching relaying (TSR) protocol, where the fraction is optimized so that the sum rate of the primary and secondary systems is maximized. In the first phase, the ST and FDR harvest energy from the primary signal and utilize the harvested energy to transmit their independent signals. With the given spectrum access and harvested energy, in the second phase, the FDR is enabled to assist PT and ST in forwarding the data to PR and SR, respectively. To the best of our knowledge, a thorough performance evaluation of FDR enabled by EH in CRNs has not been reported in the literature. The proposed model brings some advantages: (i) inherited from the property of relaying, the FDR utilizes distances geometrically near receivers for improving the system performance; (ii) FDR is able to exploit the benefit of the multiple-antenna technique to harvest energy and transmit data efficiently; (iii) the proposed model is particularly suitable for CRNs, in that the CRNs can sense the appearances of PT's signals in one block transmission to recycle the energy from PT's signal, instead of requesting the new energy beam from PT.

Under the proposed model, we formulate a novel sum rate maximization problem for both SR and PR, which is found to be non-convex programming. To solve the problem efficiently, we then propose a low-complexity iterative algorithm based on the inner convex approximation (ICA) framework. Numerical results are provided not only to demonstrate the effectiveness of the proposed system model and optimization algorithm, but also to show the performance trade-off between the secondary and primary systems under the multiple-antenna technique and fractional time optimization.

Notation: $\mathbf{X}^H$ and $\mathrm{tr}(\mathbf{X})$ are the Hermitian transpose and trace of a matrix $\mathbf{X}$, respectively. $\|\cdot\|$ and $|\cdot|$ denote the Euclidean norm of a vector and the absolute value of a complex scalar, respectively. $x \sim \mathcal{CN}(a, \sigma^2)$ indicates that the random variable x follows the complex normal distribution with mean a and variance σ^2.

2. System Model and Problem Formulation

As illustrated in Figure 1, we consider a CRN consisting of a primary system with one PT and one PR and a secondary system with one ST and one SR. A decode-and-forward (DF)-based FDR equipped with N-receive and M-transmit antennas is primarily employed in the secondary system, under the assumption that there is no direct link from ST to SR due to path loss and shadowing. We denote the channel vectors from PT/ST to FDR by $\mathbf{h}_z \in \mathbb{C}^{N \times 1}$, $z \in \{p, s\}$, while $\mathbf{g}_z \in \mathbb{C}^{1 \times M}$ represents the channel vectors from FDR to PR/SR. $f_{ps} \in \mathbb{C}$ and $f_{pp} \in \mathbb{C}$ stand for the channel responses from PT to ST and to PR, respectively. FDR is assumed to be a multiple-input and multiple-output (MIMO) device while PT and ST are merely equipped with one antenna (without loss of generality, the proposed system model can be extended to MIMO scenarios for PT and/or ST to further improve the system performance). Hence, $\mathbf{H}_{RR} \in \mathbb{C}^{N \times M}$ designates the self-interference (SI) channel from the transmit antennas to the receive antennas at FDR. In particular, the residual SI can be modeled as $\mathbf{H}_{RR} = \sqrt{\rho}\mathbf{H}_0$, where the entries of $\mathbf{H}_0$ follow a Rician distribution, and $0 \leq \rho \leq 1$ denotes the level of SI suppression after applying cancellation techniques as proposed in [25].

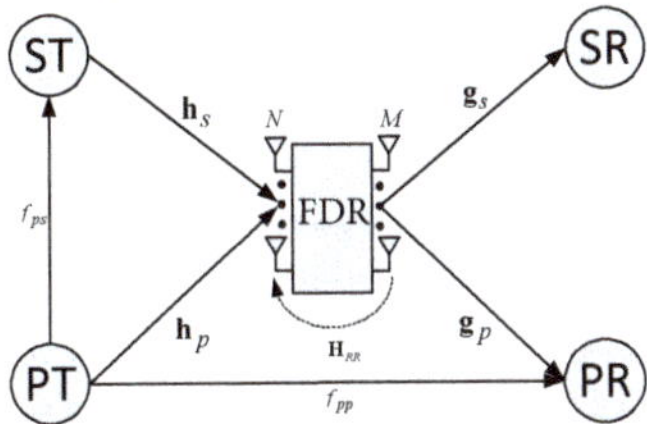

Figure 1. A cognitive radio network with a full-duplex relay (FDR).

Remark 1. *In this paper, we focus on the effect of the TSR protocol under the perfectly known channel state information (CSI). Therefore, the results of this paper will provide an upper bound of system performance. However, the following proposed model and method can be applied to a robust design under channel uncertainty by decomposing the perfect channels into the channel estimates and estimation errors [6].*

2.1. Information and Power Transfer Model

To be energy-efficient, the data transmission of CRN is considered in a time block when the link between PT and PR sharing the same spectrum is sensed to be active. Accordingly, ST and FDR are allowed to harvest energy from RF signals transmitted by PT, and then, they use the harvested energy for their own data transmission. We consider a TSR protocol for joint energy harvesting and information processing [22], as illustrated in Figure 2. According to the TSR protocol, ST and FDR harvest energy from the PT's signal for a fraction α ($0 \leq \alpha \leq 1$) of a certain time slot k of duration T. The remaining $(1 - \alpha)$ fraction of the time slot is used for information transmission. As a result, the amount of energy harvested at ST and FDR can be respectively expressed as:

$$E_s = \eta P_p \left| f_{ps} \right|^2 \alpha T \text{ and } E_r = \eta P_p \left\| \mathbf{h}_p \right\|^2 \alpha T, \tag{1}$$

where $0 < \eta < 1$ denotes the energy conversion efficiency and P_p is the transmit power of PT. Correspondingly, the maximum available power at ST and that at FDR for the time duration $(1 - \alpha)T$ are respectively calculated as:

$$P_s = \frac{\alpha}{1 - \alpha}\eta P_p \left| f_{ps} \right|^2 \text{ and } P_r = \frac{\alpha}{1 - \alpha}\eta P_p \left\| \mathbf{h}_p \right\|^2. \tag{2}$$

Herein, we suppose that the energies harvested by ST and FDR in the first phase time of αT are completely used for information transmission in the second phase time of $(1 - \alpha)T$. The received signals at FDR, SR, and PR during the time duration $(1 - \alpha)T$ at the kth time slot are respectively given as:

$$\mathbf{y}_r[k] = \mathbf{H}\mathbf{x}[k] + \sqrt{P_r\rho}\mathbf{H}_0\mathbf{W}\mathbf{s}[k] + \mathbf{n}_r[k], \tag{3}$$

$$y_s[k] = \sqrt{P_r}\mathbf{g}_s\mathbf{W}\mathbf{s}[k] + \mathbf{n}_s[k], \tag{4}$$

$$y_p[k] = \sqrt{P_r}\mathbf{g}_p\mathbf{W}\mathbf{s}[k] + \sqrt{P_p}f_{pp}x_p[k] + \mathbf{n}_p[k], \tag{5}$$

where $\mathbf{H} \triangleq [\mathbf{h}_s \; \mathbf{h}_p]$ and $\mathbf{x}[k] \triangleq [\sqrt{P_s}x_s[k] \; \sqrt{P_p}x_p[k]]^H$, with $x_s[k]$ and $x_p[k] \sim \mathcal{CN}(0,1)$ being the transmit symbols of ST and PT in the kth time slot, respectively; the elements of $\mathbf{n}_r[k]$, $\mathbf{n}_s[k]$, and $\mathbf{n}_p[k] \sim \mathcal{CN}(0,\sigma^2)$ denote the additive white Gaussian noise (AWGN) at FDR, SR and PR, respectively. At FDR, a beamforming matrix $\mathbf{W}$ is applied to combat co-channel interference, yielding the transmit signal $\mathbf{s}[k] = \left[x_{\mathbf{s}}[k - \tau] \; x_{\mathbf{p}}[k - \tau]\right]^{H}$, where τ accounts for the time delay caused by the relay processing. The system is assumed to adopt a zero-forcing (ZF) detector $\mathbf{A}^{H}$ and a ZF beamforming $\mathbf{W}$ at FDR:

$$\mathbf{A}^{H} \triangleq \left(\mathbf{H}^{H}\mathbf{H}\right)^{-1}\mathbf{H}^{H}, \text{ and } \mathbf{W} \triangleq c\mathbf{G}\left(\mathbf{G}^{H}\mathbf{G}\right)^{-1},$$

where:

$$\mathbf{G}^{H} = \left[\mathbf{g}_s^{H} \quad \mathbf{g}_p^{H}\right]^{H}, \text{ and } c = \frac{1}{\sqrt{\mathrm{tr}\left((\mathbf{G}^{H}\mathbf{G})^{-1}\right)}}.$$

Herein, the power control factor c is to ensure that the maximum transmit power at FDR does not exceed the harvested power in the corresponding transmission block. Furthermore, PT can transmit data directly to PR during the first phase, and thus, the received signal at PR via the direct link during the time αT is given as:

$$y_{\mathrm{P}}^{dir}[k] = \sqrt{P_{\mathrm{p}}} f_{pp} x_{\mathrm{P}}[k] + n_{\mathrm{p}}'[k]. \tag{6}$$

It is assumed that the processing delay is negligibly small compared to the transmission time. Moreover, the time slot k is an arbitrary slot at which PT sends the signals to PR. For simplicity, we omit τ and k, hereafter.

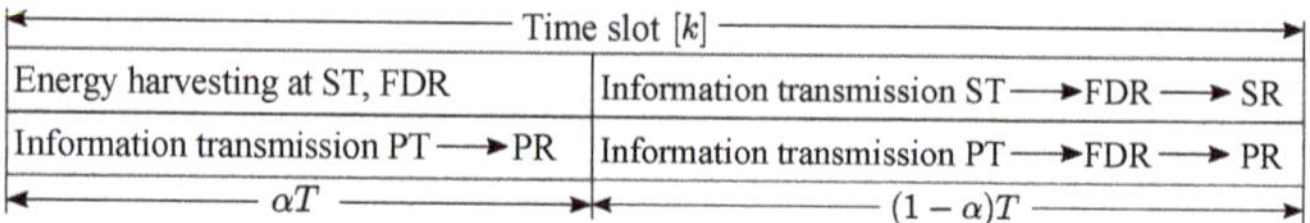

Figure 2. Time-switching relaying (TSR) protocol.

2.2. Achievable Rates and Problem Formulation

The estimated data at FDR can be expressed as $\hat{\mathbf{x}} = \mathbf{A}^{H}\mathbf{y}_r$. From (3) and (4), the signal-to-interference-plus-noise ratio (SINR) of the secondary system in each hop can be computed as:

$$\gamma_{\mathrm{ST,R}} = \frac{P_{\mathbf{s}}\left|\mathbf{a}_s^{H}\mathbf{h}_s\right|^2}{P_{\mathbf{p}}\left|\mathbf{a}_s^{H}\mathbf{h}_p\right|^2 + P_{\mathbf{r}}\rho\left\|\mathbf{a}_s^{H}\mathbf{H}_0\mathbf{W}\right\|^2 + \sigma^2\left\|\mathbf{a}_s\right\|^2}, \tag{7}$$

$$\gamma_{\mathrm{R,SR}} = \frac{P_{\mathbf{r}}\left|\mathbf{g}_s\mathbf{w}_s\right|^2}{P_{\mathbf{r}}\left|\mathbf{g}_s\mathbf{w}_p\right|^2 + \sigma^2}, \tag{8}$$

where $\mathbf{a}_s$ is the first column of $\mathbf{A}$ corresponding to $\mathbf{h}_s$; $\mathbf{w}_s$ and $\mathbf{w}_p$ are the first and second columns of $\mathbf{W}$ corresponding to $\mathbf{h}_s$ and $\mathbf{h}_p$, respectively. Similarly, the SINR of the primary system during the time $(1 - \alpha)T$ can be found as:

$$\gamma_{\mathrm{PT,R}} = \frac{P_{\mathbf{p}}\left|\mathbf{a}_p^{H}\mathbf{h}_p\right|^2}{P_{\mathbf{s}}\left|\mathbf{a}_p^{H}\mathbf{h}_s\right|^2 + P_{\mathbf{r}}\rho\left\|\mathbf{a}_p^{H}\mathbf{H}_0\mathbf{W}\right\|^2 + \sigma^2\left\|\mathbf{a}_p\right\|^2}, \tag{9}$$

$$\gamma_{\mathrm{R,PR}} = \frac{P_{\mathbf{r}}\left|\mathbf{g}_p\mathbf{w}_p\right|^2}{P_{\mathbf{r}}\left|\mathbf{g}_p\mathbf{w}_s\right|^2 + P_{\mathbf{p}}\left|f_{pp}\right|^2 + \sigma^2}, \tag{10}$$

where $\mathbf{a}_p$ is the second column of $\mathbf{A}$ corresponding to $\mathbf{h}_p$. Furthermore, the SNR of the primary system for the direct link during the time αT in (6) is found as:

$$\gamma_{\mathrm{PT,PR}} = \frac{P_{\mathrm{p}} |f_{pp}|^2}{\sigma^2}. \tag{11}$$

The end-to-end (e2e) achievable rate of the secondary system can be expressed as:

$$R_{\mathrm{s}} = (1 - \alpha) \ln\big(1 + \min\big(\gamma_{\mathrm{ST,R}}, \gamma_{\mathrm{R,SR}}\big)\big). \tag{12}$$

Similarly, the e2e achievable rate of the primary system that uses a maximal-ratio-combining (MRC) to combine (9) and (11) can be written as:

$$R_{\mathrm{p}} = R_{\alpha} + (1 - \alpha) \ln\big(1 + \min\big(\gamma_{\mathrm{PT,R}}, \gamma_{\mathrm{R,PR}}\big)\big), \tag{13}$$

where $R_{\alpha} \triangleq \alpha \ln(1 + \gamma_{\mathrm{PT,PR}})$ is a linear function with respect to α.

From (12) and (13), the optimization problem of maximizing the sum rate of the primary and secondary systems can be formulated as (To maximize the sum rate, we focus on the optimization of the time fraction α, which determines a trade-off between the energy harvesting and data transmission. The joint optimization of power allocation and time fraction will probably enhance the performance, but with much higher complexity.):

$$\underset{\alpha}{\operatorname{maximize}} \quad R_{\mathrm{s}} + R_{\mathrm{p}}, \tag{14a}$$

$$\text{subject to} \quad 0 \le \alpha \le 1. \tag{14b}$$

It can be seen that Problem (14) is non-convex, since the objective function in (14a) is non-concave. In what follows, a low-complexity approach to find the solution of (14) will be presented.

3. Proposed Solution for SRM Problem

We first transform Problem (14) into a tractable non-convex form and then apply the ICA method to devise a low-complexity algorithm.

3.1. Tractable Formulation for SRM Problem

We first introduce a new variable β with an additional constraint: $\alpha + \beta \le 1$. Accordingly, $(1 - \alpha)$ in (12) and (13) can be equivalently replaced by β. This provides a smoothing optimization instead of a strictly-splitting time slot in the TSR protocol. Then, we tackle the non-smooth functions in (12) and (13) by adding the following constraints:

$$\gamma_{\mathrm{ST,R}} \ge \frac{1}{t_{\mathrm{s}}} \text{ and } \gamma_{\mathrm{R,SR}} \ge \frac{1}{t_{\mathrm{s}}}, \tag{15a}$$

$$\gamma_{\mathrm{PT,R}} \ge \frac{1}{t_{\mathrm{p}}} \text{ and } \gamma_{\mathrm{R,PR}} \ge \frac{1}{t_{\mathrm{p}}}. \tag{15b}$$

Therefore, R_{s} and R_{p} are respectively expressed as:

$$R_{\mathrm{s}} \ge \beta \ln\big(1 + \frac{1}{t_{\mathrm{s}}}\big) := \bar{R}_{\mathrm{s}}, \tag{16a}$$

$$R_{\mathrm{p}} \ge R_{\alpha} + \beta \ln\big(1 + \frac{1}{t_{\mathrm{p}}}\big) := \bar{R}_{\mathrm{p}}. \tag{16b}$$

By using (16), we can equivalently rewrite (14) as:

$$\underset{\alpha,\beta,t}{\text{maximize}} \quad \bar{R}_{\mathrm{s}} + \bar{R}_{\mathrm{p}}, \tag{17a}$$

$$\text{subject to} \quad 0 \le \alpha \le 1, \ 0 \le \beta \le 1, \tag{17b}$$

$$\alpha + \beta \le 1, \tag{17c}$$

$$(15), \tag{17d}$$

where $\mathbf{t} \triangleq \left[t_{\mathrm{s}}; \ t_{\mathrm{p}} \right]$. We note that Problem (17) is still non-convex due to the non-concave objective in (17a) and the non-convex constraints in (17d).

3.2. Proposed Iterative Algorithm

In this subsection, we focus on convexifying Problem (17) by applying the ICA method to (17a) and (17d). To do this, we first introduce two approximate functions as follows.

- Consider the function $f(x,y) \triangleq \frac{1}{y} \ln(1 + \frac{1}{x})$, $(x,y) \in \mathbb{R}^2_{++}$. It can be seen that $f(x,y)$ is convex, since its Hessian is a positive definite matrix. According to [6], we can obtain an upper bound of $f(x,y)$ around the point $(x^{(\kappa)}, y^{(\kappa)})$ as:

$$f(x,y) \ge \mathrm{A}(x^{(\kappa)}, y^{(\kappa)}) + \mathrm{B}(x^{(\kappa)}, y^{(\kappa)})x + \mathrm{C}(x^{(\kappa)}, y^{(\kappa)})y$$
$$:= \tilde{f}^{(\kappa)}(x,y), \tag{18}$$

where:

$$\mathrm{A}(x^{(\kappa)}, y^{(\kappa)}) \triangleq \frac{2}{y^{(\kappa)}} \ln\left(1 + \frac{1}{x^{(\kappa)}}\right) + \frac{1}{\left(x^{(\kappa)} + 1\right)y^{(\kappa)}},$$

$$\mathrm{B}(x^{(\kappa)}, y^{(\kappa)}) \triangleq -\frac{1}{x^{(\kappa)}\left(x^{(\kappa)} + 1\right)y^{(\kappa)}},$$

$$\mathrm{C}(x^{(\kappa)}, y^{(\kappa)}) \triangleq -\frac{1}{(y^{(\kappa)})^2} \ln\left(1 + \frac{1}{x^{(\kappa)}}\right).$$

- Similarly, an upper bound of a convex function $g(x,y) \triangleq \frac{x^2}{y}$, $(x,y) \in \mathbb{R}^2_{++}$ around the point $(x^{(\kappa)}, y^{(\kappa)})$ can be found as:

$$g(x,y) \ge \frac{2x^{(\kappa)}}{y^{(\kappa)}}x - \frac{(x^{(\kappa)})^2}{(y^{(\kappa)})^2}y := \tilde{g}^{(\kappa)}(x,y). \tag{19}$$

Inner approximation of (17a): We introduce a new variable δ that satisfies the following convex constraint:

$$\frac{1}{\delta} \le \beta, \ \delta \ge 1. \tag{20}$$

By substituting (20) into (16) and applying (18), the concave minorants of $\bar{R}_{\mathrm{s}}$ and $\bar{R}_{\mathrm{p}}$ at the iteration $\kappa + 1$ are respectively given as:

$$\bar{R}_{\mathrm{s}} \ge f(t_{\mathrm{s}}, \delta) \ge \tilde{f}^{(\kappa)}(t_{\mathrm{s}}, \delta) := \breve{R}_{s}, \tag{21a}$$

$$\bar{R}_{\mathrm{p}} \ge R_{\alpha} + f(t_{\mathrm{p}}, \delta) \ge R_{\alpha} + \tilde{f}^{(\kappa)}(t_{\mathrm{p}}, \delta) := \breve{R}_{p}. \tag{21b}$$

It is observed that (17a) can be iteratively replaced by $\breve{R}_s + \breve{R}_p$, which is a concave objective function. We notice that when $\kappa \to \infty$, the inequalities in (16), (20), and (21) hold with equalities.

To convexify (15), the following theorem is derived using the ICA framework.

Theorem 1. *For two arbitrary vectors* $\mathbf{x}$ *and* $\mathbf{y}$, *let* $\lambda(\mathbf{x}, \mathbf{y}) \triangleq \eta P_{\mathrm{p}} \|\mathbf{x}\|^2 \|\mathbf{y}\|^2$. *The constraints in* (15a) *are convexified as:*

$$\alpha\lambda(\mathbf{h}_p, \sqrt{\rho}\mathbf{a}_s^H \mathbf{H}_0 \mathbf{W}) + \beta\psi_{\mathrm{ST,R}} \leq \nu_{\mathrm{ST,R}}, \tag{22a}$$

$$g^{(\kappa)}(\sqrt{\alpha}, \nu_{\mathrm{ST,R}})\lambda(f_{ps}, \mathbf{a}_s^H \mathbf{h}_s) \geq \frac{1}{t_{\mathrm{s}}}, \tag{22b}$$

$$\alpha\lambda(\mathbf{h}_p, \mathbf{g}_s \mathbf{w}_p) + \beta\sigma^2 \leq \nu_{\mathrm{R,SR}}, \tag{22c}$$

$$g^{(\kappa)}(\sqrt{\alpha}, \nu_{\mathrm{R,SR}})\lambda(\mathbf{h}_p, \mathbf{g}_s \mathbf{w}_s) \geq \frac{1}{t_{\mathrm{s}}}, \tag{22d}$$

where $\nu_{\mathrm{ST,R}}$ *and* $\nu_{\mathrm{R,SR}}$ *are newly introduced variables and* $\psi_{\mathrm{ST,R}} \triangleq P_{\mathrm{p}} |\mathbf{a}_s^H \mathbf{h}_p|^2 + \|\mathbf{a}_s\|^2 \sigma^2$. *Similarly, the constraints in* (15b) *are approximated by the following convex constraints:*

$$\alpha\phi_{\mathrm{PT,R}} + \beta \|\mathbf{a}_p\|^2 \sigma^2 \leq \nu_{\mathrm{PT,R}}, \tag{23a}$$

$$P_{\mathrm{p}} |\mathbf{a}_p^H \mathbf{h}_p|^2 g^{(\kappa)}(\sqrt{\beta}, \nu_{\mathrm{PT,R}}) \geq \frac{1}{t_{\mathrm{p}}}, \tag{23b}$$

$$\alpha\lambda(\mathbf{h}_p, \mathbf{g}_p \mathbf{w}_s) + \beta(P_{\mathrm{p}} |f_{pp}|^2 + \sigma^2) \leq \nu_{\mathrm{R,PR}}, \tag{23c}$$

$$g^{(\kappa)}(\sqrt{\alpha}, \nu_{\mathrm{R,PR}})\lambda(\mathbf{h}_p, \mathbf{g}_p \mathbf{w}_p) \geq \frac{1}{t_{\mathrm{p}}}, \tag{23d}$$

where $\nu_{\mathrm{PT,R}}$ *and* $\nu_{\mathrm{R,PR}}$ *are new variables and* $\phi_{\mathrm{PT,R}} \triangleq \lambda(f_{ps}, \mathbf{a}_p^H \mathbf{h}_s) + \lambda(\mathbf{h}_p, \sqrt{\rho}\mathbf{a}_p^H \mathbf{H}_0 \mathbf{W})$.

Proof. Please see Appendix A. $\square$

For convenience, we define $\nu \triangleq [\nu_{\mathrm{ST,R}}; \nu_{\mathrm{R,SR}}; \nu_{\mathrm{PT,R}}; \nu_{\mathrm{R,PR}}]$. By using (20), (21), and Theorem 1, the successive convex program providing a minorant maximization for (14) at the iteration $\kappa + 1$ is given by:

$$\underset{S}{\text{maximize}} \;\; \ddot{R}_\Sigma^{(\kappa+1)} \triangleq \ddot{R}_{\mathrm{s}} + \ddot{R}_{\mathrm{p}}, \tag{24a}$$

$$\text{subject to } (17b), (17c), (20), (22), (23), \tag{24b}$$

where $S \triangleq \{\alpha, \beta, \delta, \mathbf{t}, \nu\}$ and, correspondingly, $S^{(\kappa)} = \{\alpha^{(\kappa)}, \beta^{(\kappa)}, \delta^{(\kappa)}, \mathbf{t}^{(\kappa)}, \nu^{(\kappa)}\}$ at iteration κ. We successively solve (24) and update the involved optimization variables after each iteration until convergence, which is guaranteed to achieve at least a locally optimal solution of (14). In summary, the proposed algorithm for solving (14) is given in Algorithm 1.

Algorithm 1 Proposed iterative algorithm for solving Problem (14).

1: **Initialization:** Set $\kappa := 0$, $\varepsilon := 10^{-3}$, $\ddot{R}_\Sigma^{(0)} := -\infty$, and randomly generate an initial point $S^{(0)}$.

2: **repeat**

3:    Solve (24) to obtain the objective value $\ddot{R}_\Sigma^{(\kappa+1)}$ and solution $S^{(\star)}$.

4:    Update $S^{(\kappa+1)} := S^{(\star)}$.

5:    Set $\kappa := \kappa + 1$.

6: **until** $\ddot{R}_\Sigma^{(\kappa)} - \ddot{R}_\Sigma^{(\kappa-1)} < \varepsilon$.

7: **Output:** The solution $\alpha^* = \alpha^{(\kappa)}$. The corresponding achievable rate is given by (14a).

Complexity analysis: The proposed algorithm takes low complexity in the scene that all the constraints in (24) are conic constraints. In particular, the per-iteration complexity of solving (24) is

$\mathcal{O}(C^{2.5}V^2 + C^{3.5})$, where C denotes the number of constraints in (24b) and V is the number of decision variables determined by the number of elements in $\mathcal{S}$.

Remark 2. *It can be foreseen that the proposed algorithm is easily applied to an SR's rate (resp., PR's rate) maximization problem under the quality-of-service (QoS) constraint for PR (resp., SR) receiver, i.e.,*

$$\underset{\mathcal{S}}{maximize} \ \tilde{R}_{\Sigma}^{(\kappa+1)} \triangleq \ddot{R}_{\mathbf{s}}, \tag{25a}$$

$$subject\ to\ (17b), (17c), (20), (22), (23), \tag{25b}$$

$$\ddot{R}_{\mathbf{p}} \geq \bar{R}_{\mathbf{p}}, \tag{25c}$$

where $\bar{R}_{\mathbf{p}}$ is a data rate requirement for PR. In fact, Constraint (25c) is linear due to the property of the ICA method, leading to a convex programming in (25). Typically, this problem is associated with a trade-off between SR and PR rates, which is examined in Figure 5.

4. Numerical Results

We consider that an FDR (denoted by R) equipped with N receive antennas and M transmit antennas is located 20 m away from PT, i.e., $d_{\mathrm{PT,R}} = 20$ m, where $d_{\mathrm{x,y}}$ is the distance from x to y. An ST is located between PT and FDR such that $d_{\mathrm{PT,ST}} = 2$ m. PR and SR are located at two points, which are symmetric with respect to the line from PT to FDR, such that $d_{\mathrm{R,PR}} = d_{\mathrm{R,SR}} = 5$ m, $d_{\mathrm{PT,PR}} = \sqrt{d_{\mathrm{PT,R}}^2 + d_{\mathrm{R,PR}}^2}$ m, and $d_{\mathrm{ST,SR}} = \sqrt{d_{\mathrm{ST,R}}^2 + d_{\mathrm{R,SR}}^2}$ m. The channel responses are determined by $\ddot{\mathbf{h}} = d_z^{-\varphi}\hat{\mathbf{h}}$, where $\ddot{\mathbf{h}} \in \{\mathbf{h}_p, \mathbf{h}_s, \mathbf{g}_p, \mathbf{g}_s, \mathbf{f}_{ps}, \mathbf{f}_{pp}\}$ corresponds to the link $z \in \{(\mathrm{PT,R}), (\mathrm{ST,R}), (\mathrm{R,PR}), (\mathrm{R,SR}), (\mathrm{PT,ST}), (\mathrm{PT,PR})\}$, respectively; the path-loss exponent is assumed to be $\varphi = 3$, while the elements of $\hat{\mathbf{h}}$ follow $\mathcal{CN}(0,1)$. Unless otherwise specified, the transmit power at PT, background noise power, and energy conversion parameter are respectively set as $P_{\mathbf{p}} = 26$ dBm, $\sigma^2 = -104$ dBm, and $\eta = 0.5$ [26–28], while the suppression level of the residual SI is set as $\rho = -90$ dB following the worst-case design in [29].

Figure 3 depicts the sum rate of SR and PR versus the maximum power budget at PT. The range of the examined power budget is from 18 to 43 dBm, which is usually used for the maximum power at BS in small-cell to macro-cell scenarios [30]. To evaluate the proposed algorithm (Algorithm 1), we consider three other schemes: (i) SR and PR directly receive the signals from ST and PT, respectively, in which the harvesting time α for ST is optimized (Opt.α w/oFDR); (ii) data transmission using the same beamforming and detection as the proposed method is performed with α fixed to 0.5 (fixed α w/FDR); (iii) we plot the performance of a baseline scheme using a half-duplex relay (HDR) (named as "Alg. 1 w/HDR"), in which the proposed algorithm is utilized to find the optimal value of α, and the later phase with time fraction $(1 - \alpha)$ is split into two sub-phases: one sub-phase for transferring signals from ST and PT to HDR and the other sub-phase for forwarding the messages from HDR to SR and PR. We can observe that the proposed method provides the best performance, due to the fact that the power is efficiently utilized by a cooperation between the harvesting time and beamforming at the FDR. In particular, for all the considered values of $P_{\mathbf{p}}$, the proposed scheme gives about 2 bps/Hz higher than the one using "fixed α w/FDR", due to only the linear difference in time fraction optimization. Meanwhile, the gain of the proposed scheme over "Opt. α w/o FDR" decreases as $P_{\mathbf{p}}$ increases. Clearly, "Opt. α w/o FDR" suffers from high attenuation, due to the long-distance propagation from PT (ST) to PR (SR), which is merely compensated by the increase in power budget at the PT. It can be observed that with a low transmit power at PT, "Alg. 1 w/HDR" slightly outperforms "Opt. α w/o FDR", which confirms the benefit of the use of relaying. However, the performance gap is seen to diminish as $P_{\mathbf{p}}$ increases. Figure 4 further investigates the system performance with various numbers of receive/transmit antennas at FDR. Obviously, the performance of the scheme without FDR is independent of the number of antennas, while the rate gain obtained by the proposed method is verified. Indeed, the performance of FDR-assisted schemes is improved more quickly when both N and M increase than when only one of them increases.

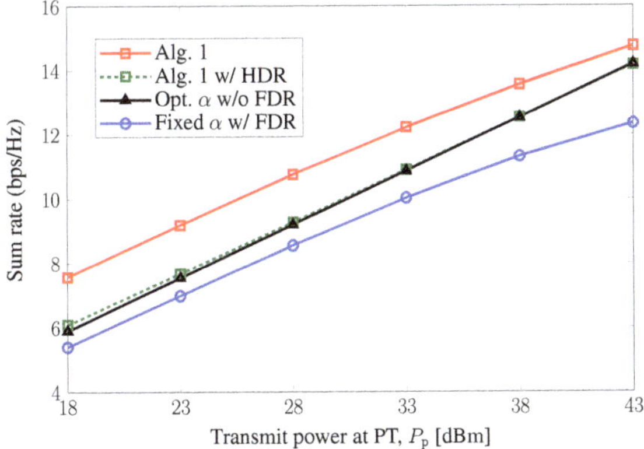

Figure 3. Sum rate of the secondary receiver (SR) and the primary receiver (PR) versus P_p with $M = N = 4$.

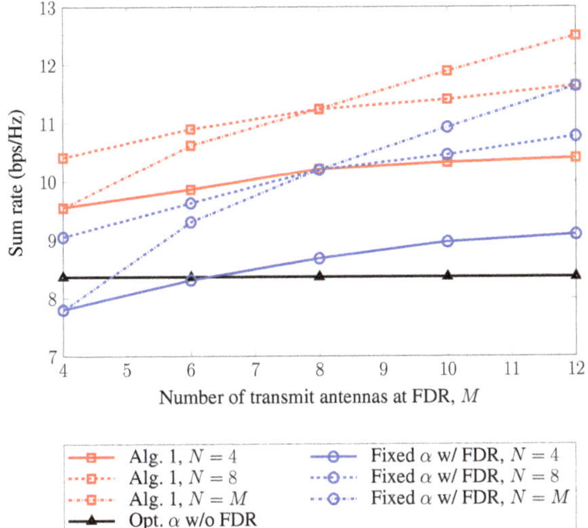

Figure 4. Sum rate of SR and PR versus the transmit power at the primary transmitter (PT) and the number of antennas at the full-duplex relay (FDR).

Figure 5 shows the trade-off between SR and PR rates with respect to α for different numbers of receive/transmit antennas. Considering practicability, we set the number of transmit/receive antennas at the FDR to the powers-of-two values. The energy harvested at the ST and FDR becomes higher as α increases, leading to a higher SR rate. However, when the value of α is higher than a certain threshold, the SR rate deteriorates due to the decrease in the information transmission time, $(1 - \alpha)$. On the contrary, the PR rate increases with α, since the time of direct information transmission for the PR increases. At the maximum sum-rate point associated with the optimal α, an increase in the number of transmit antennas at the FDR ($M = 4, 8, 16$) provides more degrees of freedom, which boosts both the SR and PR rates. On the other hand, the sum rate also increases as the number of receive antennas at FDR ($N = 4, 8, 16$) increases. This is primarily attributed to an increase in the SR rate, as FDR would be in the proximity of ST.

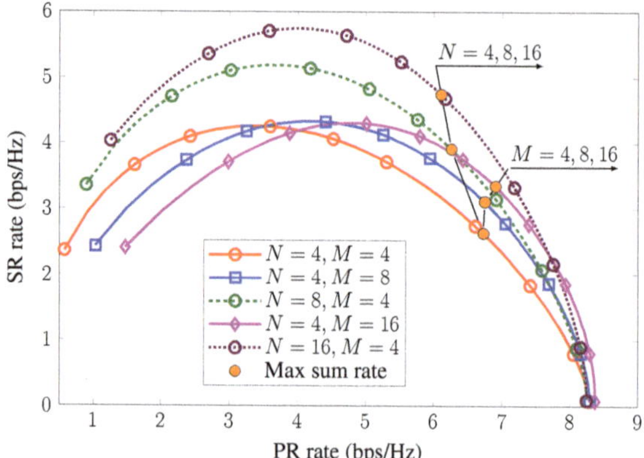

Figure 5. Trade-off between SR and PR rates.

5. Conclusions

In this paper, we considered a sum rate maximization problem for the primary and secondary systems assisted by an FDR with energy harvesting. A design problem based on the TSR protocol for joint optimization of energy harvesting and information transfer was established as a non-convex problem. By using the ICA framework, we derived the optimal solution for the problem. Remarkably, the proposed iterative algorithm was shown to provide a low computational complexity per iteration. Numerical results indicated that the proposed scheme and algorithm provided a spectral efficiency 0.5 to 2 bps/Hz higher than other schemes in the medium to high transmit power at PT by exploiting the benefits of multiple antennas and time fraction optimization. Moreover, we demonstrated a trade-off between the SR rate and the PR rate, as well as the effect of the number transmit and receive antennas at FDR on the sum rate.

Author Contributions: Conceptualization, H.V.N. and V.-D.N.; methodology, H.V.N., V.-D.N., and O.-S.S.; simulation, H.V.N.; validation, H.V.N., V.-D.N., and O.-S.S.; writing, original draft preparation, H.V.N.; writing, review and editing, H.V.N., V.-D.N., and O.-S.S.; supervision, O.-S.S.; project administration, O.-S.S.; funding acquisition, O.-S.S. All authors read and agreed to the published version of the manuscript.

Funding: This work was supported by the National Research Foundation of Korea (NRF) grant funded by the Korean government (MOE and MSIT) (No. 2017R1D1A1B03030436, No. 2017R1A5A1015596 and No. 2019R1A2C1084834).

Conflicts of Interest: The authors declare no conflict of interest.

Appendix A. Proof of Theorem 1

To convexify (15), we first address Constraint (15a). For two arbitrary vectors $\mathbf{x}$ and $\mathbf{y}$, let $\lambda(\mathbf{x}, \mathbf{y}) \triangleq \eta P_\mathrm{p} \|\mathbf{x}\|^2 \|\mathbf{y}\|^2$, and then, Constraint (15a) is equivalent to the following constraints:

$$(15a) \Leftrightarrow \begin{cases} \dfrac{\alpha\lambda(f_{ps}, \mathbf{a}_s^H \mathbf{h}_s)}{\alpha\lambda(\mathbf{h}_p, \sqrt{\rho}\mathbf{a}_s^H \mathbf{H}_0 \mathbf{W}) + \beta\psi_{\mathrm{ST,R}}} \geq \dfrac{1}{t_\mathrm{s}}, & \text{(A1a)} \\[4ex] \dfrac{\alpha\lambda(\mathbf{h}_p, \mathbf{g}_s \mathbf{w}_s)}{\alpha\lambda(\mathbf{h}_p, \mathbf{g}_s \mathbf{w}_p) + \beta\sigma^2} \geq \dfrac{1}{t_\mathrm{s}}, & \text{(A1b)} \end{cases}$$

where $\psi_{\text{ST,R}} \triangleq P_{\text{p}} \left| \mathbf{a}_s^H \mathbf{h}_p \right|^2 + \|\mathbf{a}_s\|^2 \sigma^2$. By applying (19) to (A1), Constraint (15a) is convexified as:

$$g^{(\kappa)}(\sqrt{\alpha}, \nu_{\text{ST,R}})\lambda(f_{ps}, \mathbf{a}_s^H \mathbf{h}_s) \geq \frac{1}{t_{\text{s}}}, \tag{A2a}$$

$$g^{(\kappa)}(\sqrt{\alpha}, \nu_{\text{R,SR}})\lambda(\mathbf{h}_p, \mathbf{g}_s \mathbf{w}_s) \geq \frac{1}{t_{\text{s}}}, \tag{A2b}$$

with the imposed constraints:

$$\alpha\lambda(\mathbf{h}_p, \sqrt{\rho}\,\mathbf{a}_s^H \mathbf{H}_0 \mathbf{W}) + \beta\psi_{\text{ST,R}} \leq \nu_{\text{ST,R}}, \tag{A3a}$$

$$\alpha\lambda(\mathbf{h}_p, \mathbf{g}_s \mathbf{w}_p) + \beta\sigma^2 \leq \nu_{\text{R,SR}}, \tag{A3b}$$

where $\nu_{\text{ST,R}}$ and $\nu_{\text{R,SR}}$ are new variables. When $\kappa \to \infty$, the equality of (15a) holds with the equalities in (A2) and (A3), which are given in (22). Next, we apply the same steps as above to (15b) to obtain the convex constraints in (23), which completes the proof.

References

1. Sabharwal, A.; Schniter, P.; Guo, D.; Bliss, D.W.; Rangarajan, S.; Wichman, R. In-band full-duplex wireless: Challenges and opportunities. *IEEE J. Sel. Areas Commun.* **2014**, *32*, 1637–1652. [CrossRef]
2. Yadav, A.; Tsiropoulos, G.I.; Dobre, O.A. Full-Duplex Communications: Performance in Ultradense mm-Wave Small-Cell Wireless Networks. *IEEE Veh. Technol. Mag.* **2018**, *13*, 40–47. [CrossRef]
3. Nguyen, D.; Tran, L.N.; Pirinen, P.; Latva-aho, M. On the spectral efficiency of full-duplex small cell wireless systems. *IEEE Trans. Wirel. Commun.* **2014**, *13*, 4896–4910. [CrossRef]
4. Tam, H.H.M.; Tuan, H.D.; Ngo, D.T. Successive Convex Quadratic Programming for Quality-of-Service Management in Full-Duplex MU-MIMO Multicell Networks. *IEEE Trans. Commun.* **2016**, *64*, 2340–2353. [CrossRef]
5. Nguyen, V.D.; Nguyen, H.V.; Nguyen, C.T.; Shin, O.S. Spectral Efficiency of Full-Duplex Multiuser System: Beamforming Design, User Grouping, and Time Allocation. *IEEE Access* **2017**, *5*, 5785–5797. [CrossRef]
6. Nguyen, H.V.; Nguyen, V.D.; Dobre, O.A.; Wu, Y.; Shin, O.S. Joint antenna array mode selection and user assignment for full-duplex MU-MISO systems. *IEEE Trans. Wirel. Commun.* **2019**, *18*, 2946–2963. [CrossRef]
7. Yang, K.; Cui, H.; Song, L.; Li, Y. Efficient Full-Duplex Relaying With Joint Antenna-Relay Selection and Self-Interference Suppression. *IEEE Trans. Wirel. Commun.* **2015**, *14*, 3991–4005. [CrossRef]
8. Yu, B.; Yang, L.; Cheng, X.; Cao, R. Power and Location Optimization for Full-Duplex Decode-and-Forward Relaying. *IEEE Trans. Commun.* **2015**, *63*, 4743–4753. [CrossRef]
9. Haykin, S.; Setoodeh, P. Cognitive Radio Networks: The Spectrum Supply Chain Paradigm. *IEEE Trans. Cogn. Commun. Netw.* **2015**, *1*, 3–28. [CrossRef]
10. Nguyen, H.V.; Nguyen, V.-D.; Kim, H.M.; Shin, O.-S. A Convex optimization for sum rate maximization in a MIMO cognitive radio network. In Proceedings of the 2016 Eighth International Conference on Ubiquitous and Future Networks (ICUFN), Vienna, Austria, 5–8 July 2016; pp. 495–497.
11. Erdogan, E.; Afana, A.; Ikki, S.; Yanikomeroglu, H. Antenna Selection in MIMO Cognitive AF Relay Networks with Mutual Interference and Limited Feedback. *IEEE Commun. Lett.* **2017**, *21*, 1111–1114. [CrossRef]
12. Zheng, G.; Krikidis, I.; Ottersten, B. Full-Duplex Cooperative Cognitive Radio with Transmit Imperfections. *IEEE Trans. Wirel. Commun.* **2013**, *12*, 2498–2511. [CrossRef]
13. Mohammadi, M.; Chalise, B.K.; Hakimi, A.; Mobini, Z.; Suraweera, H.A.; Ding, Z. Beamforming Design and Power Allocation for Full-Duplex Non-Orthogonal Multiple Access Cognitive Relaying. *IEEE Trans. Commun.* **2018**, *66*, 5952–5965. [CrossRef]
14. Ali, B.; Mirza, J.; Zhang, J.; Zheng, G.; Saleem, S.; Wong, K. Full-Duplex Amplify-and-Forward Relay Selection in Cooperative Cognitive Radio Networks. *IEEE Trans. Veh. Technol.* **2019**, *68*, 6142–6146. [CrossRef]
15. Lee, S.; Zhang, R.; Huang, K. Opportunistic wireless energy harvesting in cognitive radio networks. *IEEE Trans. Wirel. Commun.* **2013**, *12*, 4788–4799. [CrossRef]

16. Shi, Q.; Liu, L.; Xu, W.; Zhang, R. Joint transmit beamforming and receive power splitting for MISO SWIPT systems. *IEEE Trans. Wirel. Commun.* **2014**, *13*, 3269–3280. [CrossRef]

17. Khandaker, M.; Wong, K.K. SWIPT in MISO Multicasting Systems. *IEEE Wirel. Commun. Lett.* **2014**, *3*, 277–280. [CrossRef]

18. Bhowmick, A.; Yadav, K.; Roy, S.D.; Kundu, S. Throughput of an Energy Harvesting Cognitive Radio Network Based on Prediction of Primary User. *IEEE Trans. Veh. Technol.* **2017**, *66*, 8119–8128. [CrossRef]

19. Ponnimbaduge Perera, T.D.; Jayakody, D.N.K.; Sharma, S.K.; Chatzinotas, S.; Li, J. Simultaneous Wireless Information and Power Transfer (SWIPT): Recent Advances and Future Challenges. *IEEE Commun. Surv. Tutor.* **2018**, *20*, 264–302. [CrossRef]

20. Gao, Y.; He, H.; Deng, Z.; Zhang, X. Cognitive Radio Network with Energy-Harvesting Based on Primary and Secondary User Signals. *IEEE Access* **2018**, *6*, 9081–9090. [CrossRef]

21. Zhai, D.; Zhang, R.; Du, J.; Ding, Z.; Yu, F.R. Simultaneous Wireless Information and Power Transfer at 5G New Frequencies: Channel Measurement and Network Design. *IEEE J. Sel. Areas Commun.* **2019**, *37*, 171–186. [CrossRef]

22. Nguyen, V.D.; Duong, T.Q.; Tuan, H.D.; Shin, O.S.; Poor, H.V. Spectral and Energy Efficiencies in Full-Duplex Wireless Information and Power Transfer. *IEEE Trans. Commun.* **2017**, *65*, 2220–2233. [CrossRef]

23. Bhowmick, A.; Roy, S.D.; Kundu, S. Throughput of a Cognitive Radio Network with Energy-Harvesting Based on Primary User Signal. *IEEE Wirel. Commun. Lett.* **2016**, *5*, 136–139. [CrossRef]

24. Xing, H.; Kang, X.; Wong, K.; Nallanathan, A. Optimizing DF Cognitive Radio Networks With Full-Duplex-Enabled Energy Access Points. *IEEE Trans. Wirel. Commun.* **2017**, *16*, 4683–4697. [CrossRef]

25. Riihonen, T.; Werner, S.; Wichman, R. Mitigation of loopback self-interference in full-duplex MIMO relays. *IEEE Trans. Signal Process.* **2011**, *59*, 5983–5993. [CrossRef]

26. Duarte, M.; Dick, C.; Sabharwal, A. Experiment-driven characterization of full-duplex wireless systems. *IEEE Trans. Wirel. Commun.* **2012**, *11*, 4296–4307. [CrossRef]

27. Nguyen, V.D.; Nguyen, H.V.; Dobre, O.A.; Shin, O.S. A New Design Paradigm for Secure Full-Duplex Multiuser Systems. *IEEE J. Sel. Areas Commun.* **2018**, *36*, 1480–1498. [CrossRef]

28. Bhowmick, A.; Chatterjee, A.; Verma, T. Performance of DF Relaying in an Energy Harvesting Full Duplex Cognitive Radio Network. In Proceedings of the 2019 International Conference on Vision Towards Emerging Trends in Communication and Networking (ViTECoN), Vellore, India, 30–31 March 2019; pp. 1–4.

29. Korpi, D.; Heino, M.; Icheln, C.; Haneda, K.; Valkama, M. Compact Inband Full-Duplex Relays With Beyond 100 dB Self-Interference Suppression: Enabling Techniques and Field Measurements. *IEEE Trans. Antennas Propag.* **2017**, *65*, 960–965. [CrossRef]

30. *3GPP Technical Specification Group Radio Access Network, Evolved Universal Terrestrial Radio Access (E-UTRA): Further Advancements for E-UTRA Physical Layer Aspects (Release 9)*; Document 3GPP TS 36.814 V9.0.0; 3GPP: Valbonne, France, 2010.

MDPI

St. Alban-Anlage 66

4052 Basel

Switzerland

Tel. +41 61 683 77 34

Fax +41 61 302 89 18

www.mdpi.com

Electronics Editorial Office

E-mail: electronics@mdpi.com

www.mdpi.com/journal/electronics